Glycosylation and Glycation in Health and Diseases

Edited by

Tapan Kumar Mukherjee

Amity Institute of Biotechnology
Amity University, New Town, Kolkata
West Bengal 700156, India

Parth Malik

School of Chemical Sciences
Central University of Gujarat Gandhinagar
Gujarat-382030, India

&

Ruma Rani

ICAR-National Research Centre on Equines
Hisar-125001, Haryana, India

Glycosylation and Glycation in Health and Diseases

Editors: Tapan Kumar Mukherjee, Parth Malik & Ruma Rani

ISBN (Online): 978-981-5322-52-1

ISBN (Print): 978-981-5322-53-8

ISBN (Paperback): 978-981-5322-54-5

Published by Bentham Science Publishers Pte. Ltd. Singapore. All Rights Reserved.

First published in 2025.

need for a court order if at any point you breach any terms of this License Agreement. In no event will any delay or failure by Bentham Science Publishers in enforcing your compliance with this License Agreement constitute a waiver of any of its rights.

3. You acknowledge that you have read this License Agreement, and agree to be bound by its terms and conditions. To the extent that any other terms and conditions presented on any website of Bentham Science Publishers conflict with, or are inconsistent with, the terms and conditions set out in this License Agreement, you acknowledge that the terms and conditions set out in this License Agreement shall prevail.

Bentham Science Publishers Pte. Ltd.
80 Robinson Road #02-00
Singapore 068898
Singapore
Email: subscriptions@benthamscience.net

**BENTHAM
SCIENCE**

CONTENTS

PREFACE

The book titled, "Glycosylation and Glycation in Health and Diseases" is hereby planned to be completed in 10 chapters. While glycosylation is a physiological process engrossing glycoprotein and glycolipid synthesis via enzyme-assisted carbohydrates' addition to proteins and lipids respectively; glycation is a pathophysiological process where excess carbohydrates are added to the proteins via biochemical reactions without the involvement of any enzymes.

The book commences with an introductory chapter describing various aspects of glycosylation. Subsequently, each chapter is dedicated to the biological roles of glycosylated molecules and various aspects of congenital disorders associated with glycosylation, respectively. In the domain of glycation, one chapter describes the generation of advanced glycation end products (AGEs) and their effects on various mammalian life-sustaining events. Subsequently, five chapters illustrate the pathophysiological effects of the receptor for advanced glycation end products (RAGE) on various organs, about health concerns. Finally, one chapter is dedicated to the pro-inflammatory and pro-oxidative mechanisms through which RAGE complicates various diseased conditions. Thereby, the mechanistic and pathophysiological aspects of both glycosylation and glycation are comprehensively attempted for the first time, in this book.

The study of glycans or carbohydrates has emerged as a necessity for the undergraduate and graduate syllabi of various Life Sciences related subjects, including cell biology, biochemistry, molecular biology, biotechnology, microbiology, immunology (host-pathogen interaction), and others. The fundamental biomolecular aspects have therein propelled glycation and glycosylation as major research themes in the biopharmaceutical and biotechnological industry including new drug discoveries against various cancers. Considering the prominent functions of glycans vis-à-vis immune interactions, conjugated carbohydrates are being promptly screened as next-generation therapeutics, vaccines, and diagnostic augmenters.

Closing in on the footmarks of glycosylation, the study of glycation has swiftly emerged as an essential aspect of current research, considering the awareness of physical stress as a deteriorating health feature in developed and developing economies. The biology of glycation is one of the most evolving areas of present-day "Molecular Medicine" research. It is now widely accepted that the interaction of AGEs with RAGE enhances inflammatory and oxidative stress. This enhanced level of inflammation and oxidative stress propagates various non-communicable disorders, including diabetes, cardiovascular, pulmonary, and vascular complications, associated with major organs, viz. kidney, liver, pancreas, and nervous system. At present, the glycated haemoglobin *i.e.* HBA1c in the RBC is recognized as a decisive hallmark of diabetes, being in regular use for diagnosis. Finally, AGE-RAGE interaction aggravated extents of inflammation and oxidative stress complicates various cancers. Thus, studies probing the therapeutic usefulness of anti-AGE/anti-RAGE molecules against senescence and aging are being conducted with increasing reliability. Concerning this, every year thousands of manuscripts are published on "Glycosylation and Glycation", across the globe. Essentially, not only the students and teachers but also the glycobiology researchers too, would substantially benefit from this book, which in due course, would cement its place in many libraries as well as laboratories.

Tapan Kumar Mukherjee
Department of Biotechnology
Amity University, New Town, Kolkata

West Bengal 700156, India
Parth Malik
School of Chemical Sciences
Central University of Gujarat Gandhinagar
Gujarat-382030, India

&

Ruma Rani
ICAR-National Research Centre on Equines
Hisar-125001, Haryana, India

List of Contributors

Himel Mondal Department of Physiology, All India Institute of Medical Sciences, Deoghar, Jharkhand-814152, India

Parth Malik School of Chemical Sciences, Central University of Gujarat, Gandhinagar, Gujarat-382030, India

Swarrnim Startup & Innovation University, Bhoyan-Rathod, Gandhinagar-Gujarat, India

Ruma Rani ICAR-National Research Centre on Equines, Hisar-125001, Haryana, India

Raj Singh Department of Biotechnology, Maharishi Markandeshwar (Deemed to be University), Mullana, Ambala, Haryana 133207, India

Raman Kumar Department of Biotechnology, Maharishi Markandeshwar (Deemed to be University), Mullana, Ambala, Haryana 133207, India

Rajeev K. Singla Joint Laboratory of Artificial Intelligence for Critical Care Medicine, Department of Critical Care Medicine and Institutes for Systems Genetics, Frontiers Science Center for Disease-related Molecular Network, West China Hospital, Sichuan University, Chengdu, China

Shaikat Mondal Department of Physiology, Raiganj Government Medical College and Hospital, Raiganj, West Bengal, India

Tapan Kumar Mukherjee Amity Institute of Biotechnology, Amity University, New Town, Kolkata, West Bengal 700156, India

Vishal Haribhai Patel Institute of Biotechnology, Amity University, Noida-201301, Uttar Pradesh, India

CHAPTER 1

The Basic Concept of Glycosylation

Parth Malik[1,2,†], Ruma Rani[3,†] and **Tapan Kumar Mukherjee[4,*]**

[1] *School of Chemical Sciences, Central University of Gujarat, Gandhinagar, Gujarat-382030, India*

[2] *Swarrnim Startup & Innovation University, Bhoyan-Rathod, Gandhinagar-Gujarat, India*

[3] *ICAR-National Research Centre on Equines, Hisar-125001, Haryana, India*

[4] *Amity Institute of Biotechnology, Amity University, New Town, Kolkata, West Bengal 700156, India*

Abstract: Glycobiology aims at structure-function correlational analysis of carbohydrates (sugar or glycan). A monosaccharide is the simplest form of carbohydrate that no longer be hydrolyzed. The other forms of carbohydrates are formed by glycosidic linkages of monosaccharides, such as disaccharides, oligosaccharides, and polysaccharides, comprising two, three to ten, and more than ten monosaccharides, respectively. Carbohydrates act as one of the major energy sources (*e.g.*, ATP) and are also involved in cellular protection, stabilization, organization, and barrier functions. In the cellular system, carbohydrates are present in pure and protein-conjugated forms, which are referred to as glycoproteins. Conjugated carbohydrates are also present in the form of glycolipids and proteoglycans. Notably, *N- and O*-linked glycosylation as major forms occur in the rough surface endoplasmic reticulum (RER) and Golgi apparatus respectively, adding carbohydrates to proteins and thus making glycoproteins. Relatively fewer common types of glycosylation are the *C*-linked glycosylation, *S*-linked glycosylation, glypiation, and phosphoglycosylation. A complex interplay of two enzyme groups such as glycosyl transferases (adding carbohydrates to proteins) and glycosidases/glycosyl hydrolases (removing carbohydrates from proteins) control the glycosylation extent. Prominent cellular factors regulating glycosylation are the availability of carbohydrates, proteins, enzymes, movement of proteins from RER to Golgi, and several other environmental factors regulating post-translational modifications. This chapter describes the various aspects of glycobiology including protein glycosylation, purification, and analysis of glycans, and their role in physiology and pathophysiology.

^{*} **Corresponding author Tapan Kumar Mukherjee:** Amity Institute of Biotechnology, Amity University, New Town, Kolkata, West Bengal 700156, India; E-mail: tapan400@gmail.com
[†] These authors contributed equally to this work.

Keywords: Glycobiology, Glycosylation, Glycoproteins, Glycosyltransferases, Glypiation, Glycan analysis, Glycosidases/glycosyl hydrolases, *N*-linked glycosylation, *O*-linked glycosylation, Phosphoglycosylation, *S*-linked glycosylation.

INTRODUCTION

The study of structure, functions, and biological aspects of saccharides (carbohydrates, sugar chains, or glycans) is called "Glycobiology". The generic term for carbohydrate is interchangeable with sugar (the sweet-tasting carbohydrate) or glycans (compounds containing manifold monosaccharides). Biochemically, carbohydrates are polyhydroxy aldehydes or ketones. Based on the chemical structure, the number of monosaccharides, and length, glycans are divided into monosaccharides, disaccharides, oligosaccharides, and polysaccharides. While monosaccharides consist of a single sugar molecule (*e.g.*, glucose, fructose, and galactose), the disaccharides consist of two monosaccharides (*e.g.*, maltose: glucose + glucose; sucrose: glucose + fructose; lactose: glucose+galactose, *etc.*) joined *via* glycosidic linkage(s) (Fig. **1**). Carbohydrates that consist of three to ten monosaccharides are called oligosaccharides (*e.g.*, raffinose consists of trisaccharide) and those containing more than ten monosaccharides are called polysaccharides (*e.g.*, starch in plant cells and glycogen in animal cells). Dietary carbohydrates are the main sources of fuel (one gm of carbohydrate is equivalent to 4 calories) in most living organisms.

Carbohydrates or glycans are also involved in cellular protection, stabilization, organization, and barrier functions. At present, glycobiology is a swiftly emerging domain of biology, exhibiting relevance for biotechnology, and biomedicine, as well as basic research of carbohydrates, about physiological and various pathophysiological conditions (*e.g.*, host-pathogen interactions and subsequent adhesion, invasion, virulence, and pathogenicity of microorganisms). In the cellular system, carbohydrates exist in pure and conjugated form, with other molecules such as proteins and lipids. Following the synthesis of proteins *via* translation, several post-translational modifications are known to prevail. One of these modifications involves the addition of carbohydrates to the protein, at a specific pre-determined position using multiple enzymes. This process of carbohydrate-to-protein addition is called glycosylation. The glycosylated proteins are called glycoproteins. Not only proteins but some lipids and proteoglycans are also glycosylated. In general, the process of glycosylation occurs in specific cell organelles such as the endoplasmic reticulum (ER) and Golgi apparatus. Glycosyl transferases and glycosidases are the two enzyme groups involved in glycosylation. During glycosylation, the hemiacetyl group of a glycosyl donor reacts with the –OH or –NH$_2$ group of the protein (glycosyl accepter) forming a covalent bond. Glycoproteins are found in almost all living organisms including eubacteria, archaea, and eukaryotes. All eukaryotic living creatures from single-cellular to complex multicellular, generate glycoproteins. In most organelles of eukaryotic cells, proteins prevail as glycoproteins and are, therefore, involved in glycosylation.

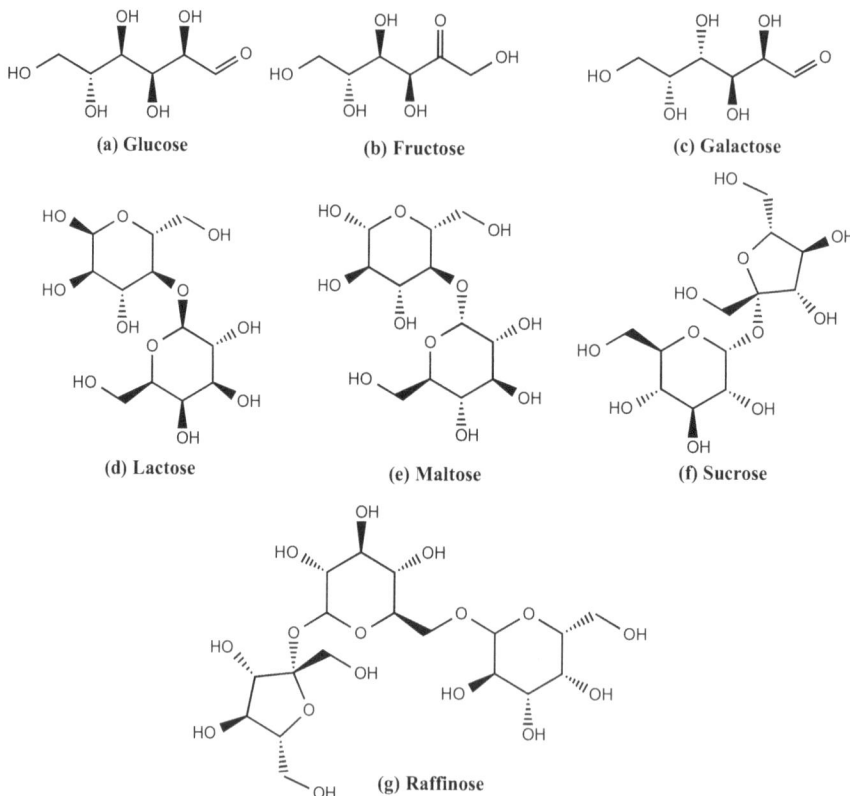

Fig. (1). Structural distinctions of some eminent mono (**a-c**), di (**d-f**), and trisaccharide (**f**). Typical features comprise similar or dissimilar monosaccharides as linking units to confer a functional diversity to the respective di, oligo, and polysaccharides.

Multiple kinds of glycosylation are known to happen within a cell, such as *N*-linked glycosylation, *O*-linked glycosylation, *C*-linked glycosylation, glypiation, and phosphoglycosylation. The *N*-linked glycosylation takes place in the ER lumen. In this mode of glycosylation, glycans bind to the $-NH_2$ group of the amino acid asparagine. On the other hand, in *O*-linked glycosylation, monosaccharides bind to the –OH group of the amino acids- serine, threonine, tyrosine, hydroxylysine, hydroxyproline side chains, or oxygens on lipids such as ceramide within the ER, Golgi apparatus, cytosol, and nucleus. To complete the glycosylation process, *N*-linked glycosylation entails the association of a special lipid, called dolichol phosphate. About erstwhile small glycosylation contingents in glypiation, the glycan core links a phospholipid and a protein. In the *C*-linked glycosylation, mannose binds to the indole ring of amino acid tryptophan, and finally, phosphoglycosylation happens at the site corresponding to glycan binding with amino acid serine *via* phosphodiester linkage. Enormously diversified proteins are formed because of glycosylation. This large diversity of glycosylation is related to glycosylation modifications, at every stage during the process of glycosylation such as glycan composition: the sugar types that are linked to a particular protein, glycan length: short- or long-chain oligosaccharides, glycosidic linkage: identification of glycan (oligosaccharide) binding location, and glycan structure: the extent of branching.

The purpose of glycosylation is to generate a functional protein, commencing from an immature protein. The process assists in adequate protein folding that enhances protein stability. The glycoproteins are involved in several physiological processes and maintenance of cellular homeostasis, which is not possible without the proper level of glycosylation of these glycoproteins. Briefly, some major utilities of glycosylation comprise: (i) Every glycoprotein is targeted to its proper destination. The functional proteins produced by glycosylation contain specialized signaling sequences called leader sequences that traffic the proteins to their specific destination, (ii) The newly synthesized glycoproteins are protected from being degraded by the various hydrolytic enzymes. Thus, glycosylation of the proteins serves as a protection from proteolytic degradation. (iii) Some membrane glycoproteins such as G-proteins act as receptors, which bind their specific cognate ligands and are involved in coordinating multiple signaling events of the cells. Finally, immunological molecules such as immunoglobulins (Igs) or antibodies (Abs) are nothing but glycoproteins. These immunological molecules safeguard our body against invading pathogens. Of late, several studies are dedicated to understanding the role of various glycoproteins (including antibodies) in targeting various immunological reactions, vaccine production, and next-generation therapy against multiple diseases such as cancers and manifold diagnostic assays (*e.g.*, human chorionic gonadotrophin [HCG] antibody is used to examine pregnancy). This introductory chapter describes the process of glycosylation, factors affecting glycosylation, analysis of glycosylation, biological importance of glycosylation, significance of various glycoproteins produced, and finally disease conditions related to improper protein glycosylation [1 - 3].

THE ENZYMES INVOLVED IN THE PROCESS OF GLYCOSYLATION

The process of glycosylation predominantly takes place in the ER lumen and inside the Golgi apparatus as part of the post-translational modifications of various secretory proteins. A detailed analysis reveals more than 50% of all cellular proteins are typically subjected to some form of glycosylation. In the process of glycosylation, carbohydrates are covalently added to the proteins. The proteins that are glycosylated through the ER-Golgi secretory pathways are the secretory proteins (*e.g.*, cytokine TNF and interleukins) and their receptors (*e.g.*, TNF receptors or TNFR), cell surface receptors including receptors for peptide/protein hormones (*e.g.*, insulin receptors), and growth factor (*e.g.*, vascular endothelial growth factor receptors or VEGFRs), antibodies (*e.g.*, IgM and IgD), ligands (*e.g.*, P selectin glycoprotein) and organelle resident proteins (Gp96, a resident protein of ER). Not only proteins, lipids (*e.g.*, rhamnolipids), and proteoglycans (*e.g.*, decorin, a leucine-rich proteoglycan that can bind to collagen) also undergo glycosylation and therefore potentially contribute to the number of substrates for this modification regime.

Many enzymes are involved in the process of glycosylation thereby making this process one of the most complex post-translational modifications. Typically, the process does not involve any template usage by the cells (unlike replication and transcription), but a host of enzymes that are added or removed as one or more sugars to proteins, producing a diverse group of glycoproteins. Despite being non-template dependent, the process of glycosylation is highly ordered, comprising stepwise reactions wherein individual enzyme activity depends on the completion of the preceding enzymatic reaction. At the cellular level, the molecular events of glycosylation involve the following three basic steps: (i) Linking monosaccharides together, (ii) Transferring sugars from one substrate to another, and finally (iii) Trimming sugars from the functional glycan structure. The enzymes involved in glycosylation may vary from cell to cell or even various compartments within a cell as also the glycan structure from one to another cell.

The process of glycosylation can be regulated by the action of the following enzymes:

• Glycosyltransferases

• Glycosidases

Here is a brief discussion about them:

Glycosyltransferases

• More than 30,000 glycosyltransferases (GTs) sequences are known for all living creatures /kingdoms (prokaryotic and eukaryotic).

• In eukaryotes present in ER and Golgi apparatus.

• Primary sequence analysis reveals these GTs consist of nearly 90 Gtf families.

• It is known that the two structural folds GT-A and GT-B act as nucleotide sugar-dependent enzymes.

• GTs add sugar residue to the oligosaccharide chain, protein, or lipid in the proper order.

• The GTs are vast in scope since glycosidic linkages have been discovered in practically every protein functional group, with glycosylation being implicated to a considerable level in the incorporation of most regularly occurring monosaccharides.

• Each GT has specificity for glycosidic linkage of a particular sugar with a donor (sugar nucleotide or dolichol) to a substrate (the –OH or –NH$_2$ group of a protein, called glycosyl accepter), functioning independently of other Gtfs.

• Thus, the transfer of mono- or oligosaccharide moieties from an activated nucleotide sugar to a nucleophilic glycosyl acceptor molecule is involved in the catalytic process of GTs. In this case, the nucleophile might be based on oxygen, carbon, or nitrogen. In addition, several transferases modify glycans by the addition of acetyl, methyl, phosphate, sulfate, and other groups.

• The GTs transfer the glycosyl groups to a nucleophilic acceptor with either retention or inversion of configuration at the anomeric center. This allows the classification of GTs as either retaining or inverting enzymes.

• The transglycosylation reaction requires the involvement of divalent cations such as Mg^{++} or Mn^{++} as cofactors and are active in the pH range that persists in the ER lumen and Golgi apparatus, *i.e.*, of pH 5.0 to 7.0.

• Several oligosaccharides can be produced using the available GTs and activated nucleotide sugars. Some of the most suited examples are, UDP-glucose (UDP-Glc), UDP-*N*-acetylglucosamine (UDP-GlcNAc), UDP-*N*-acetylgalactosamine (UDP-GalNAc), UDP-galactose (UDP-Gal), UDP-glucuronic acid (UDP-GlcUA), GDP-fucose (GDP-Fuc), GDP-mannose (GDP-Man) and CMP-sialic acid (CMP-NeuAc) [4].

• The glycosyl transferase inhibition induced by nucleotide diphosphates formed during the reaction is a significant limitation to enzyme-catalyzed glycosylation.

• GTs have been widely employed in drug discovery and development for the targeted synthesis of glycol-conjugates, as well as the synthesis of distinctively glycosylated libraries of medicines, biological probes, or natural products (a process referred to as glycol randomization).

The following two techniques are known to subside enzymatic inhibition:

1. The phosphatase enzyme, which removes the phosphate group and degrades the nucleotide diphosphates (Fig. **2A**).

2. Using multi-enzyme regeneration methods, nucleotide diphosphates are recycled to the appropriate nucleotide triphosphates. Although multiple enzymes and cofactors are involved in these *in situ* regenerations, the approach does not need the usage of stoichiometric sugar nucleotide concentrations (Fig. **2B**).

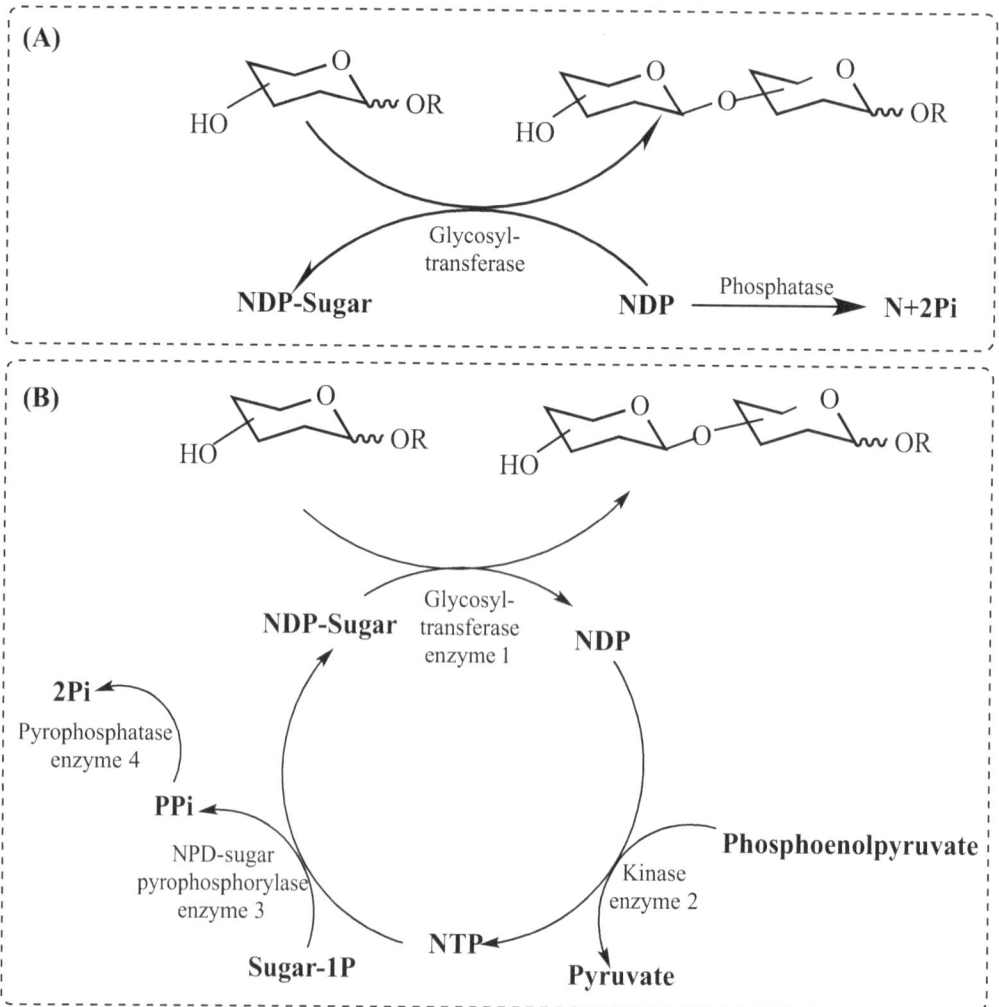

Fig. (2). Methods for preventing enzyme inhibition in glycosyltransferase-catalyzed synthesis. (**A**) Phosphatase addition. (**B**) Sugar nucleotide recycling (NDP = nucleotide diphosphates, NTP = nucleotide triphosphates, N = nucleotide, Pi = phosphate).

Glycosidases

• Except for some Archaeans and a few unicellular parasites, glycosidases, or glycoside hydrolases (GHs) are present in almost all living organisms.

• GHs are used to retrieve information about carbohydrate groups conjugated to glycopeptides and glycoproteins.

• GHs are critical for glycan processing in the ER, Golgi apparatus, and lysosomes. These enzymes are also found in saliva and the gastrointestinal tract.

• These enzymes hydrolyze the glycosidic bonds to eliminate sugars from proteins. Each glycosidase is specialized for eliminating a particular sugar (for example, mannosidase).

• Broadly, glycosidases are divided into exoglycosidases and endoglycosidases.

• While exoglycosidases release a single monosaccharide from the non-reducing terminus of an oligosaccharide, the endoglycosidases cleave the internal glycosidic bonds [5].

• For research purposes, endoglycosidases are promptly used followed by one or more exoglycosidases. In the next step, the product is analyzed using various modes of liquid chromatography and SDS-PAGE.

THE PROCESS OF VARIOUS TYPES OF GLYCOSYLATION

The following types of glycosylation occur in the cells:

1. *N*-Linked Glycosylation

2. *O*-Linked Glycosylation

3. *C*-Linked Glycosylation

4. *S*-Linked Glycosylation

5. Glypiation

6. Phosphoglycosylation

N-linked and *O*-linked glycosylation are the major types of glycosylation that predominantly, but not exclusively occur in RER and Golgi apparatus respectively (Table **1**). Fundamentally, both RER and Golgi are involved in both *N*-linked and *O*-linked glycosylation. The oligosaccharide GlcNAc2Man9Glc3 is cotranslationally transferred to specific asparagine residues of nascent glycoproteins in the RER for *N*-linked Glycosylation, where it is trimmed by glucosidase and mannosidase enzymes to GlcNAc2Man. The GlcNAc2Man reaches the Golgi apparatus, where it undergoes further trimming and elongation. Modifications in the Golgi apparatus contribute to the structural variety of *N*-linked glycosylation.

Table 1. The table summarizes the various types of glycosylation *via* their biological significance.

Types of Glycosylation	Distinctive Aspect
N-linked	Glycan binds with the -NH$_2$ group of asparagine in the rough surface of the endoplasmic reticulum (RER).
O-linked	Monosaccharides bind to the -OH group of serine or threonine in the RER, Golgi apparatus, cytosol, and nucleus.
C-linked	Mannose binds with a tryptophan indole ring.
S-linked	Conjugation of oligosaccharides to the sulfur atom of cysteine.
Glypiation	Glycan core links a phospholipid with a protein.
Phosphoglycosylation	Glycan binds to serine *via* a phosphodiester linkage.

O-linked glycosylation occurs on folded proteins and is triggered by the transfer of monosaccharides to serine and threonine residues. Some kinds of *O*-glycosylation begin in the RER, such as *O*-fucosylation [6] and *O*-mannosylation [7], whereas mucin-type *O*-glycosylation [8] and the glycosaminoglycan chain biosynthesis [9] commence within the Golgi apparatus.

The following paragraphs briefly describe various types of glycosylation processes:

N-Linked Glycosylation

• Approximately 90% of glycoproteins are *N*-glycosylated and therefore it is called the most common type of glycosylation. The glycans are covalently bonded to the carboxamido nitrogen on asparagine (Asn or N) residues, hence the name *"N"* linked glycosylation (Fig. 3).

• In this case, the glycan is bound to the nascent protein that is being translated and transported into the RER through sec61, a 61kDa pore-forming protein complex residing on the RER surface. This protein is involved in the activation and secretion of various secretory proteins and hence the name "sec" originated. Carbohydrates after being bound to the proteins undergo multiple processing steps and finally often produce large branched glycans.

• The proteins are released from RER and enter the Golgi apparatus through multiple vehicles (*e.g.*, COP-1, COP-II, *etc.*) and undergo manifold modifications to complete the process of glycosylation.

• Several lines of evidence indicate that both in archaea and in various eukaryotic species, a considerable number of enzymes as well as the glycosylation process remain conserved.

• Different enzymes are required for each phase of glycosylation, manifesting in glycans diversity.

• However, *N*-glycosylation is initially similar for all proteins, and the diversity is not established until later trimming and glycan maturation.

• These types of glycoproteins contain vast, typically heavily branched glycans that go through exhaustive processing after protein binding.

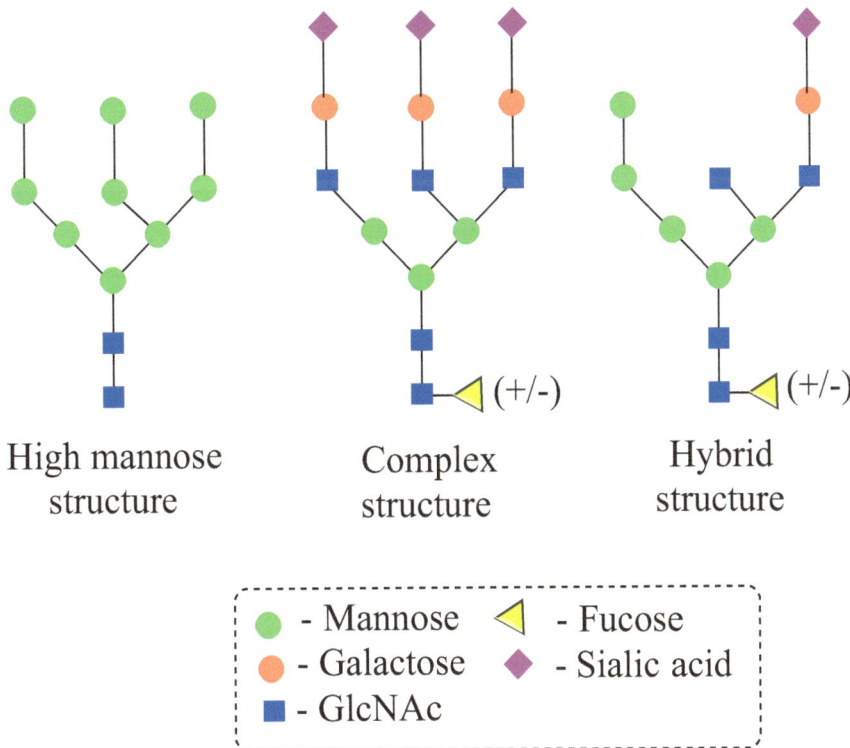

Fig. (3). Basic core structure of N- glycans.

N-glycosylation can be better understood *via* categorizing into the following events:

• Precursor Glycan Assembly and Attachment of the Assembled Glycans

• Glycan Trimming in the lumen of rough surface Endoplasmic Reticulum

• Glycan Maturation in the Golgi Apparatus

Here is a brief discussion of the above events:

Precursor Glycan Assembly and Attachment of the Assembled Glycans

The ribosomes are attached to the cytosolic face of RER. The *N*-terminus leader sequence of a *de-novo* generated linear sequence of amino acids or primary structure of a protein enters the RER lumen and this is the location where *N*-linked glycosylation takes place. So, the direction of the newly synthesized primary structured protein (newly generated nascent polypeptide chain which consists of amino terminus linear chain of amino acids) binds to the signal recognizing particles (SRPs) and subsequently enters RER.

Additionally, the oxidizing environment in the ER lumen assists in the protein folding *i.e.*, the formation of secondary, tertiary, and if needed, quaternary structure (only the proteins with multiple subunits such as heme protein in hemoglobin require a quaternary structure). Subsequently, amino acids are continuously added one by one to the newly generated nascent proteins, whereby the length of the ER lumen peptide is increased until protein synthesis is completed. Thus, it is in the ER lumen, where the simultaneous addition of sugars and protein folding happens.

Since access to amino acids of a continuously folding nascent peptide is necessary for the sugar addition by glycosylation process, sugar addition occurs before protein folding, particularly for those amino acids that should be located deep inside the completely folded protein structure. However, the amino acids of a newly synthesized protein that is completely exposed have easy access with the ability to undergo glycosylation in the post-folded state. The preferred amino acids for *N*-linked glycosylation are the Asn residues. Because of the inherent complexity of protein synthesis (translation) and folding, not all Asn residues of the predicted consensus sequence undergo glycosylation.

For glycosylation to occur, the precursor glycans must be assembled within the cytoplasm, followed by attachment on the RER surface, subsequent internalization into the RER lumen, and finally attachment with the specific amino acids of the protein that are newly generated and folded.

• In the first step, a 14-sugar precursor molecule composed of N-acetylglucosamine (GlcNAc), mannose (Man), and glucose (Glc) produces oligosaccharides linked *via N*-glycosidic linkages.

• Dolichol is a poly-isoprenoid lipid carrier that is incorporated in the RER membrane. It is used to transfer sugar molecules in the RER. In the process, 7 sugars are donated from sugar nucleotides (UDP- and GDP-sugars) within the cytoplasm, remaining bound to dolichol *via* pyrophosphate linkage (-PP-). The synthesis of a Man5GlcNAc2-PP-dolichol intermediate completes the process.

• In the final step, the seven sugars are conferred by Man- and Glc-P-dolichol molecules to make the Glc3Man9GlcNAc2-PP-dolichol precursor glycan.

• Eventually, the entire complex is flipped into the RER lumen. The precursor glycan is then transferred to the Asn residues on the nascent protein by oligosaccharide transferase (OSTase).

• The consensus sequence of the *N*-linked glycosylation is Asn–X–Ser/Thr where X is any amino acid except proline (pro). After *N*-linked oligosaccharides are transferred to nascent proteins by the OSTase, ER-resident glucosidases, and mannosidases generate a series of glycan-trimming intermediates that are specifically recognized by ER-localized lectins to direct the nascent proteins into protein folding.

• As protein folding increases, the ability of OSTase to access the consensus sequence for glycan transfer, becomes impaired. It has been observed that more *N*-terminal Asn residues are glycosylated than C-terminal amino acids [10].

Glycan Trimming in the Endoplasmic Reticulum

• Both RER and Golgi are the glycosidase targets, where oligosaccharides are trimmed *via* the process of hydrolysis. However, glycan trimming in RER differs from the glycan trimming in the Golgi apparatus and serves multiple purposes.

• Sugar hydrolysis is employed in the RER to both monitor protein folding and to identify the proteins that should be degraded.

• Glucosidases I and II remove two Glc terminals from the precursor glycan, and then calnexin and calreticulin (the membrane-bound and soluble sugar-binding lectins, respectively), attach to the nascent glycoprotein *via* remaining Glc and function as chaperones to aid in appropriate protein folding.

• Glucosidase II quickly hydrolyzes the last Glc, releasing the glycoprotein from the chaperone. UDP-glucose glycoprotein glucosyltransferase recognizes non-native folded proteins by transferring a Glc unit to the glycoprotein and subsequently binding the protein to the lectin chaperones to promote correct protein folding.

• Until the protein is correctly folded the Glc addition and removal are cyclically repeated. At that point, no protein is reglycosylated and the glycoprotein is trafficked to the Golgi for additional processing.

• As discussed previously, all the glycans are accurately folded glycoproteins that translocate to the Golgi apparatus, like Man9GlcNAc2 in higher eukaryotes.

• The RER-resident mannosidase (RERManI) decisively identifies the accurately unfolded proteins. Proteins that lose 3-4 Man residues in the RER *via* RERMANI activity are moved out of the RER and deglycosylated by glycanase N (losing entire glycan *en bloc*), before being eliminated *via* RER-associated degradation (RERAD).

NB

Because of its slow rate of Man hydrolysis, RERManI acts as a sort of timer that enables multiple reglycosylation of nascent proteins for adequate folding before folding.

Maturation of Glycans in the Golgi Apparatus

• During the glycosylation process, all *N*-linked glycoproteins exhibit similar precursor glycan structure.

• To distinguish the glycans on individual glycoproteins, the glycan processing in the Golgi apparatus is completed *via* combinative trimming and sugar addition.

• The process of distinct glycan synthesis is a highly ordered sequential event involving multiple enzymes that reside in the specific compartment or cisternae of the Golgi apparatus. The compartmentalization of Golgi enzymes is highly essential to complete the glycosylation.

The final glycan structures can be broadly divided into the following groups:

• Complex oligosaccharides are made up of several different kinds of sugar.

• Multiple Man residues found in high-mannose oligosaccharides.

• Hybrid molecules are constituted of branches of high Man and complex oligosaccharides.

• Glycans designed to be complex oligosaccharides are trimmed by Golgi mannosidases I and II, being glycosylated by GlcNAc transferase, with a common core region.

• The core then functions as the substrate for multiple Gtfs which successively transfer sugar moieties from sugar nucleotides to build variable-length and -branched oligosaccharides such as GlcNAc, galactose (Gal), *N*-acetylneuraminic acid (NANA or sialic acid) chains and fucose.

• Any glycoprotein that progresses through this processing from the common core stage becomes resistant to glycan removal by enzyme endoglycosidase H (endo H, cleaves the glycosidic bond between two N-acetylglucosamine (GlcNAc) residues in the core region of N-linked glycans), that used to experimentally determine the high-Man or complex oligosaccharide composition and structure of glycoproteins.

• In contrast to complex oligosaccharides, high MAN oligosaccharides lack additional sugar moieties, though certain Man residues may be trimmed by Golgi mannosidase I.

• The determination of whether a glycan is transformed into a complex oligosaccharide or prevails as a high Man-containing oligosaccharide, depends on how easily processing enzymes can access the glycan, which may remain delayed by the conformation of the glycoprotein.

• Some glycoproteins possess hybrid oligosaccharides with a mixture of complex and high-MAN glycans.

O-Linked Glycosylation

• This mode of glycosylation occurs in the Golgi apparatus.Cytosols and nuclei are also possible sites of *O*-glycosylation.

• *O*-glycosylation occurs frequently in the secretory pathway on glycoproteins that have already been *N*-glycosylated within the ER.

• However, *O*-linked glycans usually have simpler oligosaccharide structures than *N*-linked glycans (Fig. **4**).

• *O*-linked glycosylation does not require a consensus sequence, and no oligosaccharide precursor is required for protein transfer.

• In this process, carbohydrates are added post-translationally on serine and threonine amino acid protein residues.

• The oxidized forms of lysine and proline *i.e.*, hydroxylysine and hydroxyproline are also involved in the *O*-glycosylation process.

• Mucin, a family of highly *O*-glycosylated, high-molecular-weight proteins that make up mucus secretions, is the best example of an *O*-glycosylated protein. Proteoglycan core proteins, which prevent the prevalence of other proteins that are substantially glycosylated including various antibodies, are also formed by *O*-glycosylation.

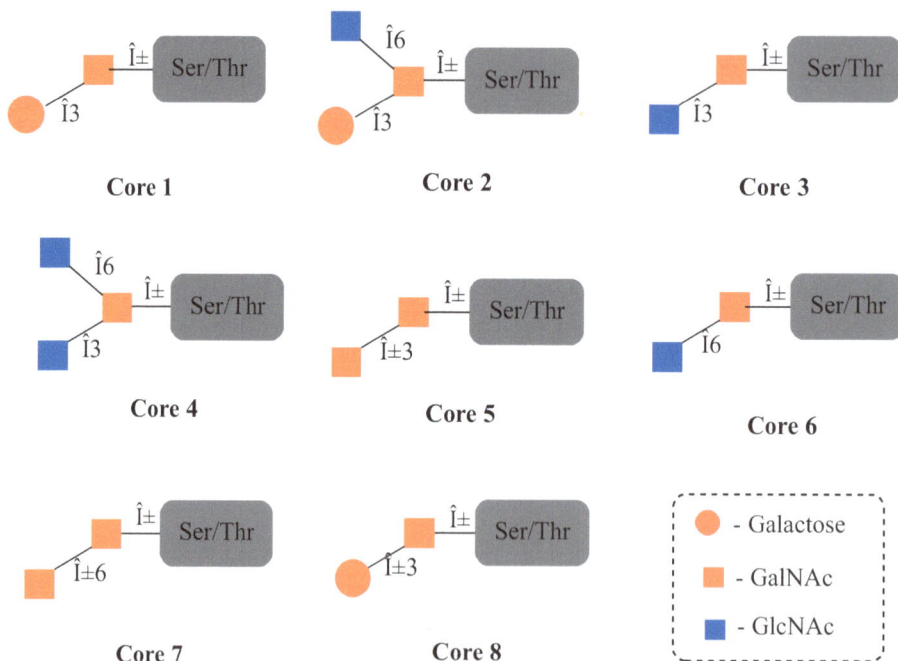

Fig. (4). Basic core structure of *O*-glycans.

Mechanism of *O*-linked glycosylation

• The *O*-linked glycosylation process is simpler than N-linked glycosylation.

• The sugar nucleotides act as monosaccharide donors during *O*-glycosylation.

• Depending on the cell and species, some proteins, such as fucose, xylose, Gal, or Man are *O*-glycosylated with N-acetylgalactosamine (GlcNAc).

• In this process, the proteins are *O*-glycosylated by GalNAc transferase (GalNAc-Ts), which transfers a single GalNAc residue to the serine or threonine β-OH group.

• While the structural motif of this enzyme has been characterized, a consensus sequence for this enzyme has yet to be discovered.

• Following the first sugar addition, a highly variable number of sugars (from only a few to more than 10) are consecutively added to the growing glycan chain [11].

C-Linked Glycosylation

• In this mode of glycosylation, the tryptophan residues of an extracellular protein bind with a Man subunit.

• To date, two recognition signals of glycosylation have been characterized. They are W-X-X-W (the first or both tryptophan residues become mannosylated) and W-S/T-X-C.

• In the case of C-mannosylation, C-C bonds rather than C-N or C-O bonds are formed.

• Further, *C*-mannosyl transferase (c-Mtf) links C1 of Man to C2 of the tryptophan indole ring.

• The enzyme recognizes the specific sequence Trp-X-X-Trp and transfers a Man residue from dolichol-P-Man to the first Trp subunit in the protein sequence.

• The rat liver microsomes as well as multiple cell lines undergo C-mannosylation.

• Some examples of *C*-glycosylated proteins are IL-12B, the erythropoietin receptor, and Trp2 in RNase.

NB

The C-glycosylated proteins are resistant to metabolic hydrolysis and therefore these proteins are the subjects of intense research interest across various species of bacteria, insects, and plants.

S-Linked Glycosylation

• This is a rare form of glycosylation documented in bacteria and humans.

• In this process of glycosylation oligosaccharides are attached to the sulfur atom of cysteine.

Glypiation

• This type of glycoprotein is localized in the membranes of archaea, protozoa, fungi, and several other eukaryotic cells.

• Thus, the specific purpose of this type of glycosylation is to anchor proteins to the biological membrane.

• In the process of glypiation, there is a covalent attachment of a glycosylphosphatidylinositol (GPI) anchor into the membrane as a stage of post-translational modification of proteins as well as targeting of membrane proteins to the specific locations.

• Enzymes like phospholipase C can cleave the GPI anchors and allow them to control the localization of the plasma membrane-anchored proteins.

• It is claimed that GPI anchor is almost ubiquitous across eukaryotes.

• A large number of functionally diverse proteins are GPI anchored including receptor proteins, prions, hydrolytic enzymes, protozoan coat proteins, complement regulatory proteins, adhesion molecules, and others.

• Interestingly, in the mammalian system multiple variants of GPI-anchored proteins are produced, either as membrane-anchored or soluble proteins. These proteins are generated by alternative mRNA splicing. For example, neural cell adhesion molecule (NCAM) is present in the GPI-anchored protein, both in soluble (in the muscle) and membrane-anchored forms (in the brain).

A GPI anchor consists of the following:

• Phosphoethanolamine linker that binds to the C-terminus of target proteins.

• The structure of the glycan core.

• A Phospholipid tail anchors the structure to the membrane.

NB: The lipid moiety of the tail and the sugar residues in the glycan core both exhibit significant variation, demonstrating substantial functional diversity including immune recognition, cell adhesion, and signal transduction.

Mechanism of GPI Anchored Protein Synthesis

The following steps are involved in the biosynthesis of GPI-anchored proteins:

1. Pre-assembly of a GPI precursor in the ER membrane.

2. Attachment of the GPI to newly produced protein in the ER lumen, while simultaneously removing a carboxy-terminal GPI-addition signal peptide.

3. Lipid remodeling and/or carbohydrate side-chain modification in the ER and post-Golgi transport.

The biosynthesis of GPI anchors, the precursor glycan for *N*-glycosylation, commences on the cytoplasmic side of the ER and ends on its luminal side. During this process, sugars obtained from sugar nucleotides and dolichol--mannose, both inside and outside the ER, to build 3-4 Man and various other sugars likeGlcNAc and Gal on a phosphatidylinositol (PI) molecule that is embedded within the membrane. Additionally, to facilitate anchor binding to proteins2-3 phosphoethanolamine (EtN-P) linker residues are contributed from phosphatidylethanolamine in the ER lumen.

Proteins that are glypiated have two distinct signal sequences. The first is found at the N-terminus of the protein and directs its transport within the ER during translation. The second signal sequence is located at the C-terminus of the protein and is recognized by GPI transamidase (GPIT). Instead of relying on a predefined consensus sequence, GPIT identifies a specific C-terminal motif. After translation, this motif allows GPIT to covalently attach a GPI anchor to the protein, as it embeds itself in the ER membrane. Then the protein is separated from the sequence and linked to a pre-existing GPI anchor.

The development of a cell-free system in *Trypanosoma brucei*(a species of parasite) enabled the study of GPI precursor biosynthesis. On the surface of each trypanosome are approximately 10 million molecules of variant-specific surface glycoprotein (VSG)with GPI anchors on the surface. Consequently, the enzymes and intermediate compounds involved in the GPI biosynthetic pathway are found in relatively large quantities in microsomal membrane preparations from this organism. Research into the sequence of events related to GPI biosynthesis has been conducted in various organisms, including *T. cruzi*, *T. brucei*, *Plasmodium falciparum*, *Leishmania* major, *Toxoplasma gondii*, *Paramecium*, *Cryptococcus neoformans*, *Saccharomycescerevisiae*, and mammalian cells. The emphasis on eukaryotic microbes is due to the abundance of GPI-anchored proteins (GPI-APs) in these organisms and the therapeutic significance of inhibiting GPI. This concept has been confirmed genetically in yeast, *Candida albicans,*and in the bloodstream of *T. brucei* [12].

Phosphoglycosylation

Phosphoglycosyltransferase (PTase, usually referred to as cyclic nucleotide phosphodiesterases) is an enzyme that transfers a prefabricatedphosphoglycan from a membrane-bound molecule; however, the exact structure and enzyme vary from species to species. This is a special type of glycosylation procedure wherein glycans are linked by serine or threonine *via* phosphodiester bonds. However, this is not a generalized glycosylation procedure across all the species since it is confined to parasites (*e.g.*, *Leishmania* and *Trypanosoma*) and slime molds (*e.g.*, *Dictyostelium*).Extensive studies demonstrate that phosphoglycosylation occurs in parasites such as *Leishmania*, where it prevails as the most abundant post-translational modification and is used to make proteophosphoglycans (PPGs), which are critical for host complement protection and promote parasite aggregation in the host [13].

THE POST-GLYCOSYLATION MODIFICATIONS

Multiple kinds of glycosylation could prevail with the same protein. Additionally, as part of the post-glycosylation events glycans can be modified and diversified by the various biochemical reactions.

The following biochemical reactions are involved in post-glycosylation modifications:

• **Sulfation:**

This is the addition of sulfur group to the Man and GlcNAc residues during the glycosaminoglycans (GAGs) production, the typical proteoglycan components in the extracellular matrix (ECM).

• **Acetylation:**

This modification involves an addition of acetyl group at sialic acid to facilitate protein-protein interactions.

• **Phosphorylation:**

This type of post-glycosylation involves the addition of phosphorus to Man residues on precursor lysosomal proteins (mannose 6-phosphate) to ensure their trafficking to lysosomes *via* binding to mannose 6-phosphate receptor (M6PR) in the Golgi.

FACTORS AFFECTING GLYCOSYLATION

A single protein not only undergoes glycosylation at multiple sites but also involves the addition of glycans with diverse glycosidic linkages at distinct sites [14].

Factors that affect the process of glycosylation are as follows:

Protein Conformation

The amino acids of a protein that need to be glycosylated, must be available for glycosylation. While glycosylation onsets as soon as the *N*-terminus of the newly synthesized protein (primary structure) enters the RER lumen, it continues even after the completion of protein folding. In a folded protein, only the exposed amino acids are the accessible sites of glycosylation.

Amino Acid Sequence

Specific amino acids and specific consensus sequences or motifs are required for glycosylation. For example, Asn is required for *N*-linked glycosylation, and Ser and Thr residues are needed for *O*-linked glycosylation.

Enzyme Availability

Available enzymes and their specific concentration at certain localized cellular compartments are another important requirement for the glycosylation process.

Example: Following completion of the glycosylation process, the RER releases the glycoprotein in the Golgi apparatus where various vehicles and trafficking proteins (*e.g.*, COP-1 and CPO-2) catch, transport, and release the glycoproteins. The manifold cisternae of Golgi have GTs and GHs which differ from the ones that act on the proteins within the RER lumen.

Movement of the Proteins Through the Rough Surface ER and Golgi Apparatus

• The ribosomes at the surface of ER synthesize the secretory proteins.

• While protein synthesis continues at the N-terminus, the newly synthesized proteins enter the ER lumen through the leader sequence or signaling sequence. Sec61, a pore-forming protein at the RER surface helps in the N-terminus entry of the newly synthesized proteins into the ER lumen. It is claimed that the oxidizing environment of the RER lumen aids in chaperon-assisted protein folding, including disulfide bond formation between cysteine amino acids. Consequently, or simultaneously, glycosylationoccurs in the RER lumen.

• Following competition, all kinds of post-translational modifications including glycosylation, phosphorylation, sulfation, addition of lipids, metals, *etc.*, the proteins move out of RER.

• Outside the RER, the protein vehicular traffic such as COP-1, and COP-2 brings them inside the cis-Golgi apparatus.

• Additional protein modification, including *O*-linked glycosylation happens in the Golgi apparatus (various cisternae of Golgi apparatus), following which proteins move from cis to trans-Golgi apparatus *via* medial Golgi apparatus and are finally secreted from the trans-Golgi apparatus into the cytoplasm.

• In the next step, completely modified proteins are subjected to "protein sorting". The specific signaling sequences, called leader sequences present at the N-terminus of a protein determine the destination of a protein. During this process, depending upon the signaling sequence, the Golgi secreted properly folded and modified proteins would move to their proper destination (for example plasma membrane-targeted proteins should reach the plasma membrane, nucleus-targeted proteins should reach the nucleus, *etc.*).

• However, the unfolded/misfolded proteins would reach lysosomes and endosomes for their degradation.

• Thus, the trafficking of proteins from the ribosomes on the RER surface into the RER and from the RER to Golgi is important for complete and proper folding as well as for the completion of various post-translational modifications including the several kinds of glycosylation.

Availability of Sugars

The sugar content in a cell culture medium can influence the glycoforms of proteins, according to experimental evidence from *in vitro* cell culture studies. In a comprehensive review article in 1918, *Blondeel and colleagues* discussed the following: "Glycosylation metabolism is affected when nucleotide-sugar precursors and associated substances are introduced to the growth medium in cell cultures. Though the results can differ, these supplements raise the nucleotide-sugar pools within cells and influence the distribution of glycoforms. Five primary factors account for these variations: (i) Cell type differences, which can affect the expression levels of enzymes and transporters; (ii) Variations in the recombinant proteins being produced, especially about glycan-site accessibility; (iii) Factors related to the fermentation and sampling timeline, such as glucose availability and exoglycosidases building up; (iv) Glutamate levels, which can affect ammonia levels, which in turn affects Golgi pH and UDP-GlcNAc pools; and (v) Finally, there are no standard metrics for consistently evaluating changes in glycol form distributions (glycosylation indices) across various experiments [15].

Environmental Factors

• Several factors that affect the proliferation and maintenance of the *in vitro* cell culture also affect the process of glycosylation.

• The major factors that affect the proliferation and maintenance of cells include cell culture medium components, pH, temperature of the culture medium, aeration, CO_2 content (generally 5%), and moisture content (95%). These factors also affect glycosylation.

• The process of glycosylation is affected by the integrity, health, and physiology of RER and ER membranes.

NB

When bioprocessing scientists optimize the culture medium for protein productivity and quality, they may modify the above factors to achieve the desired glycoform.

ANALYSIS OF GLYCANS

As discussed previously, the process of glycosylation and the specific glycoproteins produced therein affect the integrity, health, and physiology of the cells. It is also pertinent to mention that impaired glycosylation may cause multiple diseased conditions. Therefore, it is essential to determine the cellular level of glycosylation, and the specific glycoproteins produced by them. Besides checking the abundance of specific glycoproteins, the glycosylation site(s) and the glycoproteins' structure also need rigorous screening for monitoring the overall glycosylation status of cellular proteins [16, 17].

• The growing impact of glycoproteins on biological processes and pathological conditions has spurred the advancement of high-throughput and highly sensitive detection and analytical methods to gain a better understanding of the various structures and biochemical characteristics of glycoproteins.

• The great diversity in glycan structure and composition further adds to the complexity of glycoprotein analysis. The complexity arises from the fact that glycoproteins, being a fusion of proteins and oligosaccharides, are more challenging to scrutinize compared to non-glycosylated proteins.

The standard procedure of glycosylation involves the following:

• Site-specific enzymatic proteolysis (usually with proteolytic enzyme trypsin).

• Fractionation of glycopeptides (most often by liquid or affinity chromatography).

• Glycopeptide analysis using mass spectrometry (MS).

The following paragraphs briefly describe the various procedures that help to analyze glycoproteins. The protocols are herein standardized and are retrieved from 2012 review article by Roth and Khalaila [18].

Glycan Staining and Labeling

• Because of their structure, glycan sugar moieties are not responsive to staining or labeling molecules.

• Therefore, a chromogenic gel staining method called periodic acid staining (PAS) is used to stain glycosylated proteins in various biological samples.

• In addition to PAS staining, some other stains are alcian blue and Stains-all, which are employed for staining glycans, particularly for screening proteoglycans, glycosaminoglycans, and negatively charged glycoproteins.

• Glycans can be stained using periodic acid to make sugars reactive, allowing them to form covalent bonds with labeling molecules like biotin or immobilized support such as streptavidin for detection or purification purposes.

• To perform this staining, glycoprotein samples are first separated on an SDS-PAGE gel. Subsequently, the gel is treated with aperiodic acid-Schiff (PAS) stain to detect glycoproteins.

• The PAS reaction results in the development of a magenta color. During this color development, periodic acid oxidizes two adjacent diol groups, forming an aldehyde that then reacts with the Schiff reagent to create the magenta color.

• This staining method is typically done with fuchsin acid, which can be detected fluorescently at 535 nm, making it two to four times more sensitive than traditional visible staining methods.

• Certain fluorescent stains available in the market utilize periodate oxidation to attach a fluorescent hydrazide directly to glycoproteins, eliminating the need for further reduction steps. These fluorescent dyes are sensitive and capable of detecting even tiny amounts of glycoproteins down to 18 ng.

NB

However, it is important to note that periodate oxidation methods can lead to non-specific reactions with proteins of inherent oxidizable functional groups like aldehydes or ketones. Furthermore, the staining intensity may be lower in proteins with fewer glycosylation sites or less complex glycan structures than in heavily glycosylated proteins. Likewise, stains such as alcian blue and Stains-All can cause non-specific reactions with phosphoproteins and other negatively charged proteins. When glycosylation is required the removal of carbohydrates from a protein reduces its molecular weight, allowing SDS-PAGE to screen for glycosylation. Deglycosylases are used to remove glycans from glycoproteins, followed by SDS-PAGE to determine whether a protein is glycosylated.

Affinity-Based Procedures

The more specific methods for the determination of the glycosylation regime are given below:

Saccharide Binding Proteins

• Various living creatures including microbes, plants, and animals express a specialized carbohydrate (mono/oligosaccharide) binding protein, called lectin (plant proteins that bind carbohydrates).

• At present, several companies such as Roche, Applied Bioscience, Sigma, *etc.* commercially sell various lectins. These lectins are widely used for carbohydrate detection and characterization.

• To complete the Western blot, the denatured proteins/samples resolved in denatured protein gel (SDS-PAGE), transferred onto nitrocellulose or polyvinylidene fluoride (PVDF) membrane and then incubated with a specific lectin.

• In the next step, the lectin-treated membrane is labeled with either digoxigenin (DIG) or biotin.

• Finally, the membrane surface is coated with alkaline phosphatase/horseradish peroxidase (HRP)-conjugated secondary antibody or avidin.

• The color change intensity of luminescence would be assessed to understand and know about the saccharide binding proteins.

Enzyme-Based Techniques

• Enzymes in Click chemistry can be employed to label and separate glycosylated proteins.

• A prime example of glycosylated protein carries O-β-GlcNAcylation or features O-α- and N-glycans with a terminal GlcNAc.

• This protein, whether in pure form or within a protein mixture, is exposed to GALT (β-1,4-galactosyltransferase), which specifically adds an azidogalactose to GlcNAc.

• In the subsequent step, a fluorescent alkyne [a dye that can be used to label azide-tagged molecules through a copper-catalyzed click reaction (CuAAC)] is used to react with the modified O-GlcNAc.

• Following this, the sample is subjected to SDS-PAGE, allowing for the visualization of the modified protein under UV light using a gel imager.

• Likewise, mass spectrometry (MS) can be employed to label the O-GlcNAcylated peptides for further analysis.

Antibody-Based Methods

• To identify O-β-GlcNAc, a unique glycan structure, a method with a specific affinity for this glycan moiety can be employed. This is possible because primary antibodies, like commercially available CTD 110.1 and RL2 antibodies, have been developed specifically against O-β-GlcNAc.

• To detect O-GlcNAcylated proteins by Western blotting, antibodies are utilized in conjunction with secondary antibodies that are enzyme-conjugated (such as HRP or catalases). These enzymes catalyze reactions that generate color or luminescence, and this can be leveraged through techniques like blotting to identify O-GlcNAcylated proteins. Fig. (**5**) illustrates the step-by-step process of using antibodies to analyze post-translational modifications of proteins.

Glycoprotein Purification and Enrichment

For the purification of glycoproteins, lectins are widely used. As discussed in the previous paragraphs, lectins are specialized sugar molecules produced by microorganisms, plants, and animals. These proteins assist in protein folding inside the RER. Other studies reveal that lectins are associated with host-pathogen interaction. The specialized sugar lectins are the choice sugars for glycoprotein purification for the following reasons:

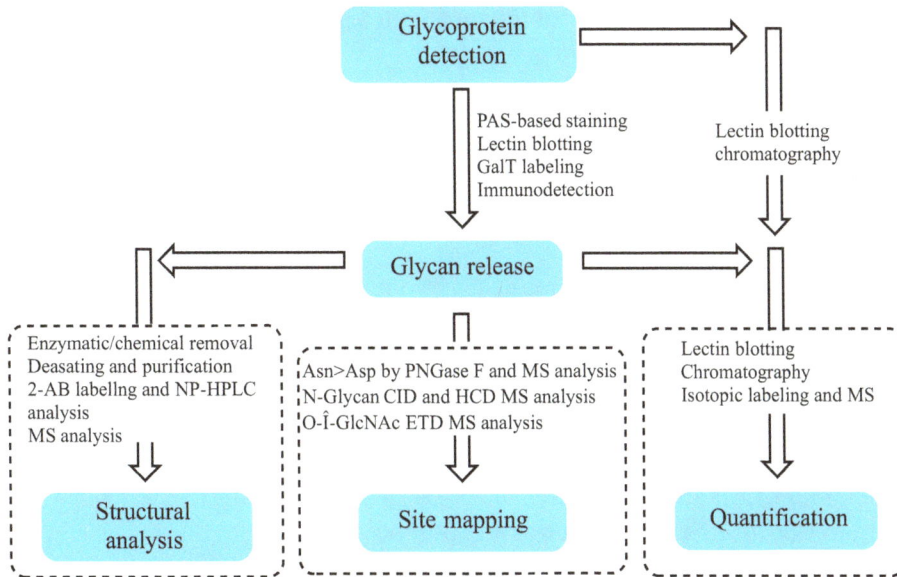

Fig. (5). Identification and Quantification of Protein Glycosylation.

Stability of Lectins

Compared totemperature-sensitive antibodies, lectins are stable over a wide range of temperatures.

Affinity towards Various Sugars

Lectins exhibit a specific affinity towards distinct sugar molecules and are, therefore, widely used for glycoprotein purification and their subsequent biochemical characterization.

Affinity towards Fluorophores, Horseradish Peroxidase, And Biotin

Just like antibodies, lectins can also bind with fluorophores, HRP, and biotin, *via* immobilization over a solid support including avidin or streptavidin.

Commercial Liability of Lectin

Compared to antibodies, lectins are cheaper and are, therefore, a preferential choice of interest. Examples of important lectins available as diversified commercial kits are Concanavalin A (ConA) from the Jackbean, Jacalin, wheat germ agglutinin (WGA), and lentil lectin (LCA).

Lectins are used for the following purposes:

• Characterization of cell surface glycoconjugates (*e.g.*, glycoproteins, glycolipids, GPI-anchored molecules).

• Screen a relative abundance of specific glycoprotein(s).

• Identify tissue and cellular localization of glycoprotein(s).

• The generation of mutants against glycoprotein(s).

• Activity analysis of GTs, GHs,*etc.*

Glycome and Glycoproteome Analysis

• Carbohydrate analysis requires sophisticated instruments since glycans prevail either as neutral molecules or charged molecules with no intrinsic UV absorbance. Although some of the carbohydrates do possess a surface charge, they are without any intrinsic UV absorbance or auto-fluorescence.

• The first phase of glycan analysis comprises the confirmation of glycan presence in a protein, *i.e.*, to confirm whether the experimental protein is a pure protein or a glycoprotein. While it may be preferred to analyze the glycan molecules of a glycoprotein on its release, the glycans can also be screened whilst being bound with the native protein or in its glycoprotein form.

• To liberate glycans from the glycoprotein, the choice between using endoglycanase H (endo H) or peptide-N4-(N-acetyl-beta-glucosaminyl) asparagine amidase (PNGase) treatment depends on the specific experiment type.

• The structural makeup of the glycan moiety within a glycoprotein can be investigated through chromatography or mass spectrometry.

• For quantitative comparative analysis of glycoproteomes, researchers can employ stable isotope labeling of amino acids in cell culture (SILAC) reagents.

• Furthermore, absolute quantification of targeted glycoproteins can be achieved using selected reaction monitoring (SRM) with isotopically labeled "heavy" reference peptides.

• Thin layer chromatography (TLC)is generally used for screening glycans where manifold glycans are abundant. However, as observed from the TLC, results the resolution is poor for complex mixtures and large molecules.

The basic steps through which glycoproteins are analyzed are as ahead:

• Enrichment of glycoproteins/glycopeptides.

• Multidimensional separation by Liquid Chromatography (LC) and Tandem Mass Spectrometry.

• Data analysis using bioinformatics.

Here the first two methods for glycans analysis are briefed, the excerpts could be traced in the 2009 contribution of Mulloy and colleagues in the second edition of "Essentials of Glycobiology".

Glycan Analysis by Mass Spectrometry

• Mass spectrometry (MS) analysis relies on glycan ionization, fragmentation, and the subsequent identification of mass fragments. Suitable MS methods for glycan analysis include electrospray ionization (ESI) and matrix-assisted laser desorption/ionization (MALDI).

• To prepare the glycan sample for analysis and prevent ionization interference during ESI and MALDI, it is necessary to desalt it. In MALDI analysis, the analyte is ionized using a matrix, often resulting in the formation of a single sodiated ion $[M^+Na]^+$.

• Various matrices are suitable for glycan screening, with 2,5-dihydroxybenzoic acid (DHB) being the most used. However, due to the diverse polarities of glycans, 2,4,6-trihydroxyacetophenone (THAP) is a more appropriate matrix in certain cases.

• In ESI and nanospray methods, the solvated analyte is converted into an aerosol using the electrospray technique. In these methods, ions are typically multiply charged, which can add complexity to the analysis of glycans.

Glycan Analysis Using Chromatography

• Chromatography is a laboratory technique for the separation of various biological molecules or mixtures of solutes based on the relative distribution of each solute between a moving fluid stream, called the mobile phase, and a contiguous stationary phase.

• The fluorescent moiety is conjugated by reductive amination to the glycan-reducing end.In this study, 2-aminobenzoic acid (2-AA), 2-aminobenzamide (2-AB), and 2-aminopyridine (2-AP) are the most frequently used fluorophores.

• On the other hand, conjugation may result in significant desialylation with aminopyridine-based fluorophores, whereas the relative extent of desialylation is low with the other two fluorophores, 2-AA and 2-AB. This can hinder a specific glycan characterization.

• Because of its negative charge, 2-AA is more suitable for electrophoresis but less suitable for chromatography and MS.

• The 2-AB fluorophore labels glycans in a non-selectively 1:1 stoichiometry, allowing sub-picomolar detection while preserving adequate molar proportions.

• For glycans separation, the most widely used chromatographic methods are normal-phase high-performance liquid chromatography (NP-HPLC), gel filtration (on Bio-Gel P4), weak anion exchange (WAX), and high-performance anion-exchange chromatography with pulsed amperometric detection (HPAEC-PAD).

• Though all methods are efficient, the appropriate method should be carefully selected to meet the specific requirements and properties of the glycans being analyzed. Gel filtration chromatography and WAX necessitate many oligosaccharides, whereas HPAECPAD necessitates high-pH and high-salt buffers (which must be removed before analyzing the separated glycans, as in the exoglycosidase sequencing).

• With its high resolution and repeatability, NP-HPLC is the most effective method for glycan analysis. Using a partially hydrolyzed 2-AB-labeled dextran as an external standard, the retention time of the unknown glycan (about the standard) is converted to glucose units (GU).

• The GU estimate is compared to an experimental value database to derive an approximate glycan structure. For some glycans, however, the NP-HPLC resolution might not be sufficient. Consequently, a supplementary technique like reverse-phase-HPLC (RP-HPLC) is needed.

• Because the dextran standard does not resolve well on RP-HPLC, arabinose is used as an alternate standard. An array of exoglycosidases can be used to validate the glycan sequence further, which involves sequentially applying specific exoglycosidases to cleave terminal monosaccharides from the non-reducing glycoprotein terminal. This chromatographic method also allows for the quantification of relative glycans by computing the peak area, which estimates the percentage of a specific oligosaccharide in the total glycan repertoire. Furthermore, the same oligosaccharide can be quantified relative to other samples.

DEGLYCOSYLATION AND ITS IMPORTANCE

For analysis of a protein existing as a glycoprotein, deglycosylation is highly essential since a protein prevails as a collection of glycoforms. The presently available proteomic analysis software is not capable of handling the myriads of possibilities that arise. Thus, the specific glycosylated protein fragment(s) may remain unrecognized. Because of this reason, deglycosylation is often recommended before performing any protein analysis experiments, in case the experimental protein is a glycosylated one and not a pure protein.

Two methods that are used for protein deglycosylation are as follows:

• Enzymatic Cleavage of Glycans

• Chemical Removal of Glycans

Here is a brief description of these methods:

Enzymatic Cleavage of Glycans

Under mild conditions, glycosidases remove sugars while maintaining the integrity of peptide bonds. Under native conditions, enzymes can also be utilized for folded proteins, though their effectiveness may be reduced by steric hindrance and other limitations.

• Most of the commercially available glycan-cleaving enzymes are specific for *N*-glycans.

• Glycoamidase, especially glycoamidase F (can also use PNGase F), is the most effective cleaving enzyme for the N-glycan moiety. It cleaves the bond between GlcNAc and an Asn residue, changing Asn to Asp. Most N-linked oligosaccharides, including complex structures like high Man, hybrid, and multi sialylated structures, as well as oligosaccharides with SO_4^{2-} substituted residues and tetra-antennary trees, are broken down by PNGase F.

• On the other hand, high PNGase F concentrations might be needed for native proteins. PNGase F will not cleave the N-glycan moiety of an oligosaccharide containing α-1,3 core fucosylation or a single N-acetylglucosamine attached to asparagine; in such an instance, PNGase A, the other glycoamidase, must be utilized. The cleavage of large peptides or intact proteins by PNGase A, however, may pose a challenge because of its high molecular weight (MW) of 80 kDa, which restricts its accessibility to the cleavage site.

• Furthermore, PNGase F and A activity is decreased when the *N*-glycan is adjacent to the N- or C-terminus of the protein.

• Therefore, endoglycosidases that cleave the *N*-glycan tree between the two core GlcNAcs (GlcNAc1-4GlcNAc) are preferred for native glycoproteins or known glycan moieties.

• There are fewer enzymes available for *O*-glycans than for *N*-glycans. Due to the wide range of O-glycans, it may require multiple enzymes to analyze a single sample, which limits their usefulness. Therefore, *O*-glycans are better suited for chemically assisted elimination.

NB

Different glycoproteins are targeted by the various endoglycosidases that are currently in use. For instance, Endo F1 cleaves the high-mannose and hybrid-type glycans; Endo F2 cleaves the high-mannose and biantennary glycans; and Endo F3 cleaves the bi- and tri-antennary glycans. The images suggest that endoglycosidases function in a very particular way. Conversely, Endo-H cleaves most hybrid-type oligosaccharides, including core-fucose, bisecting glycans, and high-mannose.

• A few glycosidases are combined in one reaction mixture; one such mixture is called deglycosylation Mix II (NEB #P6044), and it contains five different glycosidases.

• Other combinations are possible until the optimized pH range allows for the use of a common reaction buffer.

• When exoglycosidases and endoglycosidases are combined, the glycans that are released may be trimmed and thus not recovered whole. Furthermore, because pH changes can impact performance, a longer incubation time may be required when an enzyme is functioning as a non-typical buffer.

• Any specific affinity method capable of selectively capturing the protein of interest (*e.g.*, protein G for immunoglobulins) is suitable for protein purification.

• Size exclusion chromatography is not recommended unless the protein's size differs significantly from that of the glycosidase. Finally, some NEB glycosidases (Remove-iT™PNGase F, Endo S, Endo D) have an affinity tag to aid in enzyme removal from the reaction mixture *via* chitin beads assisted capture of tagged glycosidase.

Chemical Removal of Glycans

β-elimination and hydrazinolysis are the main chemical methods for removing *O*- and *N*-linked oligosaccharides.

A brief description of these is as follows:

β-Elimination

• In this method, glycan release is facilitated by exposing a glycoprotein to an alkaline environment (*e.g.*, NaOH).

• Additional *β*-elimination reactions can be used to degrade free glycans into monosaccharides until complete degradation, referred to as a "peeling reaction," occurs. However, degradation of glycan can be prevented if the reaction is carried out with NaOH, which lowers the remaining aldehydes [19].

• Although this is an efficient method, but also has a significant lacuna as the oligosaccharides have only one group that can be labeled, *i.e.*, the reduced end aldehyde. After glycan release, labeling the terminal group is not possible because this group is eliminated during the reaction.

Hydrazinolysis

• In a hydrazinolysis reaction, the hydrolysis reaction is started by adding anhydrous hydrazine to a lyophilized, salt-free glycoprotein. By altering the reaction conditions, this method may be used for both *N*- and *O*-linked carbohydrates, which can be released separately even after being mixed. For instance, at 60°C, *O*-linked oligosaccharides are released, whereas *N*-linked oligosaccharides are released at 95°C.

• The released glycan moiety remains intact, whereas the protein may break down [20].

BIOLOGICAL IMPORTANCE OF GLYCOSYLATION

• Prokaryotic and eukaryotic cells exhibit glycosylation of proteins to carry out multiple biological functions.

• Glycosylation is a quality control mechanism in the ER, monitoring protein folding status and ensuring that only correctly folded proteins are trafficked within the Golgi.

• Specific receptors in the trans-Golgi network can bind sugar moieties on soluble proteins, facilitating their trafficking to the appropriate destination.

• Additionally, these sugars can mediate attachment or activate signal transduction pathways by acting as ligands for the cell surface receptors.

• Oligosaccharides can either facilitate or inhibit protein binding to cognate interaction domains in protein-protein interactions because they can be huge and bulky. They can change a protein's solubility due to their hydrophilic nature [21].

• Thus, glycosylation of proteins is necessary because of its role in various biological processes.

The biological importance of glycosylation is briefly described in the following subtitles:

• Importance of glycosylation in protein targeting.

• Importance of glycosylation in preventing protein degradation.

• Importance of glycosylation in recognition and receptor functions.

• Importance of glycosylation in sustaining immune functioning.

Here is a brief discussion of the above points:

Importance of Glycosylation in Protein Targeting

• For a better understanding of the effects of glycosylation on protein folding and maturation measurements such as oligosaccharide structure and glycosylation site tenancy requires a quantitative analysis.

• The RER membrane is where newly synthesized proteins are attached and transported to different locations of a cell. The ribosome only synthesizes the primary structure of proteins. Certain proteins are translocated into the nucleus, and others are translocated to the plasma membrane. In the ECM, certain proteins are exported or transported [22].

• Because of this, protein transportation necessitates the presence of a tag (leader sequence) on the protein to target it at the precise locations where it is required. The protein's glycosylation acts as a tag, enabling it to be identified by a particular receptor and subsequently directed to specific locations (protein sorting).

Importance of Glycosylation in Preventing Protein Degradation

• Proteins need to be protected from degradation by various hydrolytic enzymes. Therefore, proteins that have undergone glycosylation provide defense against proteolytic degradation [23].

• Glycosidic bonds are also known as ether bondsand are present in the glycan (oligosaccharides) component of the glycoproteins. These ether bondsserve as a barrier against potential hydrolytic enzyme degradation because they are highly resistant to these enzymes. Therefore, freshly synthesized proteins are glycosylated to prevent them from being broken down by the hydrolytic enzymes [24].

• On the other hand, aged proteins are targeted for degradation by proteases. The aged glycoproteins lose parts of their glycan chains. When this happens, the protein is targeted for degradation and will eventually break down to preserve the biomolecules.

• Citing the role of glycosylation in the thermodynamic stability of a protein, *Bechor, and Levy* modeled the native protein surface texture by representing each amino acid and sugar ring by a single bead. The investigators monitored the folding of 63 engineered SH3 domain variants, which were glycosylated with varying extents of conjugated polysaccharide chains at multiple locations on the protein's surface. Analysis revealed a polysaccharide chain conferred thermal stabilization of the protein's relative extent which exhaustively depended on specific glycosylation site and not on the relative glycans size. Screening for thermodynamic stability revealed the unfolded proteins are in a higher energy state and glycosylation mediates the lower energy state and enhanced stability of a folded protein. The enthalpic sensitivity of unfolded protein was due to its higher free energy as the bulk polysaccharide chains force the unfolded ensembles to attain extended confirmations *via* preventing a residual structure formation. Observed thermodynamic stability of glycosylated proteins contributed to the corresponding kinetic stability, suggesting a relevance of incurred effects on protein biophysical stability over the closely linked polymeric conjugate systems, which prevail consistently as cell's post-translational modifications [25].

Importance of Glycosylation in Recognition and Receptor Functions

• Certain glycosylated proteins-like the G-protein linked cell surface receptor (GPCR) act as recognition sites and, as a result, function as cell surface receptors. Here, the glycan (oligosaccharide) moiety is found outside the lipid bilayer membrane, while the glycoprotein's protein subunit is embedded into it.

• After a specific signaling molecule interacts with the receptor proteins, the oligosaccharide component identifies it, and the signaling cascade starts.

• Therefore, oligosaccharides attached to the protein are crucial for recognition.

Importance of Glycosylation in Sustaining Immune Function

• Antibodies (Abs) or immunoglobulins (Igs) are glycosylated proteins.

• An immunoglobulin is a Y-shaped globulin protein with four polypeptide chains (two similar types of light chains and heavy chains respectively). Thus, a pair of one heavy and one light are adhered to one another by disulfide bonds. Subsequently, all four polypeptides joined together to form a complete Y-shaped antibody structure.

• Immunoglobulins are glycoproteins, containing *N*-linked carbohydrates in the heavy chain constant regions of all isotypes and *O*-linked carbohydrates in the hinge regions of the F_c fragment of human IgA1 and IgD. The carbohydrate content of IgM is high, about 12%. Carbohydrates are frequently attached to the V region of an IgG [26].

• An oligosaccharide moiety of a specific length that is involved in antigen recognition is located at the tip of the variable region of both the light and heavy chain, referred to as an antigen-binding site or fragment antigen binding (FAB). Thus, the oligosaccharide at the FAB region of Ig molecules performs manifold essential immunological functions.

• Certain specific glycoforms are involved in the optimum folding and assembly of peptide-loaded major histocompatibility complex (MHC) antigens and the concurrent complexation with T-cell receptors. Despite some glycopeptide antigens being presented by MHC, retrieving peptide antigens from glycoproteins inevitably involves enzyme-assisted sugar removal before the protein cleavage. Oligosaccharides conjugated to glycoproteins in the junction intervening T-cells and antigen-presenting cells (APCs) assist the proper orientation of binding faces, guarding against proteases, and arresting non-specific lateral protein-protein interactions.

• All Ig and most of the complement constituents within the humoral immune system, are glycosylated. Apart from ensuring the conjugated protein's stability, the attached glycoforms are associated with molecular recognition. For instance, in rheumatoid arthritis, galactosylated isoforms of aggregated IgG could manifest an association with Man-binding lectin, getting associated with pathology [27].

• *N*-glycans exhibit manifold structural functions, the foremost of which involves stabilization of the $-CH_2$ domain of IgG which is otherwise, thermally unstable, vulnerable to unfolding and aggregation after deglycosylation. Conjugation with *N*-glycans improves the IgG functionality, modulating the folding of the F_c domain *vis-à-vis* homogeneous glycosylations. Larger *N*-glycans, such as bi-antennary complex bind with terminal galactosylation, activate the F_c domain in the $-CH_2$ region to a horseshoe morphology, and the smaller of the conjugated *N*-glycans (such as core structure) generally attain a more closed F_c conformation. These open and closed configurations substantially influence the effector functions, arising from the interactions of the F_c domain with F_c receptors. Thus, the *N*-glycan conjugation modulates the $-CH_2$ domain flexibility to mediate the crystallization of deglycosylated IgG [28, 29].

• Glycosylation significantly affects the bioavailability and pharmacokinetics of multiple biopharmaceuticals. The F_c domain comprises fusion proteins, bearing an intricate relationship with monoclonal antibodies, mAbs. The F_c-flanked fusion proteins exhibit multiple *N*-glycosylation loci in the non-IgG fusion protein region. The noted efforts by Keck and colleagues describe a comparison of F_c fusion protein batches with distinct *N*-glycosylation patterns, for terminal GlcNAc, galactosylation, and sialylation. Although no site-specific *N*-glycan inspection was undertaken to correlate the effects with the F_c domain, the investigators witnessed a prompt and swift elimination of terminal GlcNAc [30].

• The reason to believe the participation of MAN receptors for this clearance was the inference from the 3D structure of receptors, which deciphered a receptor binding with terminal GlcNAc. Taking note of these observations, Jones and accomplices designed a fusion protein lenercept comprising an Fc domain and 2 extracellular components of the TNF-α receptor. The scientists observed that *N*-glycans with terminal GlcNAc residues had a higher clearance efficacy besides no substantial altered pharmacokinetics *via* sialic acids and terminal galactosylation [31].

• On the same lines, *Kogelberg and associates* demonstrated a clearance strategy for excessively mannosylated proteins after screening an antibody F_c-enzyme fusion protein generated in the yeast *Pichia pastoris*. The investigators noticed a glycan-specific receptor expression in the sinusoidal-liver endothelial cells [32]. Erstwhile findings by Liu and teammates (amidst their screening of distinctly glycosylated fusion proteins and mAbs generated in glycol-engineered *P. pastoris* or CHO cells) revealed that sialic acid content of fusion proteins alters their pharmacokinetics with low sialic acid extent attributing to a decreased serum half-life [33].

CONCLUSION

Glycosylation is a vital biological process that occurs in both the RER and the Golgi apparatus. It is primarily a post-translational modification that controls the formation of a functional protein from an immature protein. As a result, glycosylation promotes proper protein folding, which increases protein stability. For example, glycosylation is an enzyme-catalyzed process that adds a specific carbohydrate to specific proteins at determined locations. The process can be controlled by keeping the enzymes active. The fundamental of glycosylation is the reaction of a sugar's carbonyl group (glycosyl donor) with the protein's -OH or -

NH$_2$ group (glycosyl accepter). Several types of glycosylation occur in the cell, including *N*-linked glycosylation where glycans are attached to the nitrogen of asparagine side chains, *O*-linked glycosylation where glycans are attached to the -OH oxygen of serine, tyrosine, threonine, hydroxylysine or hydroxyproline sidechains, or to oxygens on lipids such as ceramide. Phosphoserine glycosylation implies the linkage of phosphoglycans including mannose, xylose, or fucose through the phosphate of a phosphoserine, *C*-manosylation where sugar is added to carbon on a tryptophan side chain, and finally glypiation wherein, the addition of a GPI anchor links the proteins to lipids through glycan connections. The factors affecting glycosylation include the availability of amino acids and various sugars, optimum protein conformation and availability of different enzymes, movement of the proteins through the RER, Golgi apparatus(RER-Golgi secretory pathway), and various environmental factors. At present, various methods are in use to detect and analyze glycoproteins. These comprise the multiple glycan staining methods, enzyme, and antibody-based methods. This chapter briefly describes the various methods for glycan analysis, namely MS techniques, and chromatography. The chapter also sheds light on the deglycosylation reaction. Of note, the specific glycosylated protein fragment(s) will remain unrecognized. This is the reason deglycosylation is often recommended to be done before any protein analysis. Finally, this chapter ends with a description of protein glycosylation importance such as the role of glycosylation in protein targeting, prevention of protein degradation, recognition of cell surface receptor function, and the relevance of protein glycosylation in immunological regulation. Collectively, this chapter enhances the reader's knowledge about various aspects of protein glycosylation.

REFERENCES

[1]   Varki A, Cummings RD, Esko JD, *et al.* Essentials of Glycobiology. 4th ed., Cold Spring Harbor Laboratory 2022.

[2]   Huang ML, Kiessling LL. Special Issue on Chemical Glycobiology. ACS Chem Biol 2021; 16(10): 1793-4.
[http://dx.doi.org/10.1021/acschembio.1c00764] [PMID: 34649434]

[3]   Spiro RG. Protein glycosylation: nature, distribution, enzymatic formation, and disease implications of glycopeptide bonds. Glycobiology 2002; 12(4): 43R-56R.
[http://dx.doi.org/10.1093/glycob/12.4.43R] [PMID: 12042244]

[4]   Rini JM, Moremen KW, Davis BG, Esko JD, Eds. Glycosyltransferases and glycan-processing enzymes (Ch-6).Essentials of Glycobiology. Cold Spring Harbor, NY: Laboratory Press 2022.

[5]   Bojarová P, Křen V. Glycosidases: a key to tailored carbohydrates. Trends Biotechnol 2009; 27(4): 199-209.
[http://dx.doi.org/10.1016/j.tibtech.2008.12.003] [PMID: 19250692]

[6]   Harris RJ, Spellman MW. O-Linked fucose and other post-translational modifications unique to EGF modules. Glycobiology 1993; 3(3): 219-24.
[http://dx.doi.org/10.1093/glycob/3.3.219] [PMID: 8358148]

[7]   Lommel M, Strahl S. Protein O-mannosylation: Conserved from bacteria to humans. Glycobiology
        2009; 19(8): 816-28.
        [http://dx.doi.org/10.1093/glycob/cwp066] [PMID: 19429925]

[8]   Tian E, Ten Hagen KG. Recent insights into the biological roles of mucin-type O-glycosylation.
        Glycoconj J 2009; 26(3): 325-34.
        [http://dx.doi.org/10.1007/s10719-008-9162-4] [PMID: 18695988]

[9]   Bishop JR, Schuksz M, Esko JD. Heparan sulphate proteoglycans fine-tune mammalian physiology.
        Nature 2007; 446(7139): 1030-7.
        [http://dx.doi.org/10.1038/nature05817] [PMID: 17460664]

[10]   Burda P, Aebi M. The dolichol pathway of *N*-linked glycosylation. Biochim Biophys Acta, Gen Subj
        1999; 1426(2): 239-57.
        [http://dx.doi.org/10.1016/S0304-4165(98)00127-5] [PMID: 9878760]

[11]   Steen PV, Rudd PM, Dwek RA, Opdenakker G. Concepts and principles of O-linked glycosylation.
        Crit Rev Biochem Mol Biol 1998; 33(3): 151-208.
        [http://dx.doi.org/10.1080/10409239891204198] [PMID: 9673446]

[12]   Ferguson, Michael AJ, Taroh Kinoshita, and Gerald W. Hart. Glycosylphosphatidylinositol anchors. In
        Varki A, Cummings RD, Esko JD, *et al.*, editors. Essentials of Glycobiology. 2nd edition 2009. Cold
        Spring Harbor Laboratory 2009.
        [PMID: 20301281]

[13]   Haynes PA. Phosphoglycosylation: A new structural class of glycosylation? Glycobiology 1998; 8(1):
        1-5.
        [http://dx.doi.org/10.1093/glycob/8.1.1] [PMID: 9451009]

[14]   Andersen DC, Bridges T, Gawlitzek M, Hoy C. Multiple cell culture factors can affect the
        glycosylation of Asn-184 in CHO-produced tissue-type plasminogen activator. Biotechnol Bioeng
        2000; 70(1): 25-31.
        [http://dx.doi.org/10.1002/1097-0290(20001005)70:1<25::AID-BIT4>3.0.CO;2-Q]   [PMID:
        10940860]

[15]   Blondeel EJM, Aucoin MG. Supplementing glycosylation: A review of applying nucleotide-sugar
        precursors to growth medium to affect therapeutic recombinant protein glycoform distributions.
        Biotechnol Adv 2018; 36(5): 1505-23.
        [http://dx.doi.org/10.1016/j.biotechadv.2018.06.008] [PMID: 29913209]

[16]   Ellgaard L, Helenius A. Quality control in the endoplasmic reticulum. Nat Rev Mol Cell Biol 2003;
        4(3): 181-91.
        [http://dx.doi.org/10.1038/nrm1052] [PMID: 12612637]

[17]   Dell A, Morris HR. Glycoprotein structure determination by mass spectrometry. Science 2001;
        291(5512): 2351-6.
        [http://dx.doi.org/10.1126/science.1058890] [PMID: 11269315]

[18]   Roth Z, Yehezkel Khalaila I. Identification and quantification of protein glycosylation. Int JCarbohyd
        Chem 2012. Article ID 640923.
        [http://dx.doi.org/10.1155/2012/640923]

[19]   Mulloy B, Hart GW, Stanley P. Structural analysis of glycans (Ch-47) In: Essentials of Glycobiology,
        2nd Edition, Editors: Varki A, Cummings RD, Esko JD *et al.* ColdSPring Harbor, Laboratory Press,
        2009. ISBN-13: 9780879697709.

[20]   Nakakita S, Sumiyoshi W, Miyanishi N, Hirabayashi J. A practical approach to N-glycan production
        by hydrazinolysis using hydrazine monohydrate. Biochem Biophys Res Commun 2007; 362(3): 639-
        45.
        [http://dx.doi.org/10.1016/j.bbrc.2007.08.032] [PMID: 17727814]

[21]   Dwek RA. Biological importance of glycosylation. Dev Biol Stand 1998; 96: 43-7.

[PMID: 9890515]

[22]   Schoberer J, Shin YJ, Vavra U, Veit C, Strasser R. Protein glyosylation in the ER. Methods Mol Biol 2018; 1691: 205-22.
[http://dx.doi.org/10.1007/978-1-4939-7389-7_16] [PMID: 29043680]

[23]   Jayaprakash NG, Surolia A. Role of glycosylation in nucleating protein folding and stability. Biochem J 2017; 474(14): 2333-47.
[http://dx.doi.org/10.1042/BCJ20170111] [PMID: 28673927]

[24]   Ohtsubo K, Marth JD. Glycosylation in cellular mechanisms of health and disease. Cell 2006; 126(5): 855-67.
[http://dx.doi.org/10.1016/j.cell.2006.08.019] [PMID: 16959566]

[25]   Shental-Bechor D, Levy Y. Effect of glycosylation on protein folding: A close look at thermodynamic stabilization. Proc Natl Acad Sci USA 2008; 105(24): 8256-61.
[http://dx.doi.org/10.1073/pnas.0801340105] [PMID: 18550810]

[26]   Marth JD, Grewal PK. Mammalian glycosylation in immunity. Nat Rev Immunol 2008; 8(11): 874-87.
[http://dx.doi.org/10.1038/nri2417] [PMID: 18846099]

[27]   Rudd PM, Elliott T, Cresswell P, Wilson IA, Dwek RA. Glycosylation and the immune system. Science 2001; 291(5512): 2370-6.
[http://dx.doi.org/10.1126/science.291.5512.2370] [PMID: 11269318]

[28]   Zheng K, Bantog C, Bayer R. The impact of glycosylation on monoclonal antibody conformation and stability. MAbs 2011; 3(6): 568-76.
[http://dx.doi.org/10.4161/mabs.3.6.17922] [PMID: 22123061]

[29]   Mimura Y, Church S, Ghirlando R, *et al.* The influence of glycosylation on the thermal stability and effector function expression of human IgG1-Fc: properties of a series of truncated glycoforms. Mol Immunol 2000; 37(12-13): 697-706.
[http://dx.doi.org/10.1016/S0161-5890(00)00105-X] [PMID: 11275255]

[30]   Keck R, Nayak N, Lerner L, *et al.* Characterization of a complex glycoprotein whose variable metabolic clearance in humans is dependent on terminal N-acetylglucosamine content. Biologicals 2008; 36(1): 49-60.
[http://dx.doi.org/10.1016/j.biologicals.2007.05.004] [PMID: 17728143]

[31]   Jones AJS, Papac DI, Chin EH, *et al.* Selective clearance of glycoforms of a complex glycoprotein pharmaceutical caused by terminal N-acetylglucosamine is similar in humans and cynomolgus monkeys. Glycobiology 2007; 17(5): 529-40.
[http://dx.doi.org/10.1093/glycob/cwm017] [PMID: 17331977]

[32]   Kogelberg H, Tolner B, Sharma SK, *et al.* Clearance mechanism of a mannosylated antibody–enzyme fusion protein used in experimental cancer therapy. Glycobiology 2007; 17(1): 36-45.
[http://dx.doi.org/10.1093/glycob/cwl053] [PMID: 17000699]

[33]   Liu L, Gomathinayagam S, Hamuro L, *et al.* The impact of glycosylation on the pharmacokinetics of a TNFR2:Fc fusion protein expressed in Glycoengineered *Pichia Pastoris.* Pharm Res 2013; 30(3): 803-12.
[http://dx.doi.org/10.1007/s11095-012-0921-3] [PMID: 23135825]

CHAPTER 2

Vital Functions of Glycans in the Biological Systems

Ruma Rani[1], Parth Malik[2,3], Raj Singh[4], Raman Kumar[4], Vishal Haribhai Patel[5] and Tapan Kumar Mukherjee[6,*]

[1] *ICAR-National Research Centre on Equines, Hisar-125001, Haryana, India*

[2] *School of Chemical Sciences, Central University of Gujarat, Gandhinagar, Gujarat-382030, India*

[3] *Swarrnim Startup & Innovation University, Bhoyan-Rathod, Gandhinagar-Gujarat, India*

[4] *Department of Biotechnology, Maharishi Markandeshwar (Deemed to be University), Mullana, Ambala, Haryana 133207, India*

[5] *Institute of Biotechnology, Amity University, Noida-201301, Uttar Pradesh, India*

[6] *Amity Institute of Biotechnology, Amity University, New Town, Kolkata, West Bengal 700156, India*

Abstract: Glycans and their various conjugates namely glycoproteins, glycolipids, and proteoglycans not only coat all the cells in nature and interact with the extracellular matrix (ECM) molecules but are also located in the intracellular regions of every living organism. Glycans mediate or modulate numerous biological roles that are essential for life. This most abundant cellular molecule is necessary to maintain various general and specialized functions of the cells. Some of the major vital roles of glycans include maintenance of the structural integrity and protection of the cells, cell adhesion, cell-t--cell communication, crosstalk, and bidirectional cell signaling (both inside out and outside in). Briefly, this chapter predominantly focuses on the role of glycans and their various conjugates in maintaining the structural integrity of biological membranes and the overall cells, the different modulatory functions of glycans, and their implication in nutrient sequestration. Additionally, a brief outline of the role of glycans on intrinsic or intra-species recognition and extrinsic or interspecies recognition is discussed. Overall, the biological importance of glycans and their conjugates is elaborated.

Keywords: Extrinsic/Inter-species recognition of Glycans, Glycans, Glycoproteins, Glycolipids, Intra-species/intrinsic recognition of Glycans, Proteoglycans, Structural/Modulatory role of Glycans.

---

* **Corresponding author Tapan Kumar Mukherjee:** Amity Institute of Biotechnology, Amity University, New Town, Kolkata, West Bengal 700156, India; E-mail: tapan400@gmail.com

INTRODUCTION

Glycans (carbohydrates: monosaccharides, disaccharides, oligosaccharides, and polysaccharides) are widely distributed and abundantly present in every cell of a living organism including all compartments and organelles, extracellular spaces (the extracellular matrix, glycocalyx) and even in the body fluids such as animal cell's blood. Thus, glycans are recognized as one of the most important building blocks of the living system. Different organisms are equipped to make unique complements of glycan structures by themselves, using dissimilar and sometimes, unusual monosaccharides. Enzymatic addition of glycans to other biological molecules such as proteins (*e.g.* glycoproteins, proteoglycans) and lipids (glycolipids) is famously recognized as glycosylation. The process of glycosylation happens in a cell and tissue-specific manner. In eukaryotes, the intracellular organelles namely endoplasmic reticulum (ER) and Golgi apparatus control the process of glycosylation at multiple levels. Unlike genome, exome, and proteome, it works in a non-template manner. Free glycans and various glycan conjugates including glycoproteins, glycolipids, and proteoglycans are involved in a vast array of biological functions.

The complexity and heterogeneity of glycan structures have impaired the rapid progress of glycoscience, delaying the understanding of the various roles of glycans in nature. However, in the last two or three decades, discoveries of a series of genetic and biochemical analysis tools and techniques have broadened the knowledge of the biological roles of glycans. Significant developments in nuclease-based gene editing, proteomics, and quantitative transcriptomics are now utilized to investigate protein glycosylation by identifying and focusing on glycosylation-related enzymes. Newly developed *in silico* models forecast different cell's capacity for glycosylation. The accurate and refined mapping of glycosylation pathways makes it easier to use genetic methods to address the various activities of the large glycoproteome. These strategies make use of widely accessible cell biology tools, and it is believed that the most notable developments towards a more comprehensive integration of glycosylation in general cell biology will come from the application of (single cell) transcriptomics, genetic screens, genetic engineering of cellular glycosylation capacities, and custom glycoprotein therapeutic design.

Today, it is well known that the biological functions of glycans vary from relatively subtle to crucial for the growth and development of the embryo, maintenance, and survival of a living organism, commencing from the unicellular prokaryotes(microorganisms) to complex multicellular eukaryotes, such as humans. The inceptive studies focused on the structural and modulatory roles of glycans under various physiological and pathophysiological conditions including intracellular folding and stability of proteins and lipids and the ability to interfere with carbohydrate-protein, carbohydrate-carbohydrate, and glycoprotein-glycoprotein interactions that affect the proliferation, differentiation, migration and even invasion of cancer cells. However, glycans and their conjugates prevail widely in the periphery of the plasma membrane and glycocalyx. Of note, glycocalyx is defined as a cell surrounded by dense, gel-like meshwork, and constitutes a physical barrier for any material to enter a cell. The cell surroundings or peripheral glycans are critical mediators of cell adhesion, cell-to-cell communication, and various cell signaling mechanisms. These actions of glycans are necessary for the survival and maintenance of living organisms. In the subject of cell communications, glycans, and their conjugates are intricately related to the "inside-out" and "outside-in" bidirectional signaling of a cellular system. Additionally, secretory glycoproteins are involved in various biological functions of a living body. Several studies are now dedicated to understanding the role of glycans and their conjugates on the host (*e.g.* immunological cells)-pathogen (viruses. bacteria *etc.*) interactions, and the subsequent attachment, entry, and pathogenicity. In contrast, the recognition of pathogens by the immunological cells and molecules helps to eliminate these pathogens, and glycans take significant roles in these actions. Thus, glycans are part of an elaborate communication system vital for cellular recognition, cell-cell interactions, protein transport, immune defense, and more.

The biological functions of glycans are categorized into two prominent domains, relying on their (i) structure-modulatory traits, and (ii) explicit identification by erstwhile molecules, majorly glycan-binding proteins (GBPs). Of note, the GBP actions could be intrinsic (facilitating glycans from the same organism) or extrinsic (recognizing glycans from a different organism). The intrinsic GBPs are capable of recognizing glycans on the same cell and assist in cell-cell interactions (majorly responding to extracellular molecules). Contrary to this, the extrinsic GBPs exhaustively comprise pathogenic microbial adhesions, agglutinins, or toxins though some of these roles maintain a symbiotic association. Fig. (1) depicts the explicit functions of glycans molecular aspects, and their mutual correlations, manifesting for their explicit recognition. Additionally, a broad classification of glycans is highlighted, demonstrating the roles of intrinsic and extrinsic GBPs in the screening of glycans.

This chapter predominantly focuses on the role of glycans and their various conjugates in structural-modulatory functions and nutrient sequestration. Additionally, a very brief outline of the role of glycans on intrinsic or intraspecies recognition and extrinsic or interspecies recognition is discussed. The structural and modulatory roles of glycans section briefly describe their functions in maintaining the structural integrity of the biological membranes, the involvement of glycans in extracellular matrix (ECM) organization, cell-ECM interaction, and cell adhesion. This section also talks about the role of glycans in the epigenetic modifications of nuclear proteins, particularly the histone proteins, and the subsequent role of glycans in the functioning of genetic material (DNA) and the expression of proteins through mRNA synthesis.

Biological Roles Of Glycans

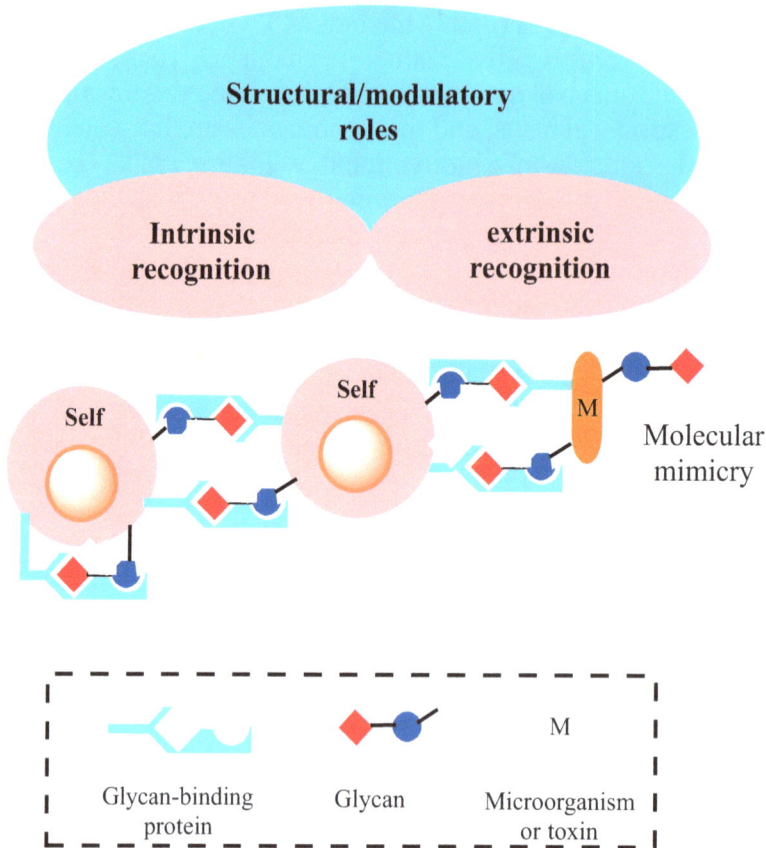

Fig. (1). Biological role of glycans.

Additionally, through their organization in the biological membranes and specific interaction with the ECM proteins, glycans maintain the solubility of various biological molecules, establishing a gradient across the biological membranes byserving as a diffusion barrier. To intra-species or intrinsic recognition of glycans, their roles in protein folding, degradation, trafficking, endocytosis, and phagocytosis are mentioned. The extrinsic or extracellular recognition of glycans briefly outlined their role in the host (*e.g.* immunological cells)-pathogen (microbes such as viruses and bacteria), and various aspects of cell signaling, both "inside out" and "outside-in" [1 - 4]. A detailed understanding of the biological roles of glycans is discussed in a series of publications by *Varki and colleagues* [5 - 8]. A specific glycan may have diversified roles in varying physiological environments, manifested from a combination of specific development stages or particular kinds of extracellular conditions. A consensus, however, deciphers the terminal sequences, uncommon structural frameworks, and altered forms of glycans as more sensitively contributing to the specific biological functions in an organism. Nevertheless, such configurations of glycans exhibit vulnerability to pathogenic and toxins attacks. The intrinsic or extrinsic variations in the functional modulation of glycans (*via* glycosylation) are ubiquitous and manifest certainty about the present diverse forms of glycans being the remnants of the yester characteristics or being the outcomes of late host-pathogen proximity [9, 10]. The working chemistry of glycosylation is highly intricate, sensitively modulating protein-protein interactions. Certain growth factor receptors are reported to bind their targets *via* glycosylation while being trafficked through the Golgi apparatus. This could be useful for restricting the undesired interactions of a newly formed receptor with its analogous protein in the same cell. For instance, despite being deglycosylated, the hormone β-human chorionic gonadotrophin (HCG) binds its receptor with unchanged affinity, though it is unable to activate adenylate cyclase (sole enzyme generating cyclic AMP from ATP) [11].

Conventionally, the outcomes of glycosylation are rather incomplete and that's why it is better suited for tuning the primary function of a protein instead of entirely activating or impairing it. This artifact is amicably illustrated by the varied functioning of select glycosylated growth factors and hormones over a wide range based on the relative extent and kind of glycosylation. The distinct glycosylation of biomolecules is aptly validated by studying the recombinant proteins, hormones, or growth factors. A befitting example explaining this scenario is the function performed by manifold polysialic acid (PSA) chains conjugated to neural cell adhesion molecule (NCAM), the adhesion receptor regulating the hemophilic proximity amongst the neuronal cells.

Fig. (2) depicts the chemical structure of sialic acid wherein manifold O and –OH at the terminal sites confer a hydrophilic sensitivity with an inclination for hydrogen bonding, although adjacent in and out of the plane orientations manifest a stressed native state. In embryonic and other neural plasticity conditions, the anionically sensitive PSA chains exert a chronic impact and perturb the hemophilic binding. Glycans are also known to affect the native protein functions by binding to erstwhile molecules in the vicinity. The PSA of embryonic NCAM, for instance, can affect the proximity of other ligand-receptor pairs in a manner as simple as *via* physical separation.

Another illustration relates to tyrosine phosphorylation functioning of epidermal growth factor (EGF) and insulin receptors, which are optimized *via* endogenous cell-surface gangliosides *via* their assortment into membrane microdomains. Despite persisting uncertainty for the manifold terminal changes in this regard, the explicit functioning of glycan sequence in the ganglioside remains the characteristically unique aspect. The net effects of glycosylation are still not well understood owing to their partial modulating essence. Nevertheless, the integration of these small outcomes confers significant contributions to the varied biological consequences, rendering glycosylation the feasibility for the genesis of essential functional diversity from the feasible, localized receptor-ligand proximities. It is important to note here that all glycans do not require or mandate glycosylation and several peptide ligand functions are not affected at all by glycosylation.

Fig. (2). Chemical structure of sialic acid.

The impaired activity of glycans or altered glycosylation may lead to various pathophysiological conditions and disorders, which have been discussed in detail in chapter 3 of this book.

STRUCTURAL AND MODULATORY ROLE OF GLYCANS

Pure glycans, glycan-conjugated proteins named glycoproteins, proteoglycan, and glycolipids have their specific structures and are involved in a large number of biological functions.

Comprehending the biological functions of glycans comprises the termination of inceptive glycosylation with a must to abolish the glycan chain elongation, altered patterns of glycans processing, enzymatic or chemically induced deglycosylation of complete chains, genetic loss of glycosylation loci besides a prospective aspect of naturally prevalent genetic variants and mutants in glycosylation. The outcomes of these changes may remain undetected or lead to drastic complete loss of specific functions. The feasibility is illustrated by cell-surface receptors wherein random consequences of altered glycosylation are exhibited. It is ironic to know that the same kind of glycosylation could result in distinctions in varied cell types, as revealed *viain vivo* or *in vitro* screening.

Fig. (3) describes the structural and regulatory tasks of glycans, discussing their utility in cell signaling, drug delivery, pathogenic response, immunological modulation, and selectivity control of plasma membrane. Prominent factors here include the characteristic glycan structure, the interaction type (extrinsic or intrinsic genesis), and the relatable biological conditions. Thus, the biological effects of glycans are mediated by their primary structural properties, and by modulating functions of proteins and lipids to which they are attached.

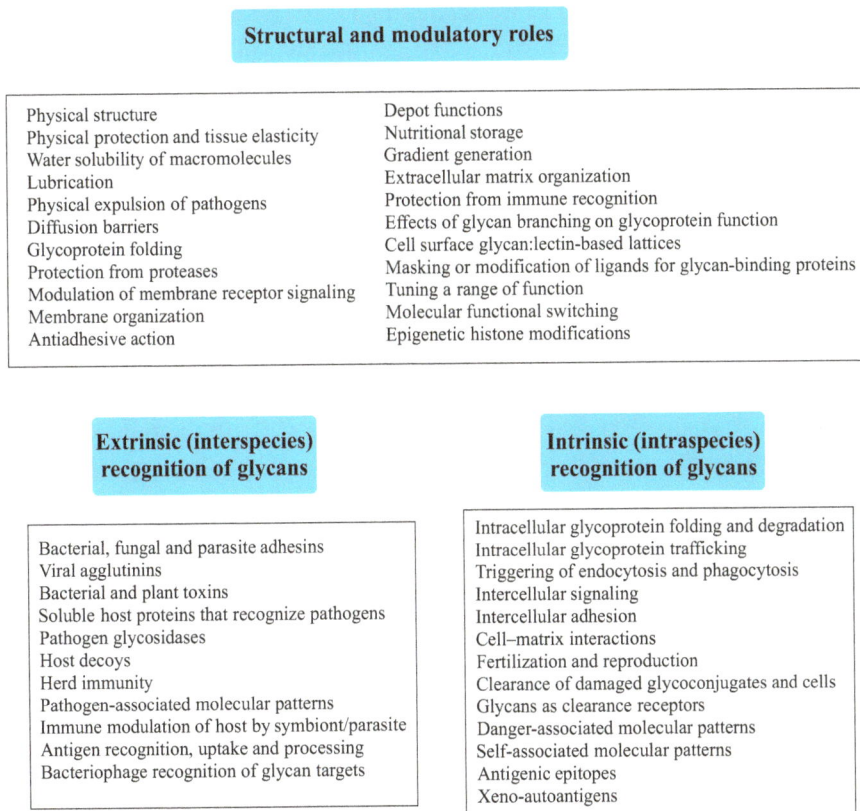

Structural and modulatory roles

Physical structure	Depot functions
Physical protection and tissue elasticity	Nutritional storage
Water solubility of macromolecules	Gradient generation
Lubrication	Extracellular matrix organization
Physical expulsion of pathogens	Protection from immune recognition
Diffusion barriers	Effects of glycan branching on glycoprotein function
Glycoprotein folding	Cell surface glycan:lectin-based lattices
Protection from proteases	Masking or modification of ligands for glycan-binding proteins
Modulation of membrane receptor signaling	Tuning a range of function
Membrane organization	Molecular functional switching
Antiadhesive action	Epigenetic histone modifications

Extrinsic (interspecies) recognition of glycans

Intrinsic (intraspecies) recognition of glycans

Bacterial, fungal and parasite adhesins	Intracellular glycoprotein folding and degradation
Viral agglutinins	Intracellular glycoprotein trafficking
Bacterial and plant toxins	Triggering of endocytosis and phagocytosis
Soluble host proteins that recognize pathogens	Intercellular signaling
Pathogen glycosidases	Intercellular adhesion
Host decoys	Cell–matrix interactions
Herd immunity	Fertilization and reproduction
Pathogen-associated molecular patterns	Clearance of damaged glycoconjugates and cells
Immune modulation of host by symbiont/parasite	Glycans as clearance receptors
Antigen recognition, uptake and processing	Danger-associated molecular patterns
Bacteriophage recognition of glycan targets	Self-associated molecular patterns
	Antigenic epitopes
	Xeno-autoantigens

Fig. (3). Schematic representation of manifold structure-function coordinated activities of glycans, describing their biological significance *via* structural compatibility.

The following paragraphs describe various aspects of the structural and modulatory actions of glycans:

ROLE OF GLYCANS IN MEMBRANE ORGANIZATION AND FUNCTIONS

The role of glycans in membrane organization and functions is multifaceted and critical, spanning various biological processes. From contributing to the physical properties of membranes to mediating cell communication and immune responses, glycans are integral to the proper functioning of cells *via* modulating the extent of clustering, mobility, and molecular interactions of various membrane receptors, decisive for their explicit functions [12].

Membrane glycans are generally present in chemically bound forms as conjugated with proteins (glycoproteins, proteoglycans) or lipids (glycolipids). These moieties do not have any hydrophobic region and are exclusively located on the cell membrane's outer (peripheral) surface. The sugar chain in the glycans is usually linked to proteins *via* the -OH group of serine or threonine or the asparagine amide group. It stands to believe that the bulk of the surface charge on cell-surface glycoproteins can modulate the functions of membrane domain organization [13, 14]. This dynamic organization involves various elements, including cholesterol-rich nanodomains (lipid rafts), the cortical actin cytoskeleton, and the glycocalyx matrix. In the context of an organized cellular architecture, glycans contribute to the structural diversity of the cell surface. The diverse structures of glycans, including branching patterns and linkage types, influence membrane properties like thickness, curvature, and fluidity. The organization of the plasma membrane glycans reveals profound effects on the overall cell organization. Two important examples in this aspect are GPI-anchored proteins and cell surface lectins such as galectin-4. GPI-anchored proteins are targeted to the apical surface in fully polarized epithelial cells. Similarly, galectin-4 prevails as the major organizing factor of such "lipid rafts" on gastric epithelial cells [15 - 17].

One of the major kinds of membrane glycoproteins is the receptor proteins. Type I single-spanning transmembrane receptor proteinsare often highly glycosylated in their extracellular domains. Receptor-bound glycans regulate the accessibility and conformation of certain ligand binding sites by projecting from the extracellular region of the protein core. The plasma membrane glycosylated proteins are connected by both N and O bonds. Multiple N-glycans, including oligomannosidic, complicated, or bisected structures, are present in almost all receptors. Fucose is a typical terminal sugar molecule still attached to sialic acid, N-acetyl lactosamine chains, and the N-glycan core. O-glycans are found on some receptors composed of sialyl T antigen structure, T antigen (core 1), and simple Tn antigen (GalNAc). It was discovered through experimental research that not all potential glycosylation sites are occupied. The cell types, receptor types, and occupied locations may all influence the function of these glycans. It has been observed that certain glycans found in the receptor proteins can either activate or hinder receptor dimerization, which in turn can start or stop signal transmission. The final point of agreement is that one class of glycans/ glycoproteins can modify other classes of glycans/ glycoproteinsorganized on the same cell surface, creating what is known as "clustered saccharide patches" [18 - 20].

The *N*-glycan molecules attached to cell membrane proteins play crucial roles in all aspects of biology, starting from embryogenesis and various types of adhesions including that of an embryo with the uterus, development, and growth of the embryo, cell to cell recognition, communication, and cell signaling. Cells recognize and identify other cells because of cell-cell recognition and crosstalk. In cell-cell recognition, receptors on a particular cell surface bind to glycoproteins on another cell, allowing the two cells to communicate. Thus, through intrinsic cell-cell recognition, one group of cells identifies familiar cells within the body. Similarly, extrinsic recognition is when cell-cell crosstalk identifies foreign cells, like pathogenic microorganisms such as bacteria, or viruses, coming from outside the body. Glycolipids are also involved in cell-cell recognition, particularly in extrinsic recognition. Various glycosylated membrane proteins such as adhesion molecules, receptors (*e.g.* lectins), and channel proteins can promote their proper folding and three-dimensional structural configuration, ensure their stability, and impact functions.

Glycolipids constitute around 5% of the cell membrane lipids. Three glycolipids are prominently well-studied, *viz.* glycosphingolipids, glycoglycerolipids, and glycophosphatidylinositols. While gangliosides are recognized as the most complex animal glycolipids, globosides are glycosphingolipids with more than one sugar as part of the carbohydrate complex. Glycolipids serve as recognition sites for cell-cell interactions. Two neighboring cells complementarily bind through the proximity of one cell's saccharides and the carbohydrates or lectins of the other. The interaction of these cell surface molecules forms the basis of cell-cell recognition, initiating the cellular responses that contribute to growth, or apoptosis [21].

ROLE OF GLYCANS IN EXTRACELLULAR MATRIX ORGANIZATION

In a multicellular organism, the space or region between two cells is known as the extracellular matrix (ECM). Various glycans such as proteoglycans are important constituents of ECM and interact with distinct bio-membrane molecules, exhibiting profound effects on membrane organization. The matrix glycans assist in the maintenance of physiologically optimal tissue architecture, porosity, and integrity. Additionally, within the matrix, various glycans interact with each other to assist in optimal organization, even in plant ECM. In general, the matrix acts as a protective shield for guarding polypeptides against proteases (mucins) or the entire tissue surfaces from microbial attachment and blocking antibody binding (*e.g.* viral glycoproteins) [22]. Many ECM components in vertebrates prevail as polymeric glycans, such as sulfated glycosaminoglycan (GAGs) and hyaluronan, self-organizing along with specific proteins as large, aggregated structures, *vis* basement membrane, and cartilage [23, 24]. Cartilage functions as a template for primary and secondary osteogenesis, formation of growth plates, and long bone terminals apart from the resting status of a bone. It is pertinent to recall that GAGs are anionic polysaccharides comprising disaccharide units within all mammalian tissue. The structural distinction of GAGs is deciphered from their linear arrangement, having amino sugar (glucosamine) and glucuronic acid as typical constituents. These are classified among heparin, keratin sulfate, hyaluronic acid, and chondroitin sulfate (the modified textures of glucosamine subunits). Fig. (**4**) depicts the chemical structures of glucosamine and glucuronic acid, wherein –OH and *O* at terminals infer their hydrophilic proximity. Structure-function significance is highlighted here with aqueous receptivity of physiological conditions. Numerous muscular dystrophy cases caused by α-dystroglycan ligand-altered glycosylation for important matrix proteins including laminins have been linked to the evidence of crucial matrix interactions with cell surface glycans [25 - 27]. For example, inflammatory cells recruited by stressed cells synthesized hyaluronan matrices that governed various early events of different disease processes [28, 29].

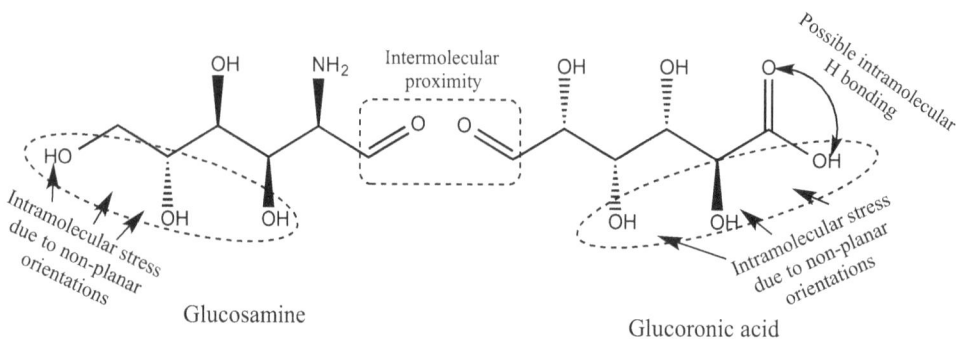

Fig. (4). Chemical structures of glucosamine and glucuronic acid, the constituent units of GAGs in the manifold cell-membrane locations.

Interestingly, this is a process undergone by most of the cells, while dividing in a hyperglycemic environment. This phenomenon presumably affects the experiments of many investigators using "standard" commercial tissue culture media, having unusually excess glucose that could result in diabetic coma [30]. The HB susceptibility of GAGs makes them feasible for mobile interactions, unlike the stronger hydrophobic-hydrophobic kind. This is why GAGs are needed for building bones, cartilage, tendons, corneas, skin, and connective tissue, besides prevailing in fluid and lubrication of the joint(s). For implant design, the specialties serve as key inputs and minimize the chance of cross-reaction. An emerging area of glycan's physiological utility being probed is their involvement in bacterial biofilm formation, assisting in the formation of discrete multi-cellular components [31, 32].

GLYCANS IMPART PHYSICAL STRENGTH AND PROTECTION OF BIOLOGICAL STRUCTURES

Glycans and their cellular conjugates are the most abundant and ubiquitous molecules present both, outside (in the ECM) and inside the cells, imparting physical strength to the overall structure. These molecules mediate the fungal and bacterial cell wall's elasticity, resilience, and compressibility [33]. The diversity of macroscopic structural changes among the planet's living forms might be far more restricted in the absence of these polymers. Important and widely distributed glycans that reinforce biological structures are glucose or N-acetylglucosamine (cellulose or chitin, respectively) β-linked homopolymers. Since these molecules are difficult to break down by physical, chemical, or enzymatic means, they give structures like the exoskeletons of arthropods and the cell walls of plants and fungi strength and rigidity [34, 35]. Several glycans confer decisive supportive roles in regulating the fungal and plant cell wall functioning in mediating selective intracellular entry [34 - 38]. For example, the hemicellulose xyloglucan not only facilitates the expansion and contraction of cellulose microfibrils but also allows a plant cell to undergo morphological changes during development and differentiation, ultimately maintaining its final form at maturation [39]. Similarly, glycans conjugated to matrix molecules including proteoglycans are crucial for optimal maintenance of the tissue structure, porosity, and overall integrity. These molecules engineer proximity for erstwhile glycans, assisting the integrated matrix arrangement.

Thick glycan coatings at the surfaces of cells play a crucial physical protective role. The thick coat of the mucus that covers many epithelial surfaces in animals, including humans, such as the inner lining of the intestines and airways, performs vital barrier functions, such as guarding against invasion by the gut microbial population [40 - 42]. Disruption of this layer by genetically altered mucin backbones, O-linked glycosyltransferases (or key chaperones like *Cosmc*) can have serious consequences, including inflammation and carcinogenesis, emanating *via* microbial invasion [43, 44]. Likewise, the thick and biochemically robust plant cell walls make it difficult for invading fungi and bacteria to reach the membrane, providing protection from pathogenic manifestation [45, 46]. In other instances, the thick layer of glycans imparts physical support, tissue strength, regenerative assistance, and a growth-aiding source, facilitating a steady and uniform energy distribution.

ROLE OF GLYCANS IN THE PROTECTION OF BIOLOGICAL STRUCTURES FROM PROTEASES

Glycans significantly prolong the stability and half-life of many proteins by guarding them from proteolysis [47, 48]. *N*-glycosylation is more frequent and therefore exhibits a prominent role in proteolytic protection contrary to that of *O*-glycosylation effective against general and specific proteolysis [49, 50]. Glycosylation regulates the binding of substrates, inhibitors, oligomer partners, activators, and cofactors, optimizing the kinetic parameters and turnover in enzymatic reactions of various proteases. Glycans protect the highly glycosylated proteins against proteolytic cleavage in the presence of steric hindrance or negative charge. Such effects are particularly shown in mucins bearing tightly packed *O*-glycans [51, 52]. Indeed, extended segments of some mucins are even resistant to broad-spectrum proteases, such as proteinase K [53]. This property might be investigated to separate mucin segments from other proteins that might break down readily into smaller pieces or even from entire tissues [54, 55]. For example, it has been shown that *N*-glycosylation in Campylobacter increases fitness in a prokaryotic cell by protecting against gastrointestinal proteases [56]. On the other hand, it has been demonstrated that single-site glycosylation can control specific cleavage events that affect protein function over the long run [57]. For example, *O*-glycosylation-provided Tango1 protection is essential for Drosophila apical secretion [58].

Biotechnological and pharmaceutical efforts aim to protect the proteases from auto-degradation and "humanize" the glycosylation expression patterns by abolishing the unfavorable immunogenic glycan epitopes [48, 59]. The prevalence of glycans on the outer periphery in most glycoproteins protects the underlying polypeptide from the recognition and degradation by proteases or antibodies. An elaborate discussion on the protease's glycosylation has been made by *Goettig P,*2016 [60]. Authors have stated considerable enhancement in the functioning of multiple proteases on the removal of glycans, such as the ADAM17 (Disintegrin and Metalloproteinase 17) expressed in insects is anchored with short *N*-glycans, and exhibits up to 30-fold increment in catalytic efficacy than a mammalian cell variant (of ADAM17) having more complex *N*-glycans interfering in the substrate binding. Strikingly, *O*-glycosylation in the vicinity of ADAM 17 substrates scissile bond considerably enhances the turnover of the molecule. Likewise, N-glycosylated kallikrein (KLKB1) exhibits a lower catalytic efficacy than deglycosylated KLKB1, which possesses a distinct cleavage regime for insulin (the substrate). Erstwhile studies in this regard noticed enhanced fibrin binding and the concomitant fibrinolytic performance emanated from plasmin on the deletion of a single *N*-glycan in the Kringle 2 domain of TPA (Tissue Plasminogen Activator) form II. Of note, the entirely deglycosylated TPA is almost 4-fold more active against small chromogenic substrates, wherein a higher mannose content or the alternative, sialic acid removal fuels the activity. Readers are suggested to refer to the 2016 article by Peter Goettig (in the International Journal of Molecular Sciences, IJMS) to ascertain the details of the referred studies herein.

The mechanistic aspects *vis-à-vis* enzyme-regulated stabilization conferred by glycosylation over conformational stabilization imparted by *N*-glycosylation, leading to enhanced specificity vis efficient interaction of substrates and regulated substrate access. As an instance, for cathepsin C and meprin, a singular *N*-glycan at the substrate binding cleft in the vicinity of catalytic subunits is crucial for substrate stabilization and binding. In many instances for the ambiguous role of individual glycans, complementary observations create a substantial realization of their augmented roles in oligomerization and interaction with erstwhile proteins. For example, *O*-glycans of recombinant carboxypeptidase exhibit synchrony with substrate engagement and tetramer formation even though their exact function is yet to be understood. Likewise, the double alanine mutant of *O*-glycosylated Ser52 and Ser 60 in the EGF locus of coagulation factor (VIIa) possessed ~14% coagulant activity concerning wild-type FVIIa. Contrary to this, amidolytic remained unchanged, deciphering the necessity of both glycans for association with tissue factor. Simultaneous screening of FVII retrieved through recombinant and wild means revealed a higher activity and belongingness with tissue factor for *N*-glycans having terminal GalNAc rather than sialic acids. Alike variations of *O*-glycans (Thr346 loci, form 1) and *N*-glycan at the Asn288 in plasmin, are known to affect the binding of inhibitor, 2-antiplasmin, and substrate fibrin. In similar works on many kallikrein-related peptidases, KLK2 possesses a single *N*-glycan in the 99-loop vicinity to the substrate-binding cleft, maintaining the substrate turnover and a potential of increased specificity for larger protein substrates. *N*-glycosylation in the surrounding 62-loop is reported for KLK3, thrombin, and human neutrophil cathepsin G, for optimized substrate binding in the prime side region. For several other long-manifested membrane proteases, *N*-glycosylation is known to forbid unspecific protein interaction (such as aggregation), wherein individual glycans may have specific functions and roles. The details of the literature cited in the above illustration are traced in Peter Goettig's IJMS article published in 2016.

ROLE OF GLYCANS IN THE PHYSICAL EXPULSION OF PATHOGENS

Excessive production of highly glycosylated secretions may be a reaction to the physical removal of imposters. For example, the removal of *Nippostrongylus brasiliensis* worms from the rat gut is associated with qualitative and quantitative changes in the goblet cell mucin composition [61]. A recent microscopic study found that bacterial products may be identified using a "sentinel" goblet cell at the mouse colonic crypt opening. This was controlled by an active Nlrp6 inflammasome, which caused nearby goblet cells in the upper crypt to secrete mucin, which helped to remove bacterial invaders that had broken through the barrier of the inner mucus layer protection [62]. The take-up of glycans by antigen-presenting cells (APCs)aids in the identification, absorption, and processing of antigens. Pathogen-produced glycans on target proteins are essential for breaking down antigens that APCs have acquired into peptides that MHC class II molecules then recognize, and T cells can recognize. For example, high densities of GlcNAc or terminal mannose residues on foreign proteins may cause phagocytosis due to the concurrent engagement of C-type lectins, which are predominant in APCs. The glycans help transport antigenic proteins to the processing compartments in this manner [63, 64].

Similarly, non-human α-Gal or Neu5Gc residues residing on injected glycoproteins could fuel immune reactions and the subsequent generation of immune complexes, enhancing the immune response against the peptide backbone [65, 66]. A bypassing strategy of self and non-self-distinction is exhibited by CD1-presented glycolipids (from foreign origin) that are detected *via* invariant T-cell receptors of NKT cells.

Another possibility of glycan's involvement in getting rid of pathogens is *via* recognition processes, which not only provide antigens for presentation to T-cells but also enable a clearance of damaged cells (or glycoproteins), which happen during microbial sialidases' entry to the circulation during sepsis, causing platelet desialylation or in the event of cancer cells generating incompletely glycosylated mucins [54]. Glycans are also recognized by innate immune cells released on the account of tissue attack in vertebrates, *vis-à-vis* hyaluronan fragments and certain matrix proteoglycans which prevail as Danger-Associated Molecular Patterns (DAMPs), generating responses alike to those manifested by exogenous Pattern-Associated Molecular Patterns (PAMPs). Studies on *Candida albicans* and *Histoplasma capsulatum* pathogenic manifestation decipher the dual functioning of fungal glycans as PAMPs and stimulate the host immune response or disguise other glycoconjugates to avoid such activation [67]. Recent studies have claimed the Self-Associated Molecular Patterns (SAMPs) functioning of glycans [68]. SAMPs are identified by intrinsic inhibitory receptors to uphold the native nonactivated state of innate immune cells, further suppressing the reactivity on encountering an immune response. The best-reported usability of glycans-based SAMPs is known to prevail as inhibitory Siglec recognition of cell surface sialoglycans (Sialic acid sugar-carrying glycans), decoding a host strategy to distinguish between infectious non-self and non-infectious self [69]. Topical attempts have consolidated the previous observations, illustrating sialoglycan recognition by factor H, which can moderate immune responses by inhibiting the swap pathway of complement stimulation [70, 71]. Such self-glycans are entities involved in pathogenic molecular mimicry, which engages these inhibiting receptors. Glycans have also been developed as significant antigenic epitopes on manifested intra and interspecies variations. Studies have demonstrated considerable neutralization impact of human Ig against the non-self-antigens having a structural similarity with glycans [72, 73]. Apart from this, structural modifications in *N-glycans* prevailing in plant and invertebrate glycoproteins induce immune responses in humans, such as in therapeutic glycoproteins [74, 75].

A suitable illustration herein pertains to IgE antibody synthesis against α-Gal epitopes (missing in humans) on being bitten by Lone Star ticks. Subsequent access to mammalian foods rich in α-Gal motifs (red meats) resulted in certain red meat allergies [76]. An important aspect related to immunological modulation by glycans works through their inflammatory responses, exhibited *via* the tendency to accumulate at the pathogenic injury undergone sites. The so-called trends are better understood as "Red Queen Effects (RQEs)", whereby glycans interactions mediate to establish a delicate balance between the maintenance of endogenous functions and averting the pathogen attack (Fig. **5**) [77]. The integrated loop of RQEs bridges the evolutionary diversification of glycans, wherein overlapping integrated pathogenic responses work out in unison to subside the swiftly evolving pathogens. To facilitate communication with glycans and avoid pathogens, hosts make the soluble glycans, referred to as "mucins" which function as pathogen-trapping entities to subvert pathogens from cell surfaces, although pathogens continue to equilibrate themselves with such defenses. The turning point of this self and non-self-recognition deciphers host recognition of pathogen-specific glycans as "non-self" entities wherein pathogens can alter their glycans for a close resemblance to host glycans.

ROLE OF GLYCANS IN MODULATING CELL ADHESION

A great diversity of glycans prevails on the mammalian cell surface, facilitating the selective adhesion of cells harboring glycan-specific receptors. Well-understood examples of cell adhesion working through receptor-engineered proximity comprise selectin-anchored leukocyte rolling on the endothelium. Other receptors with similar proximity for specific cell surface sugar epitopes are in the characterization phase [78]. Large acidic polymers such as hyaluronan and polysialic acid can inhibit cell-cell and cell-matrix interactions by bulkier sizes and negative charge. These anti-adhesive actions are visible during the development phases wherein cell migration is especially active. The "plasticity" resulting from polysialic acid surface prevalence plays a decisive role in neuronal migration and post-injury reorganization. The *N*-glycans act as a positive or negative regulator of cell adhesion [79]. Similarly, certain mucin-type *O*-glycans can either facilitate or attenuate cell adhesion depending on the core and the non-reducing termini structures [80].

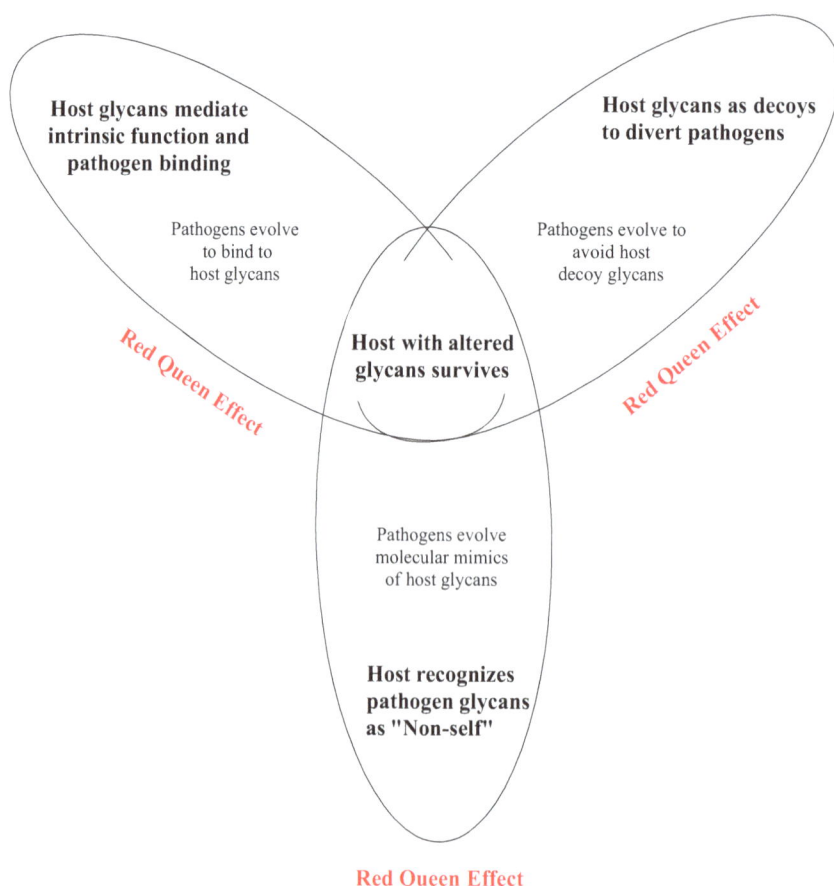

Fig. (5). Interconnected possibilities for host-pathogen interactions, mediated *via* modulated host glycans functions to recognize or subside pathogens through, (i) one-to-one interaction, (ii) bypassing the pathogens *via* decoy actions, and (iii) perceiving pathogenic glycans as antigenic.

ROLE OF GLYCANS IN DEPOT FUNCTIONS

The glycans are sensitive to the hydrophilic environment on cell surfaces and extracellular matrices and capable of attracting and assorting water molecules [81]. In addition to retaining water and cations, ECM-GAGs and polysialic acid can function as transient storehouses for localized growth factors (normally proteins or their conjugated forms) or erstwhile bioactive molecules. Such growth factors or bioactive intermediates are locally provisioned, being released as per the specific bodily requirements [82]. The long-term storage of glucose as glycogen (in animals) and starch (in plants) represents the perfect case of glycans depository functions. Evident physical scenarios for these glycan traits are noticed in marathon runners who essentially build up liver reserves of glycogen before a large race. It is pertinent to note here that in the O-linked glycosylation (the earliest form of cytosolic glycosylation), the protein glycogenin is known to undergo glycosylation on a tyrosine residue (a short 8-12 glucose residue polymer) to function as the template for glycogen formation [83, 84]. This is somewhat distinguished from the biosynthesis of potato starch, which is accomplished *via* a *de novo* route and no longer requires amylogenin primer [85].

ROLE OF GLYCANS IN GRADIENT GENERATION

Gradients of growth factors (GF) can be produced, particularly during embryonic development, when cells (at different growth stages) attach to extracellular matrix glycoproteins (GAGs) such as heparan sulfate [86]. The crucial morphogen gradients throughout development may be aided by this organization of GF by GAGs [87]. Gradient establishment is highly needed for the homogeneous distribution of drugs, nutrients, hormones, and nutraceuticals. This is essential as the hydrophobicity of most drugs impairs their bioavailability and absorption throughout the body, and consequently, the patient sensitization risk is aggravated due to continued intake.

Not limited to drugs, even multiple food ingredients like those of fats (containing fatty acids) and proteins, show a need for distributive interactive forces rather than singular hydrophobicity or hydrophilicity. The glycans by their C-C skeleton and manifold –OH substitution facilitate the interactions by hydrogen bonding (suited in the aqueous environment) and distribution of hydrophobic entities by moderating their hydrophobic-hydrophobic proximity in the physiological environment. Such attributes of glycans are finely illustrated in their utilization of nanocarriers or nanovesicles for effective trafficking in the physiological boundaries. The best illustration of nanocarriers that make use of such glycans features is lipid-based carriers, whereby the amphipathic nature itself supports the moderation of stronger self-interactions.

A recent study published in the *Journal of American Chemical Society* illustrated the interactions between SARS-CoV-2 spike protein and angiotensin-converting enzyme 2 (ACE-2) raised concerns on the functional role of glycans in mediating ACE2 dimerization and the corresponding spike protein's downstream interactions. The investigators attempted to study the interactions through combinative utilization of glycol engineering with high-resolution native mass spectrometry (MS) to ascertain the consequences on the *N*-glycan residing locations over the assembly of manifold spike-ACE2 complexes. Analysis revealed that intact spike trimers completed 66 *N*-linked sites as engaged. For monomeric ACE-2, entire 7-N-linked glycan locations are occupied to distinct extents with six locations exhibiting >90% occupancy contrary to the partially occupied (~30%) seventh site (Asn690). The investigation studied the impact of each *N*-glycan on ACE2 dimerization by resolving the ACE2 glycoforms. In an unusual trend, the investigators noticed a prominent role of Asn432 in facilitating dimerization, which was subsequently confirmed through site-directed mutagenesis. An erstwhile finding was the formation of multiple stoichiometric complexes between glycosylated dimeric ACE2 and spike trimers (spike-ACE2 and spike2-ACE2), having ~500 and < 100 nM as dissociation constants. The trends inferred the co-operated formation of ion-driven ACE2 complexes with multiple Spike trimers. Overall, the study established glycan occupancy as decisive in facilitating the interactions between ACE2 dimers and spike trimers. Knowing that soluble ACE2 maintains a native SARS-CoV-2 interaction location critical for ACE2 dimerization, so for glycosylation besides the subsequent proclivity of spike and ACE2, assembling as higher oligomers, *vis-à-vis* molecular aspects, is highly crucial for streamlining the virus-neutralization measures [88].

Another study discussed the significance of glycosylation as post-translational modification of glycans to the characteristic peptide locations, whereby Advanced Therapeutic Medicinal Products (ATMPs) developmentrequires product standardization, high-quality starting reagents and a thorough caution for safety and efficacy (bypassing the adverse reactions). Because glycosylation-mediated actions can result in adverse effects like hypersensitivity reactions, the development of antidrug antibodies, product diversity in the face of biological adverse effects, and more, regulatory agencies are swiftly initiating the strict inspection of glycan profiles on biomedical entities [89].

ROLE OF GLYCANS TO INFLUENCE DIFFUSION BARRIERS

Critical diffusion barriers are composed of highly sialylated glycoproteins and ECM-GAGs. For instance, heparan sulfate GAGs in the glomerular basement membrane [92 - 94] and the highly sialylated protein podocalyxin on glomerular podocyte foot processes [90, 91] both play important roles in preserving the integrity of blood plasma filtration by the kidney. Of note, podocytes are the cells in the kidney's Bowman's capsule that wrap around glomerulus capillaries. These cells play a critical role in establishing the filtration barrier (making up the epithelial lining of Bowman's capsule) comprising slits between interdigitating podocyte foot in the vicinity of glomerular capillaries of perforated endothelium and basement membrane thereof. Pathological or experimental damage to such glycans causes large molecules like albumin to escape into the urine and is associated with glomerular diseases [95].

ROLE OF GLYCANS IN SOLUBILIZATION AND LUBRICATION OF BIOLOGICAL MOLECULES

The body of a complex multicellular living creature constitutes various intracellular and extracellular fluids such as blood, lymph, cerebrospinal fluids (CSF), bronchoalveolar lavage (BAL), *etc.* These body fluids contain various richly glycosylated proteins, which can't function properly without being glycosylated. Importantly, glycans affect the cell membrane fluidity. The presence of glycan-containing glycolipids and glycoproteins in the membrane affects its rigidity or fluidity. For example, the high sialic acid content in gangliosides affects the negative charge and steric hindrance of the bulky glycans and increases membrane fluidity.

Some glycans with their protein and lipid conjugates are hydrophilic and acidic, making them competitive modulators of their water solubility. Unsurprisingly, glycosylation is critically responsible for the prevalence of ~ 50-70 mg•ml^{-1} protein content in human blood plasma with ~2 mM sialic acid in the bound state. Certain polysaccharides, glycoproteins, and even glycolipids possess antifreeze traits and can alter the solvent nature of water. The presence of these proteins in certain fish prevents nucleation of ice crystals in body fluids, assisting their survival in extreme cold temperatures [96, 97].

Glycosylation assists in easing out the friction-generating drag in between hollow organs, attained *via* soluble and membrane-bound mucins co-prevalence on the periphery of hollow organs. Inthe first instance, this appears as a routine task unless the observer cautiously notices variations in oral salivary mucins exhibited due to the radiations manifested in injury to salivary glands (accompanied by the ill fate of head and neck cancer) [98, 99]. This could also result from a life-threatening autoimmune disorder (Sjogren's), especially in the discomfort caused by food swallowing ability [100]. Another instance of the drag-minimizing role of glycans is the body fluid action of hyaluronan (functions as synovial fluid in joint cavities and tear fluid of eyes). The deficiencies therein could be addressed therapeutically [101, 102].

ROLE OF GLYCANS IN THE MODULATION OF MEMBRANE RECEPTOR SIGNALING

Current attempts mainly focus on the glycolipid's potential to modify the signaling actions of the membrane receptor proteins [103]. As an instance, faintly distinct regimes of sialylated ganglioside, GM3 could result in manifold outcomes on tyrosine kinase signaling actions of epidermal growth factor receptor (EGFRs) [104, 105], with the elimination consequences being reflected in changed functioning of insulin receptor [106]. A better-suited case is the co-receptor activity of heparin sulfate in fibroblast growth factor (FGF) signaling [107]. Studies have also elucidated the impact of glycosylation on carbohydrate-conjugated proteins. For instance, *N*-glycans undergoing α1-6 core fucosylation affect the signaling actions of transforming growth factor (TGF) [108]. Impaired activation of TGF-β1 affects the normal development of lungs and most of the core fucose-lacking mice die 3 days after birth, whereas the survivors gradually develop emphysematous lung variations. The working mechanism herewith is suspected to suppress the activity of TGF signaling, stimulating matrix metalloproteinase (MMP) and ultimately degrading the alveolar membranes. On similar lines, sialylation and fucosylation jointly modulate the epidermal growth factor receptor-regulated cell-cell communication [109, 110].

A relatively new regime emerged at the forefront through the establishment of a Fringe molecule as a glycosyltransferase that alters the major proteins involved in the alteration of signaling protein Notch to affect the Notch-Delta interactions [111, 112]. A notable coincidence is that optimal Fringe actions mandate Notch glycosylation with O-fucose, establishing O-fucosyltransferase 1 as an essential constituent of Notch-regulated intracellular communications [113, 114]. Subsequently, it became known that *O*-glucose alteration on Notch-mediated *via* glucosyltransferase is mandatory for Notch signaling and embryonic development [115, 116]. In a nutshell, glycosylation sensitively regulates manifold essential functions associated with Notch signaling, with the structural rationale of certain glycosylation-mediated Notch interactions with **its select** ligands, being explored recently [117 - 119]. In the view of worthy authors of this 2015 literature source, Notch receptors are instrumental in guiding the mammalian cell fates through the involvement of Jagged and Delta-like proteins. Analysis deciphered a 2.3-angstrom resolution crystal structure for the interacting domains of the Notch1-DLL4 complex, revealing a two-site, anti-parallel binding aided by Notch1 *O*-linked glycosylation.Investigation of underlying proximities revealed interaction of Notch1 epidermal growth factor repeats 11 and 12 with the DLL4

Delta/Serrate/Lag-2 (DSL) domain and the moiety at the N-terminus of Notch ligands, respectively. Scrutiny deciphered *O*-fucose and *O*-glucose functionalized threonine and serine residues (respectively) on Notch1, acting as surrogate amino acids by making specific and essential contacts to the DLL4 residues. Scarcity belittles the Notch-Delta/Jagged interactions owing to manifold constraints forbidding the gathering up of Notch receptor-ligand complexes from a structural viewpoint. Notch natively exhibits impaired ligand proximity wherein optimalbinding mandates adequate modification *via O*-glycosyltransferases in the recombinant fragments needed for structural screening. Finally, Ca^{++} binding by multiple Notch EGFs is essential for folding and ligand proximity. The study focused on*in vitro*evolution to synchronize and integrate the mutations in DLL4 stability and the concomitant Notch 1 affinity. Readers are requested to refer to the literature for the above prospects from the 2015 article by *Luca and colleagues*, published in the journal Science.

A quantitative grasp of glycans-modulated membrane receptor signaling is depicted in Fig. (**6**), wherein *in situ* added monosaccharides profusely modulate the cell-surface receptor signaling and ion-channel attributes. The figure schematically represents the scenario with α-2,3-linked sialic acid or α-1,3-linked fucose receptors impairing the signaling *via* epidermal and fibroblast growth factor receptors. In this way, *in situ*, glycan editing serves as an authentic strategy for studying the livelyroles of glycans in membrane receptor signaling and ion-channel functions.

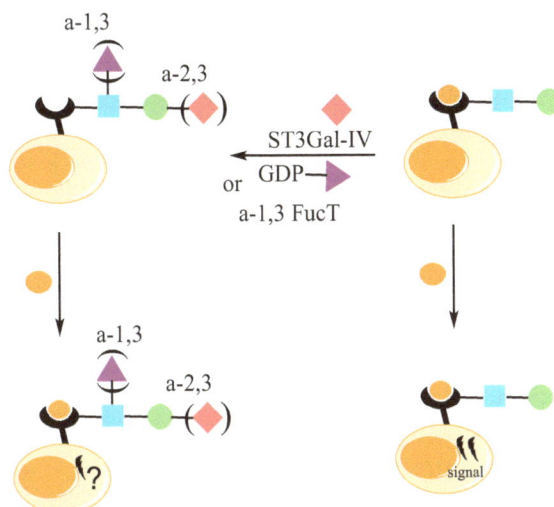

Fig. (6). The possible modulation of signaling responses *via* edited impaired signaling actions of α 2,3-linked sialic acid or α1,3-linked fucose *via* bypassed actions of epidermal and fibroblast growth factor receptors (EGFRs). The right-side portion depicts the editing-mediated signal generation.

ROLE OF GLYCANS IN EPIGENETIC MODIFICATION OF HISTONES AND NON-HISTONE NUCLEAR PROTEINS

The genetic information within chromosomal DNA is organized into nucleosomes which are further constituted of DNA and proteins and assembled into a more compact chromatin structure, eventually into very large, high-order chromosomes. In eukaryotic nuclei, the DNA is wrapped around a basic protein named histone, an octameric entity consisting of two copies, each of histones H2A, H2B, H3, and H4), to form a nucleosome, and is inhibited by histone H1. These proteins undergo a broad spectrum of post-translational modifications (PTMs) such as acetylation, methylation, phosphorylation, SUMOylation, formylation, biotinylating, butyrylation, ubiquitylation, glycosylation, *etc*. These modifications are described as epigenetic events and constitute the "histone code". The histone code modulates the recruitment of key enzymes(*e.g.* DNA methylases, histone acetyltransferases), regulating the chromatin condensation to influence the gene expression.This results in two distinct areas of chromatin, the first being slightly condensed euchromatin and the second one having highly condensed heterochromatin. While euchromatin is transcriptionally active, the highly condensed heterochromatin is transcriptionally silent [120, 121].

Epigenetics denotes a "stably heritable phenotype resulting from chromosomal changes without alterations in the DNA sequence" [122]. Two cells with the same DNA sequence, and under the influence of dissimilar epigenetic controls, exhibit distinct phenotypes. Thus, epigenetics may alter the phenotype of the cells without even altering the DNA sequence or mutating the cellular DNA. Epigenetic modifications prevail amidst the differentiation of somatic cells and after environmental changes [123]. Epigenetics happens to regulate gene expression by various mechanisms [124, 125].

The inceptive belief for glycosylation restricted its occurrence to cell surface along with the formation of secreted proteins through classical pathways of the endomembrane system until Torres and Hart (in 1984) noticed *O*-GlcNAc (*N*-acetylglucosamine) alteration *via* intracellular glycosylation in murine lymphocytes [126]. Of note, *O*-GlcNAc modification of serine and threonine aa was observed for several intracellular proteins, exhibiting 55% nuclear and 34% cytosolic prevalence, for the rat liver cells [127]. The enzymes adding (*O*-GlcNAc transferase) and removing *O*-GlcNAc (OGlcNAcase) to and from proteins have been cloned and thoroughly characterized [128]. OGT and *O*-GlcNAc target several proteins critically involved in signaling, functional gathering, nuclear pore proteins, and transcriptional domains, such as RNA polymerase II. These proteins are critically involved in sustaining cell-cell signaling, protein-protein interactions, and transcription [129]. Another noteworthy function of *O*-

GlcNAcylation pertains to its involvement in multiple metabolic pathways, regulating the metabolism of glucose, amino acids, nucleotides, and fatty acids. In zebrafish, the evolutionary diversity of OGT ascribes to its distinctly regulated six transcriptional variants [130]. The maternal transcripts, var1 and var2 prevail ubiquitously in the early stages of development while var3 and var4 are expressed subsequently, making up the OGT active forms amongst the six variants. Separate investigations have deciphered intrinsic HAT activity for *O*-GlcNacase, leading to its identification as nuclear cytoplasmic *O*-GlcNAcase and acetyltransferase (NCOAT), the noted gene transcription activators [131]. The communication between *O*-GlcNAc and erstwhile histone PTMs utilizes the phosphorylated configuration of *O*-GlcNAc. Besides, studies have deciphered an integrated function of OGT with histone deacetylase (HDAC), being recruited *via* mSin3A for gene expression [132].

GLYCANS IN INTRASPECIES RECOGNITION

Humans prevail as complex, multicellular organisms, harboring multiple cells, tissues, organs, organ systems, and the body. It is essential to maintain a coordinated cross-talk and communication between various cells within a tissue, for the survival of a complex living organism. Within a cell, proteins, and other molecules traffic from one region to the other to be processed into a functional state and aboutthe requirement of distinguished allocation. The newly synthesized proteins travel from the ribosomes through the secretory pathway (ER & Golgi apparatus) to finally reach their predetermineddestinations as designed by specific signaling sequence. For instance, receptors in the ER facilitate specific folding patterns for the incoming proteins, *vis-à-vis* their glycan epitopes. Based on the signaling sequence, the proteins are destined to reach different locations in a cell (*e.g.* membrane, nucleus, *etc.*) or even secreted outside the cells (as hormones, growth factors, cytokines, chemokines, *etc.*), the unfolded or misfolded proteins reach to endosomes or lysosomes for their degradation into the constituent aa. Of note, incorrectly folded proteins with the recognition of aberrant glycan structures can be discarded and degraded. Molecules secreted by one cell reach another cell surface, bind to their specific cell surface receptors, and send signals. Glycans have a role in protein folding and trafficking, coordinating the cellular uptake of various molecules *via* endocytosis and phagocytosis. For example, the uptake process of endocytosis and phagocytosis are often mediated by cell surface receptors that recognize macrophage terminal glycans. This section discusses the role of glycans in protein folding and degradation along with intracellular trafficking of various molecules (growth factors, signaling proteins, hormones, *etc.*)and the onset of endocytosis as well as phagocytosis [50, 133].

GLYCANS IN INTRACELLULAR GLYCOPROTEIN FOLDING AND DEGRADATION

The protein formation commences *via* binding of protein synthesis factory ribosome (30S for prokaryotes, 40S for eukaryotes) with the Kozak (eukaryotes) or Shine Dalgarno (prokaryotes) sequence of mRNA. The N-terminal leader sequence of the newly generated, aa peptide chain (primary structure) enters the lumen of the endoplasmic reticulum (ER), undergoes chaperon-assisted folding (secondary, tertiary, quaternary structure), and subsequent post-translational modifications. Some prominent aspects of these modifications include the synthesis of disulfide bonds, glycosylation (*N*-linked), phosphorylation, sulfation, addition of metal groups, *etc*. The modified proteins, thus generated, leave the ER and are transported to the cis Golgi apparatus *via* multiple vehicles, such as COP1 and COP2 (coat protein complexes). These proteins sustain a precise storing of lipids and proteins between Golgi cisternae and the subsequent retrieval from Golgi to ER. While COPII vehicles export the newly formed secretor proteins from the ER, the COPI vehicles traffic these proteins from the Golgi to the ER and intra-Golgi transport. The trafficked proteins in the Golgi apparatus undergo post-translational modifications such as *O*-linked glycosylation, crossing the medial and trans-Golgi apparatus before being released into the cytoplasm. Based on the specific signaling sequence at the amino terminus, a protein reaches its proper destination or is secreted from the cells. The misfolded proteins are the subject of targeted destruction/degradation by endosomes/lysosomes [134].

Besides the discussed biophysical roles of conjugated glycans on native glycoproteins, the identification of some glycans plays a critical role in their ER-associated degradation (ERAD). On the detection of uncommon Glc3Man9GlcNAC2-P-P dolichol sensitivity of the lipid conjugated oligosaccharide donor for *N*-glycosylation, screening deciphered it as nearly similar in nearly entire analyzed eukaryotes. Ironically, multiple aspects of this structure play a critical role in maintaining adequate *N*-glycan transfer [135, 136], variations were witnessed amidst the screening of select parasites and mutant cells [137, 138]. Persistent attempts on this aspect provide a clue, with the progressive unveiling of third glucose residue on repeated N-glycans elimination with a subsequent substitution during ER-glycoprotein folding [139, 140]. This accomplishment made way for the identification of terminal glucose units by some ER chaperones, calnexin, and calreticulin [141, 142]. Over time, the relevance of glycosylation and glycosylation flow of events in protein folding has evolved better for predictive understanding [143 - 146]. Nevertheless, despite the permanent loss of terminal glucose residue, the subsequent identification of oligomannose resembling *N*-glycans is accomplished on ER-mannosidases partial processing [147 - 149]. Such perceiving efforts are partially driven *via* mannose-

6-phosphate receptor homology in manifold chaperones, certain other mannosidase-resembling proteins, and recognition complexes [150 - 152].

Definitively, a byzantine array of glycan-modifying proteins encompasses the ER fate of glycoproteins: engineering the Golgi pathway (the terminal fate) or the ERAD involvement. With most proteins entering ER-undergoing glycosylation, significant variations are noticed for unfolded protein response-driven diseased conditions [151, 153]. Two notable events in the folding of newly generated proteins are *O*-mannosylation and *O*-fucosylation, wherein proteins being unable to fold gradually, get rid of the futile folding cycles through the cytosolic-reverse translocation. A more sophisticated pathway engineers *N*-glycan removal from the wrongly folded proteins, before the activity of proteasomes, commencing with the activity of *N*-glycanase (a cytosolic peptide) [154]. Of note, *O*-GlcNacylation of nucleo-cytoplasmic proteins could also happen *via* co-translation, guarding the nascent polypeptides from premature degradation *vis-à-vis* co-translational ubiquitination [155].

GLYCANS IN INTRACELLULAR GLYCOPROTEIN TRAFFICKING

Intracellular trafficking aims to transport proteins to the correct sub-cellular compartments such as cytoplasm, ribosomes, nucleus, membranes, and other destinations, RER to Golgi secretory pathway. The glycoproteins are trafficked to the endosomes and lysosomes for their degradation (unfolded/misfolded glycoproteins), as per the specific allocation of regulatory aspects. Glycoprotein trafficking also involves the transport of secretory proteins (*e.g.* growth factors, hormones, *etc.*) to the extracellular space through the secretory pathway. An important example of glycan functions in intracellular trafficking is mannose 6-phosphate recognition for targeting lysosomal enzymes to lysosomes. The transport of certain glycoproteins is sensitively affected by lectin-resembling moieties residing in the ER-Golgi secretory pathway. For example, the *LMAN1* gene product, ERGIC-53 within ER-Golgi intervening partition is a mannose-explicit and Ca-based human homolog of leguminous grade, L-regime lectins [156]. This molecule vitally serves as a chaperone for coagulation factors, V and VIII amidst the biosynthesis in hepatocytes and endothelial cells, modulating the bio-formation of several related glycoproteins [157 - 159]. Erstwhile noteworthy examples include VIPL and VIP36 (vasoactive intestinal polypeptides) [160 - 162]. In recent years, several glycan-recognizing proteins in the ER-Golgi pathway have been reported for protein folding (chaperons) and trafficking.

GLYCANS IN TRIGGERING OF ENDOCYTOSIS AND PHAGOCYTOSIS

Endocytosis is a cellular communication process meeting the cell's requirement from external ends through energy inputs needed to engulf the required materials across the cellular membrane. Transport of macromolecules from extracellular fluid to intracellular compartments proceeds through receptor-mediated endocytosis, initiated *via* ligand binding to the receptors. Endocytosis also assists in regulating the receptor intensity on the cell membrane through concomitant participation of post-synaptic signal communication. It could be facilitated *via* clathrin, caveolin, pinocytosis, or phagocytosis [163]. The one driven *via* clathrin is a receptor-triggered process involving the binding of clathrin proteins to certain unique membrane surface receptors. Notably, clathrin is distributed across the inner plasma membrane surface, it forms a coated pit and vesicle in the cytoplasm. The as-generated cell membrane lacking vesicles promptly eliminates clathrin and fuses with early endosomes. Most receptor proteins bound to coated vesicles across the inner membrane are recycled back to the cell membrane, and others are terminally degraded in late endosomes and lysosomes [164].

Caveolin-dependent endocytosis is yet another notable regime of cellular transport, involving caveolae as clathrin lacking buds, expressed abundantly on the plasma membrane surface. Of note, caveolin is a cholesterol-binding protein that functions *via* caveolae capture across the cholesterol and glycolipids comprising biolayer. The D2 receptors (abundant in membrane rafts) are internalized through caveolae-drivenendocytosis [165]. Endocytosis involving the transport of fluid materials (*i.e.* pinocytosis), normally occurs across the plasma membrane. The cell membrane invagination gradually results in a pocket that pinches off in the cell, forming a cyst completely saturated with extracellular fluid and some other molecules. Subsequently, the cyst moves within the cytosol and slowly amalgamates into the endosomes and lysosomes.

Phagocytosis implies a mechanism of cellular proximity and internalization of terminal extracellular wastes or invading organisms (like cell debris, microorganisms, apoptotic cells, *etc.*). Pinocytosis and phagocytosis comprise the uptake of larger membrane regions than the clathrin-driven endocytosis and caveolae pathway, whereby the cytoplasmic constituents are degraded *via* relocation to lysosomes [166]. Fig. (**7**) represents the possible phagocytic regulation by glycans, wherein the colored marks signify the classical glycan chain residues on the *N*-glycosylated cell-surface glycoprotein. The terminal sugar in this assembly is often a sialic acid (pink indication), which can bind either Factor H (for complement inhibition) or the Siglec receptors (for phagocytosis impairment). Nonetheless, neuraminidases can get rid of these terminal sialic acid molecules to picture the galactose domains (yellow spheres), having a proximity for opsonin, galectin-3, calreticulin, and C1q. Henceforth, the terminal galactose residues are deleted *via* β-galactosidase domains. These residues bind opsonin and ficolin-2 apart from complement receptor-3. The formation of sialic acid could impair phagocytosis contrary to the galactose and *N*-acetyl glucosamine-promoted phagocytosis. Hence, the manifold cell-surface receptors that identify the terminal glycans, can modulate the uptake of molecules (endocytosis), particles (phagocytosis), or even whole cells. Some of the finest illustrations herein comprise the glycoprotein receptors of hepatocytes and mannose receptors of macrophages. A significant diversity of lectins is understood to facilitate endocytosis in macrophages and dendritic cells. The recognition of this regime is important not merely for antigen presentation to T-cells but also for eliminating damaged cells, like those happening during the entryof microbial sialidases to the circulation (sepsis), culminating in platelet desialylation [167, 168] or immature glycosylated mucins secretion by cancer cells [54].

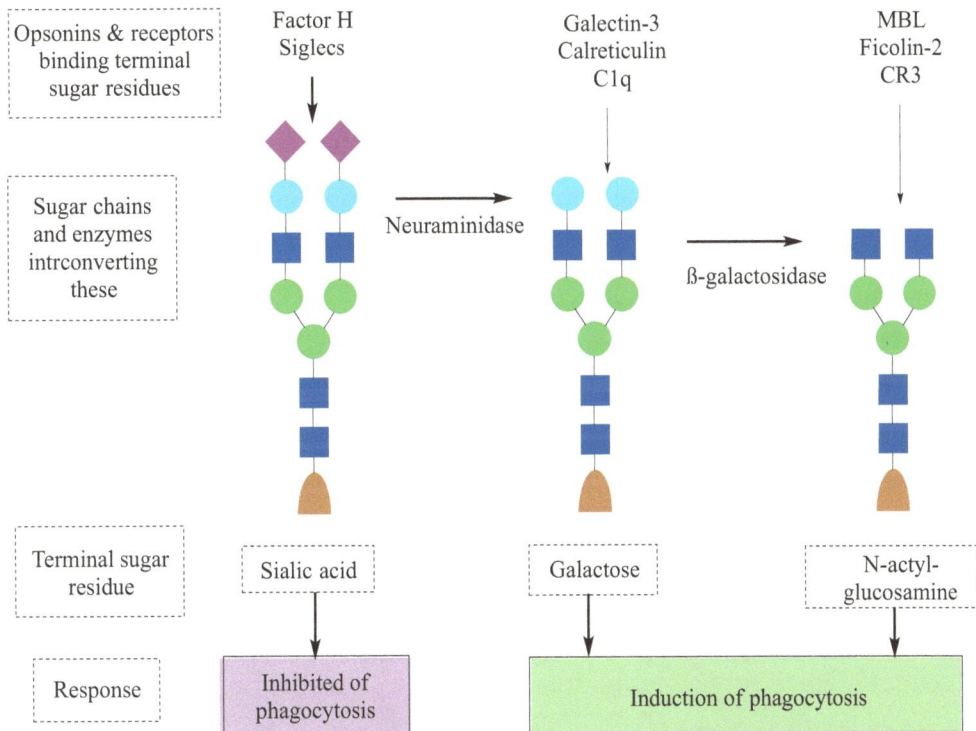

Fig. (7). Possible regulation of phagocytosis by the actions of sugar chains, wherein β-galactosidase mediated removal of terminal galactose for the coordinated actions of N-acetyl-glucosamine residues.

GLYCANS IN INTERSPECIES RECOGNITION

For various purposes, cells originate in different species interact, and communicate with each other. These interactions often involve glycan-binding proteins and can result in symbiosis, commensalism, or diseased manifestation, depending on the interaction type *via* specific biological circumstances. A large volume of studies at present, enrich our knowledge on the host-pathogen interactions or symbiotic relationships between a probiotic and intestinal host cells where cell surface glycans exhibit important roles in recognition and encountering different cells. Some important examples illustrating this include gut bacteria *E.coli* [169], and *Helicobacter* [170], which recognize host cell surface sialoglycans for adherence. In the case of *P. falciparum,* red blood cells are invaded *via* surface binding of specific sialic acid residues [171]. Another important example is the recent pandemic, *vis-à-vis* infection of the human body with the SARS-CoV 2. A growing body of literature suggests that *N-* and *O-* glycosylation of the SARS-CoV2 spike proteins strengthens its ACE2 binding affinity, assisting the evasion from the host immune system [172]. The following paragraphs briefly describe the role of glycans in microbial adhesion, interaction with the host cell's pattern recognition receptors (PRR), pathogenic uptake, and processing.

GLYCANS IN MICROBIAL ADHESIONS

The adhesion of microbes on the host cell surface is recognized as the inceptive stage of microbial colonization and biofilm formation. The host-microbe interactions subsequently result in pathogenicity (for parasitic pathogenic bacteria), commensalism (for non-pathogenic, human gut bacteria), and symbiotic relationship (for human gut probiotics). The receptor specifically targeted for bacterial adherence can be a protein/peptide or a glycoconjugate expressed on the host cell surface in connective tissue or absorbed on host tissues in contact with the abiotic surface. Many studies have elucidated that integrins residing on the surfaces of the host cells are also microbial pathogenic targets [3].

Helicobacter identification of gastric sialoglycans represents a critical case of bacterial adhesions *vis-à-vis* screened actions involved in gastric ulcers and cancer development [173, 174]. The F-pilus-supported glycans binding in *E. coli* is attributed to millions of urinary tract infections annually [175, 176], with the emergence of small molecule inhibitors as potent therapeutic agents [177]. The merozoite stage of *P. falciparum* represents a perfect example herein, resulting in malaria *via* the identification of densely sialylated glycophorins on target erythrocytes [178, 179], with the species specificity being affected by the varied nature of sialic acids [180]. For instance, in malarial infection, heparan sulfate prevailing on endothelial cells facilitates the binding of *P. falciparum*-infected erythrocytes through the DBL1α domain of PfEMP1 [181], leading to manifold complications.

The host glycan specificity squarely exhibits a key role in the binding of pathogenic fungi (*e.g. C. glabrata*) to the various target tissues [182]. Opposed to the glycan's involvement in the host-microbe interactions, multiple studies discuss the glycan-independent host-microbe interactions. With the passing evolutionary duration, a particular invading pathogen may rely exclusively on glycan-independent mechanisms for infection, as noted for the endemic, *P. falciparum* [183, 184]. Longer contact times along with the conduction in static conditions impair the impact of the initial glycan "handshake". Therefore, more studies are required to ascertain the role of glycans in host-pathogen interactions.

GLYCANS IN RECOGNITION OF PATHOGEN-ASSOCIATED MOLECULAR PATTERNS

The innate immunological cells comprise the first line of defense of a body's defense against disease-causing pathogens namely viruses, bacteria, fungus, parasites, and others. The innate immunological cells exhibit various pattern recognition receptors (PRRs) on the cell surface, such as Toll-like receptors (TLRs), *C*-type lectins, and NOD-like receptors (NLRs) [185]. Many of the PRRs are glycoprotein molecules and PRRS from the innate immunological cells react with various pathogens-associated molecular patterns (PAMPs). Some of the important PAMP molecules include bacterial lipopolysaccharides (LPS), peptidoglycan, deoxyribose (DNA), or ribose (RNA) based polymers present in distinct pathogens besides others [186]. The fungal cell wall released glucan and oligochitin oligosaccharides can also function as plant defenseelicitors [187].

VIRAL RECOGNITION OF GLYCAN TARGETS

The glycan-binding proteins of viruses are called hemagglutinins, the terminology being applied due to the ability of the viral protein to agglutinate erythrocytes. Influenza virus is one of the most common pathogenic viruses and hemagglutinin produced by it is known as influenza hemagglutinin (the "H" in "H1N1"). This molecule plays a key role in the infection process.

Adequate studies discuss the binding proximity of these pathogens, *vis-à-vis* the sialic acid ligand, exhibiting an explicit linkage to the underlying sugar chain, besides screening the viral vulnerability for avian *versus* human hosts [188 - 190]. The evolution of avian influenza viruses as human infectious agents involved selection for varied binding actions, for experimental replication [191]. Interestingly, even our closest evolutionary cousins (chimpanzees) do not have a high density of human sialic acid composition on their airway epithelium [192], justifying the lack of non-human primate models and the unlikely choice of ferret as a model for human influenza as it is believed to harbor the human-like linkage on its airway epithelium [193] and also due to its turning out ability as of humans, missing the non-human sialic acid Neu5Gc [194]. Several other instances of diligent, viral-sialic acid specificity are known, wherein the prevalence of *O*-acetyl esters at peculiar loci is exhibited. Select observations herein include a

4-O-acetyl ester residence on sialic acid targets for mouse hepatitis viral infection [195, 196] and a 9-O-acetyl ester on sialic acid side chain need for the binding of coronaviruses and erstwhile influenza C, D viruses [197 - 199]. The two specificities are distinguished only *via* few prominent aa changes in the viral receptor sequences [200].

ANTIGEN RECOGNITION, UPTAKE AND PROCESSING

Immunity against any pathogen requires coordinated interaction between innate (1st line of defense) and acquired (2nd line of defense) immune cells. In the process, the primary antigen-presenting cells (APCs) namely macrophages, dendritic cells, and B cells recognize the antigens uptake *via* phagocytosis/endocytosis to kill them, process them into small peptides and present the antigens on APC surfaces along with Major Histocompatibility Complex, MHCII. The APC surface MHC II and antigenic peptide react with T helper cell surface receptors (TCR), CD^3, and CD^4 molecules. However, endogenous antigens on the surfaces of tumor/cancer/virus-infected cells present the antigenic peptide along with MHC I on their surface. Now MHC-1-Ag interacts with TCR, CD^3, and CD^8 on the surface of T cytotoxic cells [201]. Subsequently, various glycans on the target cells facilitate the processing and presentation of the various antigens. For instance, the prevalence of terminal Man or GlcNAc residues to large extents on non-native proteins or microbes can activate phagocytosis through C-regime lectins residence on APCs, alongside the antigenic proteins trafficking to processing zones [63 - 65]. Another well-suited example is non-human α-galactose or Neu5GC fragments residing on injected glycoproteins. These entities could result in immune reactions and the subsequent generation of immune complexes that collectively stimulate the immunological response against the peptide backbone [65, 66]. A modified version of self/non-self-recognition is illustrated by the expression of non-self-glycolipids *via* CD1 molecules, recognized by hindered TCRs of NKT cells [202, 203].

CONCLUSION

Glycans are the most abundant biological molecules prevailing across all living kingdoms starting from actinomycetes (bacteria), fungus/yeast, plantae, Animalia (including humans), and even viruses (entities recognized as bridges between life and death). However, glycans are unique to each cell type and significantly different from one to other organisms. Additionally, the diversity and complexity of glycan structures including their conjugation with proteins (glycoproteins, proteoglycans) and lipids (glycolipids) makes it challenging to screen their precise involvement in diverse biological functions across various living organisms. Despite all such bottlenecks, significant progress has been made to decode the multiple biological functions of glycans. Glycans are not only essential to maintain the structural and functional integrity of the cells but are also involved in cell adhesion, bi-directional cell signaling across biological membranes, and cell-to-cell crosstalk and communication. Glycans help to maintain various physiological roles by intra-species (different members of the same species) and inter-species (members in between two different species) recognition of various biomolecules. In a nutshell, life on Earth is impossible without glycans, the most abundant biological molecule in the living universe.

REFERENCES

[1]   Moremen KW, Tiemeyer M, Nairn AV. Vertebrate protein glycosylation: diversity, synthesis and function. Nat Rev Mol Cell Biol 2012; 13(7): 448-62.
[http://dx.doi.org/10.1038/nrm3383] [PMID: 22722607]

[2]   Rabinovich GA, van Kooyk Y, Cobb BA. Glycobiology of immune responses. Ann N Y Acad Sci 2012; 1253(1): 1-15.
[http://dx.doi.org/10.1111/j.1749-6632.2012.06492.x] [PMID: 22524422]

[3]   Zhang L, Ten Hagen KG, Kelly G. The cellular microenvironment and cell adhesion: a role for O-glycosylation. Biochem Soc Trans 2011; 39(1): 378-82.
[http://dx.doi.org/10.1042/BST0390378] [PMID: 21265808]

[4]   Stanley P, Schachter H, Taniguchi N. N-Glycans.Essentials of Glycobiology. 2nd ed.

[5]   Varki A, Cummings RD, Esko JD, *et al.* Essentials of Glycobiology 2nd Editioned. Cold Spring Harbor, NY, USA: Cold Spring Harbor Laboratory Press 2009.

[6]   Varki A. Biological roles of glycans. Glycobiology 2017; 27(1): 3-49.
[http://dx.doi.org/10.1093/glycob/cww086] [PMID: 27558841]

[7]   Awofiranye AE, Dhar C, He P, Varki A, Koffas MAG, Linhardt RJ. *N* -glycolylated carbohydrates in nature. Glycobiology 2022; 32(11): 921-32.
[http://dx.doi.org/10.1093/glycob/cwac048] [PMID: 35925816]

[8]   Sackstein R, Stowell SR, Hoffmeister KM, Freeze HH, Varki A. Glycans in Systemic Physiology. In: Varki A, Cummings RD, Esko JD, Stanley P, Hart GW, Aebi M, Mohnen D, Kinoshita T, Packer NH, Prestegard JH, Schnaar RL, Seeberger PH, editors. Essentials of Glycobiology [Internet]. 4th Edition. Cold Spring Harbor (NY): Cold Spring Harbor Laboratory Press; 2022. Ch-41.

[9]   Hebert DN, Lamriben L, Powers ET, Kelly JW. The intrinsic and extrinsic effects of N-linked glycans on glycoproteostasis. Nat Chem Biol 2014; 10(11): 902-10.
[http://dx.doi.org/10.1038/nchembio.1651] [PMID: 25325701]

[10]   Varki A, Lowe JB. Biological roles of glycans Ch-6 In: Essentials of Glycobiology. Varki A, Cummings RD, Esko JD, et al., editors. Cold Spring Harbor (NY): Cold Spring Harbor Laboratory Press. Bookshelf ID: NBK1897 2009; pp. 75-88.

[11]   Rebois RV, Fishman PH. Deglycosylated human chorionic gonadotropin. An antagonist to desensitization and down-regulation of the gonadotropin receptor-adenylate cyclase system. J Biol Chem 1983; 258(21): 12775-8.
[http://dx.doi.org/10.1016/S0021-9258(17)44033-6] [PMID: 6313676]

[12]   Colley KJ, Varki A, Haltiwanger RS, Kinoshita T. Cellular Organization of Glycosylation Ch-4.Essentials of Glycobiology. 4th ed.,

[13]   Wier M, Edidin M. Constraint of the translational diffusion of a membrane glycoprotein by its external domains. Science 1988; 242(4877): 412-4.
[http://dx.doi.org/10.1126/science.3175663] [PMID: 3175663]

[14]   Barbour S, Edidin M. Cell-specific constraints to the lateral diffusion of a membrane glycoprotein. J
         Cell Physiol 1992; 150(3): 526-33.
         [http://dx.doi.org/10.1002/jcp.1041500313] [PMID: 1537882]

[15]   Brown DA, Rose JK. Sorting of GPI-anchored proteins to glycolipid-enriched membrane subdomains
         during transport to the apical cell surface. Cell 1992; 68(3): 533-44.
         [http://dx.doi.org/10.1016/0092-8674(92)90189-J] [PMID: 1531449]

[16]   Varma R, Mayor S. GPI-anchored proteins are organized in submicron domains at the cell surface.
         Nature 1998; 394(6695): 798-801.
         [http://dx.doi.org/10.1038/29563] [PMID: 9723621]

[17]   Hansen GH, Immerdal L, Thorsen E, *et al.* Lipid rafts exist as stable cholesterol-independent
         microdomains in the brush border membrane of enterocytes. J Biol Chem 2001; 276(34): 32338-44.
         [http://dx.doi.org/10.1074/jbc.M102667200] [PMID: 11389144]

[18]   Cohen M, Varki A. Modulation of glycan recognition by clustered saccharide patches. Int Rev Cell
         Mol Biol 2014; 308: 75-125.
         [http://dx.doi.org/10.1016/B978-0-12-800097-7.00003-8] [PMID: 24411170]

[19]   Varki A. Selectin ligands. Proc Natl Acad Sci USA 1994; 91(16): 7390-7.
         [http://dx.doi.org/10.1073/pnas.91.16.7390] [PMID: 7519775]

[20]   Cohen M, Hurtado-Ziola N, Varki A. ABO blood group glycans modulate sialic acid recognition on
         erythrocytes. Blood 2009; 114(17): 3668-76.
         [http://dx.doi.org/10.1182/blood-2009-06-227041] [PMID: 19704115]

[21]   Schnaar RL. Glycolipid-mediated cell–cell recognition in inflammation and nerve regeneration. Arch
         Biochem Biophys 2004; 426(2): 163-72.
         [http://dx.doi.org/10.1016/j.abb.2004.02.019] [PMID: 15158667]

[22]   Boon L, Ugarte-Berzal E, Vandooren J, Opdenakker G. Glycosylation of matrix metalloproteases and
         tissue inhibitors: present state, challenges and opportunities. Biochem J 2016; 473(11): 1471-82.
         [http://dx.doi.org/10.1042/BJ20151154] [PMID: 27234584]

[23]   Farach-Carson MC, Warren CR, Harrington DA, Carson DD. Border patrol: Insights into the unique
         role of perlecan/heparan sulfate proteoglycan 2 at cell and tissue borders. Matrix Biol 2014; 34: 64-79.
         [http://dx.doi.org/10.1016/j.matbio.2013.08.004] [PMID: 24001398]

[24]   Aspberg A. The different roles of aggrecan interaction domains. J Histochem Cytochem 2012; 60(12):
         987-96.
         [http://dx.doi.org/10.1369/0022155412464376] [PMID: 23019016]

[25]   Grewal PK, Hewitt JE. Glycosylation defects: a new mechanism for muscular dystrophy? Hum Mol
         Genet 2003; 12(Spec No 2) (Suppl. 2): R259-64.
         [http://dx.doi.org/10.1093/hmg/ddg272] [PMID: 12925572]

[26]   Haliloğlu G, Topaloğlu H. Glycosylation defects in muscular dystrophies. Curr Opin Neurol 2004;
         17(5): 521-7.
         [http://dx.doi.org/10.1097/00019052-200410000-00002] [PMID: 15367856]

[27]   Inamori K, Yoshida-Moriguchi T, Hara Y, Anderson ME, Yu L, Campbell KP. Dystroglycan function
         requires xylosyl- and glucuronyltransferase activities of LARGE. Science 2012; 335(6064): 93-6.
         [http://dx.doi.org/10.1126/science.1214115] [PMID: 22223806]

[28]   de la Motte CA, Hascall VC, Calabro A, Yen-Lieberman B, Strong SA. Mononuclear leukocytes
         preferentially bind *via* CD44 to hyaluronan on human intestinal mucosal smooth muscle cells after
         virus infection or treatment with poly(I.C). J Biol Chem 1999; 274(43): 30747-55.
         [http://dx.doi.org/10.1074/jbc.274.43.30747] [PMID: 10521464]

[29]   Zhuo L, Kanamori A, Kannagi R, *et al.* SHAP potentiates the CD44-mediated leukocyte adhesion to
         the hyaluronan substratum. J Biol Chem 2006; 281(29): 20303-14.

[http://dx.doi.org/10.1074/jbc.M506703200] [PMID: 16702221]

[30] Ren J, Hascall VC, Wang A. Cyclin D3 mediates synthesis of a hyaluronan matrix that is adhesive for monocytes in mesangial cells stimulated to divide in hyperglycemic medium. J Biol Chem 2009; 284(24): 16621-32.
[http://dx.doi.org/10.1074/jbc.M806430200] [PMID: 19276076]

[31] Davies DG, Parsek MR, Pearson JP, Iglewski BH, Costerton JW, Greenberg EP. The involvement of cell-to-cell signals in the development of a bacterial biofilm. Science 1998; 280(5361): 295-8.
[http://dx.doi.org/10.1126/science.280.5361.295] [PMID: 9535661]

[32] Neu TR, Swerhone GDW, Lawrence JR. Assessment of lectin-binding analysis for *in situ*detection of glycoconjugates in biofilm systems. Microbiology (Reading) 2001; 147(2): 299-313.
[http://dx.doi.org/10.1099/00221287-147-2-299] [PMID: 11158347]

[33] Zielinska DF, Gnad F, Wiśniewski JR, Mann M. Precision mapping of an *in vivo*N-glycoproteome reveals rigid topological and sequence constraints. Cell 2010; 141(5): 897-907.
[http://dx.doi.org/10.1016/j.cell.2010.04.012] [PMID: 20510933]

[34] McFarlane HE, Döring A, Persson S. The cell biology of cellulose synthesis. Annu Rev Plant Biol 2014; 65(1): 69-94.
[http://dx.doi.org/10.1146/annurev-arplant-050213-040240] [PMID: 24579997]

[35] Koch BE, Stougaard J, Spaink HP. Keeping track of the growing number of biological functions of chitin and its interaction partners in biomedical research. Glycobiology 2015; 25(5): 469-82.
[http://dx.doi.org/10.1093/glycob/cwv005] [PMID: 25595947]

[36] Kumar P, Yang M, Haynes BC, Skowyra ML, Doering TL. Emerging themes in cryptococcal capsule synthesis. Curr Opin Struct Biol 2011; 21(5): 597-602.
[http://dx.doi.org/10.1016/j.sbi.2011.08.006] [PMID: 21889889]

[37] Gow NAR, Hube B. Importance of the Candida albicans cell wall during commensalism and infection. Curr Opin Microbiol 2012; 15(4): 406-12.
[http://dx.doi.org/10.1016/j.mib.2012.04.005] [PMID: 22609181]

[38] Free SJ. Fungal cell wall organization and biosynthesis. Adv Genet 2013; 81: 33-82.
[http://dx.doi.org/10.1016/B978-0-12-407677-8.00002-6] [PMID: 23419716]

[39] Hayashi T, Kaida R. Functions of xyloglucan in plant cells. Mol Plant 2011; 4(1): 17-24.
[http://dx.doi.org/10.1093/mp/ssq063] [PMID: 20943810]

[40] Bergstrom KSB, Xia L. Mucin-type O-glycans and their roles in intestinal homeostasis. Glycobiology 2013; 23(9): 1026-37.
[http://dx.doi.org/10.1093/glycob/cwt045] [PMID: 23752712]

[41] Bennett EP, Mandel U, Clausen H, Gerken TA, Fritz TA, Tabak LA. Control of mucin-type O-glycosylation: A classification of the polypeptide GalNAc-transferase gene family. Glycobiology 2012; 22(6): 736-56.
[http://dx.doi.org/10.1093/glycob/cwr182] [PMID: 22183981]

[42] Hagen KGT, Tran DT. A UDP-GalNAc:polypeptide N-acetylgalactosaminyltransferase is essential for viability in Drosophila melanogaster. J Biol Chem 2002; 277(25): 22616-22.
[http://dx.doi.org/10.1074/jbc.M201807200] [PMID: 11925446]

[43] An G, Wei B, Xia B, *et al.* Increased susceptibility to colitis and colorectal tumors in mice lacking core 3–derived O-glycans. J Exp Med 2007; 204(6): 1417-29.
[http://dx.doi.org/10.1084/jem.20061929] [PMID: 17517967]

[44] Ju T, Aryal RP, Kudelka MR, Wang Y, Cummings RD. The Cosmc connection to the Tn antigen in cancer. Cancer Biomark 2014; 14(1): 63-81.
[http://dx.doi.org/10.3233/CBM-130375] [PMID: 24643043]

[45] Alberts B, Johnson A, Lewis J, *et al.* The Plant Cell Wall, In: Molecular Biology of the Cell 4th

Edition,. New York: Garland Science, 2002. Bookshelf ID: NBK26928

[46]   Underwood W. The plant cell wall: a dynamic barrier against pathogen invasion. Front Plant Sci 2012; 3: 85.
[http://dx.doi.org/10.3389/fpls.2012.00085] [PMID: 22639669]

[47]   Mitra N, Sinha S, Ramya TNC, Surolia A. N-linked oligosaccharides as outfitters for glycoprotein folding, form and function. Trends Biochem Sci 2006; 31(3): 156-63.
[http://dx.doi.org/10.1016/j.tibs.2006.01.003] [PMID: 16473013]

[48]   Russell D, Oldham NJ, Davis BG. Site-selective chemical protein glycosylation protects from autolysis and proteolytic degradation. Carbohydr Res 2009; 344(12): 1508-14.
[http://dx.doi.org/10.1016/j.carres.2009.06.033] [PMID: 19608158]

[49]   Kozarsky K, Kingsley D, Krieger M. Use of a mutant cell line to study the kinetics and function of *O*-linked glycosylation of low density lipoprotein receptors. Proc Natl Acad Sci USA 1988; 85(12): 4335-9.
[http://dx.doi.org/10.1073/pnas.85.12.4335] [PMID: 3380796]

[50]   Schjoldager KTBG, Clausen H. Site-specific protein O-glycosylation modulates proprotein processing — Deciphering specific functions of the large polypeptide GalNAc-transferase gene family. Biochim Biophys Acta, Gen Subj 2012; 1820(12): 2079-94.
[http://dx.doi.org/10.1016/j.bbagen.2012.09.014] [PMID: 23022508]

[51]   Zhong C, Li P, Argade S, *et al.* Inhibition of protein glycosylation is a novel pro-angiogenic strategy that acts *via* activation of stress pathways. Nat Commun 2020; 11(1): 6330.
[http://dx.doi.org/10.1038/s41467-020-20108-0] [PMID: 33303737]

[52]   Albrecht S, Hilliard M, Rudd P. Therapeutic proteins: facing the challenges of glycobiology. JHPOR 2014.
[http://dx.doi.org/10.7365/JHPOR.2014.5.2]

[53]   Wahrenbrock M, Borsig L, Le D, Varki N, Varki A. Selectin-mucin interactions as a probable molecular explanation for the association of Trousseau syndrome with mucinous adenocarcinomas. J Clin Invest 2003; 112(6): 853-62.
[http://dx.doi.org/10.1172/JCI200318882] [PMID: 12975470]

[54]   Malaker SA, Pedram K, Ferracane MJ, *et al.* The mucin-selective protease StcE enables molecular and functional analysis of human cancer-associated mucins. Proc Natl Acad Sci USA 2019; 116(15): 7278-87.
[http://dx.doi.org/10.1073/pnas.1813020116] [PMID: 30910957]

[55]   Wahrenbrock MG, Varki A. Multiple hepatic receptors cooperate to eliminate secretory mucins aberrantly entering the bloodstream: are circulating cancer mucins the "tip of the iceberg"? Cancer Res 2006; 66(4): 2433-41.
[http://dx.doi.org/10.1158/0008-5472.CAN-05-3851] [PMID: 16489050]

[56]   Alemka A, Nothaft H, Zheng J, Szymanski CM. N-glycosylation of Campylobacter jejuni surface proteins promotes bacterial fitness. Infect Immun 2013; 81(5): 1674-82.
[http://dx.doi.org/10.1128/IAI.01370-12] [PMID: 23460522]

[57]   Iannuzzi C, Irace G, Sirangelo I. Differential effects of glycation on protein aggregation and amyloid formation. Front Mol Biosci 2014; 1: 9.
[http://dx.doi.org/10.3389/fmolb.2014.00009] [PMID: 25988150]

[58]   Zhang L, Syed ZA, van Dijk Härd I, Lim JM, Wells L, Ten Hagen KG. O-Glycosylation regulates polarized secretion by modulating Tango1 stability. Proc Natl Acad Sci USA 2014; 111(20): 7296-301.
[http://dx.doi.org/10.1073/pnas.1322264111] [PMID: 24799692]

[59]   Jacobs P, Callewaert N. N-glycosylation engineering of biopharmaceutical expression systems. Curr Mol Med 2009; 9(7): 774-800.

[http://dx.doi.org/10.2174/156652409789105552] [PMID: 19860659]

[60]   Goettig P. Effects of Glycosylation on the Enzymatic Activity and Mechanisms of Proteases. Int J Mol Sci 2016; 17(12): 1969.
[http://dx.doi.org/10.3390/ijms17121969] [PMID: 27898009]

[61]   Koninkx JFJG, Mirck MH, Hendriks HGCJM, Mouwen JMVM, van Dijk JE. Nippostrongylus brasiliensis Histochemical changes in the composition of mucins in goblet cells during infection in rats. Exp Parasitol 1988; 65(1): 84-90.
[http://dx.doi.org/10.1016/0014-4894(88)90109-9] [PMID: 3338549]

[62]   Birchenough GMH, Nyström EEL, Johansson MEV, Hansson GC. A sentinel goblet cell guards the colonic crypt by triggering Nlrp6-dependent Muc2 secretion. Science 2016; 352(6293): 1535-42.
[http://dx.doi.org/10.1126/science.aaf7419] [PMID: 27339979]

[63]   Geijtenbeek TBH, van Vliet SJ, Engering A, 't Hart BA, van Kooyk Y. Self- and nonself-recognition by C-type lectins on dendritic cells. Annu Rev Immunol 2004; 22(1): 33-54.
[http://dx.doi.org/10.1146/annurev.immunol.22.012703.104558] [PMID: 15032573]

[64]   McGreal EP, Miller JL, Gordon S. Ligand recognition by antigen-presenting cell C-type lectin receptors. Curr Opin Immunol 2005; 17(1): 18-24.
[http://dx.doi.org/10.1016/j.coi.2004.12.001] [PMID: 15653305]

[65]   Chung CH, Mirakhur B, Chan E, *et al.* Cetuximab-induced anaphylaxis and IgE specific for galactose-alpha-1,3-galactose. N Engl J Med 2008; 358(11): 1109-17.
[http://dx.doi.org/10.1056/NEJMoa074943] [PMID: 18337601]

[66]   Ghaderi D, Taylor RE, Padler-Karavani V, Diaz S, Varki A. Implications of the presence of N-glycolylneuraminic acid in recombinant therapeutic glycoproteins. Nat Biotechnol 2010; 28(8): 863-7.
[http://dx.doi.org/10.1038/nbt.1651] [PMID: 20657583]

[67]   Mora-Montes HM, Bates S, Netea MG, *et al.* A multifunctional mannosyltransferase family in *Candida albicans* determines cell wall mannan structure and host-fungus interactions. J Biol Chem 2010; 285(16): 12087-95.
[http://dx.doi.org/10.1074/jbc.M109.081513] [PMID: 20164191]

[68]   Varki A. Letter to the Glyco-Forum: Since there are PAMPs and DAMPs, there must be SAMPs? Glycan "self-associated molecular patterns" dampen innate immunity, but pathogens can mimic them. Glycobiology 2011; 21(9): 1121-4. b
[http://dx.doi.org/10.1093/glycob/cwr087] [PMID: 21932452]

[69]   Chen GY, Brown NK, Zheng P, Liu Y. Siglec-G/10 in self-nonself discrimination of innate and adaptive immunity. Glycobiology 2014; 24(9): 800-6.
[http://dx.doi.org/10.1093/glycob/cwu068] [PMID: 24996822]

[70]   Blaum BS, Hannan JP, Herbert AP, Kavanagh D, Uhrín D, Stehle T. Structural basis for sialic acid–mediated self-recognition by complement factor H. Nat Chem Biol 2015; 11(1): 77-82.
[http://dx.doi.org/10.1038/nchembio.1696] [PMID: 25402769]

[71]   Hyvärinen S, Meri S, Jokiranta TS. Disturbed sialic acid recognition on endothelial cells and platelets in complement attack causes atypical hemolytic uremic syndrome. Blood 2016; 127(22): 2701-10.
[http://dx.doi.org/10.1182/blood-2015-11-680009] [PMID: 27006390]

[72]   Shilova N, Navakouski M, Khasbiullina N, Blixt O, Bovin N. Printed glycan array: antibodies as probed in undiluted serum and effects of dilution. Glycoconj J 2012; 29(2-3): 87-91.
[http://dx.doi.org/10.1007/s10719-011-9368-8] [PMID: 22258790]

[73]   Shilova N, Huflejt ME, Vuskovic M, *et al.* Natural antibodies against sialoglycans. Top Curr Chem 2013; 366: 169-81.
[http://dx.doi.org/10.1007/128_2013_469] [PMID: 24037491]

[74]   Fötisch K, Vieths S. N- and O-linked oligosaccharides of allergenic glycoproteins. Glycoconj J 2001; 18(5): 373-90.

[http://dx.doi.org/10.1023/A:1014860030380] [PMID: 11925505]

[75]   Wilson IBH, Zeleny R, Kolarich D, *et al.* Analysis of Asn-linked glycans from vegetable foodstuffs: widespread occurrence of Lewis a, core 1,3-linked fucose and xylose substitutions. Glycobiology 2001; 11(4): 261-74.
[http://dx.doi.org/10.1093/glycob/11.4.261] [PMID: 11358875]

[76]   Commins SP, James HR, Kelly LA, *et al.* The relevance of tick bites to the production of IgE antibodies to the mammalian oligosaccharide galactose-α-1,3-galactose. J Allergy Clin Immunol 2011; 127(5): 1286-1293.e6.
[http://dx.doi.org/10.1016/j.jaci.2011.02.019] [PMID: 21453959]

[77]   Varki A. Nothing in glycobiology makes sense, except in the light of evolution. Cell 2006; 126(5): 841-5.
[http://dx.doi.org/10.1016/j.cell.2006.08.022] [PMID: 16959563]

[78]   Taylor ME, Drickamer K. Paradigms for glycan-binding receptors in cell adhesion. Curr Opin Cell Biol 2007; 19(5): 572-7.
[http://dx.doi.org/10.1016/j.ceb.2007.09.004] [PMID: 17942297]

[79]   Gu J, Isaji T, Xu Q, *et al.* Potential roles of N-glycosylation in cell adhesion. Glycoconj J 2012; 29(8-9): 599-607.
[http://dx.doi.org/10.1007/s10719-012-9386-1] [PMID: 22565826]

[80]   Fukuda M. Roles of mucin-type *O*-glycans in cell adhesion. BBA general. 2002; 1573(3): 394-405.
[http://dx.doi.org/10.1016/S0304-4165(02)00409-9]

[81]   Espinosa-Marzal RM, Fontani G, Reusch FB, Roba M, Spencer ND, Crockett R. Sugars communicate through water: oriented glycans induce water structuring. Biophys J 2013; 104(12): 2686-94.
[http://dx.doi.org/10.1016/j.bpj.2013.05.017] [PMID: 23790377]

[82]   Cruz-Chu ER, Malafeev A, Pajarskas T, Pivkin IV, Koumoutsakos P. Structure and response to flow of the glycocalyx layer. Biophys J 2014; 106(1): 232-43.
[http://dx.doi.org/10.1016/j.bpj.2013.09.060] [PMID: 24411255]

[83]   Alonso MD, Lomako J, Lomako WM, Whelan WJ. A new look at the biogenesis of glycogen. FASEB J 1995; 9(12): 1126-37. b
[http://dx.doi.org/10.1096/fasebj.9.12.7672505] [PMID: 7672505]

[84]   Carrizo ME, Miozzo MC, Goldraij A, Curtino JA. Purification of rabbit skeletal muscle proteoglycogen: studies on the glucosyltransferase activity of polysaccharide-free and -bound glycogenin. Glycobiology 1997; 7(4): 571-8.
[http://dx.doi.org/10.1093/glycob/7.4.571] [PMID: 9184839]

[85]   Mukerjea R, Robyt JF. Tests for the mechanism of starch biosynthesis: *de novo* synthesis or an amylogenin primer synthesis. Carbohydr Res 2013; 372: 55-9.
[http://dx.doi.org/10.1016/j.carres.2013.01.020] [PMID: 23524111]

[86]   Esko JD, Selleck SB. Order out of chaos: assembly of ligand binding sites in heparan sulfate. Annu Rev Biochem 2002; 71(1): 435-71.
[http://dx.doi.org/10.1146/annurev.biochem.71.110601.135458] [PMID: 12045103]

[87]   Chen E, Stringer SE, Rusch MA, Selleck SB, Ekker SC. A unique role for 6-O sulfation modification in zebrafish vascular development. Dev Biol 2005; 284(2): 364-76.
[http://dx.doi.org/10.1016/j.ydbio.2005.05.032] [PMID: 16009360]

[88]   El-Baba TJ, Lutomski CA, Burnap SA, *et al.* Uncovering the role pf N-glycan occupancy on the cooperative assembly of spike and angiotension converting enzyme 2 complexes: Insights from Glycoengineering and native Mass Spectrometry. J Am Chem Soc 2023; 145(14): 8021-32.
[http://dx.doi.org/10.1021/jacs.3c00291] [PMID: 37000485]

[89]   Joyce K, Pandit A. The role of glycan characterisation in ATMP development and evaluation. Interdisciplinary Medicine 2023; 1(2): e20230005.

[http://dx.doi.org/10.1002/INMD.20230005]

[90]   Sawada H, Stukenbrok H, Kerjaschki D, Farquhar MG. Epithelial polyanion (podocalyxin) is found on the sides but not the soles of the foot processes of the glomerular epithelium. Am J Pathol 1986; 125(2): 309-18.
[PMID: 3538890]

[91]   Dekan G, Gabel C, Farquhar MG. Sulfate contributes to the negative charge of podocalyxin, the major sialoglycoprotein of the glomerular filtration slits. Proc Natl Acad Sci USA 1991; 88(12): 5398-402.
[http://dx.doi.org/10.1073/pnas.88.12.5398] [PMID: 2052617]

[92]   Miettinen A, Stow JL, Mentone S, Farquhar MG. Antibodies to basement membrane heparan sulfate proteoglycans bind to the laminae rarae of the glomerular basement membrane (GBM) and induce subepithelial GBM thickening. J Exp Med 1986; 163(5): 1064-84.
[http://dx.doi.org/10.1084/jem.163.5.1064] [PMID: 2939168]

[93]   Stow JL, Soroka CJ, MacKay K, Striker L, Striker G, Farquhar MG. Basement membrane heparan sulfate proteoglycan is the main proteoglycan synthesized by glomerular epithelial cells in culture. Am J Pathol 1989; 135(4): 637-46.
[PMID: 2529772]

[94]   Harvey SJ, Miner JH. Revisiting the glomerular charge barrier in the molecular era. Curr Opin Nephrol Hypertens 2008; 17(4): 393-8.
[http://dx.doi.org/10.1097/MNH.0b013e32830464de] [PMID: 18660676]

[95]   Weinhold B, Sellmeier M, Schaper W, *et al.* Deficits in sialylation impair podocyte maturation. J Am Soc Nephrol 2012; 23(8): 1319-28.
[http://dx.doi.org/10.1681/ASN.2011090947] [PMID: 22745475]

[96]   Bouvet V, Ben RN. Antifreeze glycoproteins: structure, conformation, and biological applications. Cell Biochem Biophys 2003; 39(2): 133-44.
[http://dx.doi.org/10.1385/CBB:39:2:133] [PMID: 14515019]

[97]   Duman JG. Animal ice-binding (antifreeze) proteins and glycolipids: an overview with emphasis on physiological function. J Exp Biol 2015; 218(12): 1846-55.
[http://dx.doi.org/10.1242/jeb.116905] [PMID: 26085662]

[98]   Jasmer KJ, Gilman KE, Muñoz Forti K, Weisman GA, Limesand KH. Radiation-induced salivary gland dysfunction: Mechanisms, therapeutics and future directions. J Clin Med 2020; 9(12): 4095.
[http://dx.doi.org/10.3390/jcm9124095] [PMID: 33353023]

[99]   Ziyad S H. Radiation-induced salivary gland damage/dysfunction in head and neck cancer: Nano-bioengineering strategies and artificial intelligence for prevention, therapy and reparation. Journal of Radiology and Oncology 2022; 6(3): 027-44.
[http://dx.doi.org/10.29328/journal.jro.1001044]

[100] Kamiński B. Laryngological manifestations of Sjögren's syndrome. Reumatologia 2019; 57(1): 37-44.
[http://dx.doi.org/10.5114/reum.2019.83237] [PMID: 30858629]

[101] Tamer TM. Hyaluronan and synovial joint: function, distribution and healing. Interdiscip Toxicol 2013; 6(3): 111-25.
[http://dx.doi.org/10.2478/intox-2013-0019] [PMID: 24678248]

[102] Coleman PJ, Scott D, Ray J, Mason RM, Levick JR. Hyaluronan secretion into the synovial cavity of rabbit knees and comparison with albumin turnover. J Physiol 1997; 503(3): 645-56.
[http://dx.doi.org/10.1111/j.1469-7793.1997.645bg.x] [PMID: 9379418]

[103] Mattjus P, Pike HM, Molotkovsky JG, Brown RE. Charged membrane surfaces impede the protein-mediated transfer of glycosphingolipids between phospholipid bilayers. Biochemistry 2000; 39(5): 1067-75.
[http://dx.doi.org/10.1021/bi991810u] [PMID: 10653652]

[104] Liu F, Liu H, Sollogoub M, Zhang Y. Recent advances on glycosphingolipid GM3. Carbohydr Chem

2020; 44: 230-49.
[http://dx.doi.org/10.1039/9781788013864-00230]

[105] Cavdarli S, Groux-Degroote S, Delannoy P. Gangliosides: The double-edge sword of neuro-ectodermal derived tumors. Biomolecules 2019; 9(8): 311.
[http://dx.doi.org/10.3390/biom9080311] [PMID: 31357634]

[106] Wang XQ, Lee S, Wilson H, *et al.* Ganglioside GM3 depletion reverses impaired wound healing in diabetic mice by activating IGF-1 and insulin receptors. J Invest Dermatol 2014; 134(5): 1446-55.
[http://dx.doi.org/10.1038/jid.2013.532] [PMID: 24326453]

[107] Zhu H, Duchesne L, Rudland PS, Fernig DG. The heparan sulfate co-receptor and the concentration of fibroblast growth factor-2 independently elicit different signalling patterns from the fibroblast growth factor receptor. Cell Commun Signal 2010; 8(1): 14.
[http://dx.doi.org/10.1186/1478-811X-8-14] [PMID: 20576134]

[108] Venkatachalam MA, Weinberg JM. New wrinkles in old receptors: core fucosylation is yet another target to inhibit TGF-β signaling. Kidney Int 2013; 84(1): 11-4.
[http://dx.doi.org/10.1038/ki.2013.95] [PMID: 23812359]

[109] Wang X, Gu J, Ihara H, Miyoshi E, Honke K, Taniguchi N. Core fucosylation regulates epidermal growth factor receptor-mediated intracellular signaling. J Biol Chem 2006; 281(5): 2572-7.
[http://dx.doi.org/10.1074/jbc.M510893200] [PMID: 16316986]

[110] Liu YC, Yen HY, Chen CY, *et al.* Sialylation and fucosylation of epidermal growth factor receptor suppress its dimerization and activation in lung cancer cells. Proc Natl Acad Sci USA 2011; 108(28): 11332-7.
[http://dx.doi.org/10.1073/pnas.1107385108] [PMID: 21709263]

[111] Brückner K, Perez L, Clausen H, Cohen S. Glycosyltransferase activity of Fringe modulates Notch–Delta interactions. Nature 2000; 406(6794): 411-5.
[http://dx.doi.org/10.1038/35019075] [PMID: 10935637]

[112] Moloney DJ, Panin VM, Johnston SH, *et al.* Fringe is a glycosyltransferase that modifies Notch. Nature 2000; 406(6794): 369-75. b
[http://dx.doi.org/10.1038/35019000] [PMID: 10935626]

[113] Okajima T, Irvine KD. Regulation of notch signaling by o-linked fucose. Cell 2002; 111(6): 893-904.
[http://dx.doi.org/10.1016/S0092-8674(02)01114-5] [PMID: 12526814]

[114] Shi S, Stanley P. Protein *O* -fucosyltransferase 1 is an essential component of Notch signaling pathways. Proc Natl Acad Sci USA 2003; 100(9): 5234-9.
[http://dx.doi.org/10.1073/pnas.0831126100] [PMID: 12697902]

[115] Acar M, Jafar-Nejad H, Takeuchi H, *et al.* Rumi is a CAP10 domain glycosyltransferase that modifies Notch and is required for Notch signaling. Cell 2008; 132(2): 247-58.
[http://dx.doi.org/10.1016/j.cell.2007.12.016] [PMID: 18243100]

[116] Fernandez-Valdivia R, Takeuchi H, Samarghandi A, *et al.* Regulation of mammalian Notch signaling and embryonic development by the protein *O* -glucosyltransferase Rumi. Development 2011; 138(10): 1925-34.
[http://dx.doi.org/10.1242/dev.060020] [PMID: 21490058]

[117] Stanley P, Okajima T. Roles of glycosylation in Notch signaling. Curr Top Dev Biol 2010; 92: 131-64.
[http://dx.doi.org/10.1016/S0070-2153(10)92004-8] [PMID: 20816394]

[118] Takeuchi H, Haltiwanger RS. Significance of glycosylation in Notch signaling. Biochem Biophys Res Commun 2014; 453(2): 235-42.
[http://dx.doi.org/10.1016/j.bbrc.2014.05.115] [PMID: 24909690]

[119] Luca VC, Jude KM, Pierce NW, Nachury MV, Fischer S, Garcia KC. Structural basis for Notch1 engagement of Delta-like 4. Science 2015; 347(6224): 847-53.
[http://dx.doi.org/10.1126/science.1261093] [PMID: 25700513]

[120] Strahl BD, Allis CD. The language of covalent histone modifications. Nature 2000; 403(6765): 41-5.
[http://dx.doi.org/10.1038/47412] [PMID: 10638745]

[121] Morrison O, Thakur J. Molecular complexes at euchromatin, heterochromatin and centromeric chromatin. Int J Mol Sci 2021; 22(13): 6922.
[http://dx.doi.org/10.3390/ijms22136922] [PMID: 34203193]

[122] Indellicato R, Trinchera M. Epigenetic Regulation of Glycosylation in Cancer and Other Diseases. Int J Mol Sci 2021; 22(6): 2980.
[http://dx.doi.org/10.3390/ijms22062980] [PMID: 33804149]

[123] Lind MI, Spagopoulou F. Evolutionary consequences of epigenetic inheritance. Heredity 2018; 121(3): 205-9.
[http://dx.doi.org/10.1038/s41437-018-0113-y] [PMID: 29976958]

[124] Bannister AJ, Kouzarides T. Regulation of chromatin by histone modifications. Cell Res 2011; 21(3): 381-95.
[http://dx.doi.org/10.1038/cr.2011.22] [PMID: 21321607]

[125] Greville G, McCann A, Rudd PM, Saldova R. Epigenetic regulation of glycosylation and the impact on chemo-resistance in breast and ovarian cancer. Epigenetics 2016; 11(12): 845-57.
[http://dx.doi.org/10.1080/15592294.2016.1241932] [PMID: 27689695]

[126] Hart GW. Dynamic O-linked glycosylation of nuclear and cytoskeletal proteins. Annu Rev Biochem 1997; 66(1): 315-35.
[http://dx.doi.org/10.1146/annurev.biochem.66.1.315] [PMID: 9242909]

[127] Hart GW, Haltiwanger RS, Holt GD, Kelly WG. Glycosylation in the nucleus and cytoplasm. Annu Rev Biochem 1989; 58(1): 841-74.
[http://dx.doi.org/10.1146/annurev.bi.58.070189.004205] [PMID: 2673024]

[128] Iyer SPN, Hart GW. Roles of the tetratricopeptide repeat domain in O-GlcNAc transferase targeting and protein substrate specificity. J Biol Chem 2003; 278(27): 24608-16.
[http://dx.doi.org/10.1074/jbc.M300036200] [PMID: 12724313]

[129] Sakabe K, Wang Z, Hart GW. β- *N* -acetylglucosamine (O-GlcNAc) is part of the histone code. Proc Natl Acad Sci USA 2010; 107(46): 19915-20.
[http://dx.doi.org/10.1073/pnas.1009023107] [PMID: 21045127]

[130] Sohn KC, Do SI. Transcriptional regulation and O-GlcNAcylation activity of zebrafish OGT during embryogenesis. Biochem Biophys Res Commun 2005; 337(1): 256-63.
[http://dx.doi.org/10.1016/j.bbrc.2005.09.049] [PMID: 16188232]

[131] Toleman C, Paterson AJ, Whisenhunt TR, Kudlow JE. Characterization of the histone acetyltransferase (HAT) domain of a bifunctional protein with activable O-GlcNAcase and HAT activities. J Biol Chem 2004; 279(51): 53665-73.
[http://dx.doi.org/10.1074/jbc.M410406200] [PMID: 15485860]

[132] Yang X, Zhang F, Kudlow JE. Recruitment of O-GlcNAc transferase to promoters by corepressor mSin3A: coupling protein O-GlcNAcylation to transcriptional repression. Cell 2002; 110(1): 69-80.
[http://dx.doi.org/10.1016/S0092-8674(02)00810-3] [PMID: 12150998]

[133] Nyathi Y, Wilkinson BM, Pool MR. Co-translational targeting and translocation of proteins to the endoplasmic reticulum. Biochim Biophys Acta Mol Cell Res 2013; 1833(11): 2392-402.
[http://dx.doi.org/10.1016/j.bbamcr.2013.02.021] [PMID: 23481039]

[134] Wujek P, Kida E, Walus M, Wisniewski KE, Golabek AA. N-glycosylation is crucial for folding, trafficking, and stability of human tripeptidyl-peptidase I. J Biol Chem 2004; 279(13): 12827-39.
[http://dx.doi.org/10.1074/jbc.M313173200] [PMID: 14702339]

[135] Bosch M, Trombetta S, Engstrom U, Parodi AJ. Characterization of dolichol diphosphate oligosaccharide: protein oligosaccharyltransferase and glycoprotein-processing glucosidases occurring

in trypanosomatid protozoa. J Biol Chem 1988; 263(33): 17360-5.
[http://dx.doi.org/10.1016/S0021-9258(19)77843-0] [PMID: 3053710]

[136] Imperiali B, Shannon KL. Differences between Asn-Xaa-Thr-containing peptides: a comparison of solution conformation and substrate behavior with oligosaccharyltransferase. Biochemistry 1991; 30(18): 4374-80.
[http://dx.doi.org/10.1021/bi00232a002] [PMID: 2021629]

[137] Parodi AJ. *N* -Glycosylation in trypanosomatid protozoa. Glycobiology 1993; 3(3): 193-9.
[http://dx.doi.org/10.1093/glycob/3.3.193] [PMID: 8358146]

[138] Samuelson J, Robbins PW. Effects of N-glycan precursor length diversity on quality control of protein folding and on protein glycosylation. Semin Cell Dev Biol 2015; 41: 121-8.
[http://dx.doi.org/10.1016/j.semcdb.2014.11.008] [PMID: 25475176]

[139] Parodi AJ, Mendelzon DH, Lederkremer GZ. Transient glucosylation of protein-bound Man9GlcNAc2, Man8GlcNAc2, and Man7GlcNAc2 in calf thyroid cells. A possible recognition signal in the processing of glycoproteins. J Biol Chem 1983; 258(13): 8260-5.
[http://dx.doi.org/10.1016/S0021-9258(20)82057-2] [PMID: 6863289]

[140] Parodi AJ, Mendelzon DH, Lederkremer GZ, Martin-Barrientos J. Evidence that transient glucosylation of protein-linked Man9GlcNAc2, Man8GlcNAc2, and Man7GlcNAc2 occurs in rat liver and Phaseolus vulgaris cells. J Biol Chem 1984; 259(10): 6351-7.
[http://dx.doi.org/10.1016/S0021-9258(20)82148-6] [PMID: 6373756]

[141] Hammond C, Braakman I, Helenius A. Role of N-linked oligosaccharide recognition, glucose trimming, and calnexin in glycoprotein folding and quality control. Proc Natl Acad Sci USA 1994; 91(3): 913-7.
[http://dx.doi.org/10.1073/pnas.91.3.913] [PMID: 8302866]

[142] Hebert DN, Foellmer B, Helenius A. Glucose trimming and reglucosylation determine glycoprotein association with calnexin in the endoplasmic reticulum. Cell 1995; 81(3): 425-33.
[http://dx.doi.org/10.1016/0092-8674(95)90395-X] [PMID: 7736594]

[143] Cabral CM, Liu Y, Sifers RN. Dissecting glycoprotein quality control in the secretory pathway. Trends Biochem Sci 2001; 26(10): 619-24.
[http://dx.doi.org/10.1016/S0968-0004(01)01942-9] [PMID: 11590015]

[144] Jakob CA, Burda P, Roth J, Aebi M. Degradation of misfolded endoplasmic reticulum glycoproteins in Saccharomyces cerevisiae is determined by a specific oligosaccharide structure. J Cell Biol 1998; 142(5): 1223-33.
[http://dx.doi.org/10.1083/jcb.142.5.1223] [PMID: 9732283]

[145] Deprez P, Gautschi M, Helenius A. More than one glycan is needed for ER glucosidase II to allow entry of glycoproteins into the calnexin/calreticulin cycle. Mol Cell 2005; 19(2): 183-95.
[http://dx.doi.org/10.1016/j.molcel.2005.05.029] [PMID: 16039588]

[146] Gauss R, Jarosch E, Sommer T, Hirsch C. A complex of Yos9p and the HRD ligase integrates endoplasmic reticulum quality control into the degradation machinery. Nat Cell Biol 2006; 8(8): 849-54.
[http://dx.doi.org/10.1038/ncb1445] [PMID: 16845381]

[147] Helenius A, Aebi M. Roles of N-linked glycans in the endoplasmic reticulum. Annu Rev Biochem 2004; 73(1): 1019-49.
[http://dx.doi.org/10.1146/annurev.biochem.73.011303.073752] [PMID: 15189166]

[148] Määttänen P, Gehring K, Bergeron JJM, Thomas DY. Protein quality control in the ER: The recognition of misfolded proteins. Semin Cell Dev Biol 2010; 21(5): 500-11.
[http://dx.doi.org/10.1016/j.semcdb.2010.03.006] [PMID: 20347046]

[149] Satoh T, Yamaguchi T, Kato K. Emerging structural insights into glycoprotein quality control coupled with N-glycan processing in the endoplasmic reticulum. Molecules 2015; 20(2): 2475-91.

[http://dx.doi.org/10.3390/molecules20022475] [PMID: 25647580]

[150] Castonguay AC, Olson LJ, Dahms NM. Mannose 6-phosphate receptor homology (MRH) domain-containing lectins in the secretory pathway. Biochim Biophys Acta, Gen Subj 2011; 1810(9): 815-26.
[http://dx.doi.org/10.1016/j.bbagen.2011.06.016] [PMID: 21723917]

[151] Caramelo JJ, Parodi AJ. A sweet code for glycoprotein folding. FEBS Lett 2015; 589(22): 3379-87.
[http://dx.doi.org/10.1016/j.febslet.2015.07.021] [PMID: 26226420]

[152] Christianson JC, Shaler TA, Tyler RE, *et al.* OS-9 and GRP94 deliver mutant α1-antitrypsin to the Hrd1–SEL1L ubiquitin ligase complex for ERAD. Nat Cell Biol 2008; 10(3): 272-82.
[http://dx.doi.org/10.1038/ncb1689] [PMID: 18264092]

[153] Rita Lecca M, Wagner U, Patrignani A, Berger EG, Hennet T. Genome-wide analysis of the unfolded protein response in fibroblasts from congenital disorders of glycosylation type-I patients. FASEB J 2005; 19(2): 1-21.
[http://dx.doi.org/10.1096/fj.04-2397fje] [PMID: 15545299]

[154] Hirayama H, Hosomi A, Suzuki T. Physiological and molecular functions of the cytosolic peptide:N-glycanase. Semin Cell Dev Biol 2015; 41: 110-20.
[http://dx.doi.org/10.1016/j.semcdb.2014.11.009] [PMID: 25475175]

[155] Zhu Y, Liu TW, Cecioni S, Eskandari R, Zandberg WF, Vocadlo DJ. O-GlcNAc occurs cotranslationally to stabilize nascent polypeptide chains. Nat Chem Biol 2015; 11(5): 319-25.
[http://dx.doi.org/10.1038/nchembio.1774] [PMID: 25774941]

[156] Itin C, Roche AC, Monsigny M, Hauri HP. ERGIC-53 is a functional mannose-selective and calcium-dependent human homologue of leguminous lectins. Mol Biol Cell 1996; 7(3): 483-93.
[http://dx.doi.org/10.1091/mbc.7.3.483] [PMID: 8868475]

[157] Zhang B, Cunningham MA, Nichols WC, *et al.* Bleeding due to disruption of a cargo-specific ER-t--Golgi transport complex. Nat Genet 2003; 34(2): 220-5.
[http://dx.doi.org/10.1038/ng1153] [PMID: 12717434]

[158] Everett LA, Cleuren ACA, Khoriaty RN, Ginsburg D. Murine coagulation factor VIII is synthesized in endothelial cells. Blood 2014; 123(24): 3697-705.
[http://dx.doi.org/10.1182/blood-2014-02-554501] [PMID: 24719406]

[159] Zhang B, Zheng C, Zhu M, *et al.* Mice deficient in LMAN1 exhibit FV and FVIII deficiencies and liver accumulation of α1-antitrypsin. Blood 2011; 118(12): 3384-91.
[http://dx.doi.org/10.1182/blood-2011-05-352815] [PMID: 21795745]

[160] Nufer O, Mitrovic S, Hauri HP. Profile-based data base scanning for animal L-type lectins and characterization of VIPL, a novel VIP36-like endoplasmic reticulum protein. J Biol Chem 2003; 278(18): 15886-96.
[http://dx.doi.org/10.1074/jbc.M211199200] [PMID: 12609988]

[161] Kamiya Y, Kamiya D, Yamamoto K, Nyfeler B, Hauri HP, Kato K. Molecular basis of sugar recognition by the human L-type lectins ERGIC-53, VIPL, and VIP36. J Biol Chem 2008; 283(4): 1857-61.
[http://dx.doi.org/10.1074/jbc.M709384200] [PMID: 18025080]

[162] Arshad N, Ballal S, Visweswariah SS. Site-specific N-linked glycosylation of receptor guanylyl cyclase C regulates ligand binding, ligand-mediated activation and interaction with vesicular integral membrane protein 36, VIP36. J Biol Chem 2013; 288(6): 3907-17.
[http://dx.doi.org/10.1074/jbc.M112.413906] [PMID: 23269669]

[163] Kumari S, Mg S, Mayor S. Endocytosis unplugged: multiple ways to enter the cell. Cell Res 2010; 20(3): 256-75.
[http://dx.doi.org/10.1038/cr.2010.19] [PMID: 20125123]

[164] Mettlen M, Chen PH, Srinivasan S, Danuser G, Schmid SL. Regulation of Clathrin-mediated endocytosis. Annu Rev Biochem 2018; 87(1): 871-96.

[http://dx.doi.org/10.1146/annurev-biochem-062917-012644] [PMID: 29661000]

[165]　Kiss AL, Botos E. Endocytosis *via* caveolae: alternative pathway with distinct cellular compartments to avoid lysosomal degradation? J Cell Mol Med 2009; 13(7): 1228-37.
[http://dx.doi.org/10.1111/j.1582-4934.2009.00754.x] [PMID: 19382909]

[166]　Alberts B, Johnson A, Lewis J, Raff M, Roberts K, Walter P. Transport into the Cell from the Plasma membrane: Endocytosis in Molecular Biology of the Cell. 4th ed., New York: Garland Science 2002.

[167]　Grewal PK, Uchiyama S, Ditto D, *et al.* The Ashwell receptor mitigates the lethal coagulopathy of sepsis. Nat Med 2008; 14(6): 648-55.
[http://dx.doi.org/10.1038/nm1760] [PMID: 18488037]

[168]　Grewal PK, Aziz PV, Uchiyama S, *et al.* Inducing host protection in pneumococcal sepsis by preactivation of the Ashwell-Morell receptor. Proc Natl Acad Sci USA 2013; 110(50): 20218-23.
[http://dx.doi.org/10.1073/pnas.1313905110] [PMID: 24284176]

[169]　Moonens K, Remaut H. Evolution and structural dynamics of bacterial glycan binding adhesins. Curr Opin Struct Biol 2017; 44: 48-58.
[http://dx.doi.org/10.1016/j.sbi.2016.12.003] [PMID: 28043017]

[170]　Walz A, Odenbreit S, Mahdavi J, Borén T, Ruhl S. Identification and characterization of binding properties of *Helicobacter pylori* by glycoconjugate arrays. Glycobiology 2005; 15(7): 700-8.
[http://dx.doi.org/10.1093/glycob/cwi049] [PMID: 15716466]

[171]　Orlandi PA, Klotz FW, Haynes JD. A malaria invasion receptor, the 175-kilodalton erythrocyte binding antigen of *Plasmodium falciparum* recognizes the terminal Neu5Ac(alpha 2-3)Gal- sequences of glycophorin A. J Cell Biol 1992; 116(4): 901-9.
[http://dx.doi.org/10.1083/jcb.116.4.901] [PMID: 1310320]

[172]　Mehdipour AR, Hummer G. Dual nature of human ACE2 glycosylation in binding to SARS-CoV-2 spike. Proc Natl Acad Sci USA 2021; 118(19): e2100425118.
[http://dx.doi.org/10.1073/pnas.2100425118] [PMID: 33903171]

[173]　Kesharwani A, Dighe OR, Lamture Y. Role of *Helicobacter pylori* in Gastric carcinoma: A review. Cureus 2023; 15(4): e37205.
[http://dx.doi.org/10.7759/cureus.37205] [PMID: 37159779]

[174]　Wroblewski LE, Peek RM Jr, Wilson KT. *Helicobacter pylori* and gastric cancer: factors that modulate disease risk. Clin Microbiol Rev 2010; 23(4): 713-39.
[http://dx.doi.org/10.1128/CMR.00011-10] [PMID: 20930071]

[175]　Terlizzi ME, Gribaudo G, Maffei ME. UroPathogenic *Escherichia coli* (UPEC) infections: Virulence factors, bladder responses, antibiotic, and non-antibiotic antimicrobial strategies. Front Microbiol 2017; 8: 1566.
[http://dx.doi.org/10.3389/fmicb.2017.01566] [PMID: 28861072]

[176]　Lupo F, Ingersoll MA, Pineda MA. The glycobiology of uropathogenic *E. coli* infection: the sweet and bitter role of sugars in urinary tract immunity. Immunology 2021; 164(1): 3-14.
[http://dx.doi.org/10.1111/imm.13330] [PMID: 33763853]

[177]　Kong Y, Jiang C, Wei G, Sun K, Wang R, Qiu T. Small molecule inhibitors as therapeutic agents targeting oncogenic fusion proteins: Current status and Clinical. Molecules 2023; 28(12): 4672.
[http://dx.doi.org/10.3390/molecules28124672] [PMID: 37375228]

[178]　Preiser P, Kaviratne M, Khan S, Bannister L, Jarra W. The apical organelles of malaria merozoites: host cell selection, invasion, host immunity and immune evasion. Microbes Infect 2000; 2(12): 1461-77.
[http://dx.doi.org/10.1016/S1286-4579(00)01301-0] [PMID: 11099933]

[179]　Molina-Franky J, Patarroyo ME, Kalkum M, Patarroyo MA. The cellular and molecular interaction between erythrocytes and *Plasmodium falciparum* merozites. Front Cell Infect Microbiol 2022; 12: 816574.

[http://dx.doi.org/10.3389/fcimb.2022.816574] [PMID: 35433504]

[180]   Triglia T, Chen L, Lopaticki S, *et al.* *Plasmodium falciparum* merozoite invasion is inhibited by antibodies that target the PfRh2a and b binding domains. PLoS Pathog 2011; 7(6): e1002075.
[http://dx.doi.org/10.1371/journal.ppat.1002075] [PMID: 21698217]

[181]   Lehmann F, Tiralongo E, Tiralongo J. Sialic acid-specific lectins: occurrence, specificity and function. Cell Mol Life Sci 2006; 63(12): 1331-54.
[http://dx.doi.org/10.1007/s00018-005-5589-y] [PMID: 16596337]

[182]   Vogt AM, Barragan A, Chen Q, Kironde F, Spillmann D, Wahlgren M. Heparan sulfate on endothelial cells mediates the binding ofPlasmodium falciparum–infected erythrocytes *via* the DBL1α domain of PfEMP1. Blood 2003; 101(6): 2405-11.
[http://dx.doi.org/10.1182/blood-2002-07-2016] [PMID: 12433689]

[183]   Zupancic ML, Frieman M, Smith D, Alvarez RA, Cummings RD, Cormack BP. Glycan microarray analysis of *Candida glabrata* adhesin ligand specificity. Mol Microbiol 2008; 68(3): 547-59.
[http://dx.doi.org/10.1111/j.1365-2958.2008.06184.x] [PMID: 18394144]

[184]   Persson KEM, McCallum FJ, Reiling L, *et al.* Variation in use of erythrocyte invasion pathways by Plasmodium falciparum mediates evasion of human inhibitory antibodies. J Clin Invest 2008; 118(1): 342-51.
[http://dx.doi.org/10.1172/JCI32138] [PMID: 18064303]

[185]   Tham WH, Wilson DW, Lopaticki S, *et al.* Complement receptor 1 is the host erythrocyte receptor for *Plasmodium falciparum* PfRh4 invasion ligand. Proc Natl Acad Sci USA 2010; 107(40): 17327-32.
[http://dx.doi.org/10.1073/pnas.1008151107] [PMID: 20855594]

[186]   Wicherska-Pawłowska K, Wróbel T, Rybka J. Toll-Like Receptors (TLRs), NOD-Like Receptors (NLRs), and RIG-I-Like Receptors 9RLRs) in innate immunity. TLRs, NLRs and RLRs ligand as immunotherapeutic agents for hematopoietic diseases. Int J Mol Sci 2021; 22(24): 13397.
[http://dx.doi.org/10.3390/ijms222413397] [PMID: 34948194]

[187]   Varki A. PAMPs, DAMPs anEndd SAMPs: Host glycans are self-associated molecular patterns but subject to microbial molecular mimicry. FASEB J 2021; 35(S1): fasebj.2021.35.S1.00015.
[http://dx.doi.org/10.1096/fasebj.2021.35.S1.00015]

[188]   Rabinovich GA, Croci DO. Regulatory circuits mediated by lectin-glycan interactions in autoimmunity and cancer. Immunity 2012; 36(3): 322-35.
[http://dx.doi.org/10.1016/j.immuni.2012.03.004] [PMID: 22444630]

[189]   Paulson JC, Sadler JE, Hill RL. Restoration of specific myxovirus receptors to asialoerythrocytes by incorporation of sialic acid with pure sialyltransferases. J Biol Chem 1979; 254(6): 2120-4.
[http://dx.doi.org/10.1016/S0021-9258(17)37774-8] [PMID: 422571]

[190]   Ito T, Suzuki Y, Mitnaul L, Vines A, Kida H, Kawaoka Y. Receptor specificity of influenza A viruses correlates with the agglutination of erythrocytes from different animal species. Virology 1997; 227(2): 493-9.
[http://dx.doi.org/10.1006/viro.1996.8323] [PMID: 9018149]

[191]   Gambaryan A, Yamnikova S, Lvov D, *et al.* Receptor specificity of influenza viruses from birds and mammals: new data on involvement of the inner fragments of the carbohydrate chain. Virology 2005; 334(2): 276-83.
[http://dx.doi.org/10.1016/j.virol.2005.02.003] [PMID: 15780877]

[192]   Wong SSY, Yuen K. Avian influenza virus infections in humans. Chest 2006; 129(1): 156-68.
[http://dx.doi.org/10.1378/chest.129.1.156] [PMID: 16424427]

[193]   Varki A. Uniquely human evolution of sialic acid genetics and biology. Proc Natl Acad Sci USA 2010; 107(Suppl 2) (Suppl. 2): 8939-46.
[http://dx.doi.org/10.1073/pnas.0914634107] [PMID: 20445087]

[194]   Moncla LH, Ross TM, Dinis JM, *et al.* A novel nonhuman primate model for influenza transmission.

PLoS One 2013; 8(11): e78750.
[http://dx.doi.org/10.1371/journal.pone.0078750] [PMID: 24244352]

[195] Regl G, Kaser A, Iwersen M, *et al.* The hemagglutinin-esterase of mouse hepatitis virus strain S is a sialate-4-O-acetylesterase. J Virol 1999; 73(6): 4721-7.
[http://dx.doi.org/10.1128/JVI.73.6.4721-4727.1999] [PMID: 10233932]

[196] Langereis MA, van Vliet ALW, Boot W, de Groot RJ. Attachment of mouse hepatitis virus to O-acetylated sialic acid is mediated by hemagglutinin-esterase and not by the spike protein. J Virol 2010; 84(17): 8970-4.
[http://dx.doi.org/10.1128/JVI.00566-10] [PMID: 20538854]

[197] Schultze B, Herrler G. Bovine coronavirus uses N-acetyl-9-O-acetylneuraminic acid as a receptor determinant to initiate the infection of cultured cells. J Gen Virol 1992; 73(4): 901-6.
[http://dx.doi.org/10.1099/0022-1317-73-4-901] [PMID: 1321878]

[198] Muchmore E, Varki A. Selective inactivation of influenza C esterase decreases infectivity without loss of binding; a probe for 9-Oacetylated sialic acids. Science. 1987; 236: 1293–95. jstor.org/stable/1699415

[199] Langereis MA, Bakkers MJG, Deng L, *et al.* Complexity and diversity of the mammalian sialome revealed by nidovirus virolectins. Cell Rep 2015; 11(12): 1966-78.
[http://dx.doi.org/10.1016/j.celrep.2015.05.044] [PMID: 26095364]

[200] Bai Y, Tao X. Comparison of COVID-19 and influenza characteristics. J Zhejiang Univ Sci B 2021; 22(2): 87-98.
[http://dx.doi.org/10.1631/jzus.B2000479] [PMID: 33615750]

[201] Mukherjee TK. Tumor Immunology (Ch 7) in Immunology: An Introductory Textbook by Sharma AK (Editor), Pan Stanford Publishing Pvt. Ltd. 2019. ISBN- 13: 978-9814774512; ISBN-10: 9814774510.

[202] Dutronc Y, Porcelli SA. The CD1 family and T cell recognition of lipid antigens. Tissue Antigens 2002; 60(5): 337-53.
[http://dx.doi.org/10.1034/j.1399-0039.2002.600501.x] [PMID: 12492810]

[203] Pereira CS, Macedo MF. Macedo MF. CD1-restricted T cells at the crossroad of innate and adaptive immunity. J Immunol Res 2016; 2016: 1-11.
[http://dx.doi.org/10.1155/2016/2876275] [PMID: 28070524]

CHAPTER 3

Understanding Congenital Glycosylation Disorders

Himel Mondal[1], **Shaikat Mondal**[2] and **Rajeev K. Singla**[3,*]

[1] *Department of Physiology, All India Institute of Medical Sciences, Deoghar, Jharkhand-814152, India*

[2] *Department of Physiology, Raiganj Government Medical College and Hospital, Raiganj, West Bengal, India*

[3] *Joint Laboratory of Artificial Intelligence for Critical Care Medicine, Department of Critical Care Medicine and Institutes for Systems Genetics, Frontiers Science Center for Disease-related Molecular Network, West China Hospital, Sichuan University, Chengdu, China*

Abstract: Congenital Disorders of Glycosylation (CDG) encompass a rare and complex group of genetic diseases characterized by abnormalities in the fundamental process of glycosylation. There is an abnormal synthesis or attachment of the glycan moiety of glycoproteins and glycolipids. CDG arises from mutations in genes responsible for various steps in glycosylation within the endoplasmic reticulum and Golgi apparatus. These mutations disrupt the synthesis and transfer of sugar moieties, resulting in the production of defective glycoproteins and glycolipids. Common symptoms of the disease include developmental delays, intellectual disabilities, hypotonia, seizures, and organ dysfunction. The array of CDG subtypes stems from the multitude of underlying genetic mutations and disturbed glycosylation processes making the diagnosis and management challenging. Diagnosis of CDG relies on a multifaceted approach. Clinical evaluation, biochemical analysis, and genetic testing are all essential components. The advent of next-generation sequencing has significantly improved our ability to identify the specific gene mutations responsible for individual CDG subtypes. The management of CDG involves primarily symptom alleviation and enhancing the quality of life. A multidisciplinary approach is fundamental, encompassing supportive care, physical and speech therapies, and medications targeting specific complications.

Keywords: Glycosylation, Glycosaminoglycans, Intellectual disability, Lipids, Mutation, Muscle hypotonia, Muscle weakness.

INTRODUCTION

In the intricate landscape of human genetics and cellular biology, there exists a group of conditions known as Congenital Disorders of Glycosylation (CDG).

* **Corresponding author Rajeev K. Singla:** Joint Laboratory of Artificial Intelligence for Critical Care Medicine, Department of Critical Care Medicine and Institutes for Systems Genetics, Frontiers Science Center for Disease-related Molecular Network, West China Hospital, Sichuan University, Chengdu, China; E-mail: rajeevsingla26@gmail.com

Tapan Kumar Mukherjee, Parth Malik & Ruma Rani (Eds.)

These disorders represent a unique and complex realm of genetic diseases, where the very building blocks of life - sugars - play a pivotal role. CDGs are a diverse group of disorders, each with its own genetic basis and clinical manifestations, yet they all share a common thread: disruptions in glycosylation processes within our cells.

Glycosylation is a fundamental cellular process involving the attachment of sugar molecules to proteins and lipids, transforming them into glycoproteins and glycolipids. This seemingly simple addition of sugars, however, underpins a myriad of crucial functions within the human body. Glycosylation is involved in protein folding, stability, and function, cellular signaling, and the proper functioning of various organs and tissues. When this process goes awry, as it does in CDG, the consequences can be profound [1].

The journey into the world of CDG begins with a deeper understanding of its genetic origins. Genetic mutations disrupt the glycosylation pathways within cells, resulting in a wide range of clinical presentations. From developmental delays to intellectual disabilities, seizures to hypotonia, the symptoms of CDG are as diverse as the genetic mutations underlying them. This clinical diversity poses a substantial diagnostic challenge, often requiring advanced genetic testing to pinpoint the exact mutations responsible for each individual's condition [2].

This chapter aims to discuss the intricacies of CDG, offering insight into its causes, the signs and symptoms it presents, the diagnostic hurdles faced by both patients and clinicians, and the ongoing efforts to develop effective treatments. This would explore defects in protein N-glycosylation, disorders of protein O-glycosylation, glycosaminoglycan (GAG) synthesis defects, lipid and GPI-anchor glycosylation defects, and multifaceted pathways that interconnect in the complex network of glycosylation processes.

The discovery and understanding of CDG are a relatively recent development in the field of medical genetics. Here is a brief history of CDG discovery and our evolving understanding of these complex disorders. The origins of CDG can be traced back to the late 20th century when clinicians and researchers began to observe a group of patients with unexplained, multi-systemic symptoms that did not fit into any known diagnostic categories. These symptoms included developmental delays, intellectual disabilities, and various organ dysfunctions. The first recognized case of CDG was reported in the 1980s. Professor Jaak Jaeken reported neurological disorders in twin girls. In the late 1980s and early 1990s, researchers began to characterize the underlying genetic and biochemical defects in CDG patients. They identified mutations in specific genes involved in glycosylation pathways as the root cause of these disorders. As research progressed, it became evident that CDG was not a single disorder but a group of disorders with distinct genetic and clinical features. Researchers started identifying and classifying various CDG subtypes based on the specific glycosylation pathways affected. Advances in genetic testing, particularly the development of next-generation sequencing technologies, greatly facilitated the diagnosis of CDG. This allowed for more accurate and efficient identification of the genetic mutations responsible for individual cases. As awareness of CDG grew within the medical community, more cases were diagnosed and reported [3]. Ongoing research into CDG has led to a deeper understanding of the molecular mechanisms involved in glycosylation and the consequences of glycosylation defects. This knowledge has spurred efforts to develop potential therapies and interventions for CDG. Today, CDG research continues to advance, offering hope for improved diagnostic techniques and therapeutic interventions. While there is no cure for CDG at present, ongoing studies aim to enhance the quality of life for individuals living with these rare disorders.

In summary, the discovery and understanding of CDG have evolved significantly over the past few decades. What once seemed like a mysterious and unexplained group of disorders is now being dissected at the genetic and molecular levels, with ongoing efforts to provide better diagnosis, management, and treatment options for affected individuals and their families.

CAUSES OF CONGENITAL DISORDER OF GLYCOSYLATION

CDGs are a group of rare genetic disorders caused by mutations in genes that encode enzymes or transporters involved in the glycosylation process. Glycosylation is a complex biological process where sugar molecules (glycans) are attached to proteins (glycoproteins) and lipids (glycolipids). When this process is disrupted due to genetic mutations, it can result in various CDG subtypes, each with distinct clinical manifestations. Here are some key causes and factors contributing to CDGs. The disorder can be broadly classified into four groups - protein N-glycosylation, protein *O*-glycosylation, lipid glycosylation, and other glycosylation pathways and multiple glycosylation pathways [4].

Protein *N*-Glycosylation

This is one of the most well-studied glycosylation pathways and involves the attachment of complex sugar chains (glycans) to specific asparagine (N) residues of proteins. CDGs associated with protein *N*-glycosylation primarily result from defects in the synthesis, assembly, or transfer of *N*-linked glycans. These disorders can manifest with a wide range of symptoms, including developmental delays, intellectual disabilities, seizures, and various organ dysfunctions. Classic examples of CDGs in this category include CDG type Ia (Phosphomannomutase 2 deficiency) and CDG type Ic (Phosphomannose isomerase deficiency) [5].

Protein *O*-Glycosylation

O-linked glycosylation involves the attachment of sugar molecules to serine or threonine residues of proteins. CDGs in this category are caused by mutations in genes responsible for the synthesis and transfer of *O*-linked glycans. These disorders may present with symptoms such as muscle weakness, intellectual disabilities, and developmental delays. One example is CDG type IId (Dolichol kinase deficiency).

Lipid Glycosylation

Lipid glycosylation involves the attachment of glycans to lipids, which play critical roles in cell membrane structure and function. CDGs affecting lipid glycosylation can have severe consequences, including neurological impairments, skeletal abnormalities, and skin problems. For instance, CDG type IIc (Man1B1-CDG) is associated with defects in lipid-linked oligosaccharide biosynthesis, leading to a range of clinical features.

Other Glycosylation Pathways

Some CDGs do not fit neatly into the protein or lipid glycosylation categories and instead involve various other glycosylation pathways. These disorders can result from mutations in genes responsible for unique glycosylation processes. One example is SLC35A2-CDG, which affects the transport of nucleotide sugars into the endoplasmic reticulum and can lead to a range of symptoms, including intellectual disabilities and movement disorders.

The underlying causes may be further classified into genetic mutation, enzyme deficiencies, transporter deficiencies, and complex molecular pathways.

Genetic Mutations

The primary cause of CDGs is genetic mutations that affect genes involved in glycosylation. These mutations can be inherited from one or both parents (autosomal recessive or autosomal dominant inheritance patterns) or arise spontaneously (*de novo* mutations). Mutations in different genes lead to various CDG subtypes, each with its own set of symptoms. There are over 130 known CDG subtypes, each associated with mutations in specific genes. These genes code for various glycosylation-related proteins, such as glycosyltransferases, sugar transporters, or chaperones that assist in glycosylation. Different mutations in the same gene can lead to distinct CDG subtypes or varying levels of severity within a single subtype.

As a result, these glycosylation-related genes can disrupt various aspects of the glycosylation process that can result in reduced or non-functional enzyme activity, preventing the proper attachment of glycans to proteins or lipids. They can also disrupt the sugar transport genes that may hinder the movement of sugar precursors across cellular membranes, affecting glycosylation. In addition, the mutations can lead to protein misfolding or impair quality control mechanisms within the endoplasmic reticulum and Golgi apparatus, where glycosylation occurs [6].

Enzyme Deficiencies

Many CDG subtypes result from deficiencies in specific enzymes required for glycosylation. For example, mutations in genes such as PMM2 (phosphomannomutase 2) or MPI (mannose phosphate isomerase) disrupt the early stages of *N*-glycosylation, leading to disorders like CDG-Ia and CDG-Ib, respectively. Another one is Dolichol Kinase (DOLK) Deficiency (CDG-Ij) where mutations in the DOLK gene result in reduced DOLK activity, leading to defects in glycosylation and causing CDG-Ij. The next enzyme deficiency is UDP-Glucose: Glycoprotein Glucosyltransferase (UGGT1) Deficiency (CDG-IIe). UGGT1 is an enzyme responsible for ensuring the correct folding of glycoproteins by monitoring their quality within the endoplasmic reticulum. Mutations in the UGGT1 gene result in defective glycoprotein quality control, leading to CDG-IIe. Another one is ALG3/ALG9 Deficiency (CDG-Id). ALG3 and ALG9 are enzymes involved in the assembly of lipid-linked oligosaccharides used in glycosylation. Mutations in the ALG3 or ALG9 genes lead to abnormal lipid-linked oligosaccharide synthesis, resulting in CDG-Id [7].

These are just a few examples of enzyme deficiencies associated with specific CDG subtypes. CDGs are highly heterogeneous, with more than 130 known subtypes, each linked to mutations affecting different genes and enzymes involved in glycosylation.

Transporter Deficiencies

Transporter deficiencies represent another category of genetic mutations that can cause CDGs. These disorders arise from mutations in genes encoding transporters responsible for the movement of sugar molecules, specifically nucleotide sugar substrates, across cellular membranes. These transporters are crucial for glycosylation, the process of sugar conjugation to proteins and lipids. Here are a few examples of transporter deficiencies in CDGs.

SLC35A1 encodes the UDP-galactose transporter, responsible for importing UDP-galactose into the Golgi apparatus for use in glycosylation reactions. Mutations in the SLC35A1 gene result in reduced or dysfunctional UDP-galactose transport, leading to CDG-IIf. SLC35C1 encodes the UDP-N-acetylglucosamine transporter, which transports UDP-N-acetylglucosamine into the Golgi apparatus for glycosylation reactions. Mutations in the SLC35C1 gene lead to impaired transport of UDP-N-acetylglucosamine, resulting in CDG-IIm. SLC35A2 encodes the UDP-galactose transporter and is also associated with CDG-IIm. SLC39A8 encodes a manganese transporter that is important for the proper function of glycosylation enzymes. Mutations in the SLC39A8 gene result in disrupted manganese transport, leading to CDG-IIn. While not directly a sugar transporter, SEC23B encodes a component of the COPII protein complex involved in protein trafficking, including glycosylation. Mutations in the SEC23B gene result in defective protein transport from the endoplasmic reticulum to the Golgi apparatus, affecting glycosylation and causing CDG-IIc [8].

These transporter deficiencies highlight the critical role that sugar molecule transport plays in glycosylation. Disruptions in these transporters can lead to glycosylation defects, which in turn result in the diverse clinical manifestations associated with CDGs.

Complex Molecular Pathways

CDGs may result from disruptions in complex molecular pathways involved in glycosylation. These disorders are characterized by genetic mutations that affect various stages of glycosylation, leading to defects in the attachment of sugar molecules to proteins and lipids.

One key aspect of glycosylation is the synthesis of sugar precursors like UDP-glucose, UDP-galactose, UDP-N-acetylglucosamine, and others. Mutations in genes involved in sugar precursor synthesis can lead to CDGs, as these precursors are essential for glycan formation. Sugars are attached to carrier molecules, such as dolichol phosphate or lipid-linked oligosaccharides, before being transferred to target proteins or lipids. Mutations affecting this step can disrupt glycosylation. The ER has quality control mechanisms to ensure that glycoproteins are properly folded and glycosylated. Mutations in genes involved in these quality control processes can result in the accumulation of misfolded or misglycosylated proteins, leading to CDGs. Glycosylation occurs in different cellular compartments, including the ER and Golgi apparatus. Mutations in genes that impact the function and communication between these compartments can disrupt glycosylation.

Understanding these complex molecular pathways in glycosylation is essential for diagnosing and researching CDGs. Genetic mutations that disrupt these pathways can lead to a wide range of clinical manifestations associated with different CDG subtypes, emphasizing the importance of ongoing research into these rare genetic disorders.

SIGNS AND SYMPTOMS OF CONGENITAL DISORDER OF GLYCOSYLATION

Due to the wide range of CDG subtypes and affected genes, the signs and symptoms can vary significantly. However, there are common clinical features and manifestations associated with CDGs. Some signs and symptoms are shown in Fig. (**1**) and described below [9, 10].

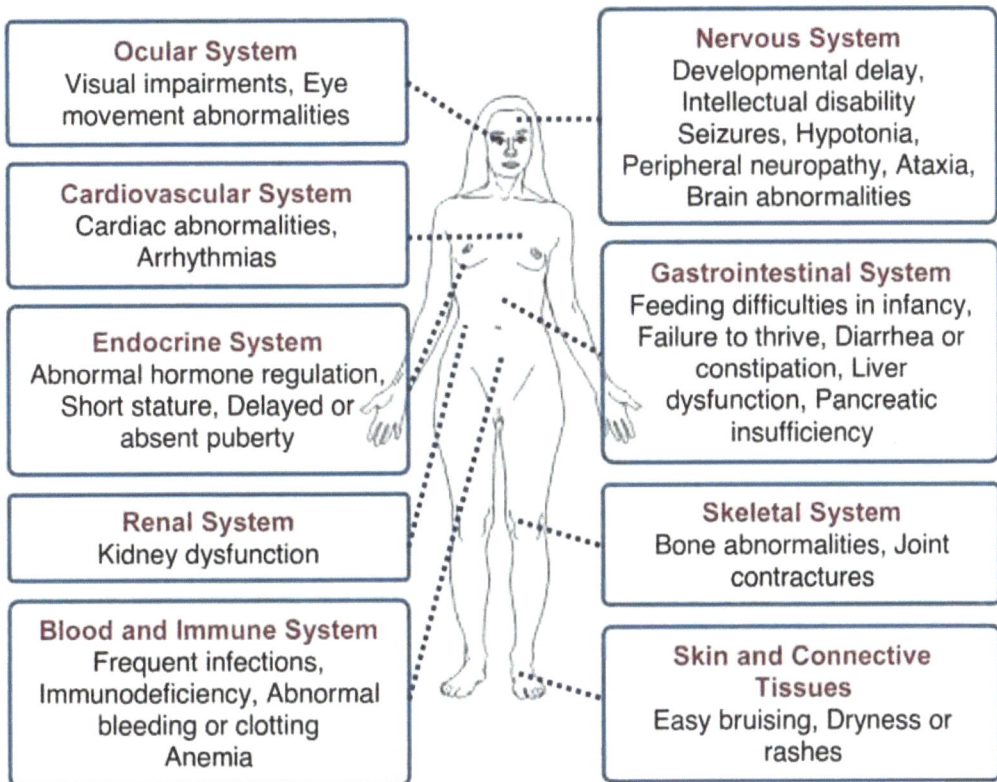

Fig. (1). Signs and symptoms of congenital disorder of glycosylation.

Developmental Delays and Intellectual Disabilities

Developmental delays and intellectual disabilities in individuals with CDGs stem from disruptions in the essential cellular process of glycosylation, which profoundly impacts brain development and function. Glycosylation plays a critical role in neural cell formation, energy provision, synaptic plasticity, and neurotransmitter regulation. When glycosylation is compromised due to genetic mutations in CDG-related genes, neural processes such as neuronal migration, axon guidance, and synaptic transmission are affected. This results in cognitive impairment, learning difficulties, and varying degrees of intellectual disability. Additionally, CDGs can lead to neuroinflammation, oxidative stress, and neuronal dysfunction, further contributing to developmental challenges. The extent of these issues varies among CDG subtypes and individuals, underscoring the complexity and heterogeneity of these rare genetic disorders.

Seizures

Seizures in individuals with CDGs arise from the intricate consequences of disrupted glycosylation processes on brain function. Glycosylation abnormalities can lead to imbalances in neurotransmitters, dysfunctional ion channels, altered synaptic plasticity, and heightened neuronal excitability, all of which increase the susceptibility to abnormal electrical activity and seizures. Furthermore, neuroinflammation, structural brain abnormalities, and the specific genetic mutations underlying each CDG subtype contribute to the variability in seizure occurrence and characteristics among affected individuals. As CDGs encompass a diverse group of rare genetic disorders, seizures may be more prevalent in certain subtypes due to the unique molecular mechanisms involved. Effective management typically involves antiepileptic medications and comprehensive assessments to tailor therapeutic approaches for individual's specific CDG subtype and seizure profile.

Hypotonia and Muscle Weakness

Hypotonia (low muscle tone) and muscle weakness in CDGs result from disruptions in glycosylation, which impacts various cellular processes crucial for muscle development and function. Glycosylation plays a role in muscle fiber structure, energy metabolism, and the formation of glycoproteins which are essential for muscle contraction and neuromuscular communication. When glycosylation is compromised due to genetic mutations in CDG-related genes, muscle cells may not function optimally. This can lead to hypotonia, making it challenging for individuals to maintain muscle tone and control their movements. Muscle weakness often accompanies hypotonia due to the inability of affected muscles to generate and sustain the necessary force for normal motor activities. The severity of these symptoms can vary depending on the specific CDG subtype and the extent of glycosylation disruption, ranging from mild muscle issues to more profound motor difficulties.

Gastrointestinal Issues

Gastrointestinal (GI) issues in individuals with CDGs stem from the vital role of glycosylation in the normal functioning of the digestive system. Glycosylation abnormalities can lead to deficiencies in enzymes critical for digestion, impair mucosal integrity, disrupt neuromuscular function, and trigger inflammation within the GI tract. These disruptions result in a range of GI symptoms, including constipation, diarrhea, feeding difficulties, abdominal pain, and malnutrition. Furthermore, liver involvement in some CDGs can exacerbate GI problems. The specific manifestations and severity of GI issues can vary among CDG subtypes and individuals. Therefore, a comprehensive approach to management, involving dietary modifications, nutritional support, and symptom-targeted therapies, is crucial to address the unique needs of each affected individual and improve their overall well-being.

Coagulation Abnormalities

CDGs can affect the structure and function of blood clotting factors. When clotting factors are improperly glycosylated due to genetic mutations in CDG-related genes, their activity becomes compromised, leading to a bleeding tendency. This can manifest as prolonged bleeding after minor injuries, easy bruising, and spontaneous bleeding events. Additionally, CDGs may cause thrombocytopenia, where there is a reduced platelet count, and disrupt platelet function, further contributing to coagulation issues. The liver's role in synthesizing clotting factors and potential hepatic involvement in some CDG subtypes can exacerbate these abnormalities. Early diagnosis and appropriate management are crucial to mitigate the risk of excessive bleeding and ensure the well-being of individuals with CDGs. Treatment strategies may involve clotting factor replacement therapies and regular monitoring of clotting parameters.

Dysmorphic Features

The CDG subtypes can present with craniofacial development. Glycosylation is integral to the proper formation and function of glycoproteins and glycolipids, which play pivotal roles in cellular signaling, cell adhesion, and tissue development, including that of the face. Mutations in genes related to glycosylation can lead to structural and functional abnormalities in craniofacial tissues, affecting the growth and patterning of facial features. Consequently, individuals with CDGs may exhibit characteristic facial characteristics, contributing to a recognizable clinical phenotype. The specific facial dysmorphism can vary among CDG subtypes, reflecting the heterogeneity of glycosylation abnormalities and their distinct impacts on craniofacial development. These distinctive facial features often serve as diagnostic clues for clinicians evaluating individuals with suspected CDGs.

Failure to Thrive

CDGs can lead to issues such as gastrointestinal problems, malabsorption of nutrients, muscle weakness, and metabolic abnormalities of which contribute to poor growth and weight gain. Abnormal glycosylation can impair the absorption and utilization of essential nutrients, leaving individuals with CDGs at risk of malnutrition and insufficient calorie intake. Additionally, the energy demands of a body coping with glycosylation abnormalities can be higher, further exacerbating the challenge of maintaining adequate growth. Early intervention, including nutritional support and tailored therapies, is essential to address the specific underlying causes of failure to thrive in CDGs and optimize the overall health and development of affected individuals.

Visceral Organ Involvement

Some CDGs can affect the function of visceral organs like the heart, kidneys, and lungs due to the widespread presence of glycosylation in these vital organ systems. Glycosylation is crucial for the proper functioning of many proteins, including those involved in organ development and maintenance. Mutations in genes responsible for glycosylation can disrupt the glycosylation of specific proteins critical for organ health. For example, in CDGs affecting the heart, glycosylation abnormalities can impair the function of cardiac proteins, leading to cardiomyopathies or arrhythmias. Similarly, in CDGs involving the kidneys, glycosylation disruptions can affect the filtration and reabsorption processes, potentially leading to kidney dysfunction. In the lungs, glycosylation plays a role in maintaining the integrity of the respiratory epithelium, and glycosylation abnormalities can contribute to respiratory issues. These orga*N*-specific complications highlight the systemic nature of CDGs and the diverse clinical manifestations that can result from glycosylation disruptions in various organ systems.

Other Neurological Symptoms

Neurological symptoms are a common feature in CDGs because glycosylation plays a fundamental role in the development and function of the nervous system. Glycosylation abnormalities can affect the function of neurotransmitter receptors and transporters, leading to imbalances in synaptic transmission and further contributing to neurological dysfunction. The specific neurological manifestations in CDGs can vary depending on the subtype, affected genes, and the extent of glycosylation disruption, underscoring the complexity of these rare genetic disorders.

Hearing and Vision Impairments

The intricate involvement of glycosylation in the development and function of sensory organs may be the basis of hearing and visual impairment in CDGs. In the auditory system, glycosylation abnormalities can affect the structure and function of the inner ear's sensory hair cells and the auditory pathway, resulting in hearing loss. In the visual system, glycosylation plays a role in the development of photoreceptor cells and the integrity of the retinal structure. Disruptions in glycosylation can lead to retinal degeneration and visual impairment. CDGs affecting these processes can lead to a range of hearing and vision impairments, which can vary in severity depending on the specific CDG subtype and the extent to which glycosylation is disrupted.

Immune System Dysregulation

Glycosylation abnormalities can affect various components of the immune system, including immune cell signaling, antigen recognition, and immune response regulation. CDGs can lead to immunodeficiency, making individuals more susceptible to recurrent infections due to impaired immune cell function. Furthermore, glycosylation disruptions can trigger immune system-related inflammation and autoimmune responses, leading to a state of chronic immune activation. These immune system dysregulations contribute to the wide spectrum of clinical symptoms observed in CDGs, including increased susceptibility to infections, autoimmunity, and immune-mediated complications affecting various organ systems. The specific immune system manifestations in CDGs depend on the subtype.

The glycosylation pathways affected can differ among CDG subtypes, further contributing to variability in clinical manifestations. Moreover, individual genetic makeup and genetic modifiers may influence symptom severity. Environmental factors, such as nutrition and infections, can also play a role. Lastly, the timing of diagnosis and the availability of appropriate medical care and interventions can significantly affect the management and prognosis of individuals with CDGs. Consequently, the combination of genetic, molecular, and environmental factors contributes to the broad spectrum of symptom severity observed in CDGs, from mild and manageable symptoms to profound disabilities.

It is important to note that CDGs are a highly heterogeneous group of disorders, and not all individuals with CDGs will exhibit all these symptoms. The specific signs and symptoms depend on the subtype of CDG, the affected genes, and the extent to which glycosylation is disrupted. Accurate diagnosis and comprehensive medical evaluation are essential to determine the specific CDG subtype and provide appropriate medical care and management for affected individuals. Early intervention and supportive therapies can improve the quality of life for individuals with CDGs.

TESTING AND DIAGNOSIS FOR CONGENITAL DISORDER OF GLYCOSYLATION

If we have a case vignette like the following:

Brishti is a 3-year-old girl brought to the pediatric clinic by her parents due to concerns about her developmental delays and unusual physical features. Emily has consistently missed developmental milestones, such as walking and speaking, and exhibits generalized hypotonia, making her appear floppy. On physical examination, she presents with distinctive facial features, including a high forehead, widely spaced eyes, and a small nose with anteverted nares. There are no signs of organomegaly or dysmorphic features.

From these clinical features, it is very difficult to provisionally diagnose that it may be a case of CDG. In addition, any unexplained multisystem disorder may be a case of CDG. Furthermore, genetic tests are very costly, especially in connection to the per capita income in developing countries. This makes the diagnosis further difficult.

Hence, a comprehensive diagnostic approach involving a combination of clinical assessment, biochemical testing, and genetic analysis is needed for diagnosing a case of CDG. Fig. (2) shows a generalized schema of diagnosis of CDG and described below [11].

Clinical Evaluation

A detailed medical history is essential to identify developmental delays, neurological symptoms, growth issues, and other clinical features associated with CDGs. A thorough physical examination may reveal dysmorphic facial features, organ-specific abnormalities, or signs of neurological involvement.

Biochemical Testing

Transferrin Isoelectric Focusing (TIEF) is a widely used initial screening test for CDGs. It assesses the glycosylation pattern of serum transferrin. Abnormal patterns, including underglycosylation, may suggest CDGs. Mass spectrometry of serum or urine glycoproteins can identify abnormal glycan structures associated with CDGs. Abnormal glycolipid profiles may indicate specific CDG subtypes.

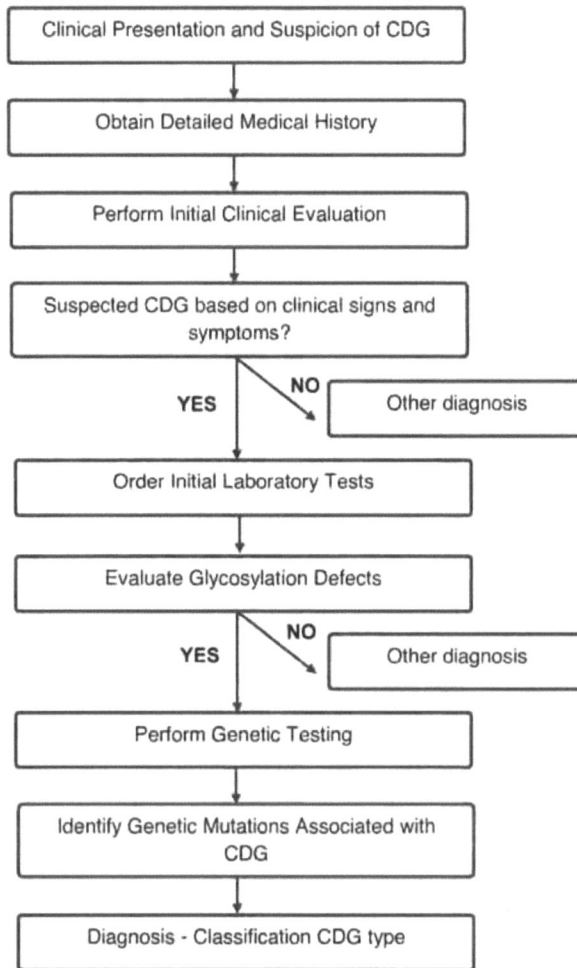

Fig. (2). Steps that are commonly followed for diagnosis of congenital disorder of glycosylation.

Genetic Testing

Whole-Exome Sequencing (WES) or Whole-Genome Sequencing (WGS) is essential to identify specific CDG-related gene mutations. WES analyzes the protein-coding regions of an individual's genome, known as the exome. It is a comprehensive approach that can identify a wide range of genetic mutations, including those responsible for CDGs. WGS examines an individual's entire genome, including non-coding regions. It provides an even more comprehensive view of genetic variations. Targeted testing focuses on specific genes known to be associated with CDGs based on clinical presentation or initial biochemical screening results. This approach is employed when there is a high suspicion of a particular CDG subtype.

Genetic testing typically involves collecting a sample of DNA from the individual being tested. This can be done through various methods, including blood samples, saliva, cheek swabs, or tissue biopsies. For CDGs, blood samples are often preferred as they contain DNA from various tissues and are relatively easy to collect, especially in pediatric patients. Once the genetic sample is collected, the DNA is extracted and sequenced using advanced laboratory techniques. The obtained DNA sequences are compared to a reference genome to identify variations, including mutations responsible for CDGs.

Neuroimaging and Organ-specific Evaluations

Neuroimaging, such as MRI and CT scans, can reveal structural abnormalities in the brain. Depending on clinical manifestations, additional tests may be necessary to evaluate organ involvement, such as cardiac evaluation, liver function tests, and renal function assessments.

Early and accurate diagnosis of CDGs is critical for initiating appropriate interventions, such as supportive therapies, dietary modifications, and targeted treatments when available. Timely diagnosis can also aid in genetic counseling and family planning decisions for affected individuals and their families. Due to the evolving understanding of CDGs and the development of new diagnostic techniques, healthcare professionals need to stay updated with the latest research and testing methodologies in this field.

DISORDERS OF PROTEIN *N*-GLYCOSYLATION

Dysregulation of *N*-glycosylation can lead to a wide spectrum of clinical manifestations, including developmental delays, intellectual disabilities, growth deficiencies, neurological impairments, and a range of organ-specific symptoms. Advances in genetic testing and our understanding of glycosylation pathways have facilitated the diagnosis and management of these rare conditions, providing hope for improved patient care and potential therapeutic interventions.

Features of *N*-glycosylation Defects

N-glycosylation serves several essential functions, including protein folding, stability, and cellular transport. It also plays a role in cell signaling, immune system responses, and cell adhesion. Properly glycosylated proteins are crucial for various biological processes, such as neuronal development, immune response regulation, and enzymatic activity. *N*-glycosylation defects in CDG are characterized by a range of molecular features that arise from genetic mutations affecting the normal glycosylation processes. These molecular features contribute to the diverse clinical presentations observed in CDG patients [12].

One prominent molecular feature of *N*-glycosylation defects in CDG is the incomplete assembly of glycans (oligosaccharides) on target proteins. Mutations in gene-encoding enzymes involved in glycan synthesis can result in truncated or improperly structured glycans. The incomplete glycan assembly can hinder the proper functioning of glycoproteins, affecting their stability, structure, and function.

Due to incomplete glycosylation, newly synthesized proteins may not fold correctly. Inadequately glycosylated proteins can adopt abnormal conformations, rendering them misfolded and less stable. Misfolded proteins often accumulate within the endoplasmic reticulum (ER), leading to ER stress and the activation of cellular stress responses.

In CDG, the molecular feature of accumulating aberrant proteins is a consequence of the impaired glycosylation process. Misfolded or dysfunctional glycoproteins tend to accumulate in the ER, as they cannot pass quality control mechanisms effectively. This accumulation can disrupt ER homeostasis and trigger the unfolded protein response (UPR).

Another molecular feature of *N*-glycosylation defects is the activation of the UPR. The UPR is a cellular stress response that aims to restore ER function and alleviate protein-folding stress. It involves the upregulation of specific genes and signaling pathways to reduce the burden of misfolded proteins. However, chronic UPR activation can have detrimental effects on cell function and contribute to the clinical symptoms seen in CDG patients.

CDG is a heterogeneous group of disorders with various subtypes, each associated with specific genetic mutations. This leads to a wide range of glycosylation patterns, depending on the affected gene. The molecular features of CDG can vary greatly, resulting in different clinical presentations, from mild to severe, affecting various organs and systems.

N-glycosylation defects often lead to molecular features of multi-organ involvement. As glycosylation plays a crucial role in various biological processes, including cell signaling, immune function, and neural development, defects in glycosylation can affect multiple organ systems, resulting in a wide spectrum of clinical symptoms.

Proper glycosylation is essential for the correct trafficking and localization of glycoproteins within the cell. Defects in *N*-glycosylation can disrupt the normal trafficking of proteins to their intended cellular compartments, further contributing to the dysfunction of affected cells and tissues.

In a nutshell, *N*-glycosylation defects in CDG are associated with a range of molecular features, including incomplete glycan assembly, misfolded and unstable proteins, the accumulation of aberrant proteins, activation of the UPR, variability in glycosylation patterns, multi-organ involvement, and impaired protein trafficking. These molecular abnormalities collectively underlie the clinical manifestations observed in CDG patients and highlight the importance of glycosylation in maintaining normal cellular function and homeostasis.

Defects in Endoplasmic Reticulum *N*-glycosylation

Defect in the ER-*N*-glycosylation is a critical component of CDG. Understanding how these defects occur at the molecular level involves examining the events that take place within the ER during *N*-glycosylation [13].

The process begins with the synthesis of proteins in the ribosomes attached to the ER. These proteins are then translocated into the ER lumen, where they undergo various post-translational modifications, including *N*-glycosylation. In the ER, a critical molecular event in *N*-glycosylation involves the formation of the dolichol pyrophosphate-linked oligosaccharide (DLO). This precursor molecule is synthesized through a series of enzymatic reactions that involve the sequential addition of sugar residues, such as glucose and mannose, to a lipid carrier called dolichol phosphate. Once the DLO molecule is fully assembled, it is translocated across the ER membrane to the cytoplasmic side. Here, another critical molecular event occurs, which is the transfer of the DLO from the cytoplasmic side to the ER lumen, where the nascent protein resides. This transfer is facilitated by a complex known as the DLO flippase.

The DLO molecule is then used to transfer the oligosaccharide to specific asparagine residues on the nascent protein. This step involves the action of an enzyme called oligo-saccharyltransferase (OST). The oligosaccharide is attached to the protein as a whole unit in a process known as en bloc transfer. The ER has quality control mechanisms in place to ensure that the attached oligosaccharides are correctly assembled. Chaperone proteins, such as calnexin and calreticulin, assist in the folding of glycoproteins and help identify misfolded or incompletely glycosylated proteins.

In congenital disorders of glycosylation, defects can occur at various points in this process. Mutations in genes encoding enzymes responsible for DLO synthesis, DLO flipping, OST activity, or other glycosylation-related proteins can lead to abnormalities in *N*-glycosylation. These molecular defects result in the incomplete or improper glycosylation of proteins, leading to the accumulation of misfolded or dysfunctional glycoproteins within the ER. When ER *N*-glycosylation is impaired, the misfolded or inadequately glycosylated proteins may trigger UPR and ER stress. These cellular stress responses can lead to a wide range of clinical symptoms observed in CDG patients, affecting various organ systems.

Defects in Golgi *N*-glycosylation

Defects in Golgi *N*-glycosylation represent a subset of CDG, characterized by molecular events that disrupt the crucial final stages of protein glycosylation within the Golgi apparatus.

After initial glycosylation in the ER, glycoproteins are transported to the Golgi apparatus, a series of stacked membranous compartments. Within the Golgi, glycan structures are further elaborated and modified through a series of enzymatic reactions involving various glycosyltransferases. A primary molecular event in Golgi *N*-glycosylation defects is mutations in genes encoding Golgi-resident glycosyltransferases. These enzymes are responsible for adding specific sugar residues to the growing glycan chains on glycoproteins. Mutations in these genes can result in the loss of enzyme activity, substrate specificity, or the addition of incorrect sugar residues, leading to abnormal glycan structures [14].

Due to the enzymatic defects, glycoproteins passing through the Golgi may acquire aberrant glycan structures. These irregular glycans can affect the stability, solubility, and functionality of glycoproteins, compromising their biological roles. Along with it, misfolded or improperly glycosylated proteins emerging from the Golgi may not pass the quality control mechanisms in the ER. This results in their retention within the ER and can lead to protein misfolding and ER stress. The UPR may be activated as a compensatory mechanism.

Golgi *N*-glycosylation defects can affect the proper trafficking of glycoproteins within the cell. Proteins that are not correctly glycosylated may not be transported to their designated cellular locations, disrupting essential cellular functions. As a result, the molecular events in Golgi *N*-glycosylation defects contribute to a range of clinical manifestations observed in CDG patients. The severity and specific clinical features depend on the gene affected and the extent of glycosylation impairment.

Disorders of *N*-glycoprotein Deglycosylation

Disorders of *N*-glycoprotein deglycosylation represent a group of conditions characterized by molecular abnormalities that affect the removal of carbohydrate moieties from glycoproteins. Glycosylation and deglycosylation are fundamental processes in protein maturation and regulation, and defects in deglycosylation can have far-reaching consequences for cellular function and health.

In the cell, glycoproteins undergo glycosylation in the ER and Golgi apparatus, where carbohydrate moieties are added to specific asparagine residues. Afterward, glycoproteins may be targeted for deglycosylation, primarily within lysosomes or the cytoplasm. Deglycosylation is vital for protein quality control, degradation, and regulation of cellular processes. Disorders of N-glycoprotein deglycosylation often result from mutations in genes encoding enzymes or transporters involved in the removal of glycan chains. These mutations can lead to impaired deglycosylation, causing glycoproteins to retain their carbohydrate moieties, either partially or entirely. The molecular hallmark of deglycosylation disorders is the accumulation of glycosylated proteins within cells. These glycoproteins are typically targeted for degradation, but when deglycosylation is defective, they persist in the cell, leading to cellular dysfunction [15]

Deglycosylation disorders can affect different cellular compartments. In lysosomal storage disorders, mutations in genes encoding lysosomal enzymes result in the inability to break down glycosylated proteins properly. In cytoplasmic deglycosylation defects, mutations may affect enzymes responsible for removing glycan chains in the cytoplasm. The consequences of disorders of N-glycoprotein deglycosylation are broad and can affect multiple organ systems. Accumulation of glycosylated proteins in lysosomes or the cytoplasm can lead to cellular dysfunction, tissue damage, and organ failure. Clinical presentations vary widely but often include neurological symptoms, skeletal abnormalities, developmental delays, and visceral organ involvement. Like other congenital disorders, N-glycoprotein deglycosylation disorders are heterogeneous, with different genetic mutations leading to distinct molecular events and clinical manifestations. Each specific disorder is typically associated with mutations in a particular gene or genes involved in deglycosylation.

Treatment options for deglycosylation disorders are often limited and primarily focus on managing the symptoms and complications. Enzyme replacement therapy, substrate reduction therapy, and other approaches may be explored depending on the specific disorder and its underlying molecular defect. Gene therapy and other emerging treatments hold promise for addressing the root cause of these disorders.

DISORDERS OF PROTEIN O-GLYCOSYLATION

O-glycosylation is a crucial post-translational modification that plays a fundamental role in maintaining proper protein structure and function. When defects occur in this pathway, it can have far-reaching consequences on cellular processes, leading to a wide range of clinical manifestations.

O-glycosylation is a complex and highly regulated cellular process involving the attachment of glycans to specific amino acid residues, primarily serine (Ser) or threonine (Thr), within proteins. This process takes place in the ER and Golgi apparatus and plays a crucial role in protein folding, stability, and function. Unlike *N*-glycosylation, which attaches glycans to asparagine (Asn) residues, *O*-glycosylation adds glycans to hydroxyl groups of Ser or Thr. The process typically begins with the synthesis of a glycan precursor on a lipid carrier molecule in the ER. The precursor is then transferred to specific Ser or Thr residues of target proteins in the Golgi apparatus, where further modification and processing occur. Proper *O*-glycosylation is essential for the functionality of many proteins, including those involved in cell adhesion, immune response, signaling, and mucin formation [16].

Disorders of protein *O*-glycosylation encompass a group of genetic conditions characterized by a range of clinical features stemming from abnormalities in the addition of *O*-linked glycans to proteins. These clinical manifestations can vary widely depending on the specific *O*-glycosylation defect, but they often involve multi-system involvement. Common clinical features include developmental and intellectual disabilities, facial dysmorphism, skeletal abnormalities, and various organ system dysfunctions. Additionally, individuals with *O*-glycosylation disorders may experience growth retardation, seizures, and neurological impairments. The severity and combination of clinical symptoms can differ significantly among patients, making the diagnosis and management of these complex disorders challenging. A multidisciplinary approach, including genetic testing and specialized medical care, is often necessary to provide appropriate support and treatment for affected individuals.

Defects in *O*-man Synthesis (Congenital Muscular Dystrophies)

Defects in *O*-mannosyl glycan synthesis are characterized by progressive muscle weakness and wasting, often presenting early in infancy or childhood. Congenital Muscular Dystrophies (CMD) are part of the broader spectrum of muscular dystrophies, a group of genetic conditions primarily affecting muscle structure and function. Understanding the molecular mechanisms and clinical consequences of *O*-mannosyl glycan synthesis defects in CMDs is essential for diagnosis and potential therapeutic interventions [17].

O-mannosylation is a post-translational modification where mannose sugars are added to serine or threonine residues of target proteins, including α-dystroglycan. This glycosylation is crucial for the proper binding of α-dystroglycan to extracellular matrix proteins, such as laminin, which is necessary for muscle cell stability. Molecular events leading to defects in *O*-mannosylation are primarily driven by genetic mutations in genes encoding enzymes involved in the synthesis and addition of *O*-linked mannose sugars. Notably, mutations in genes like POMT1, POMT2, POMGNT1, and LARGE are associated with CMDs. Mutations in these genes can lead to the loss or reduced activity of glycosyltransferases responsible for adding mannose residues to α-dystroglycan. As a result, α-dystroglycan remains underglycosylated or hypoglycosylated. The molecular consequence of hypoglycosylation is the disruption of the interactions between α-dystroglycan and its extracellular binding partners, causing muscle cells to lose their structural integrity. This results in muscle fiber degeneration and progressive muscle weakness [18].

CMDs due to defects in *O*-mannosylation typically manifest as severe muscle weakness, particularly in the proximal muscles. This weakness often presents from infancy or early childhood and progresses over time. Affected individuals frequently exhibit delayed motor milestones, such as delayed crawling or walking, reflecting the underlying muscle dysfunction. Some CMD subtypes, particularly those associated with mutations in genes like POMGNT1 and POMT1, can be accompanied by cognitive impairment, intellectual disability, and developmental delays. Ocular abnormalities, including retinal abnormalities and congenital cataracts, may occur in some CMD subtypes. Additionally, brain involvement can lead to structural brain abnormalities, such as cobblestone lissencephaly. A common clinical feature in CMDs is elevated serum CK levels, which result from muscle degeneration and leakage of CK from damaged muscle fibers. As CMDs progress, individuals may develop respiratory and cardiac complications, including respiratory muscle weakness and cardiomyopathy, which can be life-threatening. Contractures, scoliosis, and joint deformities are common orthopedic problems seen in CMD patients due to muscle weakness and imbalance.

In summary, defects in *O*-mannosylation, leading to congenital muscular dystrophies, result from genetic mutations affecting the glycosylation of α-dystroglycan. These molecular events disrupt muscle cell stability, leading to a range of clinical features, including muscle weakness, cognitive impairment, eye and brain abnormalities, and orthopedic and respiratory complications.

Defects in *O*-GalNAc Synthesis

O-linked *N*-acetylgalactosamine (*O*-GalNAc) glycosylation is a critical post-translational modification that plays a vital role in various cellular processes, including cell adhesion, signaling, and immune response regulation. Dysregulation of *O*-GalNAc synthesis can lead to a range of clinical manifestations, highlighting the significance of this process in maintaining normal cellular and physiological functions [19].

O-GalNAcylation is initiated in the Golgi apparatus, where a sugar molecule, UDP-GalNAc, is transferred onto serine or threonine residues of target proteins. This reaction is catalyzed by a family of enzymes known as GalNAc-transferases (GalNAc-Ts). Defects in *O*-GalNAc synthesis primarily result from genetic mutations affecting the GalNAc-T genes. Mutations can lead to a loss of enzyme function or a reduction in substrate specificity, causing alterations in *O*-GalNAc attachment to proteins. Mutations in GalNAc-T genes may result in incomplete *O*-GalNAcylation of target proteins, leading to hypoglycosylation or altered glycosylation patterns. This can disrupt protein function and affect their stability, localization, and interactions with other molecules. *O*-GalNAcylation plays a crucial role in modulating cell signaling pathways. Defects in *O*-GalNAc synthesis can disrupt this regulation, leading to aberrant signaling events that contribute to the pathogenesis of various disorders.

Defects in *O*-GalNAc synthesis can manifest as congenital disorders, with symptoms often present from birth or infancy. The clinical features vary depending on the specific genetic mutation and its impact on *O*-GalNAcylation. Many individuals with defects in *O*-GalNAc synthesis experience neurological impairments, including developmental delays, intellectual disabilities, and seizures. The altered glycosylation of neuronal proteins can disrupt neural development and function. Some individuals may exhibit muscle weakness or hypotonia, similar to muscular dystrophies, due to the defective glycosylation of muscle-related proteins. Facial dysmorphism, characterized by distinctive facial features, can be a clinical hallmark of certain *O*-GalNAcylation disorders. Gastrointestinal symptoms, such as feeding difficulties and swallowing problems, are observed in some cases, likely due to glycosylation abnormalities affecting proteins in the digestive system.

The variable severity of these disorders underscores the importance of precise diagnosis, early intervention, and tailored management strategies for affected individuals.

Defects in other *O*-glycosylation Families

Defects in *O*-glycosylation can affect various families of glycosyltransferases and lead to a range of congenital disorders, each characterized by specific clinical consequences. While it is discussed defects in *O*-GalNAc synthesis previously, there are other *O*-glycosylation families, each with its set of associated genetic disorders [20, 21].

C-Mannosylation (C-Galactosylation): C-mannosylation involves the attachment of mannose to tryptophan residues in proteins. It's catalyzed by protein *O*-linked mannose beta-1,2-*N*-acetylglucosaminyltransferase (POMGnT1). Defects in C-mannosylation can lead to a rare genetic disorder known as muscular dystrophy-dystroglycanopathy with mental retardation (MDDG). Individuals with MDDG typically exhibit muscle weakness, intellectual disabilities, and structural brain abnormalities. Mutations in genes like POMGnT1 disrupt C-mannosylation, affecting the glycosylation of alpha-dystroglycan and its function in maintaining muscle cell integrity.

Sialylation: Sialylation involves the addition of sialic acid to terminal sugar residues of glycoproteins. This process is crucial for regulating cell adhesion, signaling, and immune response. Defects in sialylation can lead to various disorders collectively known as congenital disorders of glycosylation (CDGs). For example, Sialic Acid Storage Disease (SASD) results from mutations in SLC17A5, which encodes a sialic acid transporter. SASD is characterized by developmental delays, hepatomegaly, and neurological symptoms due to the accumulation of free sialic acid in lysosomes.

T-synthase and core 1 synthase: T-synthase and Core 1 synthase are responsible for adding Galactose (Gal) and *N*-acetylgalactosamine (GalNAc), respectively, to proteins, forming core 1 *O*-glycans. These glycans are essential for mucin-type *O*-glycosylation. Mutations in genes encoding these enzymes can lead to various congenital disorders, including Tn syndrome and Congenital Disorder of Glycosylation Type 2 (CDG-II). These disorders often present with intellectual disabilities, skeletal abnormalities, and coagulation issues.

Sulfation: Sulfation involves the addition of sulfate groups to sugar molecules in glycoproteins and glycolipids. Defects in sulfation can lead to disorders known as Sulfation Deficiencies. These disorders often result in developmental delays, intellectual disabilities, and skeletal abnormalities. Maroteaux-Lamy syndrome, a mucopolysaccharidosis caused by a deficiency in sulfatase enzymes, is an example of a sulfation disorder.

O-Fucose and *O*-Glucose glycosylation: *O*-fucose and *O*-glucose glycosylation involves the addition of fucose or glucose, respectively, to serine or threonine residues of target proteins. These glycosylations are important for signaling pathways. Mutations affecting these pathways can lead to Alagille syndrome, a disorder characterized by liver, cardiac, skeletal, and facial abnormalities. Alagille syndrome is caused by mutations in genes such as JAG1 or NOTCH2, affecting *O*-fucose glycosylation and disrupting Notch signaling.

Defects in various *O*-glycosylation families can result in a wide range of congenital disorders, each with distinct clinical consequences. These disorders underscore the critical role of *O*-glycosylation in maintaining normal cellular and physiological functions, and ongoing research in this field is essential for understanding the molecular mechanisms of these diseases and developing potential therapies.

DEFECTS IN GLYCOSAMINOGLYCAN (GAG) SYNTHESIS

Defects in Glycosaminoglycan (GAG) synthesis are associated with a group of rare genetic disorders known as mucopolysaccharidoses (MPS) and mucolipidoses. GAGs are long, linear carbohydrates that play essential roles in extracellular matrix structure, cellular signaling, and various biological processes. These disorders result from mutations in genes encoding enzymes responsible for GAG synthesis and degradation. Because of these mutations, individuals with GAG synthesis defects accumulate incompletely degraded GAGs within lysosomes, leading to cellular dysfunction and a range of clinical manifestations. MPS and mucolipidoses exhibit a wide spectrum of symptoms, including developmental delays, skeletal abnormalities, organ enlargement, and neurological impairments [22].

GAGs are long, unbranched polysaccharides composed of repeating disaccharide units. The major types of GAGs include chondroitin sulfate, dermatan sulfate, heparan sulfate, keratan sulfate, and hyaluronic acid. Each type of GAG has a distinct structure and plays specific roles in different tissues and biological processes. The biosynthesis of GAGs begins in the endoplasmic reticulum and continues in the Golgi apparatus. Enzymes responsible for elongating and modifying the GAG chains sequentially add specific sugar molecules to the growing polysaccharide. These enzymes include xylosyltransferases, galactosyltransferases, glucuronosyltransferases, and sulfotransferases, among others. Proper GAG synthesis is essential for the formation of biologically active proteoglycans, which are key components of the extracellular matrix. On the other hand, GAG degradation is mediated by lysosomal enzymes called glycosidases. These enzymes break down GAGs into smaller components that can be further metabolized and excreted. In individuals with GAG synthesis defects, these lysosomal glycosidases are often normal, but their function is compromised due to the accumulation of undegraded GAGs in lysosomes.

The clinical consequences of GAG synthesis defects primarily result from the accumulation of undegraded GAGs within lysosomes and their subsequent impact on cellular function. Some common clinical features and consequences of GAG synthesis are discussed below.

Skeletal deformities, joint contractures, and short stature are common in many GAG storage disorders. The accumulated GAGs interfere with bone and cartilage development. Enlargement of organs such as the liver and spleen can occur due to GAG accumulation in these tissues. Some GAG storage disorders can lead to heart valve abnormalities, cardiomyopathy, and other cardiac issues. Cognitive decline, developmental delays, and intellectual disabilities are frequent features in severe forms of GAG storage disorders. Accumulated GAGs can affect the central nervous system. Obstructive airway disease and sleep apnea can occur, often due to the enlargement of tonsils and adenoids. Opacification of the cornea is a characteristic feature in some GAG storage disorders, affecting vision.

Specific GAG Synthesis Disorders

Several well-defined GAG storage disorders fall under the broader category of mucopolysaccharidoses as shown in Table **1** [23]. Some of the most prominent examples include the following.

Table 1. Type of mucopolysaccharidoses and their characteristics.

Disorder	Characteristics
Mucopolysaccharidosis Type I (MPS I)	MPS I is caused by a deficiency of the enzyme alpha-L-iduronidase (IDUA), which is involved in the degradation of heparan sulfate and dermatan sulfate. There are two subtypes of MPS I: Hurler syndrome (severe) and Hurler-Scheie syndrome (intermediate).
Mucopolysaccharidosis Type II (MPS II or Hunter syndrome)	MPS II is caused by a deficiency of iduronate-2-sulfatase (IDS), leading to the accumulation of heparan sulfate and dermatan sulfate.
Mucopolysaccharidosis Type III (MPS III or Sanfilippo syndrome)	MPS III includes four subtypes (A, B, C, and D), each caused by a deficiency of a specific enzyme involved in heparan sulfate metabolism. These disorders primarily affect the central nervous system, leading to severe neurological impairment.
Mucopolysaccharidosis Type IV (MPS IV or Morquio syndrome)	MPS IV includes two subtypes (A and B) and is characterized by the accumulation of keratan sulfate and chondroitin sulfate. Skeletal abnormalities are prominent in this disorder.
Mucopolysaccharidosis Type VI (MPS VI or Maroteaux-Lamy syndrome)	MPS VI results from a deficiency of arylsulfatase B, leading to the accumulation of dermatan sulfate.
Mucolipidosis Type II (I-cell disease) and Type III	These disorders result from defects in the transport of enzymes required for lysosomal function. They are characterized by the accumulation of both GAGs and lipids in lysosomes.

Diagnosing GAG synthesis defects involves clinical assessments, biochemical testing to detect elevated levels of GAGs in urine or blood, and genetic testing to identify specific gene mutations responsible for the disorder. Advances in molecular genetics have significantly improved diagnostic accuracy.

While there are currently no curative treatments for GAG storage disorders, several therapeutic approaches aim to alleviate symptoms and improve the quality of life for affected individuals. These include ERT involves the administration of recombinant enzymes to replace the deficient or dysfunctional enzymes responsible for GAG degradation. HSCT can be considered in some MPS disorders, particularly when initiated early, to provide a source of healthy enzyme-producing cells. SRT aims to reduce the production of GAGs within the body. Research in gene therapy holds promise for the development of treatments aimed at correcting the genetic mutations underlying GAG storage disorders.

DEFECTS IN LIPID AND GPI-ANCHOR GLYCOSYLATION

Defects in lipid and GPI-anchor glycosylation disrupt the synthesis and attachment of carbohydrate molecules to lipids and glycosylphosphatidylinositol (GPI)-anchored proteins. Defects can occur at different stages, including the biosynthesis of lipid-linked oligosaccharides (LLOs) or the attachment of GPI anchors to proteins. Mutations in genes encoding enzymes involved in these processes can lead to the accumulation of abnormal lipid intermediates and affect the function of GPI-anchored proteins. Consequently, individuals with defects in lipid and GPI-anchor glycosylation may experience a wide range of clinical manifestations, including neurological impairments, skeletal abnormalities, and organ dysfunction.

Defects in GPI-anchored Proteins

GPI-anchored proteins are a diverse group of cell surface proteins that play essential roles in various cellular functions, including signal transduction, immune response regulation, and protection against complement-mediated cell lysis. GPI anchors serve as lipid tethers that attach these proteins to the cell membrane. Defects in the biosynthesis or attachment of GPI anchors can have far-reaching consequences, leading to a range of clinical manifestations [24, 25].

GPI anchors are complex glycolipids composed of phosphoethanolamine, inositol, mannose, glucosamine, and other sugar moieties, attached to a phosphatidylinositol lipid tail. These anchors are synthesized in the ER and attached to specific proteins in the Golgi apparatus. The attachment process involves the transfer of the GPI anchor to the C-terminus of target proteins, facilitating their insertion into the lipid bilayer of the cell membrane.

Defects in GPI anchoring result from mutations in genes encoding various enzymes involved in GPI anchor biosynthesis or attachment. These mutations disrupt the production of functional GPI anchors and lead to the release of GPI-anchored proteins into the bloodstream, where they can contribute to the pathophysiology of GPI anchor deficiencies and Paroxysmal Nocturnal Hemoglobinuria (PNH). PNH is characterized by chronic intravascular hemolysis, where red blood cells are destroyed in the bloodstream. This can lead to anemia, fatigue, and an increased risk of thrombosis. PNH patients may experience bone marrow failure, resulting in decreased production of red blood cells, white blood cells, and platelets. This can lead to an increased risk of infections, bleeding, and compromised immune function. PNH is associated with an increased risk of thrombosis (blood clot formation), which can be life-threatening if it occurs in critical organs such as the brain (cerebral venous thrombosis) or the liver (hepatic vein thrombosis). Hematuria (blood in the urine) is a common symptom in PNH due to the presence of GPI-deficient red blood cells in the urine.

It is important to note that PNH can be either inherited (caused by germline mutations) or acquired (developed during one's lifetime). Acquired PNH typically results from somatic mutations in the PIG-A gene, leading to the biosynthesis of GPI anchors. These mutations lead to the expansion of a clone of blood cells lacking GPI-anchored proteins. Acquired PNH is often associated with other bone marrow disorders, such as aplastic anemia and myelodysplastic syndromes.

Defects in Glycosphingolipid (GSL) Synthesis

Glycosphingolipids (GSL) are complex lipids that play crucial roles in cell membranes, cell signaling, and cellular recognition. These lipids are composed of a hydrophilic carbohydrate (glycan) head group linked to a hydrophobic ceramide tail. GSLs are synthesized in the endoplasmic reticulum and Golgi apparatus through a series of enzymatic reactions. Defects in GSL synthesis can occur due to mutations in genes encoding these enzymes, leading to the accumulation of GSL intermediates and disrupting cellular functions. Several sphingolipidoses, including Gaucher disease, Tay-Sachs disease, and Fabry disease, are associated with defects in GSL synthesis [26, 27].

GSL synthesis involves several steps, beginning with the formation of ceramide, the core structure of GSLs, and the subsequent addition of various carbohydrate residues. The specific GSL produced depends on the type of glycosyltransferases present and their substrate specificity. Key enzymes involved in GSL synthesis include glucosylceramide synthase (GCS), galactosylceramide synthase (GalCerS), and ceramide glucosyltransferase (CGT), among others.

Molecular events leading to GSL synthesis defects can vary depending on the specific mutation and the enzyme involved. For instance, mutations in genes coding for glycosyltransferases can result in incomplete or improper GSL structures, disrupting their normal functions. Additionally, defects in transporters responsible for GSL trafficking within the cell can lead to the mislocalization of GSLs. The consequences of GSL synthesis defects are diverse and can manifest as lysosomal storage disorders, neurological impairments, skin abnormalities, and developmental delays, underscoring the critical role of GSLs in various cellular processes. Understanding the molecular basis of these defects is pivotal for diagnosing GSL-related disorders and exploring potential therapeutic strategies to mitigate their effects.

Defects in GSL synthesis disrupt the normal production of GSLs and lead to the accumulation of GSL intermediates in lysosomes, resulting in cellular dysfunction. The clinical consequences of GSL synthesis defects are highly variable and depend on the specific GSL affected, the extent of accumulation, and the tissues and organs involved. Many GSL storage disorders manifest with progressive neurological symptoms, including developmental delays, intellectual disabilities, seizures, and motor impairments. Accumulated GSLs can affect neuronal membranes and neurotransmitter systems. Some GSL storage disorders lead to the enlargement of visceral organs, such as the liver and spleen. This hepatosplenomegaly can cause abdominal distension and discomfort. In addition, skeletal, ocular, and cardiac involvements may also be there.

Several sphingolipidoses are associated with defects in GSL synthesis, each characterized by the accumulation of specific GSL intermediates. Some notable examples are shown in Table **2**.

Table 2. Disorders due to deficiencies in enzymes responsible for the synthesis or degradation of glycosphingolipids.

Disease	Characteristics
Tay-Sachs Disease	Tay-Sachs disease is caused by mutations in the HEXA gene, leading to a deficiency of hexosaminidase A (Hex A). This enzyme is crucial for the breakdown of GM2 ganglioside, a GSL. The accumulation of GM2 ganglioside in neurons leads to severe neurological deterioration.
Gaucher Disease	Gaucher disease results from mutations in the GBA gene, leading to a deficiency of glucocerebrosidase (GBA enzyme). This deficiency leads to the accumulation of glucocerebroside, a GSL, primarily in macrophages. The disease can manifest with hepatosplenomegaly, skeletal abnormalities, and neurological complications.
Fabry Disease	Fabry disease is caused by mutations in the GLA gene, resulting in a deficiency of alpha-galactosidase A (α-Gal A). This deficiency leads to the accumulation of globotriaosylceramide (Gb3), a GSL, in various tissues. Symptoms can include skin lesions, kidney dysfunction, cardiac issues, and neuropathic pain.
Krabbe Disease	Krabbe disease is characterized by mutations in the GALC gene, leading to a deficiency of galactocerebrosidase (GALC enzyme). This deficiency results in the accumulation of galactocerebroside, a GSL, primarily in oligodendrocytes. The disease affects the central nervous system and peripheral nerves, leading to severe neurological symptoms.

DEFECTS OF MULTIPLE GLYCOSYLATION AND OTHER PATHWAYS

Defects in multiple glycosylation pathways and other related cellular processes can result in a group of complex and often severe genetic disorders. These disorders may involve disruptions in various aspects of glycosylation, including *N*-glycosylation, *O*-glycosylation, glycosaminoglycan synthesis, glycolipid synthesis, and more. Such defects can lead to a wide array of clinical symptoms and complications, ranging from developmental delays and intellectual disabilities to skeletal abnormalities, organ dysfunction, and neurological impairments. The heterogeneity and variability in these disorders make diagnosis and treatment challenging. Nonetheless, advances in molecular genetics and research into these pathways continue to expand our understanding of these conditions, offering hope for improved diagnostic tools and potential therapeutic interventions to alleviate the burden on affected individuals and their families.

Defects in the Synthesis of Sugar Precursors

The molecular pathway and clinical features of defects in sugar precursor synthesis are closely linked to the pathogenesis of CDGs [28, 29].

The synthesis of sugar precursors starts with the production of nucleotide sugars in various cellular compartments. These nucleotide sugars serve as donors for glycosyltransferases, the enzymes responsible for attaching sugar residues to proteins and lipids. Different nucleotide sugars, such as UDP-glucose, UDP-galactose, and GDP-mannose, are synthesized through a series of enzymatic reactions, each involving specific enzymes. Activated monosaccharides, like UDP-glucose or GDP-mannose, are generated by transferring sugar moieties onto nucleotides. These activated monosaccharides are crucial substrates for glycosylation reactions that occur in the ER and Golgi apparatus.

Defects in sugar precursor synthesis lead to hypoglycosylation, where glycoproteins and glycolipids are inadequately or incorrectly glycosylated. This results in misfolded, unstable, or non-functional glycoconjugates. Many CDG subtypes are associated with neurological impairments, ranging from mild motor and cognitive delays to severe intellectual disability. The brain's high demand for glycosylation makes it particularly susceptible to defects in sugar precursor synthesis. Some CDG subtypes result in lysosomal storage disorders (LSDs) due to the accumulation of undegraded glycoconjugates within lysosomes. This can lead to additional clinical features, such as organomegaly and skeletal abnormalities.

The sialic acid pathway is a critical metabolic pathway responsible for the biosynthesis and regulation of sialic acids, a family of nine-carbon monosaccharides primarily found at the termini of glycan chains on cell surface glycoproteins and glycolipids. Sialic acids play essential roles in various cellular processes, including cell-cell interactions, immune responses, and signaling. The molecular basis of the sialic acid pathway involves a series of enzymatic reactions that take place mainly in the cytoplasm and the nucleus of eukaryotic cells.

The pathway begins with the conversion of glucose-6-phosphate to *N*-acetylglucosamine-6-phosphate (GlcNAc-6-P) in the cytoplasm. This reaction is catalyzed by the enzyme Glucosamine-6-phosphate synthase (GlmS). Subsequently, GlcNAc-6-P is converted to *N*-acetylmannosamine-6-phosphate (ManNAc-6-P) by Glucosamine-6-phosphate *N*-acetyltransferase (Gnpat). ManNAc-6-P undergoes epimerization to form *N*-acetylneuraminic acid--phosphate (Neu5Ac-9-P). This reaction is catalyzed by the enzyme ManNAc-6-P epimerase (MNX). Neu5Ac-9-P is a key intermediate in the pathway and represents the most common sialic acid in humans, known as *N*-acetylneuraminic acid (Neu5Ac). Neu5Ac-9-P is then converted to cytidine monophosphate (CMP)-Neu5Ac by the enzyme CMP-Neu5Ac synthase (CMAS). This reaction involves the transfer of a CMP moiety from cytidine triphosphate (CTP) to Neu5Ac-9-P, producing the activated sialic acid. CMP-Neu5Ac serves as the substrate for various glycosyltransferases in the Golgi apparatus. These enzymes catalyze the transfer of sialic acids to glycoproteins and glycolipids, where they participate in cell adhesion, immune recognition, and other cellular functions. Sialic acids can be removed from glycoconjugates by sialidases, and the released Neu5Ac can be recycled into the pathway. Neu5Ac can also be converted to other sialic acid forms, such as *N*-glycolylneuraminic acid (Neu5Gc), in some species [30, 31].

The sialic acid pathway is tightly regulated at multiple levels, including gene expression and enzyme activity. The expression of enzymes in the pathway is regulated by various factors, including transcription factors and hormones. Additionally, feedback inhibition mechanisms help maintain appropriate levels of sialic acids within cells.

Mutations in genes encoding enzymes in the sialic acid pathway can lead to congenital disorders known as SASD. These disorders result in the accumulation of sialic acid-containing compounds within lysosomes and can lead to neurological impairments, skeletal abnormalities, and other clinical features.

Defects in the Biosynthesis of Dolichol-monosaccharides

Defects in the biosynthesis of dolichol-monosaccharides are associated with rare genetic disorders that disrupt the formation of DLOs. These DLOs are critical intermediates in the *N*-linked glycosylation pathway, which is essential for the proper glycosylation of proteins in the ER. Understanding the molecular basis and clinical features of defects in dolichol-monosaccharide biosynthesis is vital for diagnosing and managing these complex conditions.

Dolichol-monosaccharide biosynthesis begins with the synthesis of dolichol, a long-chain isoprenoid lipid. This process involves several enzymatic reactions that occur in the ER membrane. Once dolichol is synthesized, multiple sugar residues are sequentially added to it to form a DLO. The precise structure of the DLO varies among species, but it typically contains 14 to 20 sugar residues, including *N*-acetylglucosamine (GlcNAc), mannose, and glucose. After assembly, the DLO is flipped from the cytoplasmic side to the luminal side of the ER membrane, making it accessible for glycosylation of nascent proteins [32].

Defects in the biosynthesis of dolichol-monosaccharides lead to CDG characterized by abnormal protein glycosylation. Clinical features of CDG encompass a wide range of symptoms, primarily affecting the nervous system, growth and development, and various organ systems. Understanding these molecular and clinical aspects is crucial for the diagnosis and management of CDG patients.

Defects in Golgi Homeostasis

Defects in Golgi homeostasis refer to disruptions in the structure, function, and regulation of the Golgi apparatus, a critical organelle involved in post-translational modification, sorting, and trafficking of proteins and lipids within eukaryotic cells. Golgi homeostasis is essential for maintaining cellular integrity and function, and defects in this process can have profound consequences [33].

One common manifestation of Golgi homeostasis defects is the fragmentation of the Golgi apparatus into smaller, dispersed structures. This can result from impaired fusion and maintenance of Golgi cisternae and vesicles. Golgi homeostasis defects can disrupt protein glycosylation, leading to abnormal glycan structures on glycoproteins. This can impact protein stability, localization, and function. Proper Golgi homeostasis is essential for the sorting and trafficking of proteins and lipids between the Golgi compartments and other cellular organelles. Defects in this process can lead to mis-localization of key cellular components.

Mutations in genes encoding proteins involved in Golgi maintenance, fusion, or vesicular trafficking can result in Golgi homeostasis defects. For example, mutations in genes like GOLGB1 or GOLGA2 can lead to Golgi fragmentation. Aberrant signaling pathways can disrupt Golgi homeostasis. For instance, altered Ca-signaling or the activation of stress response pathways like the UPR can impact Golgi structure and function. Some viruses can manipulate the Golgi apparatus to facilitate their replication. Viral proteins or mechanisms may disrupt Golgi homeostasis to benefit viral replication.

Golgi homeostasis defects can affect neurons and result in neurological disorders. For example, some hereditary spastic paraplegias (HSPs) are caused by mutations in genes that disrupt Golgi structure and trafficking in neurons, leading to spasticity and motor dysfunction. Disrupted Golgi homeostasis can impact protein glycosylation and lead to CDGs. The CDGs may result in a wide range of clinical features, including developmental delays, intellectual disabilities, and multi-system dysfunction. Golgi homeostasis defects can activate cellular stress responses, such as the UPR, which can contribute to cell dysfunction and apoptosis if not properly regulated.

OVERCOMING THE DEFECTS/TREATMENT OF GLYCOSYLATION

Glycosylation disorders can manifest as a wide range of clinical symptoms, including developmental delays, intellectual disabilities, skeletal abnormalities, and multi-system dysfunction. The treatment of glycosylation disorders is complex and multifaceted, aiming to address the underlying molecular defects, manage clinical manifestations, and improve the quality of life for affected individuals. This essay explores the basis of treatment for glycosylation disorders, encompassing genetic counselling, supportive care, and emerging therapeutic approaches [10, 34, 35]

Genetic Counseling and Diagnosis

The foundation of treatment for glycosylation disorders begins with an accurate diagnosis. Identifying the specific genetic mutation(s) responsible for the disorder is essential for tailoring treatment strategies. Genetic counseling provides affected individuals and their families with comprehensive information about the genetic basis of the disorder. This includes recurrence risks, family planning options, and the importance of genetic testing for at-risk family members.

Supportive Care and Symptomatic Management

Many glycosylation disorders result in a variety of clinical symptoms, such as developmental delays, muscle weakness, and organ dysfunction. Symptomatic care is essential and often involves physical therapy, occupational therapy, and speech therapy to manage developmental and motor impairments. Addressing specific complications, such as seizures, cardiac issues, or immune system dysfunction, are crucial. This may require medical management, surgical interventions, or therapies tailored to the individual's needs.

Emerging Therapeutic Approaches

In some glycosylation disorders, where a specific enzyme deficiency is identified as the cause, ERT may be considered. ERT involves the administration of exogenous enzymes to compensate for the deficiency, thereby alleviating specific symptoms or metabolic imbalances. Gene therapy is an exciting and evolving field that holds promise for the treatment of glycosylation disorders. This approach involves delivering functional copies of the defective genes to cells, potentially restoring normal cellular function and addressing the underlying molecular defect. Ongoing research focuses on developing small molecule therapies that can modulate cellular processes affected by glycosylation defects. These approaches aim to correct or mitigate cellular dysfunction and improve the clinical outcome for affected individuals.

Multidisciplinary Management

A multidisciplinary approach to care, involving geneticists, pediatricians, neurologists, physical therapists, occupational therapists, and other specialists, provides comprehensive care tailored to the individual's needs. Support groups and networks for individuals and families affected by glycosylation disorders can offer emotional support, share experiences, and provide guidance on managing daily challenges.

The treatment of glycosylation disorders is a complex and evolving field, driven by advancements in genetics, molecular biology, and therapeutic approaches. Accurate diagnosis, genetic counseling, and supportive care are fundamental aspects of managing these rare genetic conditions. Emerging therapies, including gene therapy and small molecule interventions, offer hope for addressing the underlying molecular defects and improving the quality of life for affected individuals. A personalized approach to care, involving a team of healthcare professionals and genetic specialists, is essential to meet the unique needs of each individual and family affected by glycosylation disorders. As research continues, the outlook for individuals with these complex disorders becomes increasingly promising, offering the potential for more effective treatments and improved clinical outcomes.

CONCLUSION

In conclusion, the chapter on Congenital Disorders of Glycosylation (CDG) has provided a comprehensive overview of these rare genetic conditions that affect the glycosylation process, a fundamental post-translational modification crucial for cellular function. CDG encompasses a diverse group of disorders, each stemming from specific genetic mutations that disrupt various aspects of glycosylation. Throughout this chapter, the molecular mechanisms, clinical features, and diagnostic challenges associated with CDG are explored.

Our exploration of CDG has underscored the complex nature of these disorders, which can affect multiple organ systems and manifest with a wide range of clinical symptoms. From neurological impairments to developmental delays and multi-system dysfunction, CDG presents a complex clinical landscape. It has become evident that early diagnosis, often requiring sophisticated genetic testing and precise molecular analysis, is essential for optimizing patient care and management.

In addressing the treatment and management of CDG, the importance of a multidisciplinary approach, involving geneticists, pediatricians, neurologists, therapists, and other specialists has also been discussed. Symptomatic care, genetic counseling, and supportive interventions play a crucial role in improving the quality of life for affected individuals and their families. Additionally, emerging therapeutic approaches, such as gene therapy and small molecule interventions, hold promise for addressing the underlying molecular defects and offering new avenues for treatment.

As we conclude this chapter, it is important to acknowledge the ongoing research efforts aimed at unraveling the complexities of CDG and advancing our understanding of these disorders. With continued scientific exploration and the collaboration of researchers, clinicians, and affected individuals, the future holds the potential for more precise diagnostics, improved treatments, and enhanced outcomes for those living with CDGs.

REFERENCES

[1]   Schjoldager KT, Narimatsu Y, Joshi HJ, Clausen H. Global view of human protein glycosylation pathways and functions. Nat Rev Mol Cell Biol 2020; 21(12): 729-49.
[http://dx.doi.org/10.1038/s41580-020-00294-x] [PMID: 33087899]

[2]   Lipiński P, Tylki-Szymańska A. Congenital Disorders of Glycosylation: What Clinicians Need to Know? Front Pediatr 2021; 9: 715151.
[http://dx.doi.org/10.3389/fped.2021.715151] [PMID: 34540767]

[3]   Monticelli M, D'Onofrio T, Jaeken J, Morava E, Andreotti G, Cubellis MV. Congenital disorders of glycosylation: narration of a story through its patents. Orphanet J Rare Dis 2023; 18(1): 247.
[http://dx.doi.org/10.1186/s13023-023-02852-w] [PMID: 37644541]

[4]   Chang IJ, He M, Lam CT. Congenital disorders of glycosylation. Ann Transl Med 2018; 6(24): 477.
[http://dx.doi.org/10.21037/atm.2018.10.45] [PMID: 30740408]

[5]   Hirata T, Kizuka Y. N-Glycosylation. Adv Exp Med Biol 2021; 1325: 3-24.
[http://dx.doi.org/10.1007/978-3-030-70115-4_1] [PMID: 34495528]

[6]   Frappaolo A, Sechi S, Kumagai T, Karimpour-Ghahnavieh A, Tiemeyer M, Giansanti MG. Modeling Congenital Disorders of N-Linked Glycoprotein Glycosylation in *Drosophila melanogaster*. Front Genet 2018; 9: 436.
[http://dx.doi.org/10.3389/fgene.2018.00436] [PMID: 30333856]

[7]   Medrano C, Vega A, Navarrete R, *et al.* Clinical and molecular diagnosis of non-phosphomannomutase 2 N-linked congenital disorders of glycosylation in Spain. Clin Genet 2019; 95(5): 615-26.
[http://dx.doi.org/10.1111/cge.13508] [PMID: 30653653]

[8]   Quelhas D, Correia J, Jaeken J, *et al.* SLC35A2-CDG: Novel variant and review. Mol Genet Metab Rep 2021; 26: 100717.
[http://dx.doi.org/10.1016/j.ymgmr.2021.100717] [PMID: 33552911]

[9]   Paprocka J, Jezela-Stanek A, Tylki-Szymańska A, Grunewald S. Congenital Disorders of Glycosylation from a Neurological Perspective. Brain Sci 2021; 11(1): 88.
[http://dx.doi.org/10.3390/brainsci11010088] [PMID: 33440761]

[10]   Verheijen J, Tahata S, Kozicz T, Witters P, Morava E. Therapeutic approaches in Congenital Disorders of Glycosylation (CDG) involving N-linked glycosylation: an update. Genet Med 2020; 22(2): 268-79.
[http://dx.doi.org/10.1038/s41436-019-0647-2] [PMID: 31534212]

[11]   Grünewald S, Matthijs G, Jaeken J. Congenital disorders of glycosylation: a review. Pediatr Res 2002; 52(5): 618-24.
[http://dx.doi.org/10.1203/00006450-200211000-00003] [PMID: 12409504]

[12]   Jaeken J, Matthijs G. Congenital disorders of glycosylation. Annu Rev Genomics Hum Genet 2001; 2(1): 129-51.
[http://dx.doi.org/10.1146/annurev.genom.2.1.129] [PMID: 11701646]

[13]   Leroy JG. Congenital disorders of N-glycosylation including diseases associated with O- as well as N-glycosylation defects. Pediatr Res 2006; 60(6): 643-56.
[http://dx.doi.org/10.1203/01.pdr.0000246802.57692.ea] [PMID: 17065563]

[14]   Cherepanova N, Shrimal S, Gilmore R. N-linked glycosylation and homeostasis of the endoplasmic reticulum. Curr Opin Cell Biol 2016; 41: 57-65.
[http://dx.doi.org/10.1016/j.ceb.2016.03.021] [PMID: 27085638]

[15]   DeRosa CM, Weaver SD, Wang CW, Schuster-Little N, Whelan RJ. Simultaneous N-Deglycosylation and Digestion of Complex Samples on S-Traps Enables Efficient Glycosite Hypothesis Generation. ACS Omega 2023; 8(4): 4410-8.
[http://dx.doi.org/10.1021/acsomega.2c08071] [PMID: 36743002]

[16]   Dutta D, Mandal C, Mandal C. Unusual glycosylation of proteins: Beyond the universal sequon and other amino acids. Biochim Biophys Acta, Gen Subj 2017; 1861(12): 3096-108.
[http://dx.doi.org/10.1016/j.bbagen.2017.08.025] [PMID: 28887103]

[17]   Martin P. Congenital muscular dystrophies involving the O-mannose pathway. Curr Mol Med 2007; 7(4): 417-25.
[http://dx.doi.org/10.2174/156652407780831601] [PMID: 17584082]

[18]   Bertini E, D'Amico A, Gualandi F, Petrini S. Congenital muscular dystrophies: a brief review. Semin Pediatr Neurol 2011; 18(4): 277-88.
[http://dx.doi.org/10.1016/j.spen.2011.10.010] [PMID: 22172424]

[19]   Freeze HH, Eklund EA, Ng BG, Patterson MC. Neurology of inherited glycosylation disorders. Lancet Neurol 2012; 11(5): 453-66.
[http://dx.doi.org/10.1016/S1474-4422(12)70040-6] [PMID: 22516080]

[20]   Ng BG, Freeze HH. Perspectives on Glycosylation and Its Congenital Disorders. Trends Genet 2018; 34(6): 466-76.
[http://dx.doi.org/10.1016/j.tig.2018.03.002] [PMID: 29606283]

[21]   Bennett EP, Mandel U, Clausen H, Gerken TA, Fritz TA, Tabak LA. Control of mucin-type O-glycosylation: A classification of the polypeptide GalNAc-transferase gene family. Glycobiology 2012; 22(6): 736-56.
[http://dx.doi.org/10.1093/glycob/cwr182] [PMID: 22183981]

[22]   Mizumoto S, Yamada S. Congenital Disorders of Deficiency in Glycosaminoglycan Biosynthesis. Front Genet 2021; 12: 717535.
[http://dx.doi.org/10.3389/fgene.2021.717535] [PMID: 34539746]

[23]   Hampe CS, Wesley J, Lund TC, *et al.* Mucopolysaccharidosis Type I: Current Treatments, Limitations, and Prospects for Improvement. Biomolecules 2021; 11(2): 189.
[http://dx.doi.org/10.3390/biom11020189] [PMID: 33572941]

[24]   Kinoshita T. Glycosylphosphatidylinositol (GPI) Anchors: Biochemistry and Cell Biology: Introduction to a Thematic Review Series. J Lipid Res 2016; 57(1): 4-5.
[http://dx.doi.org/10.1194/jlr.E065417] [PMID: 26582962]

[25]   Carmody LC, Blau H, Danis D, *et al.* Significantly different clinical phenotypes associated with mutations in synthesis and transamidase+remodeling glycosylphosphatidylinositol (GPI)-anchor biosynthesis genes. Orphanet J Rare Dis 2020; 15(1): 40.
[http://dx.doi.org/10.1186/s13023-020-1313-0] [PMID: 32019583]

[26]   Xu YH, Barnes S, Sun Y, Grabowski GA. Multi-system disorders of glycosphingolipid and ganglioside metabolism. J Lipid Res 2010; 51(7): 1643-75.
[http://dx.doi.org/10.1194/jlr.R003996] [PMID: 20211931]

[27]   Ryckman AE, Brockhausen I, Walia JS. Metabolism of Glycosphingolipids and Their Role in the Pathophysiology of Lysosomal Storage Disorders. Int J Mol Sci 2020; 21(18): 6881.
[http://dx.doi.org/10.3390/ijms21186881] [PMID: 32961778]

[28]   Parker JL, Newstead S. Structural basis of nucleotide sugar transport across the Golgi membrane. Nature 2017; 551(7681): 521-4.
[http://dx.doi.org/10.1038/nature24464] [PMID: 29143814]

[29]   Hadley B, Maggioni A, Ashikov A, Day CJ, Haselhorst T, Tiralongo J. Structure and function of nucleotide sugar transporters: Current progress. Comput Struct Biotechnol J 2014; 10(16): 23-32.
[http://dx.doi.org/10.1016/j.csbj.2014.05.003] [PMID: 25210595]

[30]   Cotton TR, Joseph DDA, Jiao W, Parker EJ. Probing the determinants of phosphorylated sugar-substrate binding for human sialic acid synthase. Biochim Biophys Acta Proteins Proteomics 2014; 1844(12): 2257-64.
[http://dx.doi.org/10.1016/j.bbapap.2014.09.014] [PMID: 25242570]

[31]   Hinderlich S, Weidemann W, Yardeni T, Horstkorte R, Huizing M. UDP-GlcNAc 2-Epimerase/ManNAc Kinase (GNE): A Master Regulator of Sialic Acid Synthesis. Top Curr Chem 2013; 366: 97-137.
[http://dx.doi.org/10.1007/128_2013_464] [PMID: 23842869]

[32]   Banerjee DK, Zhang Z, Baksi K, Serrano-Negrón JE. Dolichol phosphate mannose synthase: a Glycosyltransferase with Unity in molecular diversities. Glycoconj J 2017; 34(4): 467-79.
[http://dx.doi.org/10.1007/s10719-017-9777-4] [PMID: 28616799]

[33]   Liu J, Huang Y, Li T, Jiang Z, Zeng L, Hu Z. The role of the Golgi apparatus in disease (Review). Int J Mol Med 2021; 47(4): 38.

[http://dx.doi.org/10.3892/ijmm.2021.4871] [PMID: 33537825]

[34]   Park JH, Marquardt T. Treatment Options in Congenital Disorders of Glycosylation. Front Genet 2021; 12: 735348.
[http://dx.doi.org/10.3389/fgene.2021.735348] [PMID: 34567084]

[35]   Boyer SW, Johnsen C, Morava E. Nutrition interventions in congenital disorders of glycosylation. Trends Mol Med 2022; 28(6): 463-81.
[http://dx.doi.org/10.1016/j.molmed.2022.04.003] [PMID: 35562242]

CHAPTER 4

The Biology of Advanced Glycation End Products

Parth Malik[1,2,†], **Ruma Rani**[3,†] and **Tapan Kumar Mukherjee**[4,*]

[1] *School of Chemical Sciences, Central University of Gujarat, Gandhinagar, Gujarat-382030, India*

[2] *Swarrnim Startup & Innovation University, Bhoyan-Rathod, Gandhinagar-Gujarat, India*

[3] *ICAR-National Research Centre on Equines, Hisar-125001, Haryana, India*

[4] *Amity Institute of Biotechnology, Amity University, New Town, Kolkata, West Bengal 700156, India*

Abstract: This chapter is dedicated to the biology of advanced glycation end products (AGEs). In 1912, AGEs were first identified by French chemist Louis-Camille Maillard. Early investigation revealed AGE generation during food preparation (cooking) at high temperatures, wherein carbohydrates (*e.g.* glucose/glycan) slowly react with various proteins *via* concomitant generation of Schiff's base and Amadori products. This non-enzymatic process of AGE generation is termed glycation. Later, subsequent investigations revealed that AGE is exogenously produced during cooking and other processing of foodsand also endogenously generated in the human body including blood, skin, and other tissues. To date, more than 20 AGEs are postulated to prevail within human blood, tissues, and food resources. AGEs are optical sensitive molecules and based on their optical sensitivity AGEs are distinguished into fluorescent and non-fluorescent categories. The most important non-fluorescent components are carboxymethyl-lysine (CML), carboxyethyl-lysine (CEL), and pyrrolidine while pentosidine and methylglyoxal-lysine dimer (MOLD) are prominent compounds having fluorescent sensitivity. AGE binds with several receptor molecules, the prominent among whichare receptors for advanced glycation end products (RAGE). Additional cell surface molecules capable of binding with AGE including macrophage scavenger receptors (MSRs) type A, B1 (CD36), oligosaccharyltransferase-48/OST48, also termed "AGE receptor 1" (AGE-R1), 80K-H phosphoprotein (AGE receptor 2, AGE-R2), and galectin-3 (AGE receptor 3, AGE-R3), the scavenger receptor family (SR-A, SR-B, SR-1, SR-E, LOX-1, FEEL-1, FEEL-2, and CD36). This chapter describes the steps of AGE synthesis, their biochemical characterization, and the implication of the AGE-RAGE interactions at the cellular platform.

[*] **Corresponding author Tapan Kumar Mukherjee:** Amity Institute of Biotechnology, Amity University, New Town, Kolkata, West Bengal 700156, India; E-mail: tapan400@gmail.com
[†] These authors contributed equally to this work.

Keywords: Advanced glycation end products (AGE), Glycation, Lipid peroxidation, Methylglyoxal, Oxidative and Inflammatory stress, Reactive oxygen species (ROS), Skin autofluorescence.

INTRODUCTION

The process of glycation was initially introduced in 1912 by French chemist Louis-Camille Maillard and is therefore also recognized as the "Maillard reaction". In the typical process, free reducing sugars spontaneously (without the involvement of any enzymes/ catalysts) gradually react with free amino groups (-NH$_2$) of proteins, DNA, and lipids, culminating in the formation of Amadori products. The Amadori products undergo diverse, irreversible dehydration and rearrangement reactions and generate "advanced glycation end products (AGEs)". The biochemical analysis of AGE formation indicates that the typical process of glycation commences *via* sugar carbonyl or aldehyde group and the amino acid nucleophilic, free -NH$_2$ group chemical reaction, swiftly generating Schiff base as an unstable adduct. The spontaneous rearrangements of this adduct reversibly form "Amadori product" as a further stable fraction. Finally, the intermediate Amadori products undergo irreversible oxidation, dehydration, polymerization, and cross-linking to form AGEs (Fig. **1**) [1]. The depicted AGE molecules in Fig. (**1**) are 3-deoxyglucosone, glyoxal, methylglyoxal, N$^\varepsilon$-(carboxymethyl) lysine, and N$^\varepsilon$-(carboxyethyl) lysine, formed as a consequence of Amadori rearrangement of the Schiff bases from the Maillard reaction.

AGE generation is mediated endogenously *i.e.* within the human body and exogenously (outside the human body). The most significant exogenous generation of AGEs is from food. *Grossin and associates* elaborately discussed the significance of dietary AGEs and their interaction with RAGE amidst aging [2]. During the processing or cooking of food on an industrial or domestic level, the Maillard reaction is frequently employed for improving the color, flavor, aroma, and consistency of foods. However, a considerable extent of AGE generation happens during sugar-protein simultaneous processing *via* cooking [3]. AGE is also produced inside the human body such as in blood, skin, and other tissues. There is a substantial possibility for the glycation of cellular proteins compared to the plasma proteins. Several studies demonstrate a familiar, intracellular generation of dicarbonyls *via* triose phosphate fragmentation and lipid peroxidation (Fig. **2**). The event of glycation culminates in lost protein function and deteriorating tissue elasticity, including those of blood vessels, skin, and tendons. All major classes of biomolecules such as DNA, proteins, and phospholipids are affected by glycation. The estimated damage level indicates that 0.1-1% of lysine and arginine fractions on proteins, 1 in 10^7 DNA nucleotides, and 0.1% alkaline phospholipids are affected by glycation [4].

In human tissues and blood, the glycation reaction is highly aggravated particularly in case of hyperglycemia and enhanced tissue oxidative stress [5]. As a result, glycation eventually mediates the pathogenesis of diabetic vascular troubles apart from cellular and tissue aging [6]. No enzymatic process is presently known to eliminate AGEs from the human system, including cells, tissues, and blood. Nevertheless, some studies claimed that AGEs are catabolized in renal proximal tubular cells. Thus, the glycation process is in agreement with the belief that metabolic waste accumulation promotes aging.

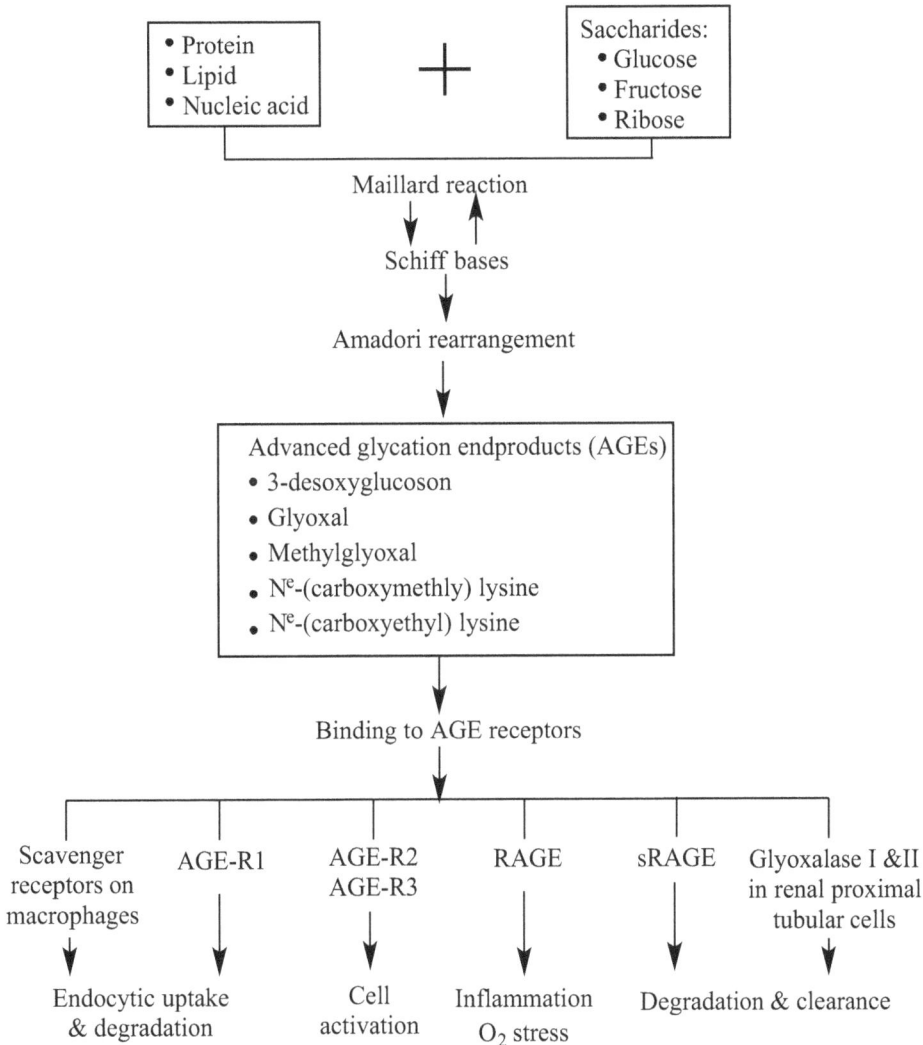

Fig. (1). Advanced-glycation end-product (AGE) formation by Maillard reaction and the fates after binding with different AGE receptors. AGE-R1, R2, and R3: AGE receptors R1, R2, and R3; RAGE: receptor for AGE; sRAGE: soluble-form RAGE.

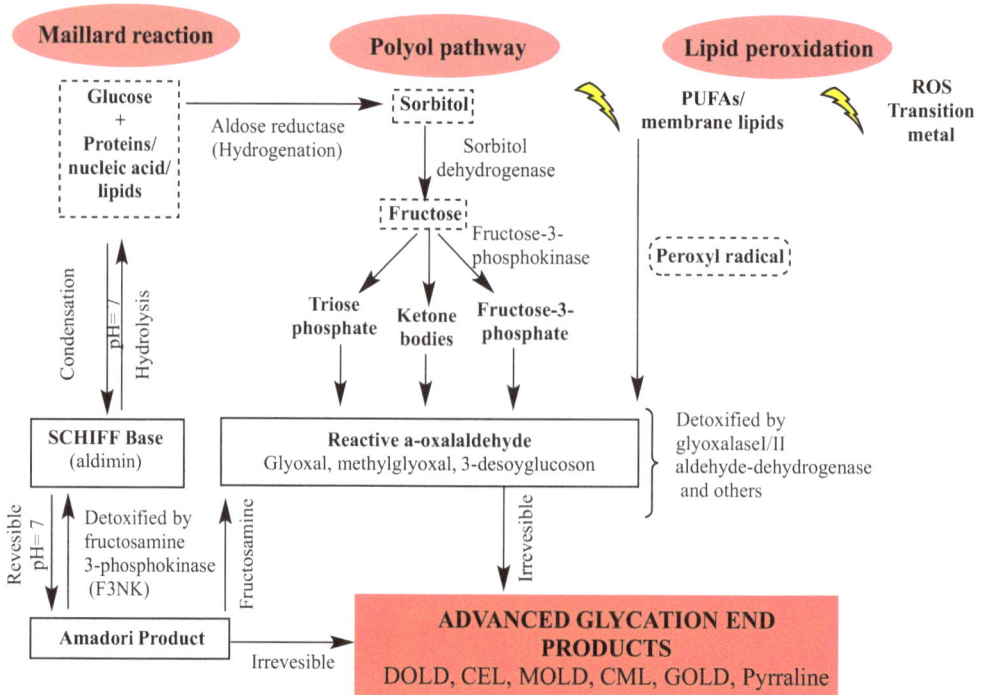

Fig. (2). *In vivo* generation of advanced glycation end products (AGEs), endogenous generation happens *via* three *in vivo* routes;(**i**)the non-enzymatic Maillard reaction, Polyol-Pathway, and lipid peroxidation. In all these processes, AGE generation happens preferentially than reactive carbonyl compounds, like glyoxal (GO), methylglyoxal (MGO), and 3-deoxyglucose. In case of impaired detoxification, the incomplete AGEs react further till irreversible AGE generation.

In cellular systems, a high level of reactive oxygen species (ROS) generation enhances AGE formation and accumulation [7 - 9]. Of note, oxidative stress is aggravated by excessive ROS generation from glucose auto-xidation and the non-enzymatic, covalent conjugation of glucose molecules with circulating proteins, resulting in AGE generation [10]. While oxidative stress induces the AGE generation, the AGE once produced reacts with its receptors *i.e.* RAGE, further generating more ROS and eventually culminating in oxidative stress. Thus, the AGE-RAGE axis leads to a continuous self-propagating process [11]. Numerous studies report the AGE-RAGE axis-dependent oxidative damage to biological macromolecules including proteins as a formidable and decisive mechanistic link in complicating diseased conditions including diabetic vascular complications, respiratory diseases, neurological disorders, cancers, and others [12].

As indicated, RAGE (*i.e.* receptor for advanced glycation end products) prevails as a prominent AGE receptor. The interaction between AGE and RAGE, which is also called the AGE-RAGE axis, not only induces oxidative stress but inflammatory stress too. The oxidative and inflammatory stresses are complementary to each other, collectively affecting several cellular activities, *vis* propagation, autophagy, apoptosis, and aging [13, 14]. Overall, the AGE-RAGE axis generated oxidative stress and inflammation resulting in the complications of several diseases including diabetes-related vascular complications, cancers, and others. Thus, the search for AGE inhibitors has unveiled manifold natural products that forbid glycation. Several medical herbs, dietary plants, and phytochemicals impair protein glycation, in the *in vitro* cell culture and physiological conditions [15]. These natural compounds with substantial antioxidant ability are gifted mediators that forbid glycation and concomitant AGE generation. Their AGE-inhibiting functions may be a potent source of human health benefits [16].

The following points are important regarding the discussion of AGE formation and its biological consequences in human systems:

- In 1912, Louis-Camille Maillard for the 1st time described the process of glycation, with the synthesis of advanced glycation end products (AGE).
- Generation of AGEs occurs *via* non-enzymatic glycation of macromolecules such as DNA, proteins (*e.g.* human serum albumin), and lipids with reducing sugar, such as aldoses (*e.g.* glucose) or ketoses (*e.g.* fructose). This accumulation can occur endogenously (*e.g.* hyperglycemia, aging, inflammation, and oxidative stress) or exogenously (*e.g.* high-fat diet, processed food, chronic alcohol consumption, and tobacco) [17]. The average glycation rate of human body glycation is quite low. Once the body is aged or continues to have high blood sugar, the glycation proceeds aggravatedly, forming an enormous number of AGEs [17].
- AGEs are the typical modifications of proteins, lipids, or nucleic acids such as DNA that undergo glycation and oxidation on encountering aldose sugars [18].
- Glycation is a spontaneous reaction and proceeds without any enzymes and cofactors. The bonding between aldose sugars and proteins is typically covalent. Thereby, glycation is an irreversible reaction. It is worth noting here with that once glycated, the stability and functional activity of proteins decreases.
- Glycations occur mainly in the bloodstream contrary to a small proportion of the absorbed simple sugars: glucose, fructose, and galactose. Glucose, including glucose-6-phosphate, is one of the predominant sources of reactive sugars for AGE but several other sugars (such as fructose, threose, and glyceraldehyde--phosphate) can condense with proteins such as fructose, threose, and glyceraldehyde-3-phosphate to form the various AGEs (Fig. **2**).
- It appears that fructose has nearly tenfold glycation activity of glucose, the primary body fuel [19]. More than 20 AGEs have been documented until recently to prevail within human blood, tissues, and dietary foods.

- Arbitrarily, these AGEs can be grouped as fluorescent and non-fluorescent with carboxymethyl-lysine (CML), carboxyethyl-lysine (CEL), and pyrrolidine, being the prominent non-fluorescent species. The major fluorescent counterparts comprise pentosidine and methylglyoxal-lysine dimer (MOLD).
- The presence of lysine residue in the peptide/protein molecules is a characteristic feature of various AGEs.
- On reaching the renal proximal tubular cells, AGEs are catabolized and are eventually eliminated from the kidneys.
- Glycation begins with a reaction between the sugar-carbonyl or aldehyde group and the free, nucleophilic $-NH_2$ amino acid moiety, swiftly forming the unstable Schiff base. Henceforth, Schiff's base rearranges to provide a reversible and comparatively steady, Amadori product.
- During Amadori rearrangement, the vigorously active intermediate carbonyl groups, (also referred to as -dicarbonyls or oxoaldehydes), form 3-deoxyglucosone and MGO, which slowly accumulate. This build-up is often recognized as "carbonyl stress."
- The -dicarbonyls can react with amino, sulfhydryl, and guanidine functionalities in proteins, resulting in denaturation, browning, and cross-linking of the targeted proteins.
- Additionally, lysine and arginine functionalities on proteins react with dicarbonyls, eventually forming stable AGE compounds, *viz.* carboxymethyl-lysine (CML), the non-fluorescent AGEs. Of note, CMLs are also formed in the *in vitro* conditions, from LDL incubated with Cu ions and glucose. For this reason, the CMLs are believed to be lipid and protein adducts.
- The intermediate products undergo subsequent irretrievable oxidation, dehydration, polymerization, and cross-linking to form AGEs over several days to weeks.
- Once formed, AGEs prevail stably and are nearly irreversible. There is evidence that enzymes, such as glyoxalase-1, can detoxify AGE precursors and inhibit AGE formation, as evidenced by the presence of deoxy fructose, a reduction product of 3-deoxyglucosone in human urine and plasma.
- AGEs may fluoresce to generate ROS, which further binds to specific cell surface receptors *via* forming cross-linkage. The interaction of AGE and RAGE leads to multiple diseased conditions [7, 20].

STEPS OF FORMATION OF ADVANCED GLYCATION END PRODUCTS

- In 1912, French food chemist Louis C. Maillard discovered that AGE formation occurs during cooking, leading to the golden browning of food with a change in taste and texture. It was also observed that during long-term storage of cooked food materials, the golden browning coloration was prominent, indicating high-level AGE formation even in stored food materials. However, several decades have passed without significant research on AGE outside the food industry.

- Around 1970, several scientists including *Bookchin & Gallop* in 1968, *Rahbar and colleagues*, 1969, *Koening and Cerami* in 1975, *Fliickiger and Winterhalter* in 1976, demonstrated that in both *in vitro*, *in vivo* as well as in diabetic conditions, carbohydrates (*e.g.* glucose/glucose-6-phosphate, fructose--phosphate) are conjugated to the *N*-terminal valine of the hemoglobin β-chain (HbAlc).

- Subsequent studies revealed that plasma and structural proteins were also glycated in diabetes. However, the relative extents of plasma protein glycation are limited due to their rapid replacement or excretion through urine.

- Another important point is that although glycation is directed predominantly against the structural proteins, other proteins such as enzymes (*e.g.* superoxide dismutase, glutathione reductase, *etc.*) are also susceptible to glycation.

- The synthesis of AGE is mediated by a step-by-step (early and late steps) chemical cascade interaction, commencing with condensation.

- The condensation reaction is non-enzymatic or reversible. It occurs between the carbonyl functionality of reducing sugar and the $-NH_2$ group of a protein or peptide. The reaction is time-consuming and is also referred to as the "Schiff base reaction", forming imines from aldehydes.

- Within a reasonable duration (few days), the Schiff base undergoes intramolecular rearrangement to attain equilibrium with more stable and slow reversible ketamine, also known as Amadori product (with glucose).

- Over several weeks, the Amadori product undergoes a series of sub sequential spontaneous inter and intra-molecular rearrangements, dehydration, and condensations to generate a diversified range of irreversible end products that prevail as a group, exhibiting fluorescence, yellow-brown color with steady inter- and intramolecular cross-links. Such end-products are referred to as "advanced glycation end products (AGEs)".

- The generated AGEs could also undergo subsequent degradation *via* multiple stages, forming furfurals, reductions, and fragmentation products.

- The Schiff base could also undergo spontaneous degradation to generate reactive α, β-dicarbonyl species such as MGO and glyoxal (GO). These dicarbonyls may also be formed from glucose lysis in the Wolff pathway, involving triose phosphate fragmentation, acetone and threonine metabolism, lipid peroxidation, and fructose-3-phosphate decomposition (Fig. **3**).

- Despite glycation being conceptualized as a reaction between sugars and $-NH_2$ groups, other macromolecule functionalities, such as protein thiols (-SH), may also participate in an analogous regime.

- Like the glycation of lysine and arginine residues on proteins, thiol functionality glycation can change the enzymatic performance. For instance, glyceraldehyde-

3-phosphate dehydrogenase is impaired by MGO-mediated elimination of -SH groups at its active loci.

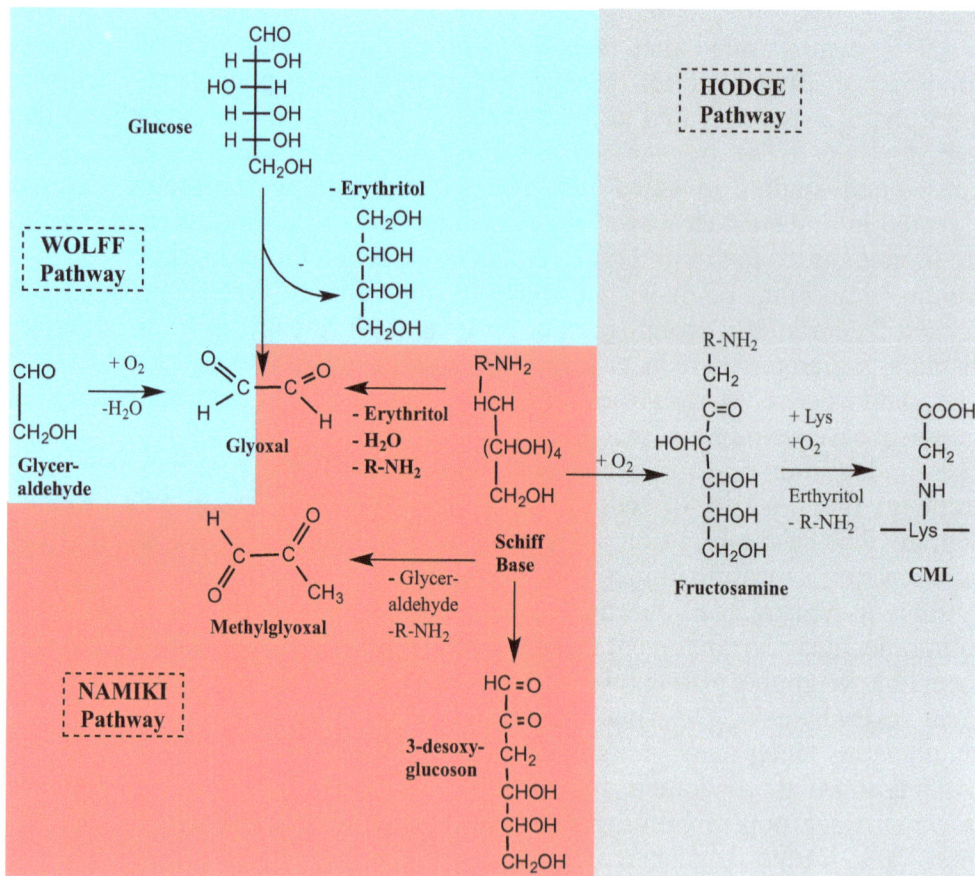

Fig. (3). Formation of AGEs *via* Wolff, Namiki, and Hodge-pathways. The autoxidation of monosaccharides or carbonyl compounds (Wolff pathway), aldimines (Namiki-pathway), or Amadori products (Hodge pathway) *via* transition metals or ROS can spontaneously form reactive carbonyl compounds *via* further reaction to AGEs.

- Glyoxal-driven glycation of creatine kinase inactivates the enzyme. Although α, β-dicarbonyls triggered thiol modification could be a sensitive factor for protein functional activity, low M_w thiols, *viz. N*-acetylcysteine and GSH (reduced glutathione) are reliable therapeutic agents for getting rid of α, β-dicarbonyls before reaction with proteins. Fig. (**1**) depicts the formation of AGEs by the above-mentioned biochemical reactions.
- The relative limit of AGE formation and gathering exhibits substantial variations with changing temperature, pH, protein and glucose amounts, and substrate

turnover. Although temperature, pH, and protein quantity remain unchanged in *vivo*, the changes in mean blood glucose extents and protein half-life ($t_{1/2}$) substantially affect the degree and assembling of AGE-modified proteins. For example, proteins with $t_{1/2}$ more than a few weeks, are likely to accumulate more AGEs than those with having a few days of $t_{1/2}$.

• Proteins that are substantially accessed by circulating glucose, like extracellular and serum proteins, also exhibit susceptibility towards non-enzymatic glycation. Besides diabetes mellitus, an illness dominated by chronic hyperglycemia promptly affects the frequency of AGE-modified protein accumulation varying as per the mean blood glucose enhancement. The distinctive structural recognition of specific AGEs has been hampered by a native variety of products formed *via* spontaneous chemical processes (mostly rearrangements) of non-enzymatic glycation pathways. As glucose is the major physiological extracellular-reducing sugar, it finds extensive use in mechanistic studies. Nevertheless, it is pertinent to note that analogous reactions could happen with erstwhile reducing sugars as well.

• The first-ever isolated AGE was 2-(2-furyl)-4(5)-(2-furanyl)-1H-imidazole (FFI), generated *via* glucose incubation with polylysine or bovine serum albumin (BSA). The remaining AGE structures including pentosidine, CML, and pyrrolidine have also been identified through model incubations and *in vivo* confirmation. Since the natural AGE formation leads to a markedly diverse range of adducts, the specific yield of any specific products is limited. For this reason, it was demonstrated that by inactivating bypassing non-enzymatic glycation pathways, one could enrich for precise intermediate compounds.

• Sulfites are well-versed moieties to inhibit non-enzymatic glycation in food materials and find significant application as preservatives. Perhaps the existence of sodium sulfite as an inhibitor in glucose and 6-amino-hexanoic acid model reaction led to the recognition of 1-alkyl-2-formyl 3,4-di glycosyl pyrrole (AFGP), an AGE intermediate.

FACTORS AFFECTING ADVANCED GLYCATION END-PRODUCT FORMATION

AGE is produced both outside and inside the human body. Exogenously, AGEs are consumed *via* food intake. The prominent environmental or exogenous factors that affect the rate of AGE formation and thereby the level of AGE in the human body are diet and smoking. Endogenously, little extent of AGEs is generated amidst the normal metabolic processes. However, endogenous AGE formation gets aggravated in pathophysiological conditions such as diabetes. Moreover, it was shown in a cohort study using healthy monozygotic and heterozygotic twins that the circulating AGE levels are genetically determined. The key internal factors implicated in AGE formation comprise the turnover frequency of proteins involved in glyco-oxidation, the relative extent of hyperglycemia, and the level of environmental oxidation. If one or more of these conditions prevail, the intracellular and extracellular proteins could be glycated and oxidized [21 - 23].

Ahead is a brief discussion of the exogenous and endogenous sources of AGE:

The Exogenous Factors Affecting Advanced Glycation End Products Formation

Manifold environmental factors, such as high-carbohydrate and high-calorie diets, high-temperature-cooked food, cigarette smoke, and other air pollutants, apart from a sedentary lifestyle may enhance AGE formation [24]. Therefore, the exogenous sources of RAGE are broadly categorized as processed food and diet, exposure to UV-light and other ionizing radiations and cigarette smoking caused air pollution (Table **1**).

Table 1. Prominent exogenous sources of Advanced Glycation End Products (AGEs), exhibiting enhanced vulnerability *via* respiratory toxins and protein or fat excess [55, 56].

Exogenous sources	Prominent constituents involved in AGE generation	Related Consequences/Evidence
Cigarette smoke	Highly reactive glycation products serve as precursors for AGE formation.	Enhanced ROS generation, pulmonary inflammation, breathing-related difficulties, worsened biochemical activities, and coordination.
Western diet	Thermal processing involving dry heating such as frying grilling, and baking (aggravates GO, MGO, and 3-DOG formation), *via* enhanced fructose content.	Aggravated oxidative and inflammatory stress, poor nutrient assimilation, age-based metabolism difficulty.
Processed foods (dry heating accounts for high levels of AGEs) prepared in inadequate oxygen using saturated oils	Ready-made cookies, biscuits, cheese, nuts, saturated and trans fats. Roasted coffee, cocoa, cereals, generated *via* baking or grilling) aggravates glycotoxins formation	Vulnerability to diabetic, cardiovascular, blood pressure vulnerable aged persons, poor absorption of foods causing deteriorated physical health.
Animal Foods rich in carbohydrates or proteins (befriended cheeses, poultry, pork, fish and eggs)	Amino acid or carbohydrate content aggravates AGE formation; composition must be stabilized with adequate antioxidants.	Slow energy release in aged diabetic sufferers, toxin accumulation, pressure on the excretory mechanisms, weakened overall coordination.

Abbreviations: GO: Glyoxal, MGO: Methylglyoxal, 3-DOG: 3-deoxyglucosone, ROS: reactive oxygen species

Processed Food and Diet

- AGEs are formed in foods and therefore diet is considered the biggest contributor to dietary AGE (dAGE) formation. Foods exposed to grilling, frying, or roasting in high temperatures generate a high level of RAGE [5, 25, 26]. In a comprehensive review article, *Nowotny and colleagues* discussed the AGE content of various foods [27]. Exogenous glycation is responsible for the brownish appearance of foods amidst cooking, popularly known as the "Maillard reaction". Accordingly, AGEs could remarkably interface with the modern diet and deteriorating human health [28].

- For a long time, the importance and functions of dAGEs in human health remained poorly understood as earlier studies claimed their poor physiological absorption. Only recently, some investigations focused on oral intake of a singular AGE-enriched diet in humans and protein-AGEs or the specific AGEs (including MG) containing diet in mice. The results convincingly established poor dAGEs absorption, signifying a potential contribution to the physiological AGE pool [29, 30]. AGE-supplemented diets in mice revealed aggravated diseased conditions emanated to be caused *via* stronger AGE generation (such as atherosclerosis) [31].

- Analysis of healthy humans revealed a convincing dAGEs correlation with circulating AGEs, *vis-à-vis* CML, MG besides the oxidative stress hallmarks [32]. Interestingly, the exclusive prevalence of dAGEs in the sufferers of diabetes and renal disorders inferred a moderation of oxidative and inflammatory stress markers [33 - 35]. The consensus realization from the results of human and animal studies established a forbiddance of dAGEs in the diet as aiding the subsiding of chronic disorders and aging in animals and humans [36].

- The subject candidates having a yester prevalence of diabetes revealed an enhanced risk of generating manifold AGEs, culminating in their gradual accumulation. Lately, findings establish a moderate dAGEs intake as attainable *via* increased consumption of fish, legumes, low-fat milk products, vegetables, fruits, and whole grains *via* reduced intake of solid fats, fatty meats, unprocessed dairy products, and excessively processed foods. These guidelines are consistent with international recommendations from the American Heart Association [37], the American Institute for Cancer Research [38], and the American Diabetes Association. Thereby, the scenario permits an upgradation of established guidelines by including recent evidence to optimize disease prevention and medical nutrition therapy for multiple conditions.

Ultraviolet Light and Ionizing Radiations

Multiple cutaneous cells, such as fibroblasts and keratinocytes [39, 40] generate AGEs, as common derivates of fibronectins, laminins, collagen, elastin, and epidermis [41, 42]. The UV light (as a source of ionizing and non-ionizing radiation) aggravates the AGE generation in dermal locations. Sunlight is perhaps the most common source of ionizing and non-ionizing radiation including UV light. The dermal composition of CML/AGE from sunlight-exposed skin has been demonstrated to be 10% greater than sunlight-guarded skin. Studies also establish that UV light aggravates the dermal accumulation of CML and pentosides by inducing oxidative stress [43, 44].

The extracellular matrix (ECM) permeability of dermal cells issubstantially affected by UV light. Several investigations have established it beyond doubt, whereby, UV light and other ionizing radiation exposure aggravates AGE extents in the skin. Thereby establishing it beyond doubt that exposure to ionizing radiation aggravates AGE formation. These findings affirm AGE as a potential target of radiation-inducedtoxicity [45]. For instance, UVA-generated ROS at the cutaneous level has been reported to aggravate the CML formation and intensify the localized degeneration mediated viaaging. Furthermore, the accumulated AGEs after UVA irradiation have been reported in dead dermal cells with damaged epidermal cells [46].

Gathered AGEs in collagen and elastin of connective tissue often culminate in softening and loss of elasticity [47]. Apart from this, a suppressed viability of UVA-exposed human dermal fibroblasts is observed in the presence of AGEs [48]. These findings collectively inferred that UV light-induced AGE accumulation in the intracellular environment aggravated the ROS generation which alters the permeability of dermal proteins aggravates inflammatory stress-mediated pathophysiological conditions and aggravates aging [49, 50].

Cigarette Smoking and Air Pollution

• The first breakthrough establishing the generation of manifold glycated products by cigarette smoke was provided by *Cerami and associates*, referring to the generated species as "glycotoxins". Analysis revealed aggressive glycotoxins inhaled into the alveoli amidst smoking, which subsequently led to their absorption into the bloodstream and following uptake by the lung parenchymal cells. Once glycotoxins were in the bloodstream, the generation of AGE moieties on serum and vascular wall proteins, was almost inevitable, thereby leading to the onset of multiple diseased conditions [51]. Consequently, cigarette smokers revealed a higher expression of AGE-lipoprotein B and AGE-albumin extents, compared to non-smokers. The studies by *Nicholl and teammates* showed enhanced AGE generation within the lungs and blood vessels of cigarette smokers.

• With an effort to further understand the underlying biochemical basis, *Dickerson and the groupmates* recognized that nornicotine (a tobacco constituent and a nicotine metabolite), could augment aldol reactions in an aqueous environment, culminating in wrongful protein glycation. The same research group demonstrated abnormally high plasma extents of nornicotine--altered glycated proteins in smokers [51]. These observations were strongly supported by the findings of *Federico and team*, whereby dermal AGE deposition in breastfed infants from smoking mothers was noticed as greater than their non-smoking counterparts [52].

• Multiple investigations also demonstrated that cigarette smoking induced and aggravated the RAGE (receptor for AGE) expression, wherein one study by *Reynolds and groupmates* reported increased RAGEexpression in pulmonary epithelium after tobacco smoke exposure. This study emphasized the importance of Egr-1, a transcription factor sustaining the RAGE expression amidst embryonic lung development in the presence of cigarette smoke [53].

• More importantly, treatment of emphysema-vulnerable AKR mice with FPS-ZM125 (a RAGE inhibitor) diminished the airspace enlarging on being exposed to cigarette smoke [54]. As only the AKR mice (and not other strains) exhibited emphysema resembling conditions on being exposed to cigarette smoke, results explained certain exclusive genetic factors as responsible for emphysema complications.

The Endogenous Factors Affecting Advanced Glycation End Products Formation

The AGE formation is mediated gradually under the physiological environment and is affected by various factors within the human body. The most prominent factors are discussed ahead:

Hyperglycemia

Despite the complex pathophysiology of type 2 diabetes mellitus, AGEs are anticipated as decisive intermediates aggravating the diabetic progression and related complications. Unattended hyperglycemia is widely reported as a principal source of diabetic complications, leading to AGE generation [57, 58].

Aging

• Hurdles in correlating AGE formation and a manifestation of age-relateddiseases have been primarily due to (**i**) multiple AGEs generating sources, (ii) a gradual accumulating tendency of AGEs, often taking several years for distinct identification (iii) inadequate accessibility andsensitive assays for quantifying the specific AGEs, (iv) increasing targets of AGEs, and (v) inadequate number of models reiterating the troubled pathologies of accumulated AGEs. These constraints have obstructed the efforts to adjudge the explicit RAGE correlation with an exclusive age-related disorder [59].

- Significant evidence prevails for aging-mediated AGE buildup in multiple species, with decisive contributions from Baynes, Thorpe, and Monnier laboratories over the past 3 decades [60]. To begin with, the correlation of characterized, manifold collagen cross-links with arterial complications undermined their importance in dietary complications and aging [61 - 63].

- A population study involving adults aged 65 years or higher revealed a positive correlation between high plasma CML extents with an aggravated susceptibility towards mortality in advancing-aged individuals. These outcomes were majorly manifested by enhanced risk of cardiovascular complications [64]. Thus, AGEs may not merely prevail as pathological markers but also as indicators of aging-driven complications.

- Typically, a heterogeneous group of compounds, AGEs, comprise more than 20 distinct molecules. Those enhancing AGE generation have been potentially illustrated as significant in multiple pathological effects [65, 66]. Manifold experimental findings have screened the AGE accumulation in aging proteins, such as crystalline in collagens and basement membranes [67, 68]. A notable study on humans herein, monitored the autofluorescence of eye lenses to screen the AGEs accumulation with advancing age [69]. Screening of diabetic sufferers' biopsies reported an accumulation of various AGEs (notably, pentosidine) in the dermis besides the collagen cross-linking [70]. Of note, pentosidine prevails as a pentose-mediated cross-linked protein, existing in human tissues such as plasma proteins and RBCs [71].

- A matching study underrated the significance of AGE in bone health with advancing age [72]. The ECM is particularly identified as a vulnerable site for forage gathering owing to a prevalence of aging proteins like collagen [73]. Accumulating AGEs have also been linked to contribute to lipofuscin manifestation with an increasing risk for advancing age [74]. Though this evidence is majorly correlative, the one for direct and causal AGEs manifesting relationship with life span is demonstrated from the studies on *C. elegans*. For instance, excessive glyoxalase (GLOD-4) expression in the mitochondria of worms is attributed to elongated lifespan which also inhibits dicarbonyl-driven altered mitochondrial proteins by AGEs [75]. Separate investigations correlate impaired GLOD-4 functioning with a shortened lifespan, which is typically exacerbated under high glucose conditions in *C. elegans* [76].

- More recently, low concentrations of MGO formed because of impaired glycine-*C*-acetyltransferase (GCAT) functioning amidst the threonine catabolism, can enhance the lifespan through proteo-hormesis [77]. A notable development herewith pertains to the FDA-approved use of rifampicin (a noted anti-tuberculosis drug) which guarded against an age-related AGE accumulation in worms *via* DAF-16/FOXO activation [78]. Studies correlating the AGE formation in differing species present crucial evidence to understand the role of AGEs in lifespan-related disorders. Earlier findings of the Monnier group revealed the pentosidine deposition in the skin samples at variable frequency, exhibiting an inverse correlation with maximum lifespan. These correlations have been screened in eight mammalian species, thereby making AGEs-mediated lifespan curtail reliability [79]. Together, the discussed studies significantly correlate the involvement of AGE in the aging process along with their causal impact on organismal ageing.

- Despite significant reports, the causal-effect association between AGEs and the aging process remains controversial. While a group of studies opens that aging accelerates AGE accumulation [80, 81], erstwhile findings support a regulatory influence of AGE onthe driven aging process [82]. However, circulating glycotoxins are implicated in oxidative and inflammatory stress-impaired cellular functioning. A notable effort by *Li and colleagues* witnessed 2.5-fold enhanced AGE formation in aged hearts contrary to the younger ones [83]. Analysis revealed a group of proteins having (50-75) kDa as the molecular weight with 4-7 as the isoelectric point, as substantially altered in aged hearts due to aggravated cardiac AGE build-up besides AGEs-driven aggravated oxidative stress and altered protein permeability.

- Erstwhile attempt by *Son and associates* illustrated an age-mediated AGE-albumin gathering as S100N along with a prominent RAGE membrane expression in visceral adipose tissues, which was subsequently screened as being involved in the pathogenesis of inflammatory aggravations in the elderly. Finally, in-depth studies by *Reynaert and Li* groups also established the AGE-RAGE signaling mediated modulations as the causative factors behind the enhanced inflammatory stress, manifesting as multiple chronic disorders such as type 2 diabetes mellitus, chronic obtrusive pulmonary disorder, several neurodegenerative diseases and osteoporosis [83, 84].

Obesity

This disease is often characterized by an aggravated vulnerability of metabolic syndrome, together with insulin resistance-driven type-2 diabetes mellitus, hypertension, fatty liver, and vascular criticalities related to an excessive generation of adipokines from fat cells. In a noted attempt herein, *Gaens and colleagues* demonstrated aggravated plasma and tissue amounts of MGO, AGEs, and advanced lipo-oxidation end products as obesity-causing factors without CML involvement [85]. In a separate attempt, *Brix and colleagues* examined obesity sufferers and noticed substantially low sRAGE generation compared to normal individuals. Interestingly, the sRAGE expression enhanced after weight loss *via* bariatric surgery which moderated the AGE-driven inflammation [86]. On the same lines, *Sanchez and accomplices* screened the AGE gathering using skin autofluorescence (SAF) in the forearm *via* an AGE reader. Analysis revealed higher SAF for obese patients suffering from metabolic syndrome compared to non-obese individuals. The SAF values remained higher even on bariatric surgery, until the failure of glycemic memory [87]. Finally, *Deo and associates* monitored the impact of reduced weight on CML extents of obesity sufferers not having diabetes. Inspection revealed a 17% decrement in CML extents after weight reduction, but the caution was not very helpful in diabetic or pre-diabetic individuals who were not overweight [88]. The findings probably inferred a likely contribution of over-weightiness and hyperglycemia in the AGE generation and accumulation within the body, with a potentially greater risk of hyperglycemia.

Chronic Renal Insufficiency

Screening the uremia sufferers, *Miyata and colleagues* observed a dramatic elevation in plasma AGE extents, irrespective of diabetes status. The investigators monitored the fate of AGEs *via* intravenous injection of pentosidine into rats. Inspection revealed the pentosidine filtration by renal glomeruli, a subsequent re-absorption in the proximal renal tubes (the catabolism locations), and finally excretion *via* urination. Henceforth, *Asano and colleagues* screened the pentosidine metabolism in proximal tubular, distal tubular, and non-renal cell lines and promptly noticed pentosidine as being promptly screened within the cytoplasm of proximal renal tubular (TJC-12) cells [89]. The findings made the investigators conclude that renal proximal tubular cells performed a key role in getting rid of plasma pentosidine.

Gaining inputs from these results, *Waanders and associates* screened a renal pentosidine gathering in the non-diabetic, chronic adriamycin-aggravated nephropathy in rats [90]. Thereby, it was concluded that AGE accumulation in chronic kidney disease is inherently responsible for the onset of chronic heart failure, cardiovascular disorders, diabetes, neurodegenerative complications, osteoarthritis, and non-diabetic atherosclerosis. Realizing this, the scientists voiced caution about consuming a high-AGE-containing diet that could be a risk factor for chronic disorders including chronic renal disease.

Glyoxalase I Deficiency

- The conventionally reactive dicarbonyl compounds (glyoxal, methyl glyoxal, 3 deoxyglucosone) are detoxified by the Glyoxylase (GLO) enzyme system. This enzyme comprises glyoxalase I and II configurations, which lyse the dicarbonyl compounds and inhibit the AGE formation.

- The impaired functioning of these enzymes destabilizes the intracellular environment, leading to carbonyl stress [91]. This unstable environment inside a cell eventually culminates in oxidative stress, cellular apoptosis, and vascular damage. Below par functioning of these enzymes builds up the toxic carbonyl compounds within the cell, which can co-exist with carrier proteins like albumin, hemoglobin, lens crystals, and low-density lipoproteins [92].

- It is important to note that despite HbA1C being a glycated intermediate, it is not considered a true AGE as it accounts for a (6-12) week ranged glycemia status while the advanced glycation involves a comparatively longer duration of glycation [93].

- Noted effort from *Miyata and colleagues* focused on examining the significance of the glyoxalase detoxification system on reactive carbonyl compounds (pentosidine precursors) while ascertaining the AGE extents of a uremia sufferer undergoing hemodialysis. The investigators noticed excessively aggravated pentosidine and CML plasma extents in the renal blood vessels of the patient than others who underwent hemodialysis. Further analysis revealed an impaired glyoxalase activity in the RBCs of this sufferer; leading to the affirmation that glyoxalase I deficiency could not detoxify the AGEs and exhibited a significant role in aggravated AGE extents in uremia sufferers [94].

- A separate investigation by *Shinohara and colleagues* noticed a suppressed intracellular AGE formation following the GLO-1 over-expression in bovine endothelial cells, which subsequently forbade the hyperglycemia-induced elevated endocytosis in the blood [95]. On the same lines, *Brouwers and associates* noticed GLO-1 over-expression-driven hyperglycemia as the source of AGE and concomitant oxidative stress in mesangial cells from diabetic rats and mice [96]. Counterverification by *Kurz and team* also revealed that GLO-1 induction mediated reduced toxicity of MGO, GO, and other AGEs to guard against cellular damage from glycation stress [97]. Erstwhile effort by *Xue and group* investigated the GLO-1 transcriptional regulation by erythroid 2-related factor 2 (transcription factor). Analysis revealed a defense mechanism that worked against decarbonylateglycation (MG) induced stress by the GLO system in high glucose concentration, causing inflammation, cell aging, and senescence [98]. Emergent efforts from *Garrido and the team* also concluded that fatty acid synthesis amalgamation with GLO-1 could guard against glycation toxicity and arrest the intracellular buildup of MGO-induced AGEs [99].

Autoimmune and Inflammatory Conditions support Advanced Glycation End-product Formation

- Aggravated AGEs generation is well-postulated about autoimmune, rheumatic inflammation, neurodegenerative, and neuropsychiatric disorders and cancers. Inflammatory responses stimulate innate immunity characterized by macrophages, dendritic cells, central nervous system prevailing microglial cells, and astrocytes to induce a metabolic trigger for glycolysis to happen in such inflammatory cells [100].

- During glycolysis, MGO is generated from glyceraldehyde-3-phosphate and dihydroxyacetone phosphate whereas is directly formed through glucose. As a result, the MGO and GO generation serves as the beginning step for the AGE generation. MGO interaction with lysine residue forms CML, while that of GO forms N^ε-(carboxyethyl) lysine (CEL) [101]. As a guarding measure from the MGO and GO toxicities, the methyl glyoxalase and glyoxalase, comprising of GLO-1 and GLO-2 are stimulated to catabolize the two AGEs antecedents.

- The molecular basis of AGE generation in neurodegenerative conditions (*e.g.* Alzheimer's and Parkinson's disease) is quite distinct from erstwhile inflammatory brain disorders. Quite unconventionally, α-synuclein accumulation is believed to be decisive for AGE formation *via in situ* mechanisms. The α-synuclein is correlated with synaptic vesicles, acting as a chaperone of the SNARE complex, which regulates vesicle formation and neurotransmitter release. Furthermore, theα-synuclein deposition impairs the GLO-1 expression to inhibit the AGE clearance from brain tissues.

STRUCTURAL DISTINCTIONS OF VARIOUS ADVANCED GLYCATION END PRODUCTS

As discussed in the earlier sections, AGEs are the typical heterogeneous molecules obtained *via* non-enzymatic intermediates from the reaction between glucose or other saccharide derivatives with proteins or lipids. At present, more than 20 distinct AGEs are reported in human blood, tissues, and foods. Based on their fluorescent traits, AGEs are divided further into fluorescent and non-fluorescent molecules, wherein the most important in the first domain include carboxymethyl-lysine (CML), carboxyethyl-lysine (CEL), pyrroline (a non-fluorescent AGE), pentosidine and methylglyoxal dimer (MOLD, a fluorescent AGEs) [105, 106]. Despite considerable diversity in their chemical structures, a common prospect for all is the prevalence of lysine residue.

- Complex mechanisms of formation, characterized by significant variability of precursors are the major reasons for the prevalence of chemically robust AGE molecules recognized to date in human blood, tissues, and foods. The respective chemical structures, traits, and physiological significance allow the AGE subdivision into multiple groups, per the specific listing criteria. Fig. (**4**) depicts the structural distinctions of potential AGE molecules, wherein H-bonding sensitivity is inferred *via* multiple H and -OH substitutions in all compounds. The various endogenous (external end) sources of AGEs are summarized in Table **2**, along with the likely toxicity threats and cautions.

- Representing a typical heterogeneity, AGEs exhibit unique structural distinctions. Inceptive studies on these compounds screened the sugars and haemoglobin A1c (HbA1c), better identified as glycated haemoglobin in diabetic sufferers. The formation mechanism of these compounds involves the conjugation of glucose molecules to the -NH₂ groups on haemoglobin β-chains, alongside the concomitant formation of a Schiff base, an unstable structure leading to Amadori rearrangement (Fig. 5). The final product, 1-deoxy-1-fructosyl residue exhibits a carbohydrate fragment attached with HbA1c (Fig. **6A**). In recent years, several other AGEs have been identified *via* cell culture and animal model studies. N-carboxymethyl lysine (CML), pentosidine, and MGO are by far, some of the well-characterized compounds (Fig. **6b-d**). Most of the recently discovered AGEs are distinguished based on their fluorescence characteristics.

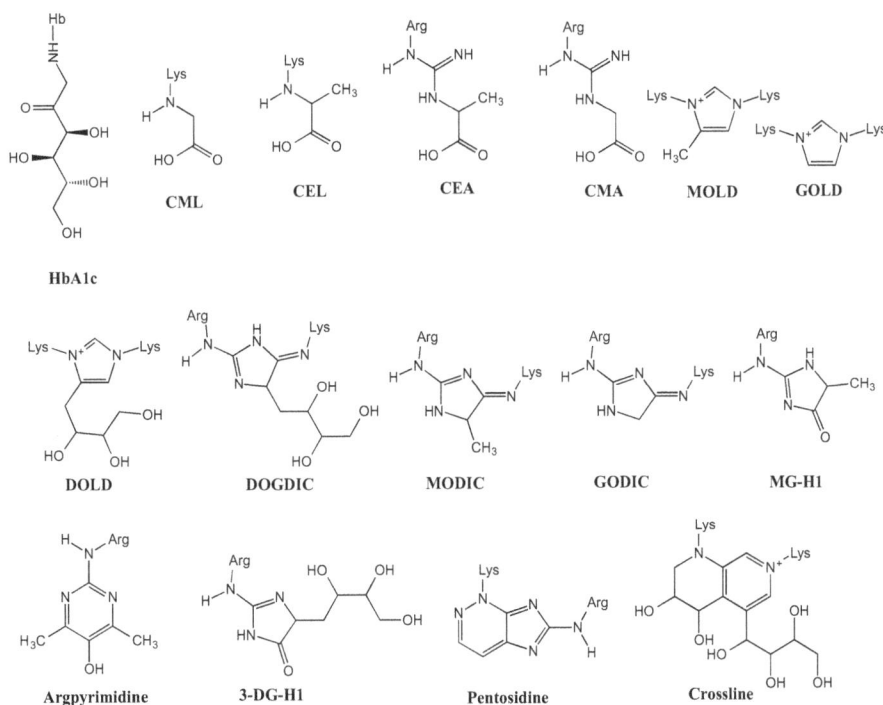

Fig. (4). Representative chemical structures of AGEs. Abbreviations: glycated hemoglobin (HbA1c), N^{ε}-(carboxymethyl)lysine (CML), Nᵉ-(1-carboxyethyl)lysine (CEL), N^{7}-(1-carboxyethyl)arginine (CEA), N^{7}-(carboxymethyl)arginine (CMA), 6-(2-formyl-5-hydroxymethyl-1-pyrrolyl)-L-norleucine (pyrraline), 6-{--[(5S)-5-ammonio-6-oxido-6-oxohexyl]-4-methyl-imidazolium-3-yl}-L-norleucine (MOLD), 6-{1-[(5S-

-5-ammonio-6-oxido-6-oxohexyl]imidazolium-3-yl}-L-norleucine   (GOLD),1,3-di(N^ε-lysino)-4-(2,-,4-trihydroxybutyl)-imidazolium   (DOLD),1,3-bis-(5-amino-5-carboxypentyl)-4-(1,2,3,4-tetrahydro-ybutyl)-3H-imidazolium   (GLUCOLD),   2-ammonio-6-({2-[4-ammonio-5-oxido-5-oxopently)am-no]4-methyl-4,5-dihydro-1H-imidazol-5-ylidene}amino)hexanoate   (MODIC), N^6-(2-{4S(-4-ammo-io-5-oxido-5-oxopentyl]amino}-3,5-dihydro-4H-imidazol-4-ylidene)-L-lysine   (GODIC), N^6-{2-{[(4S-4-ammonio-5-oxido-5-oxopentyl]amino}-5-[(2S,3R)-2,3,4-trihydroxybutyl]-3,5-dihydr--4H-imidazol-4-ylidene}-L-lysinate   (DOGDIC), N^6-{2-{[(4S)-4-ammonio-5-oxido-5-oxopentyl-amino}-5-[(2S)-2,3-dixydroxypropyl]3,5-dihydro-4H-imidazol-4-ylidene}-L-lysinate (DOPDIC), N^6-(--methyl-4-imidazolon-2-yl)-L-ornithine   (MG-H1), N^6-(45-hydro-5-(2,3,4-trihydroxybutyl)-4-imid-zolon-2-yl]-L-ornithine (3DG-H1), N^6-(5-hydroxy-4,6-dimethylpyrimidine-2-yl)-L-ornithine (argpyrimidine), and   6-[2-[[(4S)-4-amino-5-hydroxy-5-oxopentyl]amino]-4-imidazo   [4,5-b]pyridinyl]-L-norleucine (pentosidine). Modified protein surface models (light pink) are based on the structure of human hemoglobin (PDB ID 1COH) [107].

Table 2. Major endogenous sources of Advanced Glycation End products, their associated turbulences, and cautionary measures [102 - 104].

Physiological artefacts leading to enhanced AGEs generation	Missing link of normal functioning	Cautionary aspect/regulating measures
Multistage glycation *via* non-enzymatic Maillard reaction	Aggravates glycation in the hyperglycemic environment generates highly reactive intermediate AGEs, carbonyl precursors.	The rate of the reaction is enhanced *via* an increase in pH and temperature, pickling thereby lowers the pH and moderates AGEs generation.
Dicarbonyls such as α-oxoaldehydes generated *via* glucose oxidation, polyol pathway and lipid peroxidation	Increased glucose flux *via* glycolytic pathway in diabetes sufferers accumulates DHAP due to impaired glycolytic enzymes and TPI actions.	Vegetables, fruits, dairy products, high-fat, oils, and olive oils are low AGEs contributors. Cooking, steaming, and stewing without frying reduce protein glycation.
Polyunsaturated fatty acids are oxidized to form reactive carbonyl intermediates	Inadequate actions of TPI, DHAP, and GAP inter-conversion, eventually forming highly reactive dicarbonyl and MGO.	Enhanced glucose manifests oxidative stress, and glucose autooxidation *via* polyol pathway.
Enhanced lipid peroxidation in diabetic sufferers	Oxidation of PUFAs to generate malondialdehyde and methylgloxal, forming ALEs .	Dicarbonyl intermediates are vital aspects of endogenous AGEs generation, dicarbonyl stress is enhanced by ketone metabolism in hyperglycemia.

Abbreviations: DHAP: Dihydroxyacetone phosphate, GAP: Glyceraldehyde-3-phosphate, TPI: Triose Phosphate Isomerase, AGE: Advanced Glycation End Products, ALEs: Advanced Lipid Peroxidation end products.

The four major classes of AGEs based on their fluorescent responses are as follows:

1. Fluorescent and cross-linked (Fluorescent cross-linked).

2. Non-fluorescent and not cross-linked.

3. Non-fluorescent, cross-linked with proteins.

4. Fluorescent with no cross-linkage.

The first isolated and analyzed AGE is pentosidine, exhibiting a cross-linked arrangement with a fluorescent sensitivity. This AGE is retrieved from collagen, involvement of arginine and lysine subunits, and exhibiting simultaneous prevalence from hexoses and ascorbic acid [71]. The relative pentosidine extent is better considered as the fraction of major glycol-oxidative end product, the reason for its feasibility in quantifying plasma or other tissues AGE build up [107].

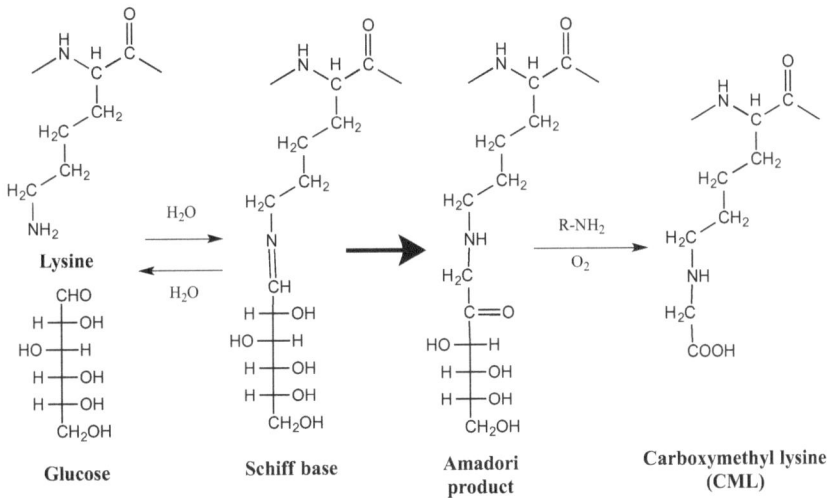

Fig. (5). Sequential reactions leading to advanced glycation end-product formation through the concomitant generation of an Amadori adduct.

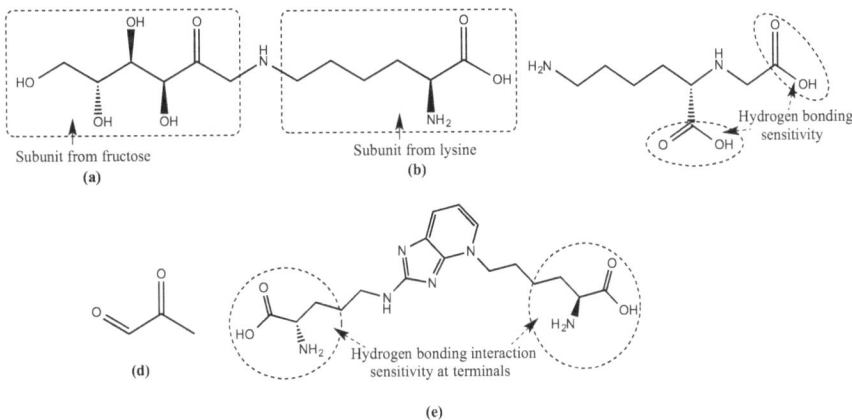

Fig. (6). Chemical structures of (**a**) 1-deoxy-1-fructosyl residue, (**b**) *N*-carboxymethyllysine (CML), (**c**)pentosidine and (**d**)methylglyoxal (MGO). The hydrogen bonding mediated proximity of each AGEs provides a clue for their qualitative detection and isolation.

Apart from pentosidine, the other fluorescent, cross-linked AGEs include pentodilysine, crossline, AGE-XI, vesper lysine A, and vesper lysine C (Fig. **7A**). Several other cross-linkers with a non-fluorescence sensitivity have been reported, having imidazolium dilysine cross-links, more familiar as glyoxal-lysine dimer (GOLD) or methylglyoxal-lysine dimer cross-links (MOLD) [108] (Fig. **8**). Both GOLD and MOLD are obtained from the reaction between two lysine side chains and two molecules of glyoxal (GO) and methylglyoxal (MGO), respectively.

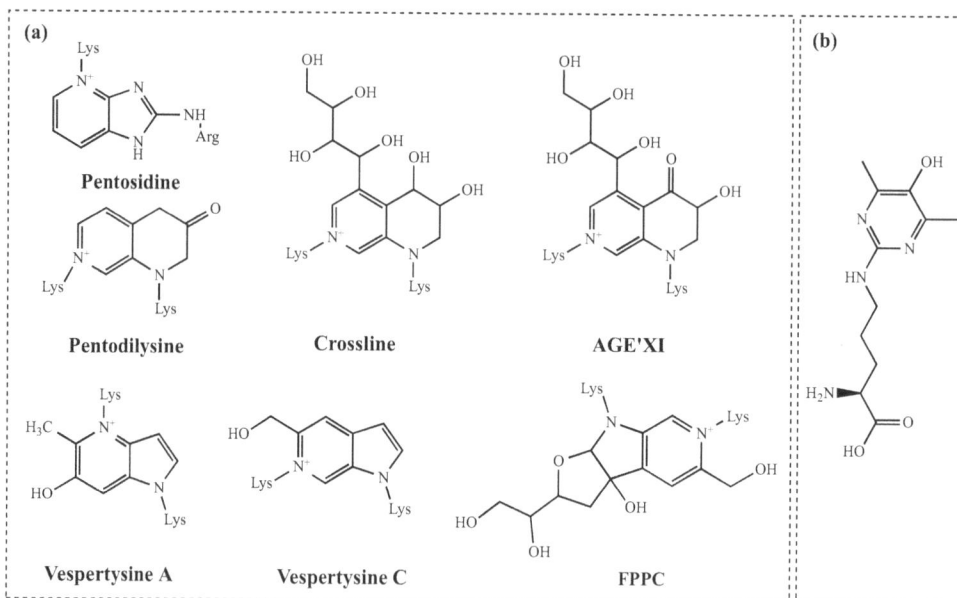

Fig. (7). Structural distinctions of (**a**) fluorescent cross-linked AGE, and (b)non-cross-linked AGE.

MEASUREMENT OF ADVANCED GLYCATION END PRODUCTS

The relative complexity of glycation with a significant diversity of chemical structures and associated physical properties often complicates a quantitative estimate of AGEs. To date, there is no standardized method based on which AGE characterization or screening from two or more laboratories could be compared on a one-to-one basis. Therefore, arriving at some conclusive observations based on the relative toxicity of diverse AGE categories is highly ambiguous. Table **3** summarizes the various methods for the characterization of AGEs along with their respective advantages and constraints. Methods for quantitative screening from biological sources are distinguished as instrumental or immunochemical types.

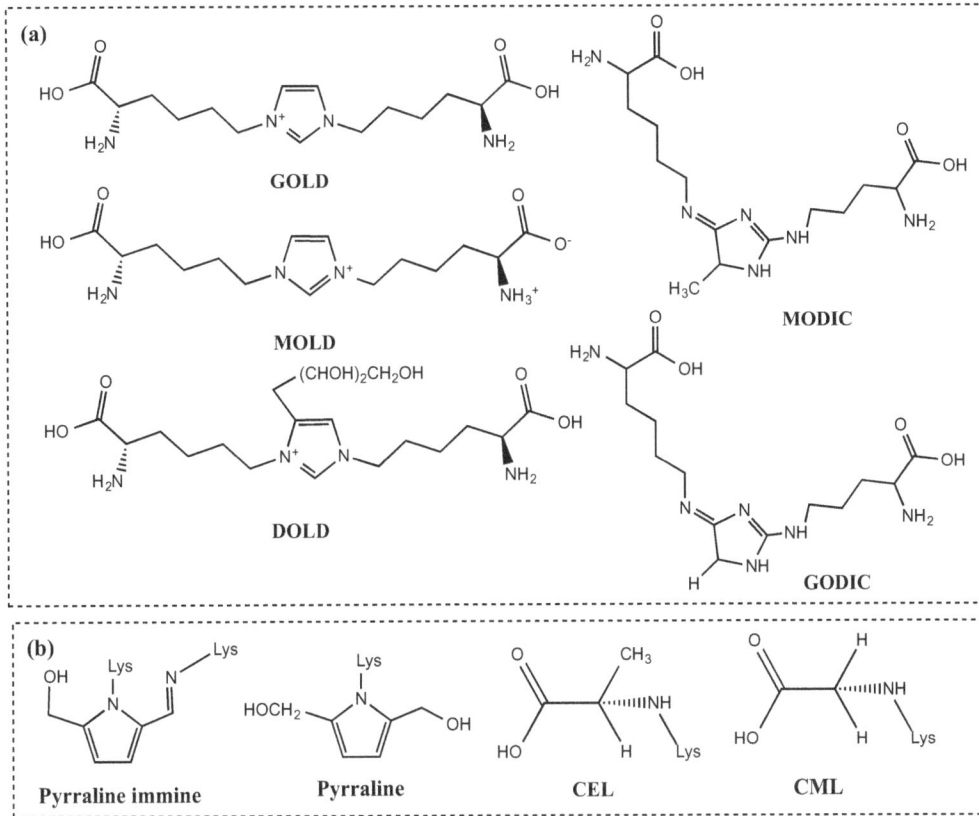

Fig. (8). Structural distinctions of non-fluorescent, (**a**) cross-linked (GOLD, MOLD, DOLD, MODIC, GODIC) and (**b**) non-crosslinked(CEL, CML, pyrraline, pyrralineimmine) AGEs.

The screening of AGEs from serum, urine, and saliva could be done using spectroscopic and fluorometric probes. Nevertheless, both these methods provide only a quantitative estimate of fluorescent AGEs. Screening of fluorescent AGEs from physiological sources is relatively modest. To be precise, samples are diluted (50-fold for serum, 10-fold for saliva, and 10 to 200-fold for urine) with phosphate-buffered saline (PBS) at 7.4 pH. The discrete fluorescence of probed AGEs is presented in arbitrary units.

A relatively robust, non-invasive method to detect AGEs from the skin could be optimized using autofluorescence spectroscopy, optimized *via* interaction of UV A radiation at 370 nm (low intensity) with the dermal AGEs. Photoemission from the skin is typically assessed at 440 nm through a moveable spectrophotometer.

Table 3. Methods for assessing the formation of AGEs in humans, along with their advantages and disadvantages.

Method name	Marker	Compartment	Advantages	Cautions
Fluorometric method	Fluorescent AGEs (pentosidine)	Serum, urine, saliva	Easy and prompt method	Cannot detect non-fluorescent AGEs, interference from the non-AGE fluorophore, results expressed in arbitrary units.
Auto-fluorescence spectroscopy	Pentosidine and remaining fluorescent AGEs	Skin	Non-invasive, easy to understand with prompt methodology. Feasible for clinical and epidemiological studies, correlates well with HPLC-ascertained AGEs.	Major fluorescence contribution is from fluorescent AGEs, which detect lower AGE in dark skin despite being worked out with similar conditions.
High Pressure Liquid Chromatography	AGEs, pentosidine, CML, CEL, MG	Plasma, tissues	Amicable for screening AGE	Costly equipment, time-consuming processing, applicable only to structurally known AGEs.
Gas Chromatography coupled with Mass Spectrometry (GC-MS)	CML, CEL and others	Urine	High sensitivity, providing lid and accurate results	Expensive equipment, elaborate skilled manpower.
Liquid Chromatography coupled with Mass Spectrometry (LC-MS)	Non-volatile compounds (CML, CEL and MG)	Plasma, urine	No derivatization is needed, and high sensitivity provides valid and accurate results.	Sophisticated method with expensive infrastructure, needs skilled manpower.
UHPLC	Plasma, tissues	Plasma, tissue	Prompt assay with adequate resolution	Expensive equipment needs trained manpower.
Enzyme-Linked Immunosorbent Assay (ELISA)	Serum, urine, tissues	Serum, urine, tissues	Simple, fast, and economical method, does not need any sophisticated laboratory equipment.	Inadequate antibody specificity, perturbation from glycation-free adducts.
Western Blotting	Any tissue	Any tissue	Highly specific and accurate method	Complex procedure with expensive conditioning.

- Several investigations have demonstrated the feasibility of AGE reader from dermal locations and its association with the net AGE count estimated *via* HPLC, although the process mandates the need for invasive skin sample collection. The conducted investigations in this regard, have illustrated that dermal AGE contents of distinct patients and healthy individuals (screened *via* AGE reader) are considerably related to the explicit extents of fluorescent (such as pentosidine) as well as non-fluorescent (CML, CEL, *etc.*) AGEs screened *via* dermal biopsies. Although an exclusive contribution in this context is retrieved *via* fluorescent AGEs, skin autofluorescence (SAF) has been examined as a reliable estimator of long-term diabetic complications (ranging from 5 to 10 years) instead of prompt glycemic memory manifested *via* hemoglobin A1c (HbA1c, typically within 3-6 months). The foremost cause of such distinctions is that SAF infers a comprehensive reminiscence of increasing metabolic stress than HbA1c or remaining normal risks (smoking, blood pressure, *etc.*). The SAF has also been noticed to be linked with progressive renal and cardiovascular criticalities, emerging as a prominent AGE marker [109, 110].

- Erstwhile studies illustrate an impact of varied light excitation or emission *vis--vis* darker skin complexion, exhibiting lower values than the individuals with fair skin color. Of late, a relatively new scheme *via* comparative assessment of the findings was validated to monitor the dermal AGEs irrespective of skin complexion [111]. The SAF overall, is a non-invasive, prompt, exact, and reliable AGEs evaluation tool for dermal AGEs, having an association with oxidative stress whereby aggravated impediments like diabetes, stroke, neuropathy, retinopathy, and nephropathy are manifested [112]. The method is user-friendly and applicable to epidemiological studies in fit individuals' subjects and those exhibiting vulnerability to disease development, despite some constraints discussed in the following paragraphs.

- Even though SAF performs a key role in ascertaining metabolic health, the autofluorescence determination procedures exhibit certain constraints that must be improvised to manifold their feasibility in clinical settings. Determination of SAFs not merely screens the quantitative dermal AGEs content since the prevalence of endogenous fluorescent responses from cutaneous fluorophores (such as nicotinamide adenine dinucleotide) exhibits similar energizing and emanation ranges (typically within 350-410 and 420-620 nm), likely to perturb the optimum and net fluorescence estimation [112].

- Currently, SAF is highly preferred for research purposes while its anticipation for future purposes pertains to the daily clinics needing thorough optimization. The prime advantages comprise a positive scenario for monitoring subsequent patients to minimize the occurrence of diabetes and chronic disorders wherever oxidative and inflammatory stress is associated. To date, however, SAF screens the manifold technical aspects of the constraints although several simultaneous attempts are being optimized to better its implementation. A new imaging system with a transmission geometry engaged in ascertaining SAF aims at improving future diagnostic performance although the subject analysis is optimized on a limited Korean population [112]. Emergent Hope projects SAF as a non-invasive probe that promises a relatively cheap and robust utility with prompt feasibility.

- Expensive and sophisticated laboratory methods like mass spectroscopy and gas/liquid chromatography enable remarkable precision and accuracy for quantitative screening of certain AGEs but the working complications do necessitate the involvement of qualified personnel along with high maintenance costs. These limitations peg back an arrest on the widespread usability of spectroscopic and chromatographic assays. For instance, liquid chromatography conjugated with mass spectrometry exhibits a greater sensitivity than HPLC-coupled fluorescent screening [113]. Similarly, the analysis of non-volatile compounds is done using liquid chromatography coupled with mass spectrometry (LC-MS), the distinction that it no longer requires a derivatization step, and the tandem MS provision allows for an increased sensitivity [114]. Likewise, GC-MS is in wide application for CML and CEL analysis in the urine and protein hydrolysates and has been used for quantitative estimation of CML in varied food samples [115]. A recent innovation in this technique has involved ultra-high pressure whereby smaller particles (< 1.7 μm) confer a prompt screening with a higher resolution.

- Conventionally, AGE biomarkers comprising CML, MG, and pentosidine have been enthusiastically screened *via* LC-MS and competitive ELISA, involving an AGE explicit monoclonal antibody to screen the AGEs extent in distinct human tissues, bio-samples, and foods. The methodology herein is comparatively modest and affordable with no specific requirement of sophisticated laboratory equipment. However, pegging restrictions attribute to lacking the required antibody specificity as per the commercial kit being employed *vis-à-Vis* interference with glycation-free adducts [116, 117]. Herewith, the AGE assessment using an immunoassay does not quantify these in absolute extents but does in random units, not necessarily with normalization to a fixed AGE glycated protein standard [118]. Here also, the handling of biological samples is challenging albeit food samples are easily manageable.

- Over the past few years, certain promising approaches for generating high throughput extent of monoclonal antibodies (mAbs) mapped with AGE-directed epitopes have come to the fore. For instance, emerging efforts have enabled the generation of novel and more efficacious mAbs directed toward CML [119, 120]. The mAbs are equipped with the ability to screen the singular glycated epitopes but effectual and exact entityprovisions are far too restricted, currently. The feasible mAbscould not elucidate the epitopes, implying an urgency to design the specific MAbs with a well-defined binding regime for better and enhanced analytical capability.

- Immunological methods encompass multiple advantages in the identification of specific compounds that could accomplish prompt attainment of results with enhanced detection sensitivity and an easy-to-use application interface [121]. Nevertheless, progress in cultivation methods alongside the making of specific monoclonal epitopes remains one of the important goals that has been accomplished. At present, no single approach is available for AGE quantification and detection. A major reason for this restriction is no litigated recognized standard measurement unit for reporting the AGE formation extents, unlike other measurable ones. The AGEs are typically accompanied by structural complexity and molecular heterogeneity, making it highly challenging to compare the outcomes for their quantification from two or more distinct experimental settings.

- ELISA has been extensively used for AGE detection from serum, biological sources, or food matrices. The methodology of ELISA uses monoclonal or polyclonal antibodies and is perhaps the primed approach for antigen-antibody quantification. Below-per-antibody specificity, high background responses (with a substantial content of protein glycation adduct), and perturbation with glycation-free adducts (caused *via* multiple pre-treatments such as heating and alkaline treatment) are some of the pertinent ELISA limitations [82, 122]. Emerging improvements have manifested enhanced faith in ELISA, making it substantial for ascertaining food, plasma, and urinal CMLextents [123, 124]. A recent study demonstrated the relationship between estimated CML extents using ELISA and HPLC-MS methods, involving ELISA use in screening CML from foods [124].

- A trusted method to quantify AGEs, RAGEs, and sRAGE is *via* Western Blotting, wherein samples are initially separated using SDS-PAGE (10% gel). Subsequently, the gel is electro-transferred to a polyvinylidene difluoride(PVDF) membrane. The non-specific protein binding sites are hereby blocked using TBST (Tris-buffered saline having Tween 20), with 5-10% powdered non-fat dry milk as the blocking agent. The immobilized membrane is incubated at room temperature for 1 h with rabbit anti-human AGE polyclonal antibody, goat anti-rabbit IgG, or polyclonal goat Ab against human RAGE. The visualized bands are screened using an enhanced chemiluminescence screening assay which is then rendered to X-ray film [125].

- A standardized familiar method to identify and quantify the specific proteins, such as AGE-modified entities, is Western blotting, befitted for biological specimens. Fluorescent-conjugated primary enzymes or antibodies having an affinity for binding to the specific antigens are generally used in this method. Despite Western blotting being recognized as a standard protein analysis technique. With a high sensitivity, Western blotting can detect up to 0.1 ng protein in a sample, making it a reliable diagnostic tool. Besides, Western blotting attributes its specificity to the gel-electrophoresis mediated arrangement of proteins based on size, charge, and conformation, providing important hints on the relative dimensions of protein or polypeptide. The second aspect of Western blotting making it an exclusive approach is the high specific antigen-antibody interaction mediated antibody detection. Despite the ability to screen a target protein from a mixture comprising manifold distinct proteins (high precision and diligence), Western blotting exhibits certain shortfalls. The method is prone to incorrect results, wherein false positives could occur in the experiment using an antibody (particularly polyclonal antibody)cross-reacting with some epitopes of a non-specific protein. Apart from this, a false negative could occur if high molecular weight proteins are improperly or even not transferred at all to the PVDF/ nitrocellulose membrane. Owing to this, inconclusive or manifold bands could be generated, leading to erring results. Some other potential concerns of Western blotting include its high operational cost, the requirement of experts for accurate analysis, and specific laboratory devices. The method warrants accuracy at each step and a chance lapse in any of the operational steps could delineate the whole process. The detection and imaging probe *vis-à-vis* chemoiluminescence, fluorescence, radioactivity, and laser detection could be expensive [124, 126].

- The execution of Western blotting is undoubtedly distinct in the manner that it permits simultaneous detection of multiple targets contrary to a singular product in ELISA. Furthermore, it also makes the determination of target size feasible. Thereby, quantification of RAGE or AGE of interest could be accomplished *via* running a gel with a target protein (as a sample) adjacent to a known standard quantity. A major concern for Western blotting over ELISA is its higher time consumption and more sophisticated operational conditioning (protein isolation, buffers, type of separation, and gel concentration).
- Taking a comprehensive note of the above aspects, the suitable method of AGE characterization is highly affected by the availability of experienced personnel (AGE concentration, number, and type of samples). The conditioning concerning optimum screening involves the retrieval of reliable data for monitoring normal subjects, observing the outcomes of specific therapeutics, and the respective pathological continuation.
- As mentioned earlier, CML, pentosidine, and MG are the noted, well-characterized AGEs. Quantitative assessment of early glycation products can ascertain the exposure duration to glucose apart from the yester metabolic control. On the contrary, screening the intermediate and late glycation products remains a decisive step in validating the relationship between glycation products and the relative tissue modifications. To have a better idea, readers are suggested

to refer to the comprehensive review article by *Putta and associates*, wherein the different methods to determine AGE extents in different samples are illustrated [127]. The prominent among these are LC-MS, HPLC, spectrofluorometric, ELISA, and *via* 1D-SDS polyacrylamide gels.

- AGE reader is a device used to quantify AGEs at the tissue level using fluorescence sensitivities. This instrument is equipped with a light source that illuminates the tissue of interest and as a consequence the incident light excites fluorescent moieties in the tissues having accumulated AGEs, emitting light with a separate wavelength. The major advantages of an AGE reader include its user-friendly, prompt, and non-invasive measurement of tissue fluorescence.

Instrumental methods used for AGE characterization are as follows:

1. Spectrofluorometer

2. High-Performance Liquid Chromatography coupled with Mass Spectrometry (HPLC/MS)

3. Gas Chromatography coupled with Mass Spectrometry (GC-MS)

4. Liquid Chromatography coupled with Tandem Mass Spectrometry (LC-MS/MS)

5. HPLC with fluorescent detection

6. A method based on Ultra-High-Pressure Liquid Chromatography (UHPLC)

7. Immunochemical methods: Enzyme-linked immunosorbent Assay (ELISA) and Western Blotting (WB), employing AGE-specific antibodies.

Spectrofluorimetric Methods

To 100 µl samples add 10 µl of 100% (w/v) TCA in each tube. The supernatant containing

The standard protocol for spectrofluorimetric analysis involves adding 10 µl, 100% (w/v) trichloroacetic acid (TCA) to 100 µl samples in distinct centrifuge tubes. The supernatant containing sugar, test sample, and interfering agents were removed afteragitation and centrifugation (at 15000 rpm, 4°C). Thereafter, the AGEs-TCA precipitate was dissolved in 400 µl buffer (PBS) for analysis. The fluorescence intensity of glycated materials was measured at 370 nm excitation and 440 nm emission using a Varian spectrofluorometer, Cary Eclipse model.

Quantification of Advanced Glycation End Products from serum

The blood was centrifuged, and the serum was diluted with a 1:50 ratio *vis-à-vis* PBS at 7.4 pH. Initially, the AGE amount was measured using a spectrofluorometer and subsequently,the fluorescence intensity was measured at 350 nm excitation and 440 nm emission wavelengths, considering PBS as the reference standard.

Detection of Serum and Urine Small-sized Advanced Glycation End products Peptides using Flow Injection Assay

The screening of AGEs from serum and urine could be done *via* Flow Injection Assay, workable *via* the HPLC system. The samples are treated with TCA and then centrifuged. Henceforth, the aqueous layer was injected at 0.5 mL•min^{-1}flow rate into the flow system, driven *via* Merck-Hitachi L-6200 pump to the fluorescence detector. The 50mg•L^{-1}AGE-BSA on proteinase K hydrolysis is used as a standard [128].

Western-Blotting Mediated Detection of Advanced Glycation End products

AGE-modified proteins in 1D-SDS gels are screened using 3 μL test samples, wherein, 20-30 μg (total protein) was mixed with 1 μl, 4× sample buffer (0.125 M Tris-HCl, 2% SDS, 40% *v/v* glycerol, 0.8% bromophenol blue, pH 6.8). On 10 min incubation at ambient temperature (AT), samples are loaded onto a pre-cast Bio-Rad 4-12% Bis-Tris 1.0 mm mini gel. Electrophoresis is done at 100 V in a running buffer (25 mM Tris base, 0.1% SDS, 192 mM glycine, pH 8.3) till the front dye accesses the gel end. Thereafter, 30 min gel soaking in the equilibrating buffer (25 mM Tris base, 192 mM glycine, 20% methanol, pH 8.3) was performed and the proteins were electrolytically shifted to nitrocellulose (NC) membrane *via* Bio-Rad mini-gel transfer apparatus in transfer buffer (250 mM Tris base, 1.92 M glycine, 20% methanol, pH 8.3) at 100 V, 4°C for 1 h [129, 130].

The membrane was rinsed twice using MilliQ water, before 2 h inactivation using a protein-devoid blocking buffer at RT. On adequate 0.05% Tween-20 in Tris-buffer saline (TBS), TBST rinsing, the membrane was incubated with rabbit anti-human AGE polyclonal antibody, goat anti-rabbit IgG (diluted 1: 40,000 in protein-free blocking buffer), and substrate like the dot immuno-binding assay. Thereafter, bands are screened and examined as per the above procedure. The membranes were stained using Ponceau S, a protein staining dye, on WB to examine prominent proteins of test samples.

Advanced Glycation End Products-Modified Proteins in Dot-Immunobinding Assay

The test samples were diluted at 1:20 using PBS (pH 7.4). The 4 µl, diluted test samples and varied glycated BSA extents, are dotted on nitrocellulose membrane at 1 cm intervals and left for 1 h drying, at RT. Unreacted membrane-protein binding sites are inactivated *via* membrane immersing the membrane in a 3% BSA, Tris-buffer saline (TBS, 10 mM Tris base, and 150 mM NaCl, pH 7.5) followed by 2 h incubation at RT. Thereafter, triplicate washing with 0.05% TBST is done. Nitrocellulose membrane membranes are thereafter incubated with rabbit anti-human AGE polyclonal antibody, diluted as 1:1,000 in 0.5% BSA (BMBA) for 2 h at RT. On triplicate TBST washing, the membrane is incubated using a secondary antibody (goat anti-rabbit IgG peroxidase-labeled; Bio-Rad, Hercules, CA) with 1:10,000 dilution, BMBA for 1 h at RT.

Luminal/enhancement and peroxidase buffer solutions are added in 1:1 proportion to the membrane after yet another triplicate rinsing and 3-5 min incubation. The chemiluminescent patterns are screened *via* a Versa Doc Imaging System (Bio-Rad), quantity-one software (Bio-Rad). Standard curves are made using AGE-BSA plots (1,000, 500, 250, 125, 62.5, 31.2, and 15.6 ng•ml^{-1}). These curves are then used to compute the titers of AGE-altered proteins in test samples after which the results are divided by total protein amount to convert as AGE-altered proteins per mg total proteins.

THE BINDING PROTEINS OF ADVANCED GLYCATION END PRODUCTS

- In 1992, *Schmidt and colleagues* initially demonstrated a cell-surface receptor exhibiting a high affinity for AGE binding. This receptor was subsequently termed a "receptor for AGE" or RAGE.
- Subsequent studies identified multiple cell surface molecules capable of AGE binding, including macrophage scavenger receptors (MSRs) type A and B1 (CD36), oligosaccharyl transferase-48/OST48, also known as "AGE receptor 1" (AGE-R1), 80K-H-phosphoprotein (AGE-R2) and galectin-3 (AGE-R3), the scavenger receptor family (SR-A, SR-B, SR-1, SR-E, LOX-1, FEEL-1, FEEL-2, and CD36).
- Among these molecules, RAGE is perhaps the best characterized and well-studied. Chapter 5 of this book is dedicated to the various molecular aspects of RAGE.
- RAGE is a trans-membrane receptor comprising 394 amino acids (AA), a single hydrophobic, 19 amino acids domain, and 43 amino acids C-terminal cytosolictail. The receptor could bind distinct AGE-adducts with high ability,

together with CML and pentosidine.

- As of now, various erstwhile AGE-receptors are recognized, comprising AGE-receptor complex (AGE-R1/OST-48, AGE-R2/80K-H, AGE-R3/galectin-3). Certain members of the scavenger receptor family are SR-A; SR-B: CD36, SR-BI, SR-E: LOX-1, FEEL-1; FEEL-2, the expression of which, varies as per the cell/tissue kind and is maintained *via* metabolic alterations caused by aging, diabetes and hyperlipidemia [130].

Receptor For Advanced Glycation End Product

- Prevalent as a type I-AGE cell surface receptor, RAGE fits in the immunoglobulin (Ig) superfamily and is illustrated as a pattern recognition entity. As demonstrated in the molecular cloning investigations, inceptive recognition of RAGE was as a polypeptide with 35 kDa molecular weight, in bovine lung-endothelial cells. The human RAGE protein sequence comprises 404 amino acids, with 42.8 kDa as molecular weight.
- Western blotting of 293 RAGE transfected cellsrevealed~50 kDa molecular mass major bands arising from post-translational modifications. RAGE is expressed in multiple cells (monocytes/macrophages, T-lymphocytes, endothelial cells, dendritic cells, fibroblasts, smooth muscle cells, neuronal cells, glia cells, chondrocytes, keratinocytes) and identifies multiple, distinct ligands (AGEs, amyloid β peptide, S100/calgranulin protein, HMGB1, LPS).
- Ligand binding is succeeded by the activation of multiple signaling pathways. The full-length RAGE (fl-RAGE) protein comprises a bulky extracellular domain, a single transmembrane anchored helix, and a short cytoplasmic domain. This fl-RAGE intracellular domain has been demonstrated as pertinent for functionally impaired signaling and NF-κB stimulation. Its extracellular domain harbors a singular, *N*-terminal V-type (variable) and two *C-type*(constant) provinces. A typical ligand-binding site exists on the V-type region whose secondary structure contains 2 β-sheets, one short α-helix, and a random coil with reasonable similarity to immunoglobulin V-type regions.
- Opposed to the C2-type region, the V- and C1 domains function as structural sites (VC1) without an independent prevalence. Apart from this, the VC1 region is conjugated to the C2 province *via* a stretchy linker. Attributed to alternative splicing, 20 distinct RAGE variations are known and two prominent fl-RAGE mRNA, are characterized. Of these, the *N*-truncated RAGE lacks the *N*-terminal V-type area which is present in the extracellular matrix (ECM)and prevails in the cytoplasmic membrane. The erstwhile variant includes the soluble form with no *C*-terminal domain, harboring the entire full-length RAGE (fl-RAGE) immunoglobulins.
- The soluble version of RAGE is generated outside the cell for which two distinct morphologies are known. The first one is the soluble RAGE, also termed

sRAGE or esRAGE, and is an outcome of alternative RAGE mRNA splicing. The fl-RAGE goes through proteolytic cleavage *via* metalloproteinases, ADAM10 and MMP9, releasing sRAGE. It has been documented that sRAGE as a decoy receptor, forbids the ligand-RAGE interaction and ligand-other cell surface receptors proximity. In coronary artery disease, lower sRAGE plasma extents deciphered aggravated AGE-RAGE interaction, enhanced ROS generation, and a subsequent inflammatory response. Subsequent sections discuss the AGE-RAGE interaction-mediated signaling events along with their physiological roles. Readers are advised to refer to the relevant literature sources for the above information on RAGE and chapter 5 of this book describing various aspects of RAGE [11, 131 - 133].

The Advanced Glycation End Products-Receptor (AGE-R) Complex

The adequate information specific to this section could be traced in the 2015 review article by *Sharma and colleagues* [134].

- The AGE-R complex comprises three constituents, typically associated with AGE signaling and endocytic uptake of altered proteins. In 1991, the initial AGE-R complex component was reported originally as cell surface receptor p60 in the context of AGE-modulated moiety on monocytes/macrophages. Henceforth, p60 was demonstrated, homologous to OST-48, as a 48 kDa fraction of oligosaccharyl-transferase complex.
- OST-48 generates a complex with ribophorin I and II, the integral membrane glycoproteins, and squarely with defender against apoptotic cell death (DAD1), housed in the rough endoplasmic reticulum (RER). Studies have demonstrated that the AGE-R1 (a homolog to OST-48), is a prominent plasma membrane protein having a single transmembrane segment and an extended extracellular domain, depicting the OST-48 N-terminal positioned in luminal RER. This recognized protein remains active during the surface binding of AGE-modified proteins and as an oligosaccharyl-transferase complex subunit, catalyzing the shift of mannose-rich oligosaccharides onto protein asparagine-acceptor sites within the RER.
- AGEs persuade an aggravated AGE-R1/OST-48 expression to suppress the RAGE expression and concomitant cellular oxidative stress besides inactivating the oxidative stress-dependent EGF-receptor signaling. In a noteworthy effort, *Vlassara and the group* reported a positive correlation for the AGE-R1/OST-48 extent in plasma PBMCs and urine AGEs, squarely for oxidative stress hallmarks in normal individuals. The association was inversely related to patients exhibiting stage 3 chronic kidney disease (CKD-3). Together, these findings maintain the belief that AGE-R1/OST-48 is guarding against ROS generation and concomitant oxidative stress-mediated troubles at cellular and

tissue platforms.

- Much like AGE-R1/OST-48, the erstwhile AGE-R complex constituent was originally termed as 80 K-H protein, a significant phosphorylation substrate for protein kinase C (PKC). This acidic 80 kDa protein was initially decontaminated from the human carcinoma Ca9-22 cell line, prevailing within cytosol and membrane, with no evident transmembrane province. In a few years, more experimental proofs emerged for the resemblance of 80 K-H protein with p90, the AGE-binding protein recognized as a definitive cell surface proximal protein on the murine macrophage RAW 264.7 cell line (membrane), exhibiting AGE 2-(2-full)-4(5)-(2-furanyl)-1H-imidazole.

- The physiological functions of AGE-R2/80K-H are still not fully understood. The phosphorylation of this protein has been demonstrated to occur on tyrosine residues, being implicated in intracellular receptor signaling, similar to the FGF receptor. The AGE-R2/80K-H protein is implicated in intracellular transport, *vis-à-vis* AGEs closely linked to VASAP-60, a vesicle trafficker. Besides, reports also suggest the 80K-H phosphoprotein interaction with PKC-ζ and munc18c, exhibiting a prominent role in the GLUT4 vesicle trafficking, whereby insulin-triggered PKC-ζ-80K-H-munc18c complex formation facilitates increased glucose transfer across GLUT4.

- The third constituent of the AGE-R complex belongs to the lectin family, being first demonstrated to exist on activated macrophage surfaces with a 32 kDa size and referral as Mac-2. This protein is distinctly reported in several species (human, dog, rat, and mouse tissues) in varied scenarios, resulting in multiple terminologies, *viz* Mac-2, CBP-35, CBP-30, and L-29. A consensus naming criterion for this protein is galectin-3 which in 1995 was reported to prevail over an AGE-R complex domain, exhibiting a significant AGE proximity. The AGE-R3/galectin-3 comprises of a carbohydrate-proximal *C*-terminal region, having 130 amino acids, with proline-glycine-alanine-tyrosine enrichment besides a smaller, *N*-terminal space (~30 residues).

- AGE binding also occurs at the *C*-terminal carbohydrate identifying province, generating a high molecular weight complex with erstwhile receptors and AGE ligands. Missing transmembrane region complements the prevalence of this protein on the cell surface occupied with remaining AGE receptors. The AGE-R3/galectin-3 protein primarily resides within the cytoplasm, although a small proportion exists within the nucleus, cell surface, and extracellular environment. The endogenous lectin is demonstrated as being extensively phosphorylated on the *N*-terminal Ser[6] residue wherein the phosphorylated version might be screened within the 3T3 fibroblasts (cytoplasm) and colon cancer cells (nucleus).

- The phosphorylation considerably modulates the ligand proximity, more than 85% of which has been examined using laminin and mucin, the prominent galectin-3 ligands. Dephosphorylation can entirely re-establish the ligand binding, deciphering Ser[6] phosphorylation as an "on/off" trigger for its sugar proximity, with a key role in understanding the AGE-R3/galectin-3 protein biological function. Nevertheless, it remains unknown whether extracellular AGE-R3/galectin-3 could be phosphorylated amidst consequent effects while extracellular engagement of AGE-altered proteins.

- AGE-R3/galectin-3 upregulation enhances the metastatic potential in cancer, with its phosphorylation being implicated in anti-apoptotic signaling. Moreover, the physiological function of AGE-R3/galectin-3 protein is well-established in cell migration and adhesion apart from immunological optimization, propagation, and demarcation. Besides, the AGE-R3/galectin-3 and RAGE involvement is demonstrated in vascular osteogenesis *via* altered Wnt/β-catenin signaling with varied outcomes on VSMC osteogenic segregation. This aids in increasing thoughtfulness regarding the atherosclerosis manifesting mechanism, demonstrating concomitant AGE involvement in the progression.

Macrophage Scavenger Receptor Family

- The scavenger receptor family includes a collection of receptors with an invariable proximity to distinguish the altered proteins such as oxidized low-density lipoprotein (OxLDL) or acetylated LDL (AcLDL). Ascertained in 1979 by Goldstein and Brown, the scavenger receptor exhibits a strong binding ability for the recognized AcLDL (not native LDL) on mouse peritoneal macrophages. Apart from AcLDL, erstwhile anionic ligands include maleyl-LDL, malelyl-BSA, and sulfated polysaccharides, all binding to this cationic receptor.

- A current bottleneck about the ABCLDL receptor has been its physiological relevance as its uptake was regulated *in vitro* but not *in vivo*. Only after a further 2 years, did it come to light that inherent human LDL incubated with rabbit aortic endothelial cells exhibited an alteration as a kind of LDL. This LDL was squarely identified by the AcLDL macrophage receptor. The formation of altered LDL entities can happen even without cells (*via* culture medium supplemented with redox metals, *viz* Cu or by free SH- functionality) along with smooth muscle cells, endothelial cells, and macrophages, elucidating the LDL oxidation *in vivo*.

- The definition of such scavenging receptor owes to the initially documented functions to facilitate the inflow and deprivation of altered proteins. At present according to their distinct structural properties, multiple scavenger receptors are known and classified.

- The scavenger receptor class A (SR-A) are trimeric transmembrane glycoproteins having 220 kDa (monomeric subunit: 77 kDa) as relative molecular weight, typically comprising five (SR-AII) or six (SR-AI) provinces. The *N*-terminal cytoplasmic sphere houses 50 amino acids, succeeded by a 25 aa transmembrane spanning domain; a 75 aa spacer region; a 121 aa α-helical coiled-coil motif, a 69 aa collagen-like triple helix, and a *C*-terminal domain. The *C*-terminal region of the SR-AI receptor harbors a cysteine-rich motif which is missing in the SR-AII receptor. The ligand binding happens at the collagenous province, leading to OxLDL or AcLDL endocytic uptake and subsequent intracellular buildup of cholesterol with concomitant generation of foam cells from macrophages during the early phase of atherosclerosis. The SR-A was illustrated as an AGE-proximal receptor, allowing endocytic inlet and AGE-altered protein damage compared to altered LDL. Since foam cells arise from the macrophages with a concomitant SR-A expression, the AGE proximity on these receptors could be decisive in atherosclerosis development.

- CD36 and SR-BI, the other receptors associated with scavenger receptor class B (SR-B), are enriched in caveolae-resembling spheres on the cell exterior. These glycoprotein receptors exhibit two transmembrane provinces, a large extracellular loop, and a short *N*- and *C*-terminal domain having substantial sequence homology in the extracellular region. Such a sequence resemblance for the two receptors is missing in *N*- and *C*-terminals. Further investigations squarely established CD36 and SR-BI proteins as being fatty-acylated.

- In the human CD36, four cysteine subunits have been recognized as the exhaustive palmitoylation sites, two each at *N*- and *C*-terminal regions. Research attempts on murine SR-BI indicate a prominent palmitoylation site at Cys^{462} and Cys^{470}, within the *C*-terminus. Investigations on CHO cells enriched human CD36, or hamster SR-BI demonstrated an explicit cell linkage to AGE-altered BSA, illustrating an endocytic AGEs uptake for CD36 and SR-BI functioning as AGE-receptors. Being reported in murine SR-BI-transfected LDLA cells (LDL receptor-negative CHO cells) SR-BI recognizes HDL, enabling a selective transfer from HDL to cells without HDL endocytic uptake. The binding of AGE ligands to SR-BI can inhibit HDL uptake, involving AGEs in controlling the SR-BI-driven cholesterol metabolism.

- Belonging to the family of SR-A receptors, CD36 exhaustively prevails on macrophages in atherosclerotic lesions, recognizing OxLDL as a ligand. A notable study by *Nakata and associates* screened CD36 within the core domain of an atherosclerotic plaque *via* immunohistochemical staining, while SR-A was exclusively localized in the vicinity. These distinctions illustrate distinct CD36 and SR-A roles toward atherogenesis manifestation. Of note, CD36 has been illustrated as AGE-receptor squarely on mouse 3T3-L1 cells and human subcutaneous adipocytes, identifying AGE-BSA complexes engineered *via* glycolaldehyde (GA) and MGO, anticipating endocytic dilapidation. Incubation of these cells with GA-BSAaggravated the intracellular oxidative stress, suppressing leptin actions *via in vivo* exacerbated insulin sensitivity, deciphering the CD36 involvement as an AGE-receptor in diabetes and atherogenesis.

- Additionally, some more constituents of the scavenger receptor family have also been demonstrated as AGEs engaged, wherein lectin-similar oxidized LDL receptor-1 (LOX-1) remains most understood. Recognized in endothelial cells, this receptor exhibits proximity towards oxidized LDL, having 50 kDa as molecular weight. LOX-1 is a type II-transmembrane glycoprotein, associated with group V C-based lectins. It prevails as a short *N*-terminal region, a singular transmembrane cytoplasmic region, having a Neck area with a succeeding *C*-terminal lectin-resembling domain (CTLD) inside extracellular space (Fig. **9**).

- Human LOX-1 CTLD harbors an inter-chain disulfide bridge at Cys^{140} to assist the homodimeric ligand-binding and the Neck regions as a homodimer, consisting of two α-helices which wrap in a left-handed configuration, forming a coiled-coil. LOX-1 is majorly restricted within vascular tissues, prevailing exhaustively in atherosclerosis or diabetes patients. Studies demonstrate NF-κB as a LOX-1-downstream signal, the expression of which is modulated by testosterone-modulated atherosclerotic plaque succession in rabbit thoracic aorta.

- Apart from oxidized LDL LOX-1 also exhibits proximity for MG or GA-altered BSA. Studies *via* cross-competitive screening infer an OxLDL or AGE-BSA binding is typically unaffected by erstwhile ligands. Matching outcomes were retrieved for HDL or AGE-BSA proximity on over-expressed SR-BI in CHO cells. A feasible explanation for such outcomes could be the prevalence of manifold ligand-sensitive loci in both LOX-1 and SR-BI. Furthermore, other notable findings of these studies also revealed that as opposed to receptor LOX-1 mediated endocytic inflow and OxLDL degradation (not of AGE-BSA), receptor SR-BI controls an endocytic inflow of AGE-modulated proteins, though not for HDL. Erstwhile investigations demonstrate the AGE-BSA interactions driven LOX-1 expression *in vitro* within the endothelial cells, not regulated by NF-κB stimulation but rather *via* RAGE-controlled mTOR pathway. Owing to mTOR's involvement in most cardiovascular disorders, the AGEs mediated LOX-1 expression could aggravate diabetic severity, through atherosclerosis.

- Fasciclin, EGF-like, laminin-similar, EGF-resembling, and link province comprising scavenger receptor-1 (FEEL-1) and FEEL-2 were cloned from HUVECs, being the first one as AcLDL receptor. Of note, FEEL-2 is a FEEL-1 (having 2570 amino acids) paralogous protein, exhibiting a 39.8% sequence homology. Both these receptors are type I transmembrane proteins, comprising seven fasciclins, sixteen EGF-like, two laminin-kind EGF-similar, and one link province in the transmembrane vicinity. First, FEEL-1/FEEL-2 were identified as stabilin-1 and stabilin-2, wherein FEEL-1 was separately recognized as an invariable lymphatic endothelial and vascular endothelial receptor-1 (CLEVER-1). Studies on FEEL-1 or FEEL-2 over-expressing CHO cells revealed an endocytic inflow and dilapidation of AGE-conjugated BSA. The AGE's proximity *vis-à-vis* FEEL-1/FEEL-2 compared to erstwhile ligands (AcLDL or OxLDL), deciphered a similar binding location on receptors. Subsequently, AGE-BSA endocytosis by choriocapillaris endothelial cells housed in Bruch's membrane *vis-à-vis* FEEL-1/FEEL-2 conduit was also demonstrated.

- Readers are suggested to refer to review articles by *Horiuchi and Taban* groups from 2003 and 2022 for the above information [135, 136].

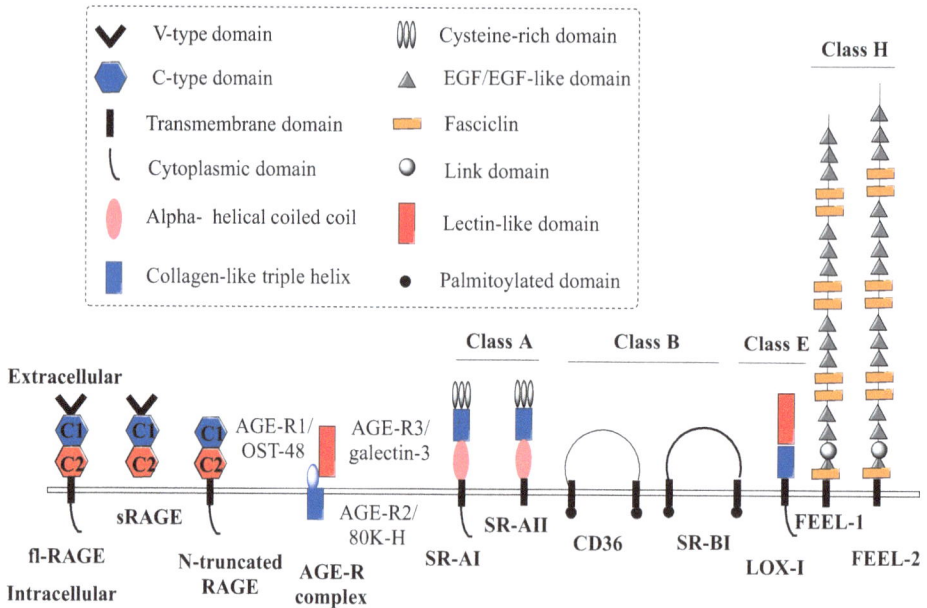

Fig. (9). Schematic representation of AGE receptors scavenging family. Major sub-categories of receptor family include class A, class B, class C, and erstwhile unclassified receptors. Select receptors can identify AGEs. The receptor complex (OST48; 80 K-H and galectin-3) and RAGE are squarely involved in AGE-binding though these are not the members of the scavenger receptor family.

THE EFFECTS OF ADVANCED GLYCATION END PRODUCTS ON CELL SIGNALING MOLECULES

- The activation of certain downstream signaling pathways by AGEs was realized for the first time following the cloning of the AGE receptor, making it evident that the possible effects are mediated *via* enhanced oxidative stress and chronic inflammation. These pathways unanimously exert their impact *via* the activation of transcription factor NF-κB to exhibit detrimental effects on localized tissues. The subsequent paragraphs discuss some prominent molecular aspects of AGE activities on signaling molecules.

- A notable aspect of AGE formation relates to a high expression of their plasma membrane receptor (reportedly, RAGE) in the immune cells, neurons, skeletal muscle cells, lungs, and heart. A consensus prevails regarding the AGE-RAGE interaction-dependent increased level of inflammatory and oxidative stress that is involved in aging and multiple pathological conditions such as atherosclerosis, arthritis, metabolic syndrome, neurodegenerative disorders, and cancer.

- The prevalence of pro-inflammatory and pro-oxidative transcription factor NF-κB binding sites on the RAGE promoter relates quite well for an encouraging feedback loop culminating in enhanced RAGE expression *via* extended NF-κB actions. Such a functional response of AGE-RAGE interaction considerably justifies a below-par RAGE expression in the adult lung tissues but aggravates in stressful conditions to manifest in troubled pathophysiology. Signal transduction *via* RAGE thereby corresponds to ligand-driven generation of RAGE oligomers, suggesting a local recruitment of manifold adaptor proteins. Topical studies demonstrate the involvement of AGE-mediated autophagy in rat vascular smooth muscle cells (VSMCs) *via* concurrent participation of extracellular signal-related kinase (ERK1/2) and Akt signaling pathways. Subsequently, the inhibition of AGE-driven apoptosis in human endothelial cells by *Ghrel* was also demonstrated *via* ERK1/2 and phosphoinositide 3-kinase/Akt pathways. Separate studies claimed AGEs aggravated migration and inflammatory actions in the adventitial fibroblasts *via* RAGE activities and concomitant MAPK, NF-κB signaling pathways. Besides, RAGE expression aggravated apoptosis in fibroblasts *via* stimulating ROS, MAPK, and the forkhead box O1 (FOXO1) transcription factor.

- Separate investigations reported the AGE-RAGE binding driven aggravated vascular endothelial growth factor (VEGF) expression within the retinal pericytes besides the mesangial TGF-β-Smad signaling *via* angiotensin II type I receptor interaction. More convincing elucidation was provided by the screening of four C57BL6 mice generations having been administered similar energy diets, with and without AGE supplementation. Analysis revealed enhanced pro-inflammatory phenotype after the AGEs inclusion in the diet, culminating with enhanced adiposity and pre-mature insulin resistance. These observations were documented by exogenously administered bovine serum albumin-AGE and lysozyme-AGE combination-driven altered gene expressions whereby exclusive gene modulations *vis-à-vis* combination stoichiometry deciphered an involvement of multiple receptor-mediated transmission pathways activated by the two combinations. An aspect still warranting attention is the little knowledge about the exact mechanism, but the closest suspicion pertains to the involvement of ROS induction following AGE binding with RAGE, aggravating the inflammation *via* multiple pathways (Fig. **10**). Related attempts to establish a role of mitochondrial electron transport chain and nicotinamide adenosine dinucleotide phosphate oxidase conduit in the RAGE aggravated oxidative stress. Thus, on a molecular level, multiple studies support the induction of inflammation and oxidative stress for a possible role of AGE-RAGE binding on the activities of signaling molecules.

- Other than RAGE, AGEs also bind to some other receptors like AGE advanced glycation receptor 1 (AGER1), which counteracts the signaling responses from RAGE activation to moderate the oxidative stress (Fig. **10**). A separate investigation on HEK293 cells revealed nearly threefold enhanced Akt phosphorylation after AGEs or CML exposure but the relative extent varied as per the redox balance of vicinity. Detailed scrutiny using p66^{shc} mutants revealed a necessary Ser-36 mediated p66^{shc} phosphorylation and the corresponding pro-oxidant outcomes were impaired in AGER1 over-expressing cells but expressed prominently in the impaired AGER1 expressing cells following small interfering RNA treatment. Henceforth, diverse receptors binding with AGEs comprise macrophages class A and B scavenger receptors, lectin-like oxidized low-density lipoprotein receptor-1 (LOX-1), megalin, TLRs, and additional scavenging receptors. Macrophage class A scavenger receptors endocytose solubilized AGE-engineered proteins along with LDL, together regulating macrophage-ECM adhesion *via* identifying collagen. Similarly, LOX-1 interacts with oxidized LDL and AGE-modified proteins.

- Contrary to macrophage class A scavenger and Lox-1 receptors, megalin is implicated in the re-absorption of inherent filtered proteins and AGE endocytosis-modulated proteins. Similarly, TLRs regulate the activation of innate immune defenses with concomitant activation being accomplished *via* distinct molecules exhibiting a similar 3-dimensional pattern. Besides AGER1, several other RAGE molecular forms can also suppress the AGE-RAGE signaling, these forms of RAGE are generated *via* erstwhile gene splicing and post-translational proteolysis of fl-RAGE. The soluble RAGE is perceived as a "decoy receptor" as these suppress the signaling activities of RAGE interaction, thereby guarding against inflammation and troubled pathophysiology [137].

THE EPIGENETIC EFFECTS OF ADVANCED GLYCATION END PRODUCTS

- Epigenetic processes typically represent the mechanisms that induce varied gene expressions and phenotypes owing to genetic changes not caused by DNA sequence alterations.

- The changes in epigenetic regime can be better understood *via* short- or long-term consequences. The former of these are characterized by a prompt response to environmental factor(s), typically transmitted to the filial generation. However, lasting consequences manifest a chronic alteration persisting as "memory," being typically inheritable, indicating the impact of a long-term, exhaustive environmental motivation. It is quite significant to note herewith that even a temporary environmental modification may instigate a long-lasting epigenetic effect.

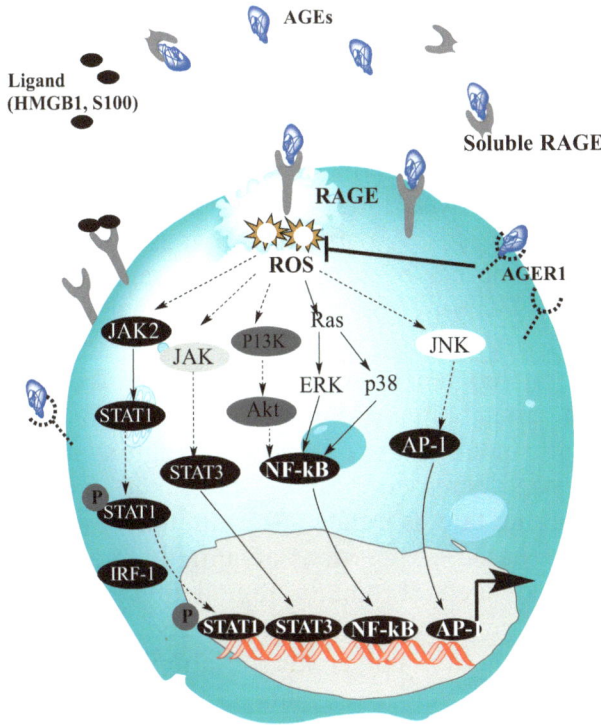

Fig. (10). Prominent signaling pathways induced in response to AGE-RAGE binding, HMGB1 and S100 being the other important RAGE binding ligands. The AGE-RAGE binding activates JAK2 which activates STAT1 phosphorylation which along with interferon regulatory factor-1 (IRF-1) binds to genomic regions of target genes. Inflammatory outcomes at the gene expression scale, are exclusively regulated *via* STAT3, NF-κB, and Activator Protein 1 (AP-1) transcription factors. Not all ligands aggravate oxidative stress and inflammation, like AGEs binding to AGER1 or soluble forms of RAGE from moderate oxidative stress, RAS signaling, and related inflammatory aggressions.

- Studies on epigenetic mechanisms demonstrate their induction *via* AGE signaling in coordination with RAGE. These responses could be moderated *via* ROS foragers like antioxidant foods, amino acids, and enzymatic antioxidant-rich compounds. Accompanying cytotoxicity with hyperglycemia can occur not only within dermal sites due to collagen-driven cross-linking with Maillard reaction products, leading to a toxic impact on cells.
- The epigenetic outcomes of AGEs are quite relevant for understanding their possible involvement in contributing factors leading to diabetes and several other constant disorders. Along with pro-inflammatory cytokine tumor necrosis factor (TNF), AGEs induce matrix metalloproteinase 9 (MMP-9) promoter demethylation though the intriguing mechanism is not yet properly understood.

DNA methylation of the MMP-9 promoter is engineered *via*GADD45a-driven suppressed propagation and formation of DNA injuring protein 45, a constituent of the miniature family of stress-responsive genes. Of these, GADD45 is associated with diabetes and diabetic cardiomyopathy, having been priorly demonstrated as stimulative *via* interacting with DNA methylation-causing proteins. The role of GADD45a in demethylating MMP-9 promoter is illustrated relatively freshly although not in completion. Studies reckon GADD45a expression is stimulative in the dermal regions of diabetic foot ulcer patients. Besides, this gene was aggravated in diabetic rats, and humans, *via* AGE action on keratinocyte (HaCaT) cells. A partial positive association between MMP-9 and GADD45a expression has been revealed, wherein GADD45a knockdown enhanced an MMP-9 transcription *via* suppressing MMP-9 promoter demethylation.

- Recent studies illustrate epigenetic mechanisms as the well-studied factors for diabetes-manifested metabolic memory. The pathological impacts of hyperglycemia are attributed to PKC stimulation, aggravated AGE generation, polyol pathway-driven increased glucose flux, and enhanced hexosamine conduit. Hyperglycemic fate is related to stimulated histone modifications characterized by increased expression of inflammatory genes. Curiously, investigations revealed such chromatin remodeling alterations and histone modulations in promoters, which could be caused even on normal glucose intake. This could manifest the epigenetic variations precisely correlated with metabolic memory. A noted effort on human podocytes herein demonstrated NAD-dependent deacetylase sirtuin 1 driven AGEs downregulated (SIRT1), provoking elevated acetylation of salient transcription factors such as STAT3, NF-κB-p65, and FOXO4.

- The STAT 3 acetylation reportedly activated the pro-inflammatory genes to initiate podocyte apoptosis and concomitant renal complications. Detailed investigations are required to recognize the epigenetically targeted genes associated with food-grade AGEs. Epigenetic mechanisms comprise the fundamentals of regulatory hyperglycemic memory which sustains a hyperglycemic state without the causative factors but is implicated with an earlier hyperglycemic condition. This pathological position is linked with oxidative stress and the AGE prevalence *via* regulating MAPK conduit. The generation of ROS in mitochondria has been related to the initiation of the hyperglycemic cycle of events like upregulated polyol pathways, protein kinase C (PKC) signaling, and aggravated AGE generation.

- Studies have correlated with NF-κB-p65 level whereby the permanent hallmarks include enhanced H3K4 expression and suppressed H3K9 methylations at the p65 gene promoter. The methylation regimens are the outcomes of enzymatic activity with contrasting impacts like histone methyltransferase, SET7, and histone demethylase, LSD1. Of note, SET7 silencing is demonstrated to be implicated in metabolic memory *via* DNMT1 methylation, causing its proteasomal degradation. Contrary to this, LSD1 likely bypasses DNMT1 degradation to get rid of methylation. Undeniably, metabolic memory should be driven *via* competitive SET7 and LSD epigenetic actions for DNMT1 loci methylation.

- A current group investigation focused on the correlation between explicit histone changes and glycemic memory *via* aggravated H3K9 acetylation at gene promoters implicated in interferon (IFN) controlling parameters, inflammation, apoptosis, and oxidative stress within monocytes. Moreover, DNA methylation is often a causative factor for diabetic vascular snags like UNC13B promoter hypermethylation in diabetic nephropathy sufferers. Thus, DNA methylation is critically linked with glycemic memory and diabetic retinopathy. For instance, the POLG1 promoter witnessed hyper-methylation linking it with glycemic memory.

- Inflammation is an archetypal pathological condition occurring as a pathogen attacking-tissue offense response. The progression is affected by multiple factors like aggravated Toll-like receptors (TLRs) expression and enhanced RAGE epigenetic changes.

- A familiar inflammatory response correlated with epigenetic outcomes is the viral meningoencephalitis-driven methyl-CpG protein 2 (MeCP2), an epigenetic issue required for DNA methylation. The early phase of this disorder involves pathogenic recognition of Purkinje cell receptors with self-critical neuronal events. Studies have demonstrated the protein RAGE as being prevalent in the adult brain and aging patients during later stages of viral meningoencephalitis, inferring inflammation modulated *via* manifold post-translational altered protein movement to the brain on activated RAGE binding.

- Inflammation could be altered *via* epigenetic modulations *vis.* DNA methyltransferase actions *vis-à-vis* methyl-CpG bound proteins (MeCP2), histone-altering enzymes, chromatin-remodeled proteins, and related complexes. Most closely related DNA methylation occurs at cytosine loci, affecting the gene expression and genomic imprinting associated with cancer and neurological impeding.

- The specific studies concerningthe above information could be traced in a 2020 review article by *Perrone and associates*, where the biochemistry, measurement methods, and epigenetic outcomes of AGEs are comprehensively discussed [24].

RECEPTOR TRAFFICKING AND INTRACELLULAR PROTEOLYSIS OF ADVANCED GLYCATION END PRODUCTS

- Irrespective of AGE association with exogenous or endogenous sources,these moieties must be eliminated to forbid long-term protein alterations including cross-linkages [137].

- Earlier for AGE removal, further processing is needed to make them amenable towards renal elimination. Thereby, the AGE degradation happens at the intracellular platform, implying their primitive intent of cellular uptake. Modification of AGE-modulated proteins into AGE-altered peptides aids in their renal elimination.

- As described earlier too, some AGE-receptors bind AGEs for henceforth intracellular modification. However, those implicated in getting rid of toxins are AGE-R1/OST-48, AGE-R3/galectin-3, and scavenger receptors (MSR-AII, MSR-BI, CD36). Strong evidence prevails concerning a high level of AGE proximity supporting basal conditions *vis-a-vis* AGE-R1/OST-48 and AGE-R3/galectin-3 actions.

- All the cells frequently exhibit most of their surface receptors to internalize signaling mediators for activating the explicit intracellular signaling flow on account of varied extracellular situations.

- The inflow of AGEs by RAGE and scavenger receptors is mediated *via* receptor-driven endocytosis.

- Binding of ligand stimulates the receptor *via* phosphorylation or ubiquitinoylation of endocytosis stimulating receptor-cytoplasmic region.

- Internalization of ligand-receptor complex onsets a regulated pattern, which could be clathrin-dependent or spontaneous. The former of these is perhaps the best studied, particularly the one facilitated *via* clathrin-modified vesicles. These vesicles are trafficked to early endosomes, being subjected to multiple sequential endosomal rearrangements. Within the recently developed endosomes, the receptor-detached AGEs moderate the endosomal pH.

- On being separated from the ligand (here, AGEs), the receptor could be reprocessed and moved reversibly to the cell surface.

- For the specific event of RAGE, studies have reported a prompt back transfer of receptors in the endothelial cells, from intracellular cargo space assemblies to the plasma membrane.

- Of course, the AGE-bound endosomes get fused with lysosomes and are thereafter, acted on by lysosomal proteases leading to AGE processing. In a noted attempt, *Grimm and associates* illustrated those lysosomal proteases, like cathepsins D, L, and B, collectively exhibit a decisive function in getting rid of AGEs.

- On being processed, small and soluble AGE-engineered peptides emanate from the cells *via* unknown mechanisms andare trafficked to the renal system through a mechanism that is still not entirely understood. Impaired renal function would result in AGE accumulation.

- Apart from the above, studies by *Sano and colleagues* showed that insulin may also result in AGE purging *via* increased uptake by macrophage scavenger receptors. Thereby, it is the activation of the phosphatidyl-inositol-3-OH kinase (PI3K) pathway that aggravates nitric oxide (NO) and insulin-driven glucose trafficking within the skeletal muscles and adipocytes.

• A separate study by *Sevillano and colleagues* demonstrated RAGE internalization after AGE binding, exhibiting a decisive role in regulating the intracellular responses.

IMPLICATION OF GLYCATION IN HUMAN PATHOPHYSIOLOGY

• The AGE products affect almost all mammalian systems. Despite the familiar glycation-glucose association, this reducing sugar itself is not very reactive. Rather, α, β-dicarbonyl species like MGO and GO, are as much as 50,000 times more reactive than glucose, believed to bemajor physiological events of glycation.

• As both MGO and GO are largely generated *via* intracellular events like triose phosphate fragmentation and lipid peroxidation, a higher probability exists for glycation of cellular proteins than those associated with plasma.

• Thereby, AGE extents are often greater in cellular proteins. For example, relative extents of MGO/arginine-retrieved hydroimidazolone, MG-H1 [Nδ-(-hydro-5-methyl-4-imidazolon-2-yl) ornithine], revealed 1.22 mmol.mol^{-1} and 0.92 mmol.mol^{-1} arginine in human blood cells and plasma proteins. Likewise, expression of another AGE, CML (N$^{\varepsilon}$-carboxymethyllysine), was three-fold greater in cellular proteins (0.068-0.233 mmol.mol^{-1} lysine) than plasma (0.021 mmol.mol^{-1} lysine).

• Disappointingly, certain AGEs are benevolent while remaining are highly reactive compared to original sugars. Consequently, these are implicated in several age-based chronic complications comprising, cardiovascular disease, Alzheimer's disease, peripheral neuropathy, and cancers.

Biological Consequences of Advanced Glycation End Products Generation in the Human Body

The sequestration of AGEs on biological macromolecules changes their structural-functional traits. Perhaps, a sluggish, progressive alteration of bio-macromolecules along with a subsequent build-up of engineered macromolecules *in vivo* , is suggested as being involved in multiple pathological complications, majorly those accompanying normal aging and chronic diabetes. The identical aspects in such complication's manifestation infer an aggravated aging similar consequence of chronic hyperglycemia.

The Intracellular Proteins

The *in vivo* occurrence of non-enzymatic glycation was initially deciphered *via* the existence of a less common hemoglobin species, A1c (HbA1c), in the erythrocytes. Thereafter, a higher frequency of HbA1c in blood from diabetic patients was screened. Additional biochemical investigations on HbA1c origin in diabetic sufferers revealed the existence of glucose-derived Amadori adduct, covalently conjugated to the *N*-terminal valine of hemoglobin β-chain. Clinical assessment of glycosylated hemoglobin in erythrocytes emerged as an authentic pointer of circulating average glucose extents, being habitually used to monitor diabetic arrest for 3-4 weeks, before sampling. As glucose prevails all through the body, the sensitivity of remaining proteins towards glucose-driven modifications was also anticipated. Lens proteins were believed to be the probable factors implicated in AGE build-up, owing to their slow turnover, if at all. Resembling hemoglobin, erythrocytes exhibit lens proteins in insulin-irrespective cells, exposing them to glucose extents analogous to extracellular sugar estimates.

Furthermore, several age-driven variations in lens proteins, like enhanced aggregation, altered assimilation peaks, and the manifestation of fluorescent chromophores deciphered their non-enzymatic glycation, particularly within the lens. In particular, gathering lens protein cross-links aggravates opacification, resulting in frequent observation of cataracts in aged humans. *in vitro* investigations assured that lens proteins may react with glucose or glucose---phosphate to form the protein cross-links, modifying the absorbance and fluorescence spectra like the noticed digests of cataractous lens proteins from aged, diabetic sufferers. The noticed invariabilities between non-enzymatic, glycated lens proteins (*in vitro*) and those from aged, diabetic individuals established for the first time non-enzymatic glycation with the manifestation of normal aging and chronic diabetes-related complications.

Extracellular Proteins

The impact of extracellular glucose on lens protein changes manifested an additional impetus to examine the non-enzymatic glycation *via* pathogenic routes comprising erstwhile long-lived proteins. During this analysis, extracellular proteins were assumed as vulnerable as proteins within normal insulin-possessing cells, the non-enzymatic glycation of extracellular proteins would be stanchly affected by inflowing glucose extents. Collagen is the principal extracellular protein in the physiological environment, prevailing in most connective tissues, like skin, tendons, and cartilage. Considering its profusion, long $t_{1/2}$, and availability, the constitutional makeup of collagen makes it vulnerable to non-enzymatic glycation. Enhanced stiffness and rigidity of this protein are usually linked to regular aging manifestations, correlating the suppressed flexibility at least to some extent with AGE-mediated cross-linkages.

The modified collagen extent *via* non-enzymatic glycation has been screened in several collagen-containing tissues such as the aorta, dura mater (a thick, tough fibrous membrane which covers and protects the brain and spinal cord) and skin. In all possibilities, the extent to which collagen exhibits the AGE modification varies directly with increasing age. Investigations on tissues from insulin-dependent diabetic patients revealed enhanced AGE-altered collagen due to chronological age, though the relative extents were analogous to healthy individuals, aged nearly twice the diabetic sufferers. Such findings re-assure a manifestation of some common aggravated aging consequences in persons with chronic diabetes history. The existence of AGEs on collagen not merely impacted its structural and constitutional whereabouts, but also mounted the risk of atherosclerosis, nephropathy, and peripheral vascular diseases. Studies demonstrate that besides collagen, AGE-modified collagen can cross-link with circulating serum proteins. Covalent conjugation and immobilization of flowing proteins like LDLs, serum albumin, and immunoglobulins to tissue collagen are likely correlated with manifested atherosclerotic lesions, renal tissues basement membrane thickening, and occlusion in the peripheral vasculature.

Nucleic Acids

Perception of non-enzymatic alteration by glucose of distinct biochemical configuration from biologically significant macromolecules complements the enveloping nature of this reaction, from a physiological viewpoint. Inceptive, the hypothesis of free -NH_2 groups on DNA bases as contributing to 638 glycationsin non-enzymatic, glucose reactions, gathered substantial reliance, in a similar regime to the protein-assisted Maillard reaction. The non-enzymatic DNA glycation could result in familiar age-driven genetic impairments, *vis.* suppressed RNA and protein formation, impaired DNA repair and replication, besides aggravated chromosomal abnormalities. Preliminary *in vitro* findings revealed-NH_2 nucleotide groups, either free or polymerized as single or double-stranded DNA, respond to glucose and glucose-6-phosphate. Such non-enzymatic, glycated DNA with unfamiliar absorbance and fluorescence spectra, in a time and sugar extent-dependent regime complemented the spectral variations accompanying non-enzymatic protein glycation. For instance, in the transformation of bacteria *via in vitro*, glycated plasmid DNA; glucose-altered DNA exhibited a considerably lower conversion than control plasmid DNA. This decreased conversion varied with incubation duration and functional glucose amount during the reaction. Plasmid DNA harvested from certain engineered colonies was observed to carry anomalous sequences. Such distinctions inferred the first-ever molecular belief that non-enzymatic DNA glycation could have unfavorable biological outcomes. The findings were almost similar in erstwhile cultured and physiological models, reassuring the mutagenic essence of DNA glycation. The outcomes of DNA glycation may alter genome integrity undyingly, culminating in varied extents of impaired cellular functions, even causing cell death in extreme cases.

Genetically engineered mice carrying Lac I as the particular mutagenesis marker gene, have been employed to ascertain the impact of non-enzymatic glycation, extrapolated to intact animal genomes. Following an assessment of accumulated Lac I mutants concerning AGE extents, the analysis revealed a linear increment of mutant frequency with time. The observed regime of DNA mutations did not remain limited to conventional base substitutions and comprised large deletions, and insertions, inferring a critical role of complex AGE-mediated DNA repair conduits. Owing to such observations for limited periods of experimental animals, there is a possibility of AGE-mediated DNA damage and mutations as the major genetic alterations noticed in elderly humans. It has been well demonstrated that the prevalence of birth disorders in the children of insulin-impaired diabetic mothers is considerably greater than in healthy counterparts. On a similar basis, the mutagenic effects of maternal hyperglycemia on lac I transgenic mice, and rising embryos could be screened. In fetuses that developed in diabetic DAM (dietary-AGEs fed mice), a two-fold increment in Lac I mutant frequency was observed contrary to thosein normal conditions. Molecular screening of such lac I mutants revealed a prototype of DNA alterations matching with the aged mice, suggesting an association between the DNA injury and henceforth, impaired revamp mechanisms for normal aging and diabetes-affected pregnancies. This investigation demonstrated first major potential association between maternal hyperglycemia, DNA injury, and innate malformations.

Likewise, DNA glycation could have multiple outcomes ranging from strand breaks, double helix unwinding, and induction of mutations *via* DNA-protein and inter-nucleotide cross-linkages. The relative alterations limit seems to vary as per the used glycating agent. For instance, MGO caused 10-fold higher DNA-protein cross-links than GO on *in vitro* incubation with *Escherichia coli-driven* DNA and DNA polymerase I. Such DNA polymerase cross-links could inhibit the imitation and further encourage frame shift mutations. The steric obstruction by DNA glycation adducts could affect the transcription *by* forbiddingtranscription factors binding. Thereby, DNA glycation is likely to influence the genome integrity by altered gene expression at transcriptional and translational stages [138].

Other than the direct biomolecule damage, AGEs, especially in extracellular regions, could aggravate the diseased pathology *via* proximity with cell surface receptors including RAGE. The ligand (AGEs)-receptor binding activates pro-inflammatory NF-κB conduit and downstream signaling molecules including p21, MAPKs, and JNKs [139].

Although multiple studies demonstrate that explicit mitochondrial glycation targets and mitochondrial impairment itself have been instrumental in aging-related diseases, the mechanisms of biomolecular glycation are still not well understood, exclusively within mitochondria *vis-à-vis* functional imparity and concomitant diseased pathology.

Other Pathological Outcomes

Multiple topical studies have deciphered the prominent role of AGE-altered proteins in numerous neuro-degenerative disorders, such as Alzheimer's disease (AD) and stroke-mediated neurological damage. Prevalence and gradual buildup of amyloid plaques within the brain is a prominent AD signature, wherein most of the work has been optimized *via* oxidative stress-mediated aggregation and buildup of β-amyloid (Aβ) plaques. Lately, amyloid plaques from abnormal and normal brains have been screened for AGEs positivity. Analysis revealed AGE positivity for samples from AD-affected brains, exhibiting nearly thrice as many AGEs per mg protein than the AGE-resembling controls. Furthermore, glycation of soluble Aβ peptides increased slightly for a greater insoluble fibrillar Aβ aggregation in test-tube incubations. Such distinctions convey a likely involvement of AGEs in Aβ plaque formation and deposition, culminating in enhanced vulnerability to AD-associated neuropathology (*e.g.*, Dementia). AGEs-driven worsening of an erstwhile neuropathological disorder has been reported in a rodent model. In a significant effort, administering physiological extents of AGE-altered proteins to healthy rats before the cerebral artery occlusion exposed a substantial volume of stroke damage and infarction than the non-treated ones. The findings infer that AGEs can capably instigate neurotoxic events, but whether AGE-induced damage is direct or driven is not known,

Pathological Effects of Advanced Glycation End Products

AGEs have several pathological effects, some prominent are as follows:

- Enhanced vascular porosity.
- Aggravated arterial rigidity.
- Inhibited vascular dilation *via* NO interference.
- Oxidizing LDL.
- Binding immune cells such as macrophages, endothelial, and mesangial, inducing the discharge of a range of cytokines.
- Aggravated oxidative stress.

In the experimental rat model, hemoglobin-AGE (HbA1c) extents are aggravated in diabetic patients while remaining AGE proteins have been demonstrated to accumulate with time, increasing from 5-50-fold over 5-20 weeks within the retina, lens, and renal cortex of diabetes-positive rats. Besides, an experimental reduced AGE activity moderated neuropathic vulnerability in diabetic rats. Thus, materials inhibiting AGE generation could arrest the disease succession, prevailing as novel therapeutic leads against AGE-mediated disorders.

AGEs exhibit multiple explicit cellular receptors; one of these well-demonstrated receptors is RAGE. Stimulation of cellular RAGE on endothelium, mononuclear phagocytes, and lymphocytes instigates free radical creation besides that of inflammatory gene regulators. These increments in oxidative stress stimulate the pro-inflammatory transcription factor NF-κB, enhancing the expression of concerned, suppressed genes implicated in atherosclerosis.

Effects of Glycation on Serological Activities, Visual Health, and Dermal Functioning

Although glycation excess affects all physiological activities to a substantial extent, the corresponding outcomes are most critically reported in blood circulation, visual health, and dermal functioning, most sensitively. The following paragraphs discuss some crucial impacts of glycation on serological activities, visual health, and dermal functioning.

(i) In the Blood

In convention, the tight junctions between the endothelial cells in the arterial wall forbid the entry of any foreign material between them. However, glycosylation weakens the intracellular junctions, making them leaky and vulnerable to tears. The body repairs these tears by plugging them with cholesterol, developing plaques in the arterial walls.

(ii) On the Eye Lens

On conjugation of glucose with proteins in the eye lenses, the lens cell morphology and appearance change from crystal clear to a little cloudy. Most of this cloudiness leads to vision impairment, commonly manifesting as "cataract formation". When glycosylation occurs in the minute blood vessels at the back of the eye, the lenses exhibit fragility and leakiness, resulting in bleeding and a consequent condition, termed "diabetic retinopathy", a leading cause of blindness.

(iii) On the Dermal Locations

After the collagen glycosylation, the elasticity of collagen in the skin decreases, making it stiffer. Conjugation of glucose with collagen in the connective tissues makes them inelastic and vulnerable to stress-induced damage. The smooth functioning of joints is mediated *via* proper collagen availability. A high blood sugar intensifies the aches and pains, leading to impaired joint movement and eventually arthritis.

Reactivityof Advanced Glycation End products

Proteins are usually glycated *via* their lysine residues. In humans, histones within thecellnucleus haveplentifullysine molecules, eventually form the glycated protein, N (6)-Carboxymethyl lysine (CML) [140]. It is notable to mention herewith that RAGEprevails on several cells, including those produced from the endothelial system, smooth muscles, the immune system, the diversified tissues such as the lungs, liver, and kidney. Amidst AGE binding, RAGE aggravates the age and diabetes-associated chronicinflammatory conditions ranging from atherosclerosis, asthma, arthritis, myocardial infarction, nephropathy, retinopathy, periodontitis, and neuropathy. Pathological manifestation of these conditions has been demonstrated to manifest *via* NF-κB activation, after AGE binding. The pathophysiology of these diseases is commonly controlled by multiple genes, implicated in hyper-inflammatory responses [141].

Clearance of Advanced Glycation End products

Clearance refers to the frequency at which any substance is eliminated (*via* natural processes) from the body. Studies have deciphered that AGE's cellular proteolysis lysed the proteins to generate the AGE peptides and concurrent free adducts (single amino acids bound AGE adducts). On being subsequently released in the plasma, these are eliminated through urine. Proteolytic resilience exhibited by ECM proteins resists AGEs as less spontaneous for getting eliminated. Though AGE-free adducts are spontaneously ejected *via* urine, AGE peptides are endocytosed *via* proximal tubule-epithelial cells and are eventually eliminated *via* the endolysosomal system, forming AGE amino acids. Some studies suspect a re-absorption of these acids *via glomerular* functioning (through the kidneys inside the region) leading to an eventual excretion. The AGE-devoid adducts prevail as prominent morphologies assisting the AGEs elimination *via* urine wherein AGE peptides prevail at a lower extent and accumulate within the plasma, in case of chronic kidney failure [142].

The macromolecular AGE proteins derived from extracellular locations are unable to move across the renal corpuscle basement membrane and should be essentially yester disgraced as AGE peptides and free adducts. Despite intensive efforts, the involvement of the hepatocellular system in this process remains ambiguous although peripheral macrophages and liver sinusoidal epithelial, Kupffer cells are indeedimplicated in this process. It is indeed established that modification of macromolecules by AGEs makes them vulnerable to elimination of *the* scavenger receptor of both liver endothelial and Kupffer cells [143].

Larger-sized AGE proteins cannot permeate the Bowman's capsule and are well-equipped for binding the endothelial and mesangial cells and mesangial matrix receptors. RAGE activation stimulates the generation of multiple cytokines including TNF-β that inhibit the activation of metalloproteinases and increase the generation of mesangial matrix, eventually resulting in glomerulosclerosis. As a result, patients with inadvertently high AGE generation exhibit impaired renal functioning. The only RAGE form that can easily excreted in the urinary process comprises AGE breakdown products, majorly peptides, and free adducts, being typically more aggressive than their parent proteins. These proteins continue to aggravate diabetes-similar pathology even after hyperglycemia has been resisted to a substantial extent by the body's physiology [144].

Certain AGEs exhibit a catalytic oxidative ability on being activated *vis-à-vis* NADPH oxidation *via* concomitant RAGE activation and concurrent damage to mitochondrial proteins, together causing impaired mitochondrial functioning with aggravated oxidative stress.A significant 2007 *in vitro* investigation in this regard reported substantially enhanced TGF-β1, CTGF, and Fn mRNA expressions in NRK-49F cells *via* AGEs on enhanced oxidative stress.The results deciphered an accurate explanation for the oxidative stress-inhibiting ability of *ginkgo biloba* (a herb rich in antioxidants) extract in diabetic patients. A consensus statement was that detrimental AGEs built up in the body could be minimized *via* antioxidant therapy. However, in extreme cases of impaired renal functioning preventing AGE elimination, kidney transplantation appears as the sole reliable option [145].

Studies establish beyond doubt that in diabetes sufferers having enhanced AGE production, kidney damage inhibits a regular and periodic urinary elimination of AGEs, leading to a positive feedback loop that aggravates the damage. A noteworthy effort from 1997 study, administered, diabetes affected and healthy candidates with a single meal of egg white (56 g protein), prepared with or without 100 g fructose. Inspection revealed an excess of 200-fold AGE immunoreactivity for the fructose-rich diet [146].

COMPARATIVE ANALYSIS OF GLYCOSYLATION VERSUS GLYCATION

Glycation and glycosylation are two mechanisms that add carbohydrates or sugars to proteins. Both processes involve the formation of carbohydrate-protein covalent bonds. Furthermore, both events also have a significant impact on native protein's functioning. While glycation refers to the binding of a sugar molecule bonding to a mature protein or lipid molecule sans enzymatic control, the glycosylation involves alteration of an organic molecule *via* regulated enzymatic modification, particularly a protein on conjugation with a sugar. Generally, protein glycosylation occurs as a part of post-translational modifications and is recognized as a normal physiological process of the cells.

Thus, the differences between glycation and glycosylation are summarized in the following paragraphs:

Materials Used in Glycosylation *Versus* Glycation

- Carbohydrates (sugars) and proteins are the major biomolecules used for both glycosylation and glycation. Besides proteins, other materials could also remain together with carbohydrates, such as lipids or even nucleic acids (DNA/RNA).
- However, the types of sugars added comprise a significant distinctive aspect of glycation and glycosylation with the former involving glucose, fructose, or galactose while the latter uses glycans, mannose, xylose, fucose, *etc*.
- For the use of proteins, glycosylation occurs as post-translational modification of proteins where protein synthesis and folding are being continued whereas glycation happens merely on mature proteins that havealready been translocated in various cellular compartments, ECMor even released in the circulation.

Functional Sites of Glycosylation *Versus* Glycation

- Glycation occurs inside and outside the cells including duringfood processing or high temperature cooking.
- Glycation is a non-enzymatic and spontaneous process that predominantly occurs in the bloodstream, where sugars and peptides/proteins prevail in abundance. The presence of more sugars in the bloodstream due to diabetes or any other pathological condition inevitably contributes to a higher extent of glycation. The diabetic status of a patient can be ascertained *via* monitoring the glycated hemoglobin, *i.e.* HbA1c. Proteins with a longer $t_{1/2}$ persist longer within the blood, exhibiting an extensive glycation probability.
- Evidence indicates glycation occurs in ECM or even at intracellular locations such as cytoplasm or in the nucleus, where nucleic acids are glycated.
- Thus, there are functional consequences of glycation, wherein basic

phospholipids may alter membrane structure and lipid per-oxidation. Similarly, *glycation* of nucleotides in DNA normally results in DNA repair induction that may induce apoptosis if the process is excessive.

- Contrary to glycation, glycosylation never occurs in extracellular spaces such as the bloodstream or ECM. Glycosylation is restricted to intracellular compartments and predominantly in the rough surface endoplasmic reticulum (RER) and Golgi apparatus.

Process of Glycosylation *Versus* Glycation

Enzymatic Versus Non-Enzymatic Reactions

- Glycation is a non-enzymatic process that majorly happens in the bloodstream. The typical event comprises a covalent conjugation of free sugars to the proteins. As the process occurs without enzymes, it can't be endogenously controlled and is spontaneously driven.
- Opposed to this, glycosylation is an entirely enzyme-assisted process that improves the functionality of translated proteins *via* post-translational modifications whereby a known carbohydrate molecule is conjugated to a pre-determined protein domain.
- Despite its intrinsic heterogeneity, protein glycosylation is a controlled response conferring optimal physiology to viable cells.
- Glycosylation is mainly a kind of post-translational modification, typically forming a functional protein from an immature counterpart. The process assists in needful protein-folding, and as a result, increases a protein's stability.
- On the other hand, glycation attenuates the functional stability of a protein, *via* decreased stability. The major concerns of glycated proteins pertain to their abnormal functioning in the blood, decreased optical resolution of eye lenses, altered skin permeability, and reduced coordination of connective tissues (discussed earlier in this chapter). Table **4** summarizes the prominent distinctions of glycation and glycosylation.

Effects of Glycosylation *Versus* Glycation on Various Human Systems

As also pointed out in section B of this chapter, glycation encompasses a chemical cascade that commences *via condensation-driven* formation of an unstable Shiff's base or aldimine that spontaneously rearranges to form a further stable ketamine, also recognized as the "Amadori product". These Amadori products are subsequently degraded *via* multiple processes, generating furfurals, reductions, and fragmentation products. Eventually, the polymorphic compounds formed because of glycation are collectively termed AGEs. While some AGEs prevail as benign, others are surprisingly livelier than their parent sugars and are thus, implicated in several age-driven chronic disorders, belonging to cardiovascular, neurodegenerative (Alzheimer's disease, peripheral neuropathy) regimes, and multiple cancers.

Table 4. Differences between glycation *versus* glycosylation.

Glycation	Glycosylation
Sugar bonding with a protein or a lipid entity without enzymatic assistance.	Controlled enzymatic alteration of an organic molecule, majorly *via* protein and sugar conjugation.
Covalent addition of a sugar molecule to a bloodstream protein.	A post-translational modification occurring in the endoplasmic reticulum and Golgi apparatus-lumen.
Not a standardized course.	It is a controllable event.
Glucose, fructose, and galactose are extensively added sugars.	Glycans, mannose, fucose, xylose, *etc.* are the mostly added sugar.
Occurs in mature proteins.	Occurs in immature as well as mature proteins.
Renders a protein non-functional.	Optimizes the protein functioning.
Decreases the protein's stability.	Increases a protein's stability.

Glycosylation is a significant biological event occurring in the endoplasmic reticulum and the Golgi apparatus. The distinctions of this process comprise the reaction of a sugar carbonyl group (glycosyl donor) with a protein's -OH or $-NH_2$ group (glycosyl accepter). Several types of glycosyloccur within the cell [146], including:

• *N-linked glycosylation*

Glycans are conjugated to the nitrogen of asparagine or arginine side chains.

• *O-linked glycosylation:*

Glycans are conjugated to the hydroxyl oxygen of serine, tyrosine, threonine, hydroxylysine, or hydroxyproline sidechains, or oxygens on lipids such as ceramide.

• *Phosphoserine glycosylation*

Phosphoglycans including mannose, xylose, or fucose could be the phosphate moiety of a phosphoserine.

• *C-manipulation*

Sugar is conjugated to carbon on a tryptophan sidechain.

• *Glypiation*

Addition of a GPI anchor, connecting the proteins to lipids *via glycan* linkages.

THE PHARMACOLOGICAL INTERVENTION OF ADVANCED GLYCATION END PRODUCTS: THE ANTI-ADVANCED GLYCATION END PRODUCT MOLECULES

To reduce the glycation-driven injury of reactive α, *and* β-dicarbonyls, multiple enzyme cascades work together to eliminate these species. The foremost of these defensive enzymes is the glyoxalase system which catalyzes MGO elimination along with some other compounds (in a minor proportion) such as glyoxal. Employing glutathione (GSH) as a cofactor, glyoxalase I forms *S*--hydroxyacylglutathione from MGO or GO. Glyoxalase II, thereafter, transforms this intermediate to an α-hydroxy acid (either D-lactate from MGO or glycolate from GO1), refurbishing GSH simultaneously. Remaining enzymes implicated in α, β-dicarbonyls detoxification include aldehyde dehydrogenases which oxidize MGO and GO to pyruvate and glyoxylate, respectively. Further, aldo-keto reductases/aldose reductases moderate the formed pyruvates and glyoxalates to alcohols (*e.g.*, MGO to acetol and lactaldehyde, GO to glycolaldehyde and ethylene glycol) [147].

Pharmacological Interventions of Advanced Glycation End Products

- Substantial interest prevails about the therapeutic agents inhibiting the AGEs formation, *via* lysing the AGE-mediated cross-links. In conventional physiological conditions, such anti-glycation guarding actions can efficiently arrest the glycation damage in coordination through revamp and elimination provisions. For instance, proteasomes and lysosomes thwart glycated protein build-up while nucleotide excision restoring action eliminates the glycated nucleotides, and lipid return aids in getting rid of glycated lipids. In the event of a disparity between the generation of glycation precursors (*viz.* α, β-dicarbonyls) and concomitant removal favouring their generation, carbonyl stress manifests and glycation damage builds up [135]. The development of anti-AGE-based therapeutic agents is discussed in subsequent paragraphs.
- Many synthetic and natural compounds are demonstrated as AGE inhibitors. Besides, a thorough understanding of AGE-forming chemical pathways has enabled the synthesis of potent and well-tuned mechanistically robust AGE-inhibitors. The first in this domain, is a small molecule, aminoguanidine (first introduced in 1996), which reacts with a post-Amadori derivative to generate subsequent intermediates but not AGEs, as demonstrated by test tube reactions and animal studies. Despite being in the initial phase of clinical assessment, pulsating studies on animals herewith demonstrate the aminoguanidine

therapeutic potential, forbidding the manifestation of AGE-related pathologies. Future anticipations rely on an improved understanding of AGEs forming underlying chemistry, leading to the genesis of added anti-AGE mechanisms based on therapeutic recourses [148].

- The *in vitro* instability of N-phenacyl thiazolium bromide resulted in its clinical failure. Another compound, alagebrium, was gradually screened as an AGE degrader, reversing the AGE build-up *in vivo*. Nevertheless, clinical investigations on such compounds are no longer being pursued and none of the identified AGE-destructing agents are being used clinically. Developing potent anti-AGE-based therapeutics currently is being worked out *via*,(i)forbidding the AGEs formation by destroying the cross-links after their formation, thereby moderating their risks. Compounds demonstrated for anti-AGE activities primarily comprise vitamin C, agmatine, benfotiamine, pyridoxamine, α-lipoic acid, taurine, pimagedine, aspirin, carnosine, metformin, pioglitazone, and pentoxifylline. Lipoic or podocarpic acid-mediated stimulation of TRPA-1 (Transient Receptor Potential Cation Channel, subfamily A) receptor, is reported as efficient towards moderating the AGEs levels *via* enhancing MGO detoxification, a noted antecedent of manifold AGEs [149]. Studies in rats and mice have demonstrated the potency of resveratrol and curcumin (prominent natural polyphenols) to subside the ill-fated outcomes of AGEs [150].

- Compared to synthetic drugs, herbal polyphenols (recommended much safer for human consumption) have been swiftly emerging as potent anti-RAGE/ anti-AGE molecules. Multiple herbal foodstuffs exhibit substantial anti-glycation traits which are either alike or sometimes greater than aminoguanidine. For instance, many polyphenols (majorly flavonoids) such as kaempferol, genistein, quercitrin, and quercetin are postulated as potent glycation inhibitors *in vitro*. Incentives for using polyphenols as antioxidant agents pertain to their edible nature, pleiotropic attributes, and robust explorations in their structure-function modulations.

- The recently demonstrated AGE degrading ability of epicatechin *in vitro* and *in vivo*, is of special interest. This compound damaged pre-formed, glycated serum albumin *in vitro*, thereby reducing the AGE build-up in AGE-injected, murine retinal tissues. The C6 and C8 on the A-ring of this compound are nucleophilic, thereby conferring the ability to attack and damage the AGE cross-links (Fig. **11e**) [151].

- Certain synthetic compounds have been studied for their AGEs destabilizing actions, *via* lysing the constitutional cross-links. Notable in this regard are Alagebrium (and related ALT-462, ALT-486, and ALT-946) and N-phenacylthiazolium bromide (Fig. **11B**). A significant study reported enhanced cell culture AGE degrading potency of rosmarinic acid over ALT-711 [152]. Disappointingly, no single agent has yet been developed to destroy glucosepane (the most familiar AGE, prevailing 10 to 1000 folds higher in humans than some other cross-linked AGE). Structural modifications have been authenticated in this regard). Structural modifications have been authenticated in this regard, whereby aminoguanidine arrested the AGEformation on reacting with 3-deoxyglucosone.

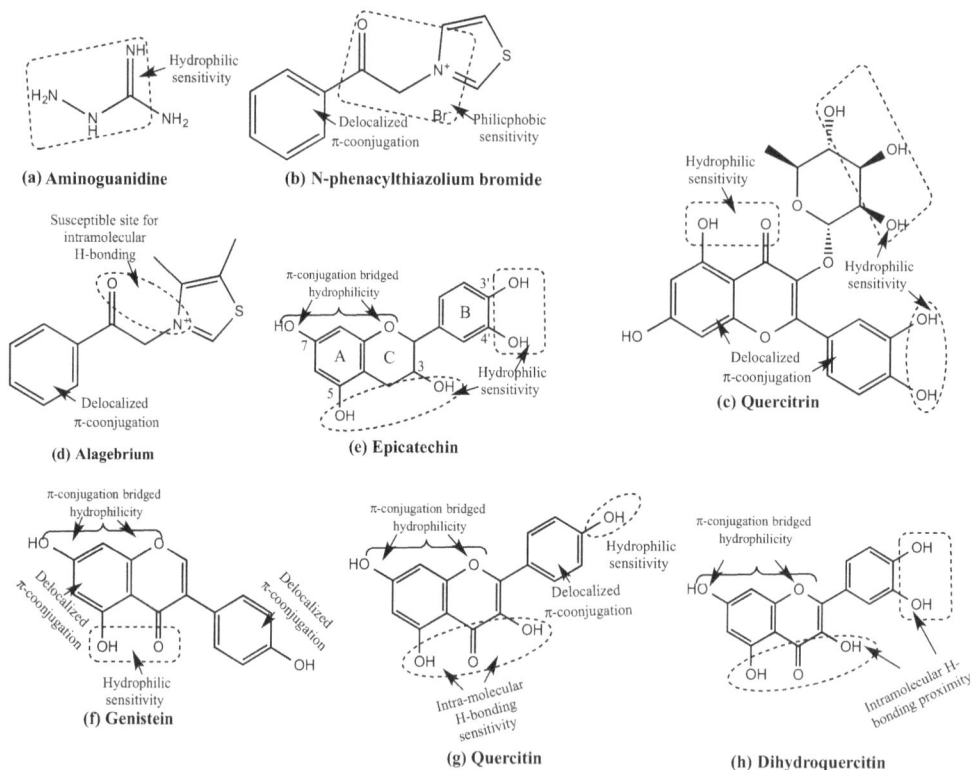

Fig. (11). Chemical structures of (**a**) aminoguanidine, (**b**) *N*-phenacylthiazolium bromide, (**c**)quercitrin, (**d**) alagebrium, (**e**) epicatechin, (**f**) genistein, (**g**) quercetin and (**h**) dihydroxyquercitin, depicting the interaction proximities *via* π-conjugation and phalicphobic sensitivities, The H⁺ releasing abilities make these polyphenols as stout antioxidant agents.

CONCLUSION

This chapter discussed various aspects of AGEs including its generation, biochemical structural variations, different ligands of AGE, pathophysiological consequences of increased level generation of AGEs, and finally the major biochemical distinctions of glycation and glycosylation *vis-à-vis* physiological sustenance and the corresponding sensitivities.

In contrast to glycosylation which is a physiologically enzymatic process of addition of carbohydrates to various proteins and lipids, glycation refers to the non-enzymatic conjugation of carbohydrates to the proteins, lipids, and other molecules, occurring *via* the Amadori reaction, Schiff base, and Maillard reactions; eventually forming AGEs.Pentosidine, a fluorescent AGE molecule is a cross-linking fragment that covalently conduits far-located lysine and arginine residues through a C5-sugar ring, forming intramolecular covalent bonds to link the different proteins. Compounds like malondialdehyde, hydroxynonenal (HNE)-lysine, or acrolein adducts obtained from oxidized hydroxy-amino acids, namely, L-serine or L-threonine, are related to the glycoxidated, terminally altered proteins. Of note, lipo-oxidative and glycol-oxidative pathways may congregate to form N^ε-(carboxymethyl)lysine (CML) as an end product contrary to pentosidine, exclusively generated *via* carbohydrate precursors. The adducts of certain AGEs, such as imidazolone and pyrroline could also be formed autonomously *via* aggravated oxygen availability. AGE reacts with multiple molecules. However, the receptor for advanced glycation end products (RAGE) is the well-characterized AGE ligand. No physiological actions of AGE were discovered. Therefore, AGEs are described as unwarranted for the physiological systems. Additionally, AGE-RAGE interaction enhances the level of oxidative stress and inflammation leading to the complications of diseased conditions including cardiovascular complications, diabetes-related vascular complications, pulmonary diseases, and cancers.

Considerable research towards understanding the biochemical regulation of glycation and glycosylation has enhanced their understanding for better control of their excessive occurrence and relief from concomitant stressful conditions. The glycation and glycosylation control mechanisms are highly distinct due to their enzyme-independent and enzyme-dependent generation, respectively. Various anti-AGE molecules are currently in clinical trials to understand their effectiveness in clinics against pathophysiological and diseased conditions. Additionally, glycation can be controlled *via* anti-RAGE therapeutic administration, requiring a proper understanding of diversified ligand interactions with RAGE. Emerging interests in AGE-RAGE crosstalk have been demonstrated in diverse physiological complications wherein antioxidants and anti-inflammatory compounds have been observed as highly potent. Enzymatic defects of the generation of glycoproteins and glycolipids by glycosylation may lead to various disease conditions. Thus, aberrant glycosylation may be arrested by controlling the inebriated enzyme actions. Out-of-proportion generation of AGE demonstrates a physiologically stressed state wherein ongoing and future interest is focused on bettering the known mechanisms of generation of AGEs and the pathophysiological consequences of excessive generation of AGEs, and finally controlling the level of AGE in the physiological system.

REFERENCES

[1]   Maillard P, Seshadri S, Beiser A, *et al.* Effects of systolic blood pressure on white-matter integrity in young adults in the Framingham Heart Study: a cross-sectional study. Lancet Neurol 2012; 11(12): 1039-47.
        [http://dx.doi.org/10.1016/S1474-4422(12)70241-7] [PMID: 23122892]

[2]   Grossin N, Auger F, Niquet-Leridon C, *et al.* Dietary CML-enriched protein induces functional arterial aging in a RAGE-dependent manner in mice. Mol Nutr Food Res 2015; 59(5): 927-38.
        [http://dx.doi.org/10.1002/mnfr.201400643] [PMID: 25655894]

[3]   Stopper H, Schinzel R, Sebekova K, Heidland A. Genotoxicity of advanced glycation end products in mammalian cells. Cancer Lett 2003; 190(2): 151-6.
        [http://dx.doi.org/10.1016/S0304-3835(02)00626-2] [PMID: 12565169]

[4]   Thornalley PJ. The enzymatic defence against glycation in health, disease and therapeutics: a symposium to examine the concept. Biochem Soc Trans 2003; 31(6): 1341-2.
        [http://dx.doi.org/10.1042/bst0311341] [PMID: 14641059]

[5]   Ahmed N, Thornalley PJ. Quantitative screening of protein biomarkers of early glycation, advanced glycation, oxidation and nitrosation in cellular and extracellular proteins by tandem mass spectrometry multiple reaction monitoring. Biochem Soc Trans 2003; 31(6): 1417-22.
        [http://dx.doi.org/10.1042/bst0311417] [PMID: 14641078]

[6]   Suji G, Sivakami S. Glucose, glycation and aging. Biogerontology 2004; 5(6): 365-73.
        [http://dx.doi.org/10.1007/s10522-004-3189-0] [PMID: 15609100]

[7]   Brownlee M. Advanced protein glycosylation in diabetes and aging. Annu Rev Med 1995; 46: 223-34.
        [http://dx.doi.org/10.1146/annurev.med.46.1.223] [PMID: 7598459]

[8]   Baynes JW. Role of oxidative stress in development of complications in diabetes. Diabetes 1991; 40(4): 405-12.
        [http://dx.doi.org/10.2337/diab.40.4.405] [PMID: 2010041]

[9]   Yan SD, Yan SF, Chen X, *et al.* Non-enzymatically glycated tau in Alzheimer's disease induces neuronal oxidant stress resulting in cytokine gene expression and release of amyloid β-peptide. Nat Med 1995; 1(7): 693-9.
        [http://dx.doi.org/10.1038/nm0795-693] [PMID: 7585153]

[10]   Giacco F, Brownlee M. Oxidative stress and diabetic complications. Circ Res 2010; 107(9): 1058-70.
        [http://dx.doi.org/10.1161/CIRCRESAHA.110.223545] [PMID: 21030723]

[11]   Mukherjee TK, Mukhopadhyay S, Hoidal JR. The role of reactive oxygen species in TNFα-dependent expression of the receptor for advanced glycation end products in human umbilical vein endothelial cells. Biochim Biophys Acta Mol Cell Res 2005; 1744(2): 213-23.
        [http://dx.doi.org/10.1016/j.bbamcr.2005.03.007] [PMID: 15893388]

[12]   Mukhopadhyay S, Mukherjee TK. Bridging advanced glycation end product, receptor for advanced glycation end product and nitric oxide with hormonal replacement/estrogen therapy in healthy *versus* diabetic postmenopausal women: A perspective. Biochim Biophys Acta Mol Cell Res 2005; 1745(2): 145-55.
        [http://dx.doi.org/10.1016/j.bbamcr.2005.03.010] [PMID: 15890418]

[13]   Boulanger E, Wautier MP, Wautier JL, *et al.* AGEs bind to mesothelial cells *via* RAGE and stimulate VCAM-1 expression. Kidney Int 2002; 61(1): 148-56.
        [http://dx.doi.org/10.1046/j.1523-1755.2002.00115.x] [PMID: 11786095]

[14]   Roca F, Grossin N, Chassagne P, Puisieux F, Boulanger E. Glycation: The angiogenic paradox in aging and age-related disorders and diseases. Ageing Res Rev 2014; 15: 146-60.
        [http://dx.doi.org/10.1016/j.arr.2014.03.009] [PMID: 24742501]

[15]   Peng X, Ma J, Chen F, Wang M. Naturally occurring inhibitors against the formation of advanced glycation end-products. Food Funct 2011; 2(6): 289-301.

[http://dx.doi.org/10.1039/c1fo10034c] [PMID: 21779567]

[16]   Sadowska-Bartosz I, Bartosz G. Effect of glycation inhibitors on aging and age-related diseases. Mech Ageing Dev 2016; 160: 1-18.
         [http://dx.doi.org/10.1016/j.mad.2016.09.006] [PMID: 27671971]

[17]   Schmidt AM. RAGE and implications for the pathogenesis and treatment of cardiometabolic disorders spotlight on the macrophage. Arterioscler Thromb Vasc Biol 2017; 37(4): 613-21.
         [http://dx.doi.org/10.1161/ATVBAHA.117.307263] [PMID: 28183700]

[18]   Goldin A, Beckman JA, Schmidt AM, Creager MA. Advanced glycation end products: sparking the development of diabetic vascular injury. Circulation 2006; 114(6): 597-605.
         [http://dx.doi.org/10.1161/CIRCULATIONAHA.106.621854] [PMID: 16894049]

[19]   McPherson JD, Shilton BH, Walton DJ. Role of fructose in glycation and cross-linking of proteins. Biochemistry 1988; 27(6): 1901-7.
         [http://dx.doi.org/10.1021/bi00406a016] [PMID: 3132203]

[20]   Schmidt AM, Yan SD, Stern DM. The dark side of glucose. Nat Med 1995; 1(10): 1002-4.
         [http://dx.doi.org/10.1038/nm1095-1002] [PMID: 7489352]

[21]   Fleming TH, Humpert PM, Nawroth PP, Bierhaus A. Reactive metabolites and AGE/RAGE-mediated cellular dysfunction affect the aging process: a mini-review. Gerontology 2011; 57(5): 435-43.
         [http://dx.doi.org/10.1159/000322087] [PMID: 20962515]

[22]   Cerami C, Founds H, Nicholl I, et al. Tobacco smoke is a source of toxic reactive glycation products. Proc Natl Acad Sci USA 1997; 94(25): 13915-20.
         [http://dx.doi.org/10.1073/pnas.94.25.13915] [PMID: 9391127]

[23]   Leslie RDG, Beyan H, Sawtell P, Boehm BO, Spector TD, Snieder H. Level of an advanced glycated end product is genetically determined: a study of normal twins. Diabetes 2003; 52(9): 2441-4.
         [http://dx.doi.org/10.2337/diabetes.52.9.2441] [PMID: 12941787]

[24]   Perrone A, Giovino A, Benny J, Martinelli F. Advanced glycation end products (AGEs): Biochemistry, signaling, analytical methods, and epigenetic effects. Oxid Med Cell Longev 2020; 2020: 1-18.
         [http://dx.doi.org/10.1155/2020/3818196] [PMID: 32256950]

[25]   Uribarri J, Woodruff S, Goodman S, et al. Advanced glycation end products in foods and a practical guide to their reduction in the diet. J Am Diet Assoc 2010; 110(6): 911-916.e12.
         [http://dx.doi.org/10.1016/j.jada.2010.03.018] [PMID: 20497781]

[26]   Goldberg T, Cai W, Peppa M, et al. Advanced glycoxidation end products in commonly consumed foods. J Am Diet Assoc 2004; 104(8): 1287-91.
         [http://dx.doi.org/10.1016/j.jada.2004.05.214] [PMID: 15281050]

[27]   Nowotny K, Schröter D, Schreiner M, Grune T. Dietary advanced glycation end products and their relevance for human health. Ageing Res Rev 2018; 47: 55-66.
         [http://dx.doi.org/10.1016/j.arr.2018.06.005] [PMID: 29969676]

[28]   Gill V, Kumar V, Singh K, Kumar A, Kim JJ. Advanced glycation end products (AGEs) may be a striking link between modern diet and health. Biomolecules 2019; 9(12): 888.
         [http://dx.doi.org/10.3390/biom9120888] [PMID: 31861217]

[29]   Koschinsky T, He CJ, Mitsuhashi T, et al. Orally absorbed reactive glycation products (glycotoxins): An environmental risk factor in diabetic nephropathy. Proc Natl Acad Sci USA 1997; 94(12): 6474-9.
         [http://dx.doi.org/10.1073/pnas.94.12.6474] [PMID: 9177242]

[30]   Cai W, He JC, Zhu L, et al. Oral glycotoxins determine the effects of calorie restriction on oxidant stress, age-related diseases, and lifespan. Am J Pathol 2008; 173(2): 327-36.
         [http://dx.doi.org/10.2353/ajpath.2008.080152] [PMID: 18599606]

[31]   Lin RY, Choudhury RP, Cai W, et al. Dietary glycotoxins promote diabetic atherosclerosis in

apolipoprotein E-deficient mice. Atherosclerosis 2003; 168(2): 213-20.
[http://dx.doi.org/10.1016/S0021-9150(03)00050-9] [PMID: 12801603]

[32]　Uribarri J, Cai W, Peppa M, *et al.* Circulating glycotoxins and dietary advanced glycation endproducts: two links to inflammatory response, oxidative stress, and aging. J Gerontol A Biol Sci Med Sci 2007; 62(4): 427-33.
[http://dx.doi.org/10.1093/gerona/62.4.427] [PMID: 17452738]

[33]　Uribarri J, Peppa M, Cai W, *et al.* Restriction of dietary glycotoxins reduces excessive advanced glycation end products in renal failure patients. J Am Soc Nephrol 2003; 14(3): 728-31.
[http://dx.doi.org/10.1097/01.ASN.0000051593.41395.B9] [PMID: 12595509]

[34]　Vlassara H, Cai W, Goodman S, *et al.* Protection against loss of innate defenses in adulthood by low AGE intake: Role of a new anti-inflammatory AGE-receptor-1. J Clin Endocrinol Metab 2009; 94: 4483-91.
[http://dx.doi.org/10.1210/jc.2009-0089] [PMID: 19820033]

[35]　Uribarri J, Peppa M, Cai W, *et al.* Dietary glycotoxins correlate with circulating advanced glycation end product levels in renal failure patients. Am J Kidney Dis 2003; 42(3): 532-8.
[http://dx.doi.org/10.1016/S0272-6386(03)00779-0] [PMID: 12955681]

[36]　Vlassara H, Uribarri J. Glycoxidation and diabetic complications: modern lessons and a warning? Rev Endocr Metab Disord 2004; 5(3): 181-8.
[http://dx.doi.org/10.1023/B:REMD.0000032406.84813.f6] [PMID: 15211089]

[37]　Lichtenstein AH, Appel LJ, Brands M, *et al.* Diet and lifestyle recommendations revision 2006: a scientific statement from the American Heart Association Nutrition Committee. Circulation 2006; 114(1): 82-96.
[http://dx.doi.org/10.1161/CIRCULATIONAHA.106.176158] [PMID: 16785338]

[38]　Food, Nutrition, Physical Activity, and the Prevention of Cancer: a Global Perspective. Washington, DC: Am. Inst. for Cancer Res 2007.

[39]　Kawabata K, Yoshikawa H, Saruwatari K, *et al.* The presence of Nε-(Carboxymethyl) lysine in the human epidermis. Biochim Biophys Acta Proteins Proteomics 2011; 1814(10): 1246-52.
[http://dx.doi.org/10.1016/j.bbapap.2011.06.006] [PMID: 21708295]

[40]　Kueper T, Grune T, Prahl S, *et al.* Vimentin is the specific target in skin glycation. Structural prerequisites, functional consequences, and role in skin aging. J Biol Chem 2007; 282(32): 23427-36.
[http://dx.doi.org/10.1074/jbc.M701586200] [PMID: 17567584]

[41]　Dyer DG, Dunn JA, Thorpe SR, *et al.* Accumulation of Maillard reaction products in skin collagen in diabetes and aging. J Clin Invest 1993; 91(6): 2463-9.
[http://dx.doi.org/10.1172/JCI116481] [PMID: 8514858]

[42]　Hashmi F, Malone-Lee J, Hounsell E. Plantar skin in type II diabetes: an investigation of protein glycation and biomechanical properties of plantar epidermis. Eur J Dermatol 2006; 16(1): 23-32.
[PMID: 16436338]

[43]　Bastien P PH, Poumés-Ballihaut C, Zucchi H, Bastien P, Tancrede E, Asselineau D. Aged human skin is more susceptible than young skin to accumulate advanced glycoxidation products induced by sun exposure. J Aging Sci 2013; 1(3): 1000112.
[http://dx.doi.org/10.4172/2329-8847.1000112]

[44]　Crisan M, Taulescu M, Crisan D, *et al.* Expression of advanced glycation end-products on sun-exposed and non-exposed cutaneous sites during the ageing process in humans. PLoS One 2013; 8(10): e75003.
[http://dx.doi.org/10.1371/journal.pone.0075003] [PMID: 24116020]

[45]　Champ CE. Advanced glycation end products: the sweet tooth of radiation toxicity. 2014. Oncology 2014; 28: 1S.

[46]　Jeanmaire C, Danoux L, Pauly G. Glycation during human dermal intrinsic and actinic ageing: an *in*

vivo and *in vitro* model study. Br J Dermatol 2001; 145(1): 10-8.
[http://dx.doi.org/10.1046/j.1365-2133.2001.04275.x] [PMID: 11453901]

[47] Bucala R, Cerami A. Advanced glycosylation: chemistry, biology, and implications for diabetes and aging. Adv Pharmacol 1992; 23: 1-34.
[http://dx.doi.org/10.1016/S1054-3589(08)60961-8] [PMID: 1540533]

[48] Masaki H, Okano Y, Sakurai H. Generation of active oxygen species from advanced glycation end-products (AGEs) during ultraviolet light A (UVA) irradiation and a possible mechanism for cell damaging. Biochim Biophys Acta, Gen Subj 1999; 1428(1): 45-56.
[http://dx.doi.org/10.1016/S0304-4165(99)00056-2] [PMID: 10366759]

[49] Kasper M, Funk RHW. Age-related changes in cells and tissues due to advanced glycation end products (AGEs). Arch Gerontol Geriatr 2001; 32(3): 233-43.
[http://dx.doi.org/10.1016/S0167-4943(01)00103-0] [PMID: 11395169]

[50] Masaki H, Okano Y, Sakurai H. Generation of active oxygen species from advanced glycation end-products (AGE) under ultraviolet light A (UVA) irradiation. Biochem Biophys Res Commun 1997; 235(2): 306-10.
[http://dx.doi.org/10.1006/bbrc.1997.6780] [PMID: 9199187]

[51] Dickerson TJ, Janda KD. A previously undescribed chemical link between smoking and metabolic disease. Proc Natl Acad Sci USA 2002; 99(23): 15084-8.
[http://dx.doi.org/10.1073/pnas.222561699] [PMID: 12403823]

[52] Federico G, Gori M, Randazzo E, Vierucci F. Skin advanced glycation end-products evaluation in infants according to the type of feeding and mother's smoking habits. SAGE Open Med 2016; 4: 2050312116682126.
[http://dx.doi.org/10.1177/2050312116682126] [PMID: 28210490]

[53] Reynolds PR, Kasteler SD, Cosio MG, Sturrock A, Huecksteadt T, Hoidal JR. RAGE: developmental expression and positive feedback regulation by Egr-1 during cigarette smoke exposure in pulmonary epithelial cells. Am J Physiol Lung Cell Mol Physiol 2008; 294(6): L1094-101.
[http://dx.doi.org/10.1152/ajplung.00318.2007] [PMID: 18390831]

[54] Guerassimov A, Hoshino Y, Takubo Y, *et al.* The development of emphysema in cigarette smoke-exposed mice is strain dependent. Am J Respir Crit Care Med 2004; 170(9): 974-80.
[http://dx.doi.org/10.1164/rccm.200309-1270OC] [PMID: 15282203]

[55] Garay-Sevilla ME, Rojas A, Portero-Otin M, Uribarri J. Dietary AGEs as exogenous boosters of inflammation. Nutrients 2021; 13(8): 2802.
[http://dx.doi.org/10.3390/nu13082802] [PMID: 34444961]

[56] Snelson M, Coughlan M. Dietary Advanced Glycation End Products: Digestion, metabolism and modulation of gut microbial ecology. Nutrients 2019; 11(2): 215.
[http://dx.doi.org/10.3390/nu11020215] [PMID: 30678161]

[57] Cole JB, Florez JC. Genetics of diabetes mellitus and diabetes complications. Nat Rev Nephrol 2020; 16(7): 377-90.
[http://dx.doi.org/10.1038/s41581-020-0278-5] [PMID: 32398868]

[58] Peppa M, Uribarri J, Vlassara H. Glucose, Advanced Glycation End Products, and Diabetes Complications: What Is New and What Works. Clin Diabetes 2003; 21(4): 186-7.
[http://dx.doi.org/10.2337/diaclin.21.4.186]

[59] Chaudhuri J, Bains Y, Guha S, *et al.* The role of advanced glycation end products in aging and metabolic diseases: bridging association and causality. Cell Metab 2018; 28(3): 337-52.
[http://dx.doi.org/10.1016/j.cmet.2018.08.014] [PMID: 30184484]

[60] Monnier VM, Taniguchi N. Advanced glycation in diabetes, aging and age-related diseases: editorial and dedication. Glycoconj J 2016; 33(4): 483-6.
[http://dx.doi.org/10.1007/s10719-016-9704-0] [PMID: 27421860]

[61]   Baynes JW. The role of AGEs in aging: causation or correlation. Exp Gerontol 2001; 36(9): 1527-37.
[http://dx.doi.org/10.1016/S0531-5565(01)00138-3] [PMID: 11525875]

[62]   Thorpe SR, Baynes JW. Maillard reaction products in tissue proteins: New products and new perspectives. Amino Acids 2003; 25(3-4): 275-81.
[http://dx.doi.org/10.1007/s00726-003-0017-9] [PMID: 14661090]

[63]   Dyer DG, Blackledge JA, Katz BM, *et al.* The Maillard reaction *in vivo*. Z Ernährungswiss 1991; 30(1): 29-45.
[http://dx.doi.org/10.1007/BF01910730] [PMID: 1858426]

[64]   Semba RD, Bandinelli S, Sun K, Guralnik JM, Ferrucci L. Plasma carboxymethyl-lysine, an advanced glycation end product, and all-cause and cardiovascular disease mortality in older community-dwelling adults. J Am Geriatr Soc 2009; 57(10): 1874-80.
[http://dx.doi.org/10.1111/j.1532-5415.2009.02438.x] [PMID: 19682127]

[65]   Monnier VM, Sun W, Sell DR, Fan X, Nemet I, Genuth S. Glucosepane: a poorly understood advanced glycation end product of growing importance for diabetes and its complications. Clin Chem Lab Med 2014; 52(1): 21-32.
[http://dx.doi.org/10.1515/cclm-2013-0174] [PMID: 23787467]

[66]   Piperi C, Adamopoulos C, Dalagiorgou G, Diamanti-Kandarakis E, Papavassiliou AG. Crosstalk between advanced glycation and endoplasmic reticulum stress: emerging therapeutic targeting for metabolic diseases. J Clin Endocrinol Metab 2012; 97(7): 2231-42.
[http://dx.doi.org/10.1210/jc.2011-3408] [PMID: 22508704]

[67]   Hammes HP, Alt A, Niwa T, *et al.* Differential accumulation of advanced glycation end products in the course of diabetic retinopathy. Diabetologia 1999; 42(6): 728-36.
[http://dx.doi.org/10.1007/s001250051221] [PMID: 10382593]

[68]   Vashishth D. Advanced glycation end-products and bone fractures. IBMS boneKEy 2009; 6(8): 268-78.
[http://dx.doi.org/10.1138/20090390] [PMID: 27158323]

[69]   Cahn F, Burd J, Ignotz K, Mishra S. Measurement of lens autofluorescence can distinguish subjects with diabetes from those without. J Diabetes Sci Technol 2014; 8(1): 43-9.
[http://dx.doi.org/10.1177/1932296813516955] [PMID: 24876536]

[70]   Monnier VM, Sell DR, Genuth S. Glycation products as markers and predictors of the progression of diabetic complications. Ann N Y Acad Sci 2005; 1043(1): 567-81.
[http://dx.doi.org/10.1196/annals.1333.065] [PMID: 16037280]

[71]   Sell DR, Monnier VM. Structure elucidation of a senescence cross-link from human extracellular matrix. Implication of pentoses in the aging process. J Biol Chem 1989; 264(36): 21597-602.
[http://dx.doi.org/10.1016/S0021-9258(20)88225-8] [PMID: 2513322]

[72]   Yamagishi S. Role of advanced glycation end products (AGEs) in osteoporosis in diabetes. Curr Drug Targets 2011; 12(14): 2096-102.
[http://dx.doi.org/10.2174/138945011798829456] [PMID: 22023404]

[73]   Singh VP, Bali A, Singh N, Jaggi AS. Advanced glycation end products and diabetic complications. Korean J Physiol Pharmacol 2014; 18(1): 1-14.
[http://dx.doi.org/10.4196/kjpp.2014.18.1.1] [PMID: 24634591]

[74]   Nowotny K, Jung T, Grune T, Höhn A. Reprint of "Accumulation of modified proteins and aggregate formation in aging". Exp Gerontol 2014; 59: 3-12.
[http://dx.doi.org/10.1016/j.exger.2014.10.001] [PMID: 25308087]

[75]   Morcos M, Du X, Pfisterer F, *et al.* Glyoxalase-1 prevents mitochondrial protein modification and enhances lifespan in *Caenorhabditis elegans*. Aging Cell 2008; 7(2): 260-9.
[http://dx.doi.org/10.1111/j.1474-9726.2008.00371.x] [PMID: 18221415]

[76]   Chaudhuri J, Bose N, Gong J, *et al.* A Caenorhabditis elegans Model Elucidates a Conserved Role for TRPA1-Nrf Signaling in Reactive α-Dicarbonyl Detoxification. Curr Biol 2016; 26(22): 3014-25.
[http://dx.doi.org/10.1016/j.cub.2016.09.024] [PMID: 27773573]

[77]   Ravichandran M, Priebe S, Grigolon G, *et al.* Impairing L-Threonine catabolism promotes healthspan through methylglyoxal-mediated proteohormesis. Cell Metab 2018; 27(4): 914-925.e5.
[http://dx.doi.org/10.1016/j.cmet.2018.02.004] [PMID: 29551589]

[78]   Golegaonkar S, Tabrez SS, Pandit A, *et al.* Rifampicin reduces advanced glycation end products and activates DAF -16 to increase lifespan in *Caenorhabditis elegans*. Aging Cell 2015; 14(3): 463-73.
[http://dx.doi.org/10.1111/acel.12327] [PMID: 25720500]

[79]   Sell DR, Lane MA, Johnson WA, *et al.* Longevity and the genetic determination of collagen glycoxidation kinetics in mammalian senescence. Proc Natl Acad Sci USA 1996; 93(1): 485-90.
[http://dx.doi.org/10.1073/pnas.93.1.485] [PMID: 8552666]

[80]   Moldogazieva NT, Mokhosoev IM, Mel'nikova TI, Porozov YB, Terentiev AA. Oxidative stress and advanced lipoxidation and glycation end products (ALEs and AGEs) in aging and age-related diseases. Oxid Med Cell Longev 2019; 2019: 1-14.
[http://dx.doi.org/10.1155/2019/3085756] [PMID: 31485289]

[81]   Münch G, Thome J, Foley P, Schinzel R, Riederer P. Advanced glycation endproducts in ageing and Alzheimer's disease. Brain Res Brain Res Rev 1997; 23(1-2): 134-43.
[http://dx.doi.org/10.1016/S0165-0173(96)00016-1] [PMID: 9063589]

[82]   Perkins RK, Miranda ER, Karstoft K, Beisswenger PJ, Solomon TPJ, Haus JM. Experimental hyperglycemia alters circulating concentrations and renal clearance of oxidative and advanced glycation end products in healthy obese humans. Nutrients 2019; 11(3): 532.
[http://dx.doi.org/10.3390/nu11030532] [PMID: 30823632]

[83]   Li SY, Du M, Dolence EK, *et al.* Aging induces cardiac diastolic dysfunction, oxidative stress, accumulation of advanced glycation endproducts and protein modification. Aging Cell 2005; 4(2): 57-64.
[http://dx.doi.org/10.1111/j.1474-9728.2005.00146.x] [PMID: 15771609]

[84]   Zawada A, Machowiak A, Rychter AM, *et al.* Accumulation of advanced glycation end-products in the body and dietary habits. Nutrients 2022; 14(19): 3982.
[http://dx.doi.org/10.3390/nu14193982] [PMID: 36235635]

[85]   Gaens KHJ, Stehouwer CDA, Schalkwijk CG. Advanced glycation endproducts and its receptor for advanced glycation endproducts in obesity. Curr Opin Lipidol 2013; 24(1): 4-11.
[http://dx.doi.org/10.1097/MOL.0b013e32835aea13] [PMID: 23298958]

[86]   Brix JM, Höllerl F, Kopp H-P, Schernthaner GH, Schernthaner G. The soluble form of the receptor of advanced glycation endproducts increases after bariatric surgery in morbid obesity. Int J Obes 2012; 36(11): 1412-7.
[http://dx.doi.org/10.1038/ijo.2012.107] [PMID: 22828946]

[87]   Sánchez E, Baena-Fustegueras JA, de la Fuente MC, *et al.* Advanced glycation end-products in morbid obesity and after bariatric surgery: When glycemic memory starts to fail. Endocrinol Diabetes Nutr (Engl Ed) 2017; 64(1): 4-10.
[http://dx.doi.org/10.1016/j.endien.2017.02.002] [PMID: 28440769]

[88]   Deo P, Keogh J, Price N, Clifton P. Effects of weight loss on advanced glycation end products in subjects with and without diabetes: A preliminary report. Int J Environ Res Public Health 2017; 14(12): 1553.
[http://dx.doi.org/10.3390/ijerph14121553] [PMID: 29232895]

[89]   Asano M, Fujita Y, Ueda Y, *et al.* Renal proximal tubular metabolism of protein-linked pentosidine, an advanced glycation end product. Nephron J 2002; 91(4): 688-94.
[http://dx.doi.org/10.1159/000065032] [PMID: 12138274]

[90] Waanders F, Greven WL, Baynes JW, *et al.* Renal accumulation of pentosidine in non-diabetic proteinuria-induced renal damage in rats. Nephrol Dial Transplant 2005; 20(10): 2060-70.
[http://dx.doi.org/10.1093/ndt/gfh939] [PMID: 15956058]

[91] Wells-Knecht KJ, Brinkmann E, Wells-Knecht MC, *et al.* New biomarkers of Maillard reaction damage to proteins. Nephrol Dial Transplant 1996; 11 (Suppl. 5): 41-7.
[http://dx.doi.org/10.1093/ndt/11.supp5.41] [PMID: 9044306]

[92] Okado A, Kawasaki Y, Hasuike Y, *et al.* Induction of apoptotic cell death by methylglyoxal and 3-deoxyglucosone in macrophage-derived cell lines. Biochem Biophys Res Commun 1996; 225(1): 219-24.
[http://dx.doi.org/10.1006/bbrc.1996.1157] [PMID: 8769121]

[93] Kennedy L. Glycation of immunoglobulins and serum proteins.International text book of Diabetes mellitus. 2nd ed. Chichester: John Wiley 2010; pp. 985-1007.

[94] Miyata T, Ueda Y, Shinzato T, *et al.* Accumulation of albumin-linked and free-form pentosidine in the circulation of uremic patients with end-stage renal failure. J Am Soc Nephrol 1996; 7(8): 1198-206.
[http://dx.doi.org/10.1681/ASN.V781198] [PMID: 8866413]

[95] Shinohara M, Thornalley PJ, Giardino I, *et al.* Overexpression of glyoxalase-I in bovine endothelial cells inhibits intracellular advanced glycation endproduct formation and prevents hyperglycemia-induced increases in macromolecular endocytosis. J Clin Invest 1998; 101(5): 1142-7.
[http://dx.doi.org/10.1172/JCI119885] [PMID: 9486985]

[96] Brouwers O, Niessen PM, Ferreira I, *et al.* Overexpression of glyoxalase-I reduces hyperglycemia-induced levels of advanced glycation end products and oxidative stress in diabetic rats. J Biol Chem 2011; 286(2): 1374-80.
[http://dx.doi.org/10.1074/jbc.M110.144097] [PMID: 21056979]

[97] Kurz A, Rabbani N, Walter M, *et al.* Alpha-synuclein deficiency leads to increased glyoxalase I expression and glycation stress. Cell Mol Life Sci 2011; 68(4): 721-33.
[http://dx.doi.org/10.1007/s00018-010-0483-7] [PMID: 20711648]

[98] Xue M, Rabbani N, Momiji H, *et al.* Transcriptional control of glyoxalase 1 by Nrf2 provides a stress-responsive defence against dicarbonyl glycation. Biochem J 2012; 443(1): 213-22.
[http://dx.doi.org/10.1042/BJ20111648] [PMID: 22188542]

[99] Garrido D, Rubin T, Poidevin M, *et al.* Fatty acid synthase cooperates with glyoxalase 1 to protect against sugar toxicity. PLoS Genet 2015; 11(2): e1004995.
[http://dx.doi.org/10.1371/journal.pgen.1004995] [PMID: 25692475]

[100] Byun K, Yoo Y, Son M, *et al.* Advanced glycation end-products produced systemically and by macrophages: A common contributor to inflammation and degenerative diseases. Pharmacol Ther 2017; 177: 44-55.
[http://dx.doi.org/10.1016/j.pharmthera.2017.02.030] [PMID: 28223234]

[101] Wetzels S, Wouters K, Schalkwijk C, Vanmierlo T, Hendriks J. Methylglyoxal-derived advanced glycation end products in multiple sclerosis. Int J Mol Sci 2017; 18(2): 421.
[http://dx.doi.org/10.3390/ijms18020421] [PMID: 28212304]

[102] Van Nguyen C. Toxicity of the AGEs generated from the Maillard reaction: On the relationship of food-AGEs and biological-AGEs. Mol Nutr Food Res 2006; 50(12): 1140-9.
[http://dx.doi.org/10.1002/mnfr.200600144] [PMID: 17131455]

[103] Lund J, Ouwens D, Wettergreen M, Bakke S, Thoresen G, Aas V. Increased glycolysis and higher lactate production in hyperglycemic myotubes. Cells 2019; 8(9): 1101.
[http://dx.doi.org/10.3390/cells8091101] [PMID: 31540443]

[104] Khalid M, Petroianu G, Adem A. Advanced glycation end products and diabetes mellitus: Mechanisms and perspectives. Biomolecules 2022; 12(4): 542.
[http://dx.doi.org/10.3390/biom12040542] [PMID: 35454131]

[105]   Luevano-Contreras C, Chapman-Novakofski K. Dietary advanced glycation end products and aging. Nutrients 2010; 2(12): 1247-65.
[http://dx.doi.org/10.3390/nu2121247] [PMID: 22254007]

[106]   Poulsen MW, Hedegaard RV, Andersen JM, *et al.* Advanced glycation endproducts in food and their effects on health. Food Chem Toxicol 2013; 60: 10-37.
[http://dx.doi.org/10.1016/j.fct.2013.06.052] [PMID: 23867544]

[107]   van Deemter M, Ponsioen TL, Bank RA, *et al.* Pentosidine accumulates in the aging vitreous body: A gender effect. Exp Eye Res 2009; 88(6): 1043-50.
[http://dx.doi.org/10.1016/j.exer.2009.01.004] [PMID: 19450456]

[108]   Miller AG, Meade SJ, Gerrard JA. New insights into protein crosslinking *via* the Maillard reaction: structural requirements, the effect on enzyme function, and predicted efficacy of crosslinking inhibitors as anti-ageing therapeutics. Bioorg Med Chem 2003; 11(6): 843-52.
[http://dx.doi.org/10.1016/S0968-0896(02)00565-5] [PMID: 12614869]

[109]   de Vos LC, Noordzij MJ, Mulder DJ, *et al.* Skin autofluorescence as a measure of advanced glycation end products deposition is elevated in peripheral artery disease. Arterioscler Thromb Vasc Biol 2013; 33(1): 131-8.
[http://dx.doi.org/10.1161/ATVBAHA.112.300016] [PMID: 23139292]

[110]   Ying L, Shen Y, Zhang Y, *et al.* Advanced glycation end products *via* skin autofluorescence as potential marker of carotid atherosclerosis in patients with type 2 diabetes. Nutr Metab Cardiovasc Dis 2021; 31(12): 3449-56.
[http://dx.doi.org/10.1016/j.numecd.2021.09.005] [PMID: 34688535]

[111]   Adl Amini D, Moser M, Chiapparelli E, *et al.* A prospective analysis of skin and fingertip advanced glycation end-product devices in healthy volunteers. J Clin Med 2022; 11(16): 4709.
[http://dx.doi.org/10.3390/jcm11164709] [PMID: 36012948]

[112]   Da Moura Semedo C, Webb MB, Waller H, Khunti K, Davies M. Skin autofluorescence, a non-invasive marker of advanced glycation end products: clinical relevance and limitations. Postgrad Med J 2017; 93(1099): 289-94.
[http://dx.doi.org/10.1136/postgradmedj-2016-134579] [PMID: 28143896]

[113]   Hameedat F, Hawamdeh S, Alnabulsi S, Zayed A. High Performance Liquid Chromatography (HPLC) with fluorescence detection for quantification of steroids in clinical, pharmaceutical, and environmental samples: A review. Molecules 2022; 27(6): 1807.
[http://dx.doi.org/10.3390/molecules27061807] [PMID: 35335170]

[114]   Pitt JJ. Principles and applications of liquid chromatography-mass spectrometry in clinical biochemistry. Clin Biochem Rev 2009; 30(1): 19-34.
[PMID: 19224008]

[115]   Agalou S, Ahmed N, Babaei-Jadidi R, Dawnay A, Thornalley PJ. Profound mishandling of protein glycation degradation products in uremia and dialysis. J Am Soc Nephrol 2005; 16(5): 1471-85.
[http://dx.doi.org/10.1681/ASN.2004080635] [PMID: 15800128]

[116]   Nagai R, Shirakawa J, Ohno R, *et al.* Antibody-based detection of advanced glycation end-products: promises *vs.* limitations. Glycoconj J 2016; 33(4): 545-52.
[http://dx.doi.org/10.1007/s10719-016-9708-9] [PMID: 27421861]

[117]   Nagai R, Horiuchi S, Unno Y. Application of monoclonal antibody libraries for the measurement of glycation adducts. Biochem Soc Trans 2003; 31(6): 1438-40.
[http://dx.doi.org/10.1042/bst0311438] [PMID: 14641083]

[118]   Nagai R, Fujiwara Y, Mera K, Yamagata K, Sakashita N, Takeya M. Immunochemical detection of Nε-(carboxyethyl)lysine using a specific antibody. J Immunol Methods 2008; 332(1-2): 112-20.
[http://dx.doi.org/10.1016/j.jim.2007.12.020] [PMID: 18242632]

[119]   Wendel U, Persson N, Risinger C, *et al.* A novel monoclonal antibody targeting carboxymethyllysine,

an advanced glycation end product in atherosclerosis and pancreatic cancer. PLoS One 2018; 13(2): e0191872.
[http://dx.doi.org/10.1371/journal.pone.0191872] [PMID: 29420566]

[120] Finco AB, Machado-de-Ávila RA, Maciel R, *et al.* Generation and characterization of monoclonal antibody against Advanced Glycation End Products in chronic kidney disease. Biochem Biophys Rep 2016; 6: 142-8.
[http://dx.doi.org/10.1016/j.bbrep.2016.03.011] [PMID: 28955871]

[121] Yuan Y, Sun H, Sun Z. Advanced glycation end products (AGEs) increase renal lipid accumulation: a pathogenic factor of diabetic nephropathy (DN). Lipids Health Dis 2017; 16(1): 126.
[http://dx.doi.org/10.1186/s12944-017-0522-6] [PMID: 28659153]

[122] Thornalley PJ. Measurement of protein glycation, glycated peptides, and glycation free adducts. Perit Dial Int 2005; 25(6): 522-33.
[http://dx.doi.org/10.1177/089686080502500603] [PMID: 16419322]

[123] Gómez-Ojeda A, Jaramillo-Ortíz S, Wrobel K, *et al.* Comparative evaluation of three different ELISA assays and HPLC-ESI-ITMS/MS for the analysis of N $^{\varepsilon}$ -carboxymethyl lysine in food samples. Food Chem 2018; 243: 11-8.
[http://dx.doi.org/10.1016/j.foodchem.2017.09.098] [PMID: 29146316]

[124] Bass JJ, Wilkinson DJ, Rankin D, *et al.* An overview of technical considerations for Western blotting applications to physiological research. Scand J Med Sci Sports 2017; 27(1): 4-25.
[http://dx.doi.org/10.1111/sms.12702] [PMID: 27263489]

[125] Walter KR, Ford ME, Gregoski MJ, *et al.* Advanced glycation end products are elevated in estrogen receptor-positive breast cancer patients, alter response to therapy, and can be targeted by lifestyle intervention. Breast Cancer Res Treat 2019; 173(3): 559-71.
[http://dx.doi.org/10.1007/s10549-018-4992-7] [PMID: 30368741]

[126] Mishra M, Tiwari S, Gomes AV. Protein purification and analysis: next generation Western blotting techniques. Expert Rev Proteomics 2017; 14(11): 1037-53.
[http://dx.doi.org/10.1080/14789450.2017.1388167] [PMID: 28974114]

[127] Putta S, Kilari EK. A review on methods of estimation of advanced glycation end products. World J Pharm Res 2015; 4(9): 689-99.

[128] Pia de la Maza M, Garrido F, Escalante N, *et al.* Fluorescent advanced glycation end-products (ages) detected by spectro-photofluorimetry, as a screening tool to detect diabetic microvascular complications. J Diabetes Mellitus 2012; 2(2): 221-6.
[http://dx.doi.org/10.4236/jdm.2012.22035]

[129] Menini S, Iacobini C, de Latouliere L, *et al.* The advanced glycation end-product N^{ε} -carboxymethyllysine promotes progression of pancreatic cancer: implications for diabetes-associated risk and its prevention. J Pathol 2018; 245(2): 197-208.
[http://dx.doi.org/10.1002/path.5072] [PMID: 29533466]

[130] Gkogkolou P, Böhm M. Advanced glycation end products. Dermatoendocrinol 2012; 4(3): 259-70.
[http://dx.doi.org/10.4161/derm.22028] [PMID: 23467327]

[131] Malik P, Chaudhry N, Mittal R, Mukherjee TK. Role of receptor for advanced glycation end products in the complication and progression of various types of cancers. Biochim Biophys Acta, Gen Subj 2015; 1850(9): 1898-904.
[http://dx.doi.org/10.1016/j.bbagen.2015.05.020] [PMID: 26028296]

[132] Malik P, Hoidal JR, Mukherjee TK. Implication of RAGE polymorphic variants in COPD complication and anti-COPD therapeutic potential of sRAGE. J. COPD 2021; 18(6): 737-48.
[http://dx.doi.org/10.1080/15412555.2021.1984417] [PMID: 34615424]

[133] Malik P, Kumar Mukherjee T. Immunological methods for the determination of AGE-RAGE axis generated glutathionylated and carbonylated proteins as oxidative stress markers. Methods 2022; 203:

354-63.
[http://dx.doi.org/10.1016/j.ymeth.2022.01.011] [PMID: 35114402]

[134] Sharma C, Kaur A, Thind SS, Singh B, Raina S. Advanced glycation End-products (AGEs): an emerging concern for processed food industries. J Food Sci Technol 2015; 52(12): 7561-76.
[http://dx.doi.org/10.1007/s13197-015-1851-y] [PMID: 26604334]

[135] Horiuchi S, Sakamoto Y, Sakai M. Scavenger receptors for oxidized and glycated proteins. Amino Acids 2003; 25(3-4): 283-92.
[http://dx.doi.org/10.1007/s00726-003-0029-5] [PMID: 14661091]

[136] Taban Q, Mumtaz PT, Masoodi KZ, Haq E, Ahmad SM. Scavenger receptors in host defense: from functional aspects to mode of action. Cell Commun Signal 2022; 20(1): 2.
[http://dx.doi.org/10.1186/s12964-021-00812-0] [PMID: 34980167]

[137] Ott C, Jacobs K, Haucke E, Navarrete Santos A, Grune T, Simm A. Role of advanced glycation end products in cellular signaling. Redox Biol 2014; 2: 411-29.
[http://dx.doi.org/10.1016/j.redox.2013.12.016] [PMID: 24624331]

[138] Pun PBL, Murphy MP. Pathological significance of mitochondrial glycation. Int J Cell Biol 2012; 2012: 1-13.
[http://dx.doi.org/10.1155/2012/843505] [PMID: 22778743]

[139] Younessi P, Yoonessi A. Advanced glycation end-products and their receptor-mediated roles: inflammation and oxidative stress. Iran J Med Sci 2011; 36(3): 154-66.
[PMID: 23358382]

[140] Soboleva A, Vikhnina M, Grishina T, Frolov A. Probing protein glycation by chromatography and mass spectrometry: analysis of glycation adducts. Int J Mol Sci 2017; 18(12): 2557.
[http://dx.doi.org/10.3390/ijms18122557] [PMID: 29182540]

[141] Martínez-García M, Hernández-Lemus E. Periodontal inflammation and systemic diseases: An overview. Front Physiol 2021; 12: 709438.
[http://dx.doi.org/10.3389/fphys.2021.709438] [PMID: 34776994]

[142] Busch M, Franke S, Rüster C, Wolf G. Advanced glycation end-products and the kidney. Eur J Clin Invest 2010; 40(8): 742-55.
[http://dx.doi.org/10.1111/j.1365-2362.2010.02317.x] [PMID: 20649640]

[143] Smedsrød B, Melkko J, Araki N, Sano H, Horiuchi S. Advanced glycation end products are eliminated by scavenger-receptor-mediated endocytosis in hepatic sinusoidal Kupffer and endothelial cells. Biochem J 1997; 322(2): 567-73.
[http://dx.doi.org/10.1042/bj3220567] [PMID: 9065778]

[144] Wihler C, Schäfer S, Schmid K, et al. Renal accumulation and clearance of advanced glycation end-products in type 2 diabetic nephropathy: effect of angiotensin-converting enzyme and vasopeptidase inhibition. Diabetologia 2005; 48(8): 1645-53.
[http://dx.doi.org/10.1007/s00125-005-1837-9] [PMID: 16010524]

[145] Sebeková K, Podracká L, Heidland A, Schinzel R. Enhanced plasma levels of advanced glycation end products (AGE) and pro-inflammatory cytokines in children/adolescents with chronic renal insufficiency and after renal replacement therapy by dialysis and transplantation--are they inter-related? Clin Nephrol 2001; 56(6): S21-6.
[PMID: 11770807]

[146] Reily C, Stewart TJ, Renfrow MB, Novak J. Glycosylation in health and disease. Nat Rev Nephrol 2019; 15(6): 346-66.
[http://dx.doi.org/10.1038/s41581-019-0129-4] [PMID: 30858582]

[147] Ko J, Kim I, Yoo S, Min B, Kim K, Park C. Conversion of methylglyoxal to acetol by Escherichia coli aldo-keto reductases. J Bacteriol 2005; 187(16): 5782-9.
[http://dx.doi.org/10.1128/JB.187.16.5782-5789.2005] [PMID: 16077126]

[148] Sellegounder D, Zafari P, Rajabinejad M, Taghadosi M, Kapahi P. Advanced glycation end products (AGEs) and its receptor, RAGE, modulate age-dependent COVID-19 morbidity and mortality. A review and hypothesis. Int Immunopharmacol 2021; 98: 107806.
[http://dx.doi.org/10.1016/j.intimp.2021.107806] [PMID: 34352471]

[149] Ohkawara S, Tanaka-Kagawa T, Furukawa Y, Jinno H. Methylglyoxal activates the human transient receptor potential ankyrin 1 channel. J Toxicol Sci 2012; 37(4): 831-5.
[http://dx.doi.org/10.2131/jts.37.831] [PMID: 22863862]

[150] Jomova K, Vondrakova D, Lawson M, Valko M. Metals, oxidative stress and neurodegenerative disorders. Mol Cell Biochem 2010; 345(1-2): 91-104.
[http://dx.doi.org/10.1007/s11010-010-0563-x] [PMID: 20730621]

[151] Kim J, Kim CS, Moon MK, Kim JS. Epicatechin breaks preformed glycated serum albumin and reverses the retinal accumulation of advanced glycation end products. Eur J Pharmacol 2015; 748: 108-14.
[http://dx.doi.org/10.1016/j.ejphar.2014.12.010] [PMID: 25530268]

[152] Oshitari T. Advanced Glycation End-Products and diabetic neuropathy of the retina. Int J Mol Sci 2023; 24(3): 2927.
[http://dx.doi.org/10.3390/ijms24032927] [PMID: 36769249]

CHAPTER 5

Receptor for Advanced Glycation End Products in Health and Physiology

Ruma Rani[1,†], **Parth Malik**[2,3,†] and **Tapan Kumar Mukherjee**[4,*]

[1] *ICAR-National Research Centre on Equines, Hisar-125001, Haryana, India*

[2] *School of Chemical Sciences, Central University of Gujarat, Gandhinagar, Gujarat-382030, India*

[3] *Swarrnim Startup & Innovation University, Bhoyan-Rathod, Gandhinagar-Gujarat, India*

[4] *Amity Institute of Biotechnology, Amity University, New Town, Kolkata, West Bengal 700156, India*

Abstract: The transmembrane protein receptor for advanced glycation end products (mRAGEs) is recognized as an immunoglobulin class of molecule. Mammalian cells produce a carboxy terminus truncated version of RAGE, either as endogenous soluble RAGE (esRAGE) or soluble RAGE (sRAGE), both being generated *via* proteolytic cleavage or alternative mRAGE-mRNA splicing. Through its extracellular domains (V, C1, and C2), RAGE interacts with seemingly unrelated ligands such as advanced glycation end products (AGEs), high mobility group box protein 1 (HMGB1), S100/calgranulin family, lysophosphatidic acid (LPA), oligomeric forms of amyloid beta peptide (Aβ-peptide), islet amyloid polypeptide (IAPP), attributing to the recognition as multi-ligand receptor. Under physiological conditions, lung tissues exhibit abundant RAGE expression compared to others, being involved in the development, spread, and homeostatic regulation, the prominent of which are lung alveolar type 1 (AT-1) epithelial cells. However, in pathophysiological conditions, supraphysiological expression of RAGE and its ligands and subsequent receptor-ligand interactions result in the aggravation of oxidative stress and inflammation, causing the propagation of various non-communicable disease conditions. The physiological RAGE expression may protect against non-small cell lung cancers (NSCLCs), as suppressed RAGE expression in lung tissues may complicate NSCLCs. The protective role of RAGE in lung tissues is surprisingly contrary to its activities in other cancers, which are unanimously characterized by its enhanced expression-driven propagation of the conditions. Anti-RAGE molecules including esRAGE/sRAGE attenuate RAGE-dependent multiple diseased conditions.

[*] **Corresponding author Tapan Kumar Mukherjee:** Amity Institute of Biotechnology, Amity University, New Town, Kolkata, West Bengal 700156, India; E-mail: tapan400@gmail.com
[†] These authors contributed equally to this work.

Keywords: Endogenous soluble RAGE (esRAGE), Inflammation, Multiligand Receptor, Oxidative stress, Pathophysiological conditions, RAGE, RAGE ligands, Receptor for advanced glycation end products (RAGE), Soluble RAGE (sRAGE).

INTRODUCTION

The year 1992 witnessed the first ever experimental evidence for receptors for advanced glycation end products *i.e.* RAGE expression in the bovine pulmonary tissues, creditable to the significant efforts of *Schmidt and associates* [1]. The gene for the RAGE molecule is located at chromosome 6, exhibiting 11 exons and 10 introns, prevailing as highly polymorphic with several alternative forms or splice variants being reported. RAGE protein is recognized as a transmembrane immunoglobulin (Ig) class of molecule, concomitantly functioning as a primitive pattern recognition receptor (PPR) molecule. Structurally, this single-spanning transmembrane protein consists of an extracellular portion having V, C1, and C2 domains. The RAGE is recognized as a "multi-ligand receptor", and interacts with these seemingly unrelated ligands, including advanced glycation end products (AGEs), high mobility group box protein 1 (HMGB1), S100/calgranulin family, lysophosphatidic acid (LPA), oligomeric amyloid beta peptide (Aβ-peptide), and islet amyloid polypeptide (IAPP).

In addition to membrane-bound RAGE (mRAGE), C-(carboxy) terminus truncated RAGE commonly prevails in bodily fluids, including bronchoalveolar lavage (BAL), the fluid surrounding lungs. Soluble RAGE (sRAGE, 50 kDa) or endogenous soluble RAGE (esRAGE, 46 kDa) are the two names given to the circulating RAGE. Although esRAGE is produced by alternative splicing of the RAGE (AGER) gene [2], sRAGE is produced by proteolytic cleavage of the receptor's extracellular domain [3, 4]. Notably, membrane-bound RAGE may use sRAGE/esRAGE as a decoy receptor.

Under completely normal physiological conditions in comparison to all other tissues of the mammalian body, RAGE is most abundantly expressed in the pulmonary tissues [5]. The basolateral membrane of pulmonary alveolar type 1 (AT-1) epithelial cells is the exclusive RAGE expression site in the lungs. RAGE may have a physiological role in lung development [6, 7]. In a significant effort on animal model experiments, *Wolf and colleagues* used RAGE knock-out mice and established that RAGE regulates the differentiation of alveolar epithelial cells, modulating the development and maintenance of lung tissue structure-function [8]. In pathophysiological conditions, supraphysiological expression and subsequent interaction of RAGE with its various ligands, such as AGE and HMGB1, lead to enhanced inflammatory and oxidative stress, eventually manifesting as multiple diseased conditions. Several lines of evidence also decipher functional similarities of the RAGE molecule with adhesion molecules such as intercellular adhesion molecule 1 (ICAM-1) and vascular cell adhesion molecule 1 (VCAM-1). VCAM-1 and ICAM-1 are cell surface immunoglobulin class of molecules and are involved in pro-inflammatory and pro-oxidative reactions. Thus, RAGE is implicated in the propagation of multiple diseases *via* enhanced inflammatory and oxidative stress. A comprehensive literature survey revealed RAGE involvement in the complication of various pulmonary diseases including asthma, chronic bronchitis (CB), cystic fibrosis (CF), chronic obstructive pulmonary disease (COPD), acute lung injury (ALI), acute respiratory distress syndrome (ARDS) and pulmonary arterial hypertension [9, 10]. The major non-pulmonary diseases complicated by RAGE are vascular complications associated with diabetes and atherosclerosis, neurological vascular complications (*e.g.* Alzheimer's disease, AD), and various cancers including breast, ovarian, endometrial, pancreatic, colon, *etc*. However, several lines of evidence indicated that an increased level of RAGE protects against non-small cell lung cancers (NSCLCs). Of note, ~85% of lung cancers are of NSCLC regime. This chapter describes the RAGE signaling mediated through its various ligands, the structure-functional relationships of RAGE, and its ligands alongside the physiological and pathophysiological roles.

CHARACTERIZATION OF RECEPTOR FOR ADVANCED GLYCATION END PRODUCTS GENE

- The RAGE gene lies on chromosome 6 (6p21.3), typically comprising 11 exons interlaced with 10 introns [11].
- The pre-B-cell leukemia transcription factor 2 (PBX2) homeobox gene and a notch homolog, located close to the human leukocyte antigen (HLA) locus are also present in proximity to the RAGE gene along with its 11 exons and 10 introns [12].
- The highly polymorphic nature of the RAGE gene has been well-characterized thus far with several alternative splice variants.
- Although the group identifies 15 transcripts with RAGE variants [13], some other research lists as many as 19 transcripts [2, 14].
- The predominant transcript, NM_001136, with 11 exons, 404 amino acids (aa), and a molecular weight of 55 kDa (cDNA: 1492 bp, DNA sequence: 4557 bp), is identified amongst the four major transcript variants for RAGE [2]. The longest isoform (NM_001206929), codes for a protein with 420 aa and has 11 exons. (Fig. **1**).
- N-terminus truncated RAGE (N-RAGE) is the subsequent transcript, having an initiation codon at exon 3 and an in-frame stop codon in the intron sequence. It encodes for 42 kDa short protein containing only 303 aa and no V-type immunoglobulin domain. This isoform can transport and localize into the

plasma membrane but cannot bind any RAGE ligands, leaving little understanding of its biological function [2, 15].

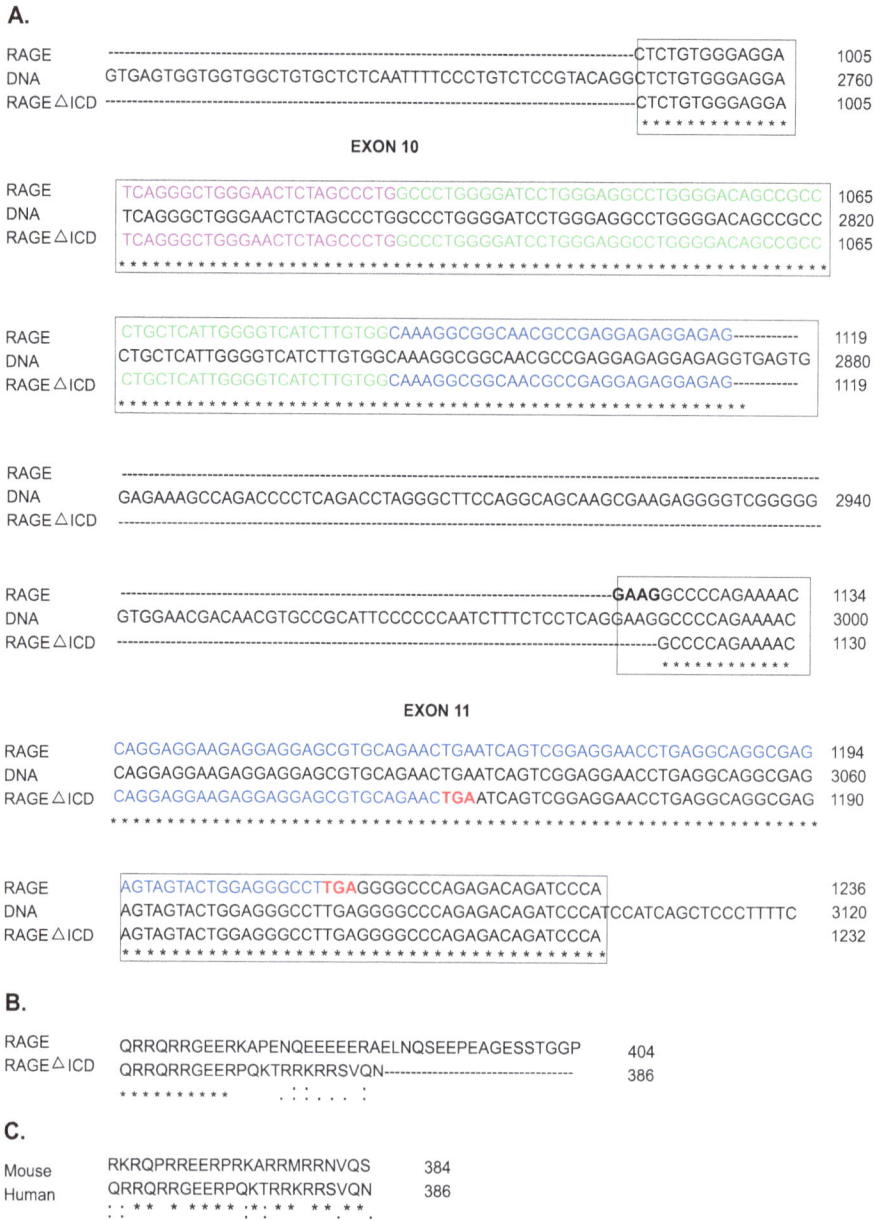

A.

```
RAGE      -----------------------------------------------------------CTCTGTGGGAGGA   1005
DNA       GTGAGTGGTGGTGGCTGTGCTCTCAATTTTCCCTGTCTCCGTACAGGCTCTGTGGGAGGA   2760
RAGE△ICD  -----------------------------------------------------------CTCTGTGGGAGGA   1005
                                                                      * * * * * * * * * * * *
```

EXON 10

```
RAGE      TCAGGGCTGGGAACTCTAGCCCTGGCCCTGGGGATCCTGGGAGGCCTGGGGACAGCCGCC   1065
DNA       TCAGGGCTGGGAACTCTAGCCCTGGCCCTGGGGATCCTGGGAGGCCTGGGGACAGCCGCC   2820
RAGE△ICD  TCAGGGCTGGGAACTCTAGCCCTGGCCCTGGGGATCCTGGGAGGCCTGGGGACAGCCGCC   1065
          * * * * * * * * * * * * * * * * * * * * * * * * * * * * * * * * * * * * * * * * * * * * * * * * * * * * * * * * * * * * *
```

```
RAGE      CTGCTCATTGGGGTCATCTTGTGTCAAAGGCGGCAACGCCGAGGAGAGGAGAG------------   1119
DNA       CTGCTCATTGGGGTCATCTTGTGGCAAAGGCGGCAACGCCGAGGAGAGGAGAGGTGAGTG   2880
RAGE△ICD  CTGCTCATTGGGGTCATCTTGTGGCAAAGGCGGCAACGCCGAGGAGAGGAGAG------------   1119
          * * * * * * * * * * * * * * * * * * * * * * * * * * * * * * * * * * * * * * * * * * * * * * * * *
```

```
RAGE      ------------------------------------------------------------------------   
DNA       GAGAAAGCCAGACCCCTCAGACCTAGGGCTTCCAGGCAGCAAGCGAAGAGGGGTCGGGGG   2940
RAGE△ICD  ------------------------------------------------------------------------   
```

```
RAGE      ---------------------------------------------------GAAGGCCCCAGAAAAC   1134
DNA       GTGGAACGACAACGTGCCGCATTCCCCCCAATCTTTCTCCTCAGGAAGGCCCCAGAAAAC   3000
RAGE△ICD  -------------------------------------------------------GCCCCAGAAAAC   1130
                                                                * * * * * * * * * * * *
```

EXON 11

```
RAGE      CAGGAGGAAGAGGAGGAGCGTGCAGAACTGAATCAGTCGGAGGAACCTGAGGCAGGCGAG   1194
DNA       CAGGAGGAAGAGGAGGAGCGTGCAGAACTGAATCAGTCGGAGGAACCTGAGGCAGGCGAG   3060
RAGE△ICD  CAGGAGGAAGAGGAGGAGCGTGCAGAACTGAATCAGTCGGAGGAACCTGAGGCAGGCGAG   1190
          * * * * * * * * * * * * * * * * * * * * * * * * * * * * * * * * * * * * * * * * * * * * * * * * * * * * * * * * * * * * *
```

```
RAGE      AGTAGTACTGGAGGGCCTTGAGGGGCCCAGAGACAGATCCCA                         1236
DNA       AGTAGTACTGGAGGGCCTTGAGGGGCCCAGAGACAGATCCCATCCATCAGCTCCCTTTTC   3120
RAGE△ICD  AGTAGTACTGGAGGGCCTTGAGGGGCCCAGAGACAGATCCCA                         1232
          * * * * * * * * * * * * * * * * * * * * * * * * * * * * * * * * * * * * *
```

B.

```
RAGE      QRRQRRGEERKAPENQEEEEERAELNQSEEPEAGESSTGGP        404
RAGE△ICD  QRRQRRGEERPQKTRRKRRSVQN-----------------------   386
          * * * * * * * * * *         .  : ! . . .  :
```

C.

```
Mouse     RKRQPRREERPRKARRMRRNVQS     384
Human     QRRQRRGEERPQKTRRKRRSVQN     386
          : . * *   * * * * * . * : . * *   * * . * * .
```

Fig. (1). Alternative RAGE splicing, giving a conserved shortened cytoplasmic domain amongst humans and mice. **A.** RAGEΔICD generated *via* alternative splicing of 5' terminal of RAGE exon 11. The DNA and cDNA sequences of human RAGE were aligned using Clustal W. Alternative 3" receptor loci at the initiation of exon 11 are depicted in bold. In red are the stop codons for canonical RAGE and the novel RAGEΔICD protein reading frames. The purple highlighting represents RAGE and RAGEΔICD regions encoding the extracellular domain, in green is the transmembrane domain while the intracellular domain is shown in blue. **B.** Alignment of human RAGE and human RAGEΔICD cytoplasmic regions protein sequences. Unvaried amino acids are depicted as "*", amino acids with strong identical properties are represented as ":" while those exhibiting weak similarity index of properties are depicted as ".". The net amino acid number of each isoform is mentioned to the right of the protein sequence.

- The third transcript, also known as "endogenous soluble RAGE or esRAGE," is a C-truncated isoform that lacks the transmembrane and cytoplasmic domains of full-length RAGE.
- The fourth encoded transcript is referred to as "soluble RAGE (sRAGE)", created when full-length RAGE is degraded by proteases, primarily to ADAM10 (a protein 10 containing a disintegrin and metalloproteinase domain).
- Hence, the soluble C-truncated RAGE proteins known as sRAGE and esRAGE are found in the bloodstream and other biological fluids like broncho-alveolar lavage (BAL) fluid. These proteins can bind AGEs and other RAGE ligands [16, 17], acting as endogenous competitive RAGE inhibitors without affecting the metabolic pathways, being subsequently recognized as "decoy receptors" [18].
- Of note, esRAGE serum extents are lower than sRAGE in healthy subjects. Both, sRAGE (50 kDa) and esRAGE (46 kDa) may not have equivalent biomarker values [16].
- Several single nucleotide polymorphisms (SNPs) of the RAGE gene have been postulated, in addition to multiple splice variants.
- Significantly, spirometry measurements of airflow obstruction and a significant RAGE gene polymorphic variant rs2070600 were revealed in two genome-wide association studies conducted in healthy persons of European ancestry [19 - 21]. RAGE polymorphic variants *i.e.*, G82S, influence pulmonary function in smokers and serum sRAGE expression in the UK population [22].
- In a recent comprehensive analysis, *Dancer and* colleagues concluded that different RAGE gene SNP frequencies based on ethnicity might account for the inconsistent findings about the RAGE involvement in complicating various diseases [23].

CHARACTERIZATION OF RECEPTOR FOR ADVANCED GLYCATION END PRODUCTS PROTEIN AND ITS SOLUBLE FORMS

- In 1992, a 35 kDa protein was identified for the first time by *Schmidt* and *colleagues* from bovine lung tissues. The AGE's binding ability of this protein led to it being named RAGE [1].
- RAGE is a single-spanning transmembrane receptor glycoprotein. In humans, RAGE is a 50-55 kDa glycosylated protein with extracellular, transmembrane, and cytoplasmic domains having sections at (aa 23-342), (aa 343-363) and (aa 363-404), respectively.

- The extracellular structure of RAGE is composed of two linked domains, a variable (V) immunoglobulin (Ig) domain (residues 23-116), C1 (residues 124-221), and C2 (residues 227-317), coupled by a flexible seven aa linker (Fig. **2**).

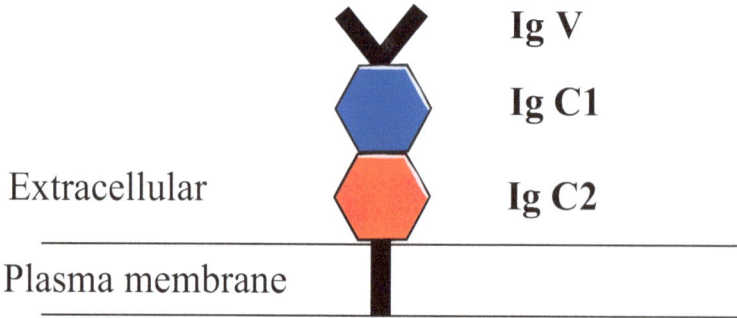

Fig. (2). Assembly of extracellular (V, C1, and C2) domains for advanced glycation end-product localization within domains within the plasma membrane.

- The eight strands (A', B, C, C', E, F, and G) that make up the V domain are joined by six loops to form two β-sheets that are interlinked by a disulfide bridge between Cys38 (strand B) and Cys99 (strand F).
- The hydrophobic cavity that anchors the molecular surfaces of V-C1 domains holds numerous highly cationic Arg and Lys residues. In contrast, the acidic aa in the C2 domain exhibits an anionic surface sensitivity. The positively charged V-C1 domain can be bound by strongly anionic regions prevailing in several RAGE ligands. The V, C1, and C2 domains are normally bound to the cell membrane but may get detached in a solubilized form of RAGE (*i.e.* the sRAGE) (Fig. **3**).
- The truncated versions of full-length RAGE generated distinct functional forms, wherein C-truncation or proteolytic cleavage account for sRAGE.
- Heparan sulfate has been shown to have a critical function in stabilizing RAGE hexamerization, getting self-associated in V-V domains, and subsequent homodimerization. Since RAGE has a low affinity for many ligands when it is monomeric, multimerization is necessary for ligand binding. Following ligand binding, RAGE oligomerizations *via* its C1-C1 domains, C2-C2 domains, and/or TM helix dimerization are crucial phases in RAGE signaling.
- The transmembrane helical structure of RAGE constitutes the meticulously conserved GxxxG motif that facilitates helix homodimerization and may assist in signal transmission.
- RAGE's cytoplasmic domain shares a high degree of sequence identity with rodents and primates, which is crucial for RAGE ligand-mediated signal transduction. There is a richly acidic area in its cytoplasmic domain that may

bind several molecules. RAGE-associated pathologic consequences are moderated when this domain is truncated as it affects downstream RAGE signaling.

- The V, C1, and C2 extracellular domains of RAGE interact and bind with several unrelated ligands, namely AGE [24], amphoteric (commonly known as high mobility group box chromosomal protein, *i.e.* HMGB1) [25], S100/calgranulins [26], Aβ-peptide, β-fibrils [27] and Mac-1 [28].

- In addition to membrane-bound RAGE (mRAGE), C-(carboxy) terminus-shortened RAGE is also commonly found in bodily fluids, including bronchoalveolar lavage (BAL), the fluid surrounding the lungs.

- The circulatory RAGE is designated either as soluble RAGE (sRAGE, 50 kDa) or endogenous soluble RAGE (esRAGE, 46 kDa).

- Although esRAGE is produced by alternate splicing of the RAGE (AGER) gene [2], sRAGE is produced by the proteolytic cleavage of the receptor's extracellular domain [3, 4] (Fig. **3**).

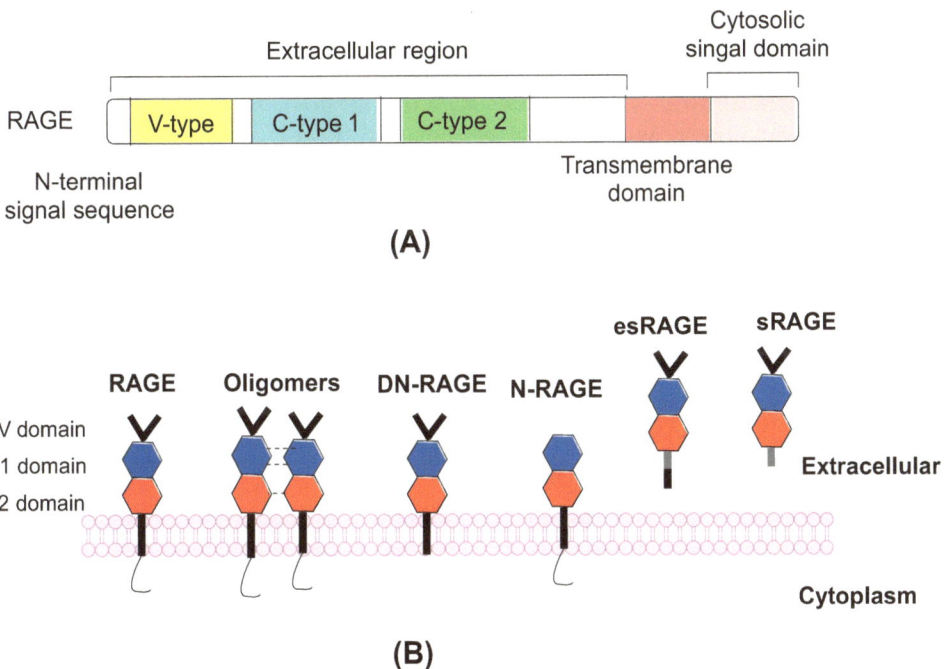

(A)

(B)

Fig. (3). Schematic representation of RAGE, extracellular, and cytoplasmic functional domains, distinguishing the working framework *via* V, C1, and C2 domains [29].

- The formation of sRAGE occurs when the mRAGE ectodomain is shed by metalloproteinases such as MMP9, A disintegrin, and metalloprotease (ADAM) 10 [4, 30]. However, it is still unclear how much proteolysis contributes to the total extent of circulating sRAGE [4, 31]. In a noted effort, *Hudson and colleagues* claimed that nearly 80% of lung RAGE functions as a membrane-associated receptor (mRAGE), the remaining RAGE (~13%) functions as sRAGE, and ~ 7% exists as esRAGE [2].

- Thus, in healthy subjects esRAGE serum levels are two to five-fold lower than sRAGE, making it inevitable for unequal sRAGE and esRAGE biomarker potency [16].
- The consensus is that inflammation is lessened and the risk of several inflammatory disorders, including COPD, is decreased when sRAGE levels are raised to the point where they overreach pro-inflammatory membrane-bound RAGE activity for AGE binding.

THE LIGANDS OF RECEPTORS FOR ADVANCED GLYCATION END PRODUCTS

- The major discovery in RAGE biology was the multiligand binding ability of this receptor, seemingly unrelated and the reason for the functional diversity of ligand-bound RAGE states.
- AGEs comprise the most prominent RAGE ligands, the first-ever decoding of AGEs is perhaps the reason for the terminology RAGE. Table 1 summarizes the characteristic biochemical traits of various RAGE ligands, including terminology, characteristic binding domain, and major clinical functions. From a specific viewpoint, chapter 2 of this book comprehensively describes the AGE ligand.

Table 1. Summary of various ligands for receptors for advanced glycation end products, their binding domains on RAGE, and the characteristic functional aspects [34 - 38].

RAGE Ligands	RAGE Binding Domain	Major Clinical Significance
Advanced Glycation End Products	V	Bind RAGE with high affinity, triggering the onset of a pro-inflammatory signaling cascade. Higher extents complicate diabetes, atherosclerosis, cardiovascular disorders, nephropathy, and chronic inflammation.
S100/calgranulins	V or VC1 or V2	Comprise >25 members with S100B, S100P, S100A2, S100A4, S100A5, S100A6, S100A7, S100A8/9, S100A12, S100A13 exhibiting varied RAGE interaction with nano to micromolar dissociation constants. S100B enhances dose based neuronal survival and axon growth; S100A7, S100A8/A9 and S100A12 are pro-inflammatory molecules; S100A2, S100A4 regulate cell growth and differentiation, S100A5, S100A6, S100P promote tumor growth.

(Table 1) cont.....

RAGE Ligands	RAGE Binding Domain	Major Clinical Significance
Human Mobility Group Box 1 Protein	VC1C2	Released actively *via* cytokine stimulation and passively amidst cell death. Involved in inflammatory disorders *via* DAMP functioning, associates with TLR ligands and cytokines, activate cells *via* distinct TLR2, TLR4, and RAGE engagement.
mDia1	Cytoplasmic	mDia1-RAGE signaling is critical in macrophages *via* Egr-1 actions under hypoxic conditions, as claimed by studies on diabetic, aged, vascular distressed animals, causing vascular dysfunction. Endothelial cells produce ROS, on activation *via* AGEs and RAGE cytoplasmic tail-DIAPH1 binding, activating NF-κB-transcription factor
Amyloid fibrils	V	A cleaved form of APP, generated on its proteolytic cleavage, constituted of acidic and hydrophobic residues susceptible to aggregation, viciously involved in AD progression.
Quinolinic acids	VC1	Interaction with RAGE is involved in Huntington's disease (a neurological disorder).
β2 integrin macrophage 1 antigen (Mac1)	CD11b	Interaction with RAGE causes leukocyte infiltration, alters S100B, S100A9, and HMGB1 levels, and exhibits immune cell infiltration.
Lysophosphatidic acid	V	Interaction with RAGE is involved in cancer, cardiovascular disorders, and diabetes, *via* STAT3, PI3K, ERK, and AKT downstream signaling pathways, causing proliferation and migration in C6 glioma and smooth muscle cells
Phosphatidylserine (PS)	VC1	RAGE interaction with PS-exhibiting apoptotic cells induces phagocytosis, and PS-RAGE binding during apoptosis on macrophages induces phagocytosis *via* mDia1-Rac1 communication.
C1q	Not yet entirely understood	Performs several complement-independent tasks in innate and acquired immunity, abnormal C1q expression may result in autoimmunity, prevents inflammation by releasing intracellular toxic materials into extracellular space, and genetic deficiency of C1q leads to SLE.

Abbreviations: SLE: Systemic Lupus Erythematosus, DAMP: Damage-Associated Molecular Pattern, TLR: Toll-Like Receptors, APP: Amyloid Precursor Protein, STAT3: Signal Transducer and activator of Transcription 3, PI3K: Phosphoinositide 3- kinase, ERK: Extracellular Regulated Kinases, AKT: A kinase (PKKA) C-terminal domain, Egr-1: Early Growth Response (Egr) gene-1, NF-κB: Nuclear Factor-kappa B, AD: Alzheimer's Disease.

- In addition to AGEs, other RAGE ligands include islet amyloid polypeptide (IAPP), proteins from the high mobility group box 1 (HMGB1), lysophosphatidic acid (LPA), and the S100/calgranulin family [26, 32 - 34].
- Additionally, Toll-like receptors (TLRs) can also increase TLR9-mediated responsiveness to improve DNA internalization.

- Most RAGE ligands bind to their extracellular domains (V, C1, C2) in a heterogeneous regimen; while the extracellular V-type immunoglobulin (Ig) domain binds to multiple ligand families. The binding sites on the V-domain are manifold with substantial spatial distinctions.
- The RAGE ligands may also bind at the extracellular C1 and C2-type Ig domains, diversifying the RAGE-ligand interaction complexity [35 - 38].

The prominent RAGE ligands are as follows:

- Advanced Glycation End Products (AGEs)
- High Mobility Group Box Protein 1 (HMGB1)
- S100/Calgranulin Family
- Lysophosphatidic Acid (LPA)
- Oligomeric Forms of Amyloid Beta Peptide (Aβ-PEPTIDE)
- Islet Amyloid Polypeptide (IAPP)

ADVANCED GLYCATION END PRODUCTS (AGES)

- Chapter 4 of this book is dedicated to the biology of advanced glycation end products (AGEs). Briefly, in 1912, AGE was first identified by French chemist *Louis-Camille Maillard*. Inceptive studies showed AGE generation amidst food preparation (cooking) at high temperatures, as carbohydrates (*e.g.* glucose) gradually react with proteins *via* generating Schiff's base, Amadori products, and finally an irreversible Maillard reaction.
- This non-enzymatic AGE generation process is called "glycation".
- Subsequently, it was demonstrated that besides the exogenous processes (majorly *via* cooking foods), the endogenous means within the human body including blood, skin, and other tissues, could also generate AGEs.
- As of now, more than 20 AGEs have been reported in human blood and tissues, and dietary foods. Based on their fluorescent sensitivity, the AGEs can be fluorescent or non-fluorescent. While major significant non-fluorescent AGEs are carboxymethyl-lysine (CML), carboxyethyl-lysine (CEL), and pyrroline; pentosidine and methylglyoxal-lysine dimer (MOLD) are the well-studied fluorescent AGEs [39].
- AGEs bind several receptors, amongst which RAGE is the most prominent. Additional cell surface entities capable of binding with AGE include macrophage scavenger receptors (MSRs) type A and B1 (CD36), oligosaccharyl transferase-48/OST48, also known as "AGE receptor 1" (AGE-R1), 80K-H phosphoprotein (AGE-R2), galectin-3 (AGE-R3), and the scavenger receptor family (SR-A, SR-B, SR-1, SR-E, LOX-1, FEEL-1, FEEL-2, and CD36) [39].

The AGE-RAGE interaction mediated aggravated inflammatory and oxidative stress results in multiple diseased conditions, the majorly noticed of which is the diabetic vascular complication. Additionally, the AGE deposition on biological macromolecules significantly affects their structure-functional properties. This is a typical, progressive modification of biological macromolecules, with concomitant involvement of accumulated, *in vivo* modified textures, being implicated in multiple pathologic abnormalities, exclusively with a faster-aging manifestation and long-term diabetes. Similar development patterns of these complications suggest a resemblance of chronic hyperglycemia with accelerated aging.

High Mobility Group Box Protein 1 (HMGB1)

The high mobility group (HMG) is a group of chromosomal non-histone proteins, involved in the regulation of DNA-dependent processes such as transcription, replication, recombination, and DNA repair.

The HMG proteins are subdivided into the following 3 superfamilies:

HMGA

It contains an AT-hook domain and has HMGA1, and HMGA2 as members.

HMGB

This is an HMG box domain, having HMGB1, HMGB2, HMB3, and HMGB4 members.

HMGN:

It contains a nucleosomal binding domain. Its members are HMGN1, HMGN2, HMGN3, HMGN4.

HMGB1:

High-mobility group box protein 1 (HMGB1), was first isolated and characterized by Goodwin GH from calf intestinal thymus. HMGB1 is an evolutionary conserved commonly prevailing member of the HMGB family. High mobility proteins are named so because of their swift mobility over the polyacrylamide gel. Certain other terminologies for HMGB1 proteins are amphoterins/alarmins/architectural factors. HMGB1 can directly bind with RAGE, toll-like receptors 2, 4, 9 (TLR2, TLR4 & TLR9), syndecan 1, phosphacan 1, heparin, and thrombomodulin.

The High-Mobility Group Gene and Protein

- While in mice, HMGB1 gene lies on chromosome 5, in humans the HMGB1 gene is localized on chromosome 13q12, further constituted of six exons. There are four major HMGB1 polymorphisms being reported, rs1045411, rs1360485, rs1412125, rs2249825.
- Every nucleated human cell articulates HMGB1. However, on invasion by pathogens or damage-associated molecules, monocytes, macrophages, natural killer cells, and myeloid/plasmacytoid dendritic cells exhibit vigorous HMGB1 expression, leading to extracellular HMGB1-recruiting myeloid dendritic cells, plasmacytoid dendritic cells, macrophages, neutrophils, and CD4+ T cells.
- On average, HMGB1 prevails to the extent of 10^6 molecules per cell, although some cells may express it to a lower degree. The total HMG proteins are ~3%, by weight of the histone or DNA content, thereby prevailing as a small group of proteins compared to histones. HMGB1 non-specifically binds to the minor DNA grooves, consisting of 215 aa, having 24.894 kDa as the molecular weight. However, the HMGB1 migrates at 30 kDa loci in sodium dodecyl sulfate-polyacrylamide gel electrophoresis gels (SDS-PAGE), possibly because of the abnormal constituent cationic aa. Studies on comparative amino acid sequence screening of HMGB1 revealed a 100% homology between mice and rats and ~99% between rodents and humans.

A typical HMGB1 polypeptide chain consists of the following:

HMG box A

It is a cationic motif of 1-79 aa.

HMG box B

It houses a cationic motif of 89-163 aa.

Acidic C Terminus

It consists of relatively higher repetitive extents of anionic, aspartic acid, and glutamic acid residues.

- Functionally, boxes A and B are DNA-binding domains, with box B exhibiting a cytokine-like activity, inducing macrophage secretion of additional pro-inflammatory cytokines. This cytokine activity is antagonized by recombinant A box. The first 21 aa residues of recombinant box B encode a minimal peptide sequence with an intact cytokine-like activity. The protein structure mediating HMGB1-RAGE binding is between 150 and 183 aa residues (Fig. **4**) [40].

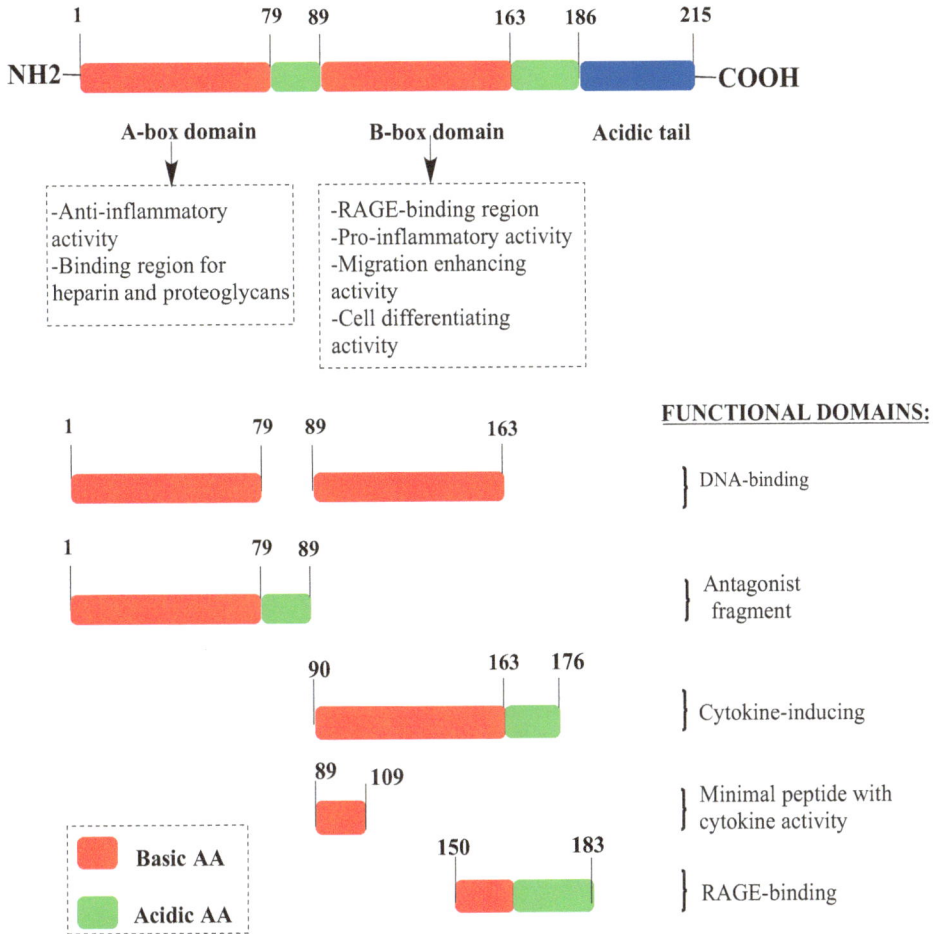

Fig. (4). Constitutional domains of a functional HMGB1-polypeptide chain, illustrating the diverse physiological controls exhibited on binding with RAGE.

- The exhaustive secretion of HMGB1 by monocytes/macrophages in response to various stimuli was first reported by *Wang and the group* [41]. Subsequently, it was elucidated that once secreted, HMGB1 activates the cells involved in immune response or inflammatory reactions, functioning as a cytokine [42].
- The protein can also be passively released by damaged or necrotic cells, leading to inflammation. HMGB1 functions as a "necrotic marker" or Damage Associated Molecular Pattern (DAMP) to facilitate cellular recognition by the innate immune system. It also acts as a surrogate for determining the relative injury limit to initiate tissue repair [42]. Several post-translational modifications

including acetylation, phosphorylation, methylation, and poly (ADP)-ribosylation are known to allocate HMGB1 to the secretory pathway.

- In convention, studies noticed tumors harboring HMGB1 as highly acetylated, contrary to those of normal cells. While HMGB1 oligomerization and conformational changes can be induced by acetylation, it was later shown that these changes are characteristic of proliferating tissues in general rather than just for tumor cells. Previous studies linked HMGB1 hyperacetylation to its 43 lysine residues, suggesting this as a mechanism to facilitate cytosol migration. This enzymatic alteration likely enables HMGB1 to "change its mask" and operate as a cytokine-like agent.

Sub-cellular Localization and Cellular Release of High-Mobility Group Box Protein

As a small protein (~24.8 kDa), HMGB1 is synthesized on the non-ER bound (Free) ribosomal surface. After ribosomal synthesis, HMGB1 is translocated to the nucleus through its nuclear localization signals 1 and 2. In the nucleus, HMGB1 binds with the DNA minor grove to sustain the transcription and other nuclear actions. Various molecules like bacterial lysophosphatidylcholine (LPS), cytokines (like IFNγ, IL-1, and TNFα), the cellular redox modulators (ROS/RNS), and necrotic/pro-apoptotic/autophagic stimulus together result in multiple HMGB1 post-translational modifications. The major biochemical responses causing HMGB1 modification are acetylation, phosphorylation, methylation, and ADP-ribosylation. These post-translational modifications decide the HMGB1 cytoplasmic to nuclear translocation or *vice versa*, its sub-cellular compartmentalization, and extracellular secretion *via* active or passive mode. In the extracellular space, HMGB1 may bind to RAGE and/or TLRs to provoke pro-inflammatory and pro-oxidative reactions, eventually complicating the multiple diseased conditions. Based on the specific biochemical environment, HMGB1 could exist freely in the extracellular space to drive multiple inflammatory reactions. In general, cells secrete HMGB1 in the extracellular region and this release can be accelerated *via* certain hormones or proteins.

The most generalized HMGB1 locations are as under:

- The Nuclear HMGB1
- The Cytoplasmic HMGB1
- The Endosomal/Lysosomal HMGB1
- The Mitochondrial HMGB1
- The Membrane-bound HMGB1
- The Extracellular HMGB1

The distinct HMGB1 locations convey their significant structurally diversified potential, prominent aspects of these locations are as discussed ahead.

The Nuclear High Mobility Group Box Protein 1

- In the basal state, HMGB1 predominantly prevails as a nuclear protein.
- HMGB1 shuttles between the cell cytoplasm and nucleus.
- This protein lacks a classical leader sequence and thereby cannot be actively secreted through the classical endoplasmic reticulum (ER)-Golgi apparatus exocytotic pathway.
- However, HMGB1 contains two non-classical nuclear localization signals. Protein-transporting molecule importin binds with non-classical HMGB1 nuclear localization signals 1 and 2, transporting it from the cytoplasm to the nucleus for DNA binding.

NB

In its non-acetylated form, HMGB1 resides within the nucleus and this state may involve deacetylases. The acetylation of lysine resides adjacent to the nuclear localization signal impairs the importin recognition and binding with nuclear localization signals 1 and 2. In response to the immunological challenge, macrophages become activated, acetylating the nuclear localization signal sites within HMGB1 to facilitate cytosolic retention. This is followed by secretory lysosomal assembly and extracellular release.

The Cytoplasmic High Mobility Group Box Protein 1

- The cytoplasmic prevalence of HMGB1 was first reported by Bustin in 1979.
- Subsequently, the same research group proved that no dissimilarity exists between the nuclear and cytoplasmic HMGB1.
- Within the cytoplasm, HMGB1 is localized outside the cell organelles, lysosomes, endosomes, and mitochondria.

Nuclear to Cytoplasmic Translocation of High Mobility Group Box Protein 1

- The nucleus-to-cytoplasm protein shuttle is sustained *via* post-translational modifications (*e.g.*, acetylation, phosphorylation, methylation) enabling the trafficked protein's nuclear localization or export sequences (NLS or NES). The export sequences, thereafter, interact specifically with import and export complexes on the nuclear membrane. In addition to the two NLS sites, HMGB1 has two non-classical NES, enabling its regular cytoplasm to nuclear movement.
- This balance is skewed toward nuclear accumulation in quiescent cells. According to new research, double-stranded RNA-activated protein kinase R (PKR)/inflammasome-mediated pyroptosis partially mediates

HMGB1extracellular release on its initial nuclear to cytoplasmic translocation controlled by JAK/STAT1-mediated acetylation.

- The JAK/STAT1 signaling is the fundamental mechanism that governs the HMGB1 nuclear-cytoplasmic translocation and is crucial for the HMGB1 hyperacetylation induced by lipopolysaccharide (LPS) or IFN within the NLS sites. The HMGB1 secretion triggered by IFN-β, IFN-γ, or LPS is interfered with by pharmacological or genetic perturbation with JAK/STAT1 signaling.

Cytoplasmic to Extracellular Space Release of High Mobility Group Box Protein 1

- Fully reduced HMGB1 can passively reach extracellular space *via* necrotic cell-assisted diffusion.
- It is actively discharged into the extracellular area as well. The secretory lysosome is a non-conventional mechanism that facilitates the secretion of numerous immune and non-immune cells, including neutrophils, dendritic cells, macrophages, monocytes, and natural killer cells, in response to cytokines and LPS.
- Immune cells stimulated with IFNγ, IL-1, and TNFα export nuclear HMGB1 to the cytoplasm and subsequently secrete it.
- HMGB1 is also released during bacterial or viral infection wherein extracellular HMGB1 can act as a chemoattractant for inflammatory cells, strongly suggesting its role as a modulator of the immune system.
- HMGB1 is a leader-devoid cytokine released from the cytosol into the extracellular environment. Post synthesis, HMGB1 is used by conventional cytokines for delivery into the rough endoplasmic reticulum (RER) and Golgi apparatus-mediated secretory pathway for a release across the cell membrane. Alternatively, HMGB1 release may be mediated *via* unique organelles associated with the endolysosomal compartment.
- Active secretion starts the activation of monocytes and macrophages, translocating them from the nucleus to cytoplasm upon autophagic stimulation.
- Release of HMGB1 from the macrophages in the extracellular milieu requires the activation of NLRC4 or NLRP3 inflammasomes.
- HMGB1 is also released from apoptotic cells, whereby active secretion from multiple immune and non-immune cells such as monocytes, macrophages, neutrophils, dendritic cells, and natural killer (NK) cells in response to stimuli like LPS and cytokines, involves non-conventional secretory process *via* secretory lysosomes.
- HMGB1 is also secreted by plasma cells in response to LPS.
- HMGB1 majorly resides on the activated platelets' surface.
- When the viral protein pVII is functional, HMGB1 exhibits an enhanced chromatin interaction.

- In response to endotoxins, ds-RNA, and other pathogen-associated molecular patterns (PAMPs), monocytes and macrophages release early pro-inflammatory cytokines (TNF, IL-1, IFN-β, and IFN-γ) and late mediators (HMGB1). IFN-β and IFN-γ are two early cytokines that can induce immune cells to release HMGB1 in a time and dose-dependent manner.

The Endosomal/Lysosomal High Mobility Group Box Protein 1

The major function of endosomes in a cell is to regulate organelle differentiation *via* sustaining amicable protein and lipid trafficking amongst the sub-cellular compartments of secretory and endocytic pathways. Contrary to this, lysosomes (frequently referred to as suicidal bags) catalyze the periodic removal of damaged cells and tissues by their hydrolytic enzymes.

- Certain lysosomes could function as secretory compartments (terminology secretory lysosomes), prominently prevailing in varied immune cells and are aided by Ca+2-regulated exocytosis [43]. Of note, cytokine secretion could be regulated without intermediates, IL1β (interleukin) and HMGB1, *via* lysosomal exocytosis [40].
- Studies using double immunofluorescence staining deciphered HMGB1 co-localization with the lysosomal marker, lysosomal-associated membrane protein 1 (LAMP1) albeit not with an early endosomal marker or early endosomes antigen 1 in LPS-treated monocytes [44].
- Further, IFI30 lysosomal thiol reductase (IF130, more commonly known as γ-interferon inducible lysosomal thiol reductase (GILT) impairs the disulfide bond formation in proteins for their unfolding and eventual degradation through lysosomes.
- The transcription factors NF-κB and STAT1 respectively sustain the GILT expression induced by LPS and IFN-γ [45, 46]. Depleting IFN-130 in T-cells and monocytes can suppress cytokine generation (*i.e.* for IL1β and TNF), releasing HMGB1 in response to LPS, antigen exposure, or mitochondrial oxidative damage [47, 48].
- Above events create the biochemical environment for HMGB1 nucleus to cytoplasm translocation, gradually aggravating the autophagy in IF130/fibroblasts [48]. Together, these observations validate the role of IF130 in mediating secretory lysosome-driven HMFB1 release.

The Mitochondrial High Mobility Group Box Protein 1

Mitochondria perform manifold functions relevant to cellular bioenergetics and apoptotic death in eukaryotes. Regarding HMGB1, it has been discussed earlier that it is an evolutionary conserved chromatin-associated protein, functioning as a critical regulator of mitochondrial functions and morphology. About the mitochondrial relevance of HMGB1, studies decode heat shock protein beta-1 (HSPB1)/HSP27 as a downstream mediator functioning of quality control HMGB1 actions.

- The findings by *Tang and colleagues* noticed HSPB1 gene disruption in embryonic fibroblasts by wild-type HMGB1 to reiterate the mitochondrial fragmentation, impairing the mitochondrial respiration and ATP formation on targeted HMGB1 deletion.
- Further, a forced expression of HSPB1 reverses this phenotype in HMGB1 knockout cells. It was noticed that a complex HSPB1 expression reverses this phenotype in HMGB1 knockout cells.
- These biochemical eventualities about mitochondrial effects *via* HMGB1 regulated HSPB1 expression function as a reliable guarding mechanism against mitochondrial malfunctioning, facilitating clearance and autophagy-driven mitigation of cellular stress [49].
- Mechanistically, the HMGB1 functioning in mitochondria for wild-type cells could be best summarized as follows: HMGB1 functions as an HSPB1 gene transcriptional regulator. It is pertinent to recall that HSPB1 phosphorylation is needed to activate the actin cytoskeleton, affecting the autophagy and mitophagy-driven cellular transport for mitochondrial insult. Collectively, impaired HMGB1 or HSPB1 results in mitophagy phenotype, characterized by mitochondrial fragmentation, suppressed aerobic respiration, and ATP generation.
- Several subsequent efforts confirmed HMGB1's role in impaired mitochondrial functioning. The first notable of these efforts is a 2015 study by *Qi and team*, whereby non-functional mitochondrial energy metabolism was examined suspecting its likely involvement in multiple neurodegenerative diseases. Screening the role of HMGB1 in striatal degeneration, investigators studied the impact of HMGB1 on autophagy and cell death activation on being exposed to 3-nitro propionic acid (3-NP). For this purpose, rat striata were intoxicated with 3-NP using stereotaxic injection following which changes in HMGB1, caspase-3, and phosphor-c-Jun amino-terminal kinases (p-JNK) expressions were monitored. Analysis revealed 3-NP triggered enhancement in p-JNK, cleaved caspase-3, and autophagic marker, LC3-II besides reduced SQSTM1 (p62). The effects were counter-confirmed as being HMGB1 driven, revealing moderation on glycyrrhizin (an HMGB1 inhibitor) administration. The HMGB1 involvement was screened critically for basal autophagy during the cellular rescue *via* the HMGB1 targeting shRNA approach. Treatment with 3-NP stimulated HMGB1, p-JNK, and LC3-II expression in striatal neurons while p-JNK expression was substantially reduced on shRNA-mediated HMGB1 knockdown, an observation that ceased to prevail on exogenous HMGB1 supplementation. These observations collectively established a prominent

HMGB1 involvement in signaling-driven autophagic and apoptotic induction for neurodegeneration *via* mitochondrial impairment [50].

- Unlike the above-discussed studies, a 2018 effort from *Huang and associates* screened RAGE-HMGB1 interaction as the source of impaired mitochondrial functioning triggered cancer cell proliferation, suppressed apoptosis, and aggravated chemoresistance in colorectal cancer. The study examined the effect of dynamin-related protein 1 (Drp1) driven mitochondrial fragmentation *via* inducing a chemo resistant response. Analysis revealed a release of HMGB1 from the dying cells in a conditioned medium on chemotherapeutic drug administration to resistant cells. The underlying mechanism was observed as Drp1 phosphorylation through a concomitant RAGE involvement.

- It was subsequently observed that RAGE signaling activates extracellular regulated kinases (ERK1/2) to phosphorylate Drp1 at residue S616, activating autophagy-mediated chemoresistance and regrowth in cancer cells. HMGB1 involvement was confirmed *via* non-phosphorylated Drp1 on treatment with HMGB1 inhibitor and RAGE blocker, improving the sensitivity to chemotherapeutic drugs *via* diminished autophagy. Besides, RAGE-HMGB1 interactions were noticed as critical owing to a high phosphor-Drp1Ser616 exhibiting rectal cancer patients with increased vulnerability on tumor relapse, abysmal 5-year disease-free and overall survival on neoadjuvant chemo-radiotherapy. This observation correlated with RAGE-G82S polymorphism, inferring an association with high phospho-Drp1Ser616 in the tumor microenvironment. The study established aggravated chemoresistance and regrowth following HMGB1 release from dying cancer cells *via* RAGE-mediated ERK/Drp1 phosphorylation [51].

- Similar observations were also made for a HMGB1 regulatory effect on hypoxia-triggered mitochondrial biogenesis in pancreatic cancer (PANC1/CFPAC1) cells, in a recent 2020 study. Analysis revealed hypoxia aggravated NRF-1/TFAM (nuclear respiratory factor 1 gene/mitochondrial transcription factor A) expressions, enhanced mitochondrial copy number, and ATP content besides increased mitochondria in pancreatic cancer cells, getting suppressed on HMGB1 knockdown. In a nutshell, HMGB1 knockdown reduced hypoxia with AICAR (a noted AMPK stimulator) triggered NRF-1, TFAM, PGC-1α, SIRT1, and complexes I, III proteins, reducing the acetylation of PGC-1α/SIRT1. Apart from this, treatment with SRT1720 (a SIRT1 stimulator) enhanced the SIRT1 functions, stimulating hypoxia-driven PGC-1α deacetylation, leaving HMGB1 inactivated cells [52].

- A more recent effort from 2022 examined the role of HMGB1 in oxidative stress-induced mitochondrial dysfunctioning. Analysis revealed significantly inhibited mitochondrial biogenesis on HMGB1 neutralization, confirmed by enhanced mitochondrial fragmentation, Drp1, and PGC-1α (peroxisome proliferator-activated receptor γ co-activator 1α, controlling multiple exercise-related aspects of muscle function) expressions with impaired Mfn2 (mitofusin 2 gene) activities. Repetitive exposure to oxidative dust over 5 days resulted in substantial impairment of transcripts encoding mitochondrial respiration and metabolism (ATP synthase, NADUF, and UQCR) and glucose uptake. Confirmation was assured following the reversal of conditions on antibody-mediated HMGB1 neutralization [53].

The observations of above above-discussed studies unanimously pinpoint HMGB1 involvement in maintaining optimum mitochondrial functioning (upright organization, morphology). Besides, HMGB1 could be a potential therapeutic target for eliminating the cancer cells, *vis-à-vis* mitochondrial insult (fragmentation and impaired metabolism).

The Membrane-bound High Mobility Group Box Protein 1

From the viewpoint of a putative RAGE ligand, it is worth noting that overall RAGE and HMGB1 are not always confirmed indicators of cancer progression as these proteins remain distributed in soluble and membrane fractions of normal tissues and tumor cells. The prominent distinctions of HMGB1 activity therein are as summarized ahead.

- A noted effort from 2012 (featuring in Oncology Letters) observed that in the rat organ specimens, HMGB1 protein was largely prevalent as the soluble fraction and could be recovered in meager amounts from insoluble membrane fractions. The observations confirmed previous findings wherein most of the protein remained in supernatant amidst high-speed centrifugation and only a small portion sedimented with microsomal membrane fraction. Screening the sub-cellular prevalence of HMGB1 and RAGE in rat organs and Guerin ascites tumor cells revealed their soluble existence in normal tissues. Contrary to this, the tumor cells carried these proteins in an insoluble, membrane-bound form, wherein HMGB1 resided as a stable RAGE-complex only in the protein extract retrieved from tumor cells as a membrane constituent.
- Distribution pattern for HMGB1 protein revealed it as prominently prevalent as an insoluble membrane protein in cancer cells contrary to its soluble form in the protein extract from normal tissues. The study recalled the earlier efforts wherein a small HMGB1 portion was screened as associated with plasma membrane [54] and an accumulated prevalence of HMGB1 in murine erythroleukemia cells, on hexamethylene bisacetamide induction [55]. However, in this attempt, the investigators noticed increased HMGB1 extents in the tumor membrane regions, recognized as an evident cancer signature. Surprisingly, a stable HMGB1-RAGE complex was, thereafter, demonstrated as a prominent

hallmark of tumor cells, having HMGB1 in the membrane form. The reasons attributed to the membrane prevalence of HMGB1 and RAGE proteins in tumor cells were the cellular redistribution needed for stable HMGB1 and RAGE interactions. The membrane existence of these proteins is not noticed in normal cells as both proteins coexist in small extents as a membrane fraction [56].

- Regarding the membrane association of HMGB1, its prevalence as a cell-surface and membrane protein on activated platelets and early neurons in neurite outgrowth amidst development and nerve regeneration is well-established [54, 57]. Secondly, HMGB1 containing secretory endolysosomes can be fused with cell membranes and secreted after cellular activation through extracellular lysophosphatidylcholine (LPC) or ATP [58]. Yet another membrane vicinity function of HMGB1 is demonstrated analogous to its LPS binding protein, wherein HMGB1 functions as a carrier for several materials. Anchored by its highly charged aa, HMGB1 exhibits sticky proximity for binding with other substances such as DNA. Herein, studies have demonstrated significantly enhanced uptake of exogenous DNA by HMGB1 *via* trafficking across the mammalian cell membranes [59].

- Highlighting the ambiguity of HMGB1 sub-cellular localization and secretion, a recent 2023 study by *Cui and colleagues* examined the HMGB1 sub-cellular localization and secretion in starved glioma cells. The study used immunofluorescence microscopy, enzyme-linked immunosorbent assay (ELISA), sub-cellular fractionation western blotting, and immunoelectron microscopy to track the HMGB1 prevalence. Analysis revealed nuclear to cytoplasmic HMGB1 translocation and ultimately to ECM in glioma cells. The observation using sub-cellular fractionation revealed HMGB1 migration to membrane-bound compartments. The distinction was HMGB1 prevalence to the specific ER and mitochondria contact regions, recognized as mitochondria-associated membranes. Visualization through immunogold electron microscopy revealed HMGB1 translocation to the nucleus, cytoplasm, and ECM of starved glioma cells. It was noticed that while being in the cytoplasm, Au particles prevailed within or in the periphery of membrane structures, getting localized around or in the mitochondria, small vesicles, endosomes, coated vesicles, and autolysosomes. The distribution of HMGB1 in the mitochondrial-associated membranes (MAM, in the vicinity of ER membranes), remains biochemically distinct from pure ER and mitochondria. In the cytoplasm, HMGB1 is distributed non-uniformly, acquiring a patchy regime, with MAM-localized proteins resembling morphology [60 - 62].

The Extracellular High Mobility Group Box Protein 1

The primary location of HMGB1 is the nucleus and its release into the extracellular environment as endogenous DAMP is triggered as a warning indication to the vicinity environment about a likely loss of homeostatic balance. Excessive quantities of extracellular HMGB1 result in the release of pro-inflammatory cytokines, such as TNF, IL-1, and IL-6 (interleukins). Active HMGB1 release commenced *via* gradual translocation of the HMGB1 nuclear pool to the cytosol [63]. The release of HMGB1 into the extracellular environment is triggered passively as a prototypical DAMP from dying cells or being profusely secreted from active or stressed cells, irrespective of the concerned tissue.

- As extracellular DAMPs, HMGB1 plays a key role in activating immune cells by interacting with multiple cell surface receptors, stimulating the macrophages and endothelial cells to generate cytokines and chemokines, sustaining a positive feedback loop of inflammatory responses. HMGB1 forms complexes with prominent PRRs, the receptors sustaining innate immunity and identifying pathogen-specific, conserved molecular patterns [64, 65]. Such actions of HMGB1 are exclusively mediated *via* interactions with RAGE and TLRs. Erstwhile prominent HMGB1 receptors comprise CXC chemokine receptor type 4 (CXCR4), macrophage antigen-1, syndecan-3, and CD24-Siglec-10 [66]. These interactions inevitably culminate in aggravated tissue inflammation and related pathologies such as diabetes, chronic sepsis, AD, and cancer, indicating the pro-inflammatory essence of HMGB1 [67, 68].
- Relating to its inflammatory involvement, a 2020 opinion review by *Andersson and the team* highlighted the significance of HMGB1 as a therapeutic target against COVID-19. In the noted view of authors, HMGB1 as endogenous DAMP, lays the foundation for an inflammatory environment through two pathways. The more common route involves disulfide-HMGB1-activated TLR4 receptors manifesting pro-inflammatory cytokine release. The extracellular HMGB1 released from perishing cells or those produced from activated innate immunity cells, forms complexes with extracellular DNA, RNA, and other DAMP molecules released after lysis-induced cell death.
- Such complexes are endocytosed *via* RAGE, expressed abundantly in the lungs, and thereafter, trafficked to the endolysosomal system, the balance being inactivated by excessive HMGB1 concentrations. Opportune redox-sensitive molecules concomitantly interact with cytosolic pro-inflammatory receptors, aggravating the inflammasome activation. Extracellular SARS-CoV-2 RNA has a definite probability of invading the cytosol through HMGB1-aided transport accompanied by the leaky texture of lysosomes. The possibility of HMGB1 interaction with SARS-CoV-2 RNA is augmented by the bound state of HMGB1 (with other molecules) as PAMPs and DAMPs. Further, an almost 40% reduction in HMGB1 plasma extents in arterial blood than that of venous, creates feasibility of these HMGB1 complexes formation in the lungs. The second factor leading to a high probability of such events is the

uncharacteristically high RAGE expression in the pulmonary tissues. This RAGE abundance in pulmonary tissues facilitates endocytosis of vulnerable molecules destined for destruction *via* lysosomes at physiological HMGB1 extents but corresponding to higher extents, detrimental inflammasome activation is inescapable. As a result, mounting stress caused by hyper-inflammation triggered by congested pulmonary physiology tempts for apoptosis in pulmonary endothelial cells from females but necrosis in cells from males. Thereby, minimizing the extracellular HMGB1 expression could moderate the inflammation-complicated interactions with SARS-CoV-2 mRNA and its trafficking beyond pulmonary regions [69].

- Recently, a 2021 study by *Lu and accomplices* examined the role of HMGB1 in macrophage inflammation amidst the systemic lupus erythematosus (SLE, a chronic autoimmune disorder) pathogenesis. Taking note of the unknown mechanism for extracellular HMGB1-mediated build-up of activated lymphocyte-derived DNA (ALD-DNA) in endosomes and aggravated macrophage inflammation, investigators screened HMGB1-driven modulation in macrophage DNA uptake pathways. Analysis revealed alone ALD-DNA uptake (in macrophages) by a weaker, unselective micropinocytosis which was potently transformed to an efficient, specific clathrin/caveolin-1 mediated receptor-driven pathway on interception by extracellular HMGB1. This transformation to a faster and more prompter DNA uptake by macrophages caused a rapid and abundant ALD-DNA aggregation in endosomes. Digging into the working mechanism, the analysis revealed this modulation in macrophage's DNA uptake being facilitated *via* DNA and TLR2/TLR4 binding ability of HMGB1. A peculiar aspect recalled by the investigators is the effect of macrophage-DNA uptake by varied HMGB1 receptors, wherein RAGE-mediated endocytosis (*via* clathrin or lipid raft independent pathway) caused macrophage pyroptosis rather than inflammation [70, 71]. So, the observation of TLR2/TLR4 bound HMGB1 driven DNA uptake was conceptualized as ALD-DNA driven macrophage inflammation, clarifying the SLE pathogenesis [72].

Thus, the multiple residence of HMGB1 in the cells establishes HMGB1 as a multifunctional, redox-sensitive protein with a concentration optimum being the basis of sustaining the native physiology with its excessive amounts that may have negative consequences to the cell survival.

Functions of High Mobility Group Box Protein 1

Owing to a nuclear, cytoplasmic, and circumstantially governed extracellular prevalence of HMGB1, it has functional activities related to all these cellular locations. Ahead is the summary of diverse HMGB1 functions:

Nuclear Functions of High Mobility Group Box Protein 1

One of the main chromatin-associated non-histone proteins that acts as a DNA chaperone in the nucleus is HMGB1. HMGB1's completely reduced form is thought to be responsible for its nuclear activities. HMGB1 has been proven as a universal biosensor for nucleic acids, being engaged in replication, transcription, chromatin remodeling, V(D)J recombination, DNA repair, and genome stability.

Role of High Mobility Group Box Protein 1 in Gene Regulation

HMGB1 is associated with chromatin, binding the DNA minor grove with higher proximity to the non-canonical DNA confirmations such as single-stranded DNA, DNA-containing cruciform or bent structures, super-coiled, and Z-DNA. Under *in vitro* circumstances, HMGB1 can remove histone H1 from severely bent DNA. It can rearrange the canonical nucleosome, lowering the structural barriers to transcription factor binding. On binding to DNA, HMGB1 loops the molecule to increase its flexibility. This process allows it to influence the activities of many gene promoters by influencing transcription factor binding and/ or bringing distant regulatory regions in proximity.

Role of High Mobility Group Box Protein 1 in Repairing the Damaged DNA

Several studies elucidate HMGB1's role in single and double-stranded DNA repair. The role of HMGB1 in nucleotide excision repair (NER) has been of significant interest although some consequences in the NER using *in vitro* systems have been reported rather conflictingly results. HMGB1 is also involved in nucleotide mismatch repair (MMR) and base excision repair (BER) pathways. Some studies have also screened HMGB1 involvement in repairing the double-strand breaks such as non-homologous end joining (NHEJ).

Role of High Mobility Group Box Protein 1 in Antibody Diversity

HMGB1 is involved in V(D)J recombination through its cofactor-like actions for the recombination-activating genes (RAG) complex, regulating the rearrangement and recombination of genes encoding immunoglobulin and T-cell receptors. The functional mechanism of HMGB1 herein involves stimulating cleavage and RAG protein binding at the 23 bp spacer of conserved recombination signal sequence (RSS).

Other Nuclear HMGB1 Functions of High Mobility Group Box Protein 1

Multiple independent studies described the following functions of HMGB1:

- Modulating the binding of sterol regulatory element-binding proteins (SREBPs) such as sterol regulatory element binding transcription factor 1 (SREBF1) to their cognate DNA sequences, increasing their transcriptional activities.
- Enables TP53 (a tumor suppressor protein) to bind with DNA.
- Though not 100% confirmed, HMGB1 has been suspected of being implicated in mitochondrial quality control and autophagy regulation in a transcription-dependent manner involving HSPB1.
- HMGB1 can modulate the telomerase activity, featuring a likely involvement in telomere maintenance.

Cytoplasmic Functions of High Mobility Group Box Protein 1

• TLR9 Mediated High Mobility Group Box Protein 1 Activities

- HMGB1 functions as a sensor and chaperone in the cytoplasm, activating the TLR9-mediated immune responses. Besides, HMGB1 is implicated in endosomal translocation and TLR9 activation in response to CpG-DNA functions *vis-à-vis* macrophages.

• Autophagic Actions of High Mobility Group Box Protein 1

- In the cytoplasm, HMGB1 is programmed to dissociate the BECN1: BcL-2 (apoptosis regulating proteins) complex (a regulator of autophagy) *via* competitive interaction with BECN1 (a novel phosphorylation substrate of the death-associated protein kinases (DAPK)), leading to autophagic activation. HMGB1 guards BECN1 and autophagy-related 5 (ATG5, a protein involved in autophagy) from calpain-mediated cleavage, thereby regulating their pro-autophagic and pro-apoptotic functions besides the extent and severity of inflammation-driven cell injury. In myeloid cells, HMGB1 has a protective role against endotoxemia and bacterial infection *via* promoting autophagy. HMGB1 is also associated with controlling oxidative stress-mediated autophagy.

Functions of HMGB1 in the Extracellular Compartment

- While necrotic cells release fully reduced HMGB1 that functions as a chemokine, the apoptotic cells viciously release disulfide HMGB1 that acts as a cytokine and sulfonyl HMGB1, promoting immunological tolerance. In the extracellular region, fully reduced HGMB1 is oxidized and in association with CXCL12, allocates the recruitment of inflammatory cells in the initial phase of tissue injury. The CXCL12-HMGB1 complex induces the CXCR4 homodimerization. Disulfide-conjugated HMGB1 binds transmembrane receptors, such as RAGE, TLR2, and TLR4, and the Triggering Receptors Expressed on Myeloid Cells (TREM1), activating the signal transduction pathways. The HMGB1 bound cells release cytokines/chemokines, TNF, IL-1, IL-6, IL-8, CCL2, CCL3, CCL4 (Chemokine (C-C) motif ligand 2,3,4), and chemokine Interferon-γ Inducible Protein 10 kDa (CXCL10), functioning as DAMP molecule for a strengthened immune response.

- HMGB1 also promotes secretion of IFN-γ by macrophage-stimulated natural killer (NK) cells in coordination with other cytokines such as IL-2 or IL-12. HMGB1 coordinates and integrates the innate and adaptive immune responses, inducing the migration of monocyte-derived immature dendritic cells and further regulation of neutrophil-adhesive and migratory functions. HMGB1 also assists the binding to class A CpG, generating cytokines in plasmacytoid dendritic cells alongside TLR9, MYD88, and RAGE-activated auto-reactive B cells.

- HMGB1 sustains the adaptive immune responses (T cell, B cells, and T regulatory cell actions) *via* conjugated chromatin immune complexes, promoting the B cell responsiveness to endogenous TLR9 ligands *via* a B-cell receptor (BCR)-dependent and RAGE-independent mechanism. Besides this, HMGB1 is also needed for tumor infiltration and activation of T-cells expressing the lymphotoxin LTA-LTB heterotrimer. Through these functions, HMGB1 promotes the progression of malignant tumors, acting without effector or regulatory T-cells. HMGB1 is also reported to limit T-cell proliferation, wherein released HMGB1-nucleosome complexes enable cytokine generation during apoptosis *via* TLR2-induced cytokine production. Finally, HMGB1 sustains the immunological tolerance of apoptotic cells, neutralizing their pro-inflammatory activities (on release by apoptotic cells) by reactive oxygen species (ROS)-driven oxidation, specifically on Cys-106 residue.

Pathophysiology of High Mobility Group Box Protein 1

- Demonstrated as contributing to the pathogenesis of multiple chronic inflammatory, autoimmune diseases and cancer.
- High serum HMGB1 extents are reported as critically implicated in multiple inflammatory events including sepsis, rheumatoid arthritis, atherosclerosis, chronic kidney disease, and SLE.
- Anticipated as being involved in multiple diseases distinguished *via* cell death and damage, including diabetes and AD.
- The nucleosome-associated HMGB1 release amidst secondary necrosis, is widely believed to have a possible role in SLE.
- HMGB1 can mediate regrowth and metastasis amidst chemotherapy of tumor cells in an AGER/RAGE expression-dependent regime.

S100/CALGRANULIN FAMILY

- In 1965 B.W. Moore first observed S100 proteins [73].
- S100 proteins represent a family of low molecular-weight proteins present in vertebrates and are distinguished by 2 Ca^{2+} binding domains, having a helix-loop-helix configuration (EF-hand kind).
- The prefix in the symbol "S100" signifies their 100% water solubility, forming saturated dispersions at neutral pH.
- The S100 protein family comprises the largest subgroup of Ca^{2+}-binding EF-hand (helix E-loop-helix F) protein group. These proteins have significantly diverse intracellular and extracellular functions *viz.* regulating Ca balance, cell apoptosis, migration, proliferation, differentiation, energy metabolism, and inflammation.
- S100 proteins are small, 10-12 kDa molecular weight acidic proteins, containing 2 distinct EF-hands, 4 α-helical segments, a central hinge region of variable length, and the N- and C- terminal variable domains.
- Contrary to the abundance of S100 genes in vertebrates, these are entirely missing in invertebrates.
- As of now, at least 25 proteins are known in the S100 protein family, wherein 21 have genes clustered at chromosome locus 1q21, termed the epidermal differentiation complex [74]. They are encoded by a family of genes with the *S100* prefix in their symbols, like, *S100A1*, *S100A2*, and *S100A3*. They are also recognized as DAMPs, causing a knockdown of aryl hydrocarbon receptors with a suppressed S100 protein expression in THP-1 cells (monocytes isolated from the peripheral blood).
- S100 staining encompasses the characterization of multiple tumors, including malignant melanoma, glial tumors, neurogenic tumors (*e.g.* schwannomas and neurofibromas), mesenchymal tumors (chondromas, chondrosarcomas, liposarcomas) and some histiocytic tumors.
- Calgranulin is an S100 Ca^{2+}-binding protein expressed on multiple cells, including renal epithelial cells and neutrophils. The proteins S100A8 and S100A9 together form a heterodimer, known as "calprotectin".
- Multiple S100 proteins, (like S100B, SA4, S100A8, S100A9, S100A12, and S100A13), are secreted to function in a cytokine-similar regime. For example, S100A8/A9 heterodimer acts as a chemotactic molecule during inflammation [75], S100B exhibits neurotrophic (at physiological concentration: nanomolar extents) or neurotoxic (micromolar levels) functions [76, 77] and S100A4 has angiogenic effects [78].
- The members of the S100 protein family are typically well-characterized multifunctional proteins expressed in a diverse spectrum of tissues. The constituent members *via* their interaction with multiple intracellular effector proteins are involved in regulating manifold cellular processes such as

contraction, motility, cell growth and differentiation, cell cycle progression, transcription, membrane structural organization, dynamics of cytoskeletal components, safeguarding from oxidative cell damage, *via* regulating protein phosphorylation and secretion [79].

S100 proteins are considerably diverse, concerning their structural distinctions and biological functions. The molecular findings observed for RAGE-S100 protein interactions are presented in Table **2** ahead [80].

Table 2. Experimental whereabouts of diverse S100-RAGE interactions concerning S100 proteins structure-function variability.

Members of the S100 family of proteins	Chromosomal loci	Prominent clinical activities known
S100A1-S100A18	1q21	**S100A1**: Interacts with V-domain of RAGE, promotes neurite outgrowth *via* activating NF-κB, in a RAGE-dependent manner. **S100A2**: Not much is reported for RAGE-S100A2 interaction, recent studies however claim S100A2-(GST-RAGE) interaction with micromolar affinity, sensitive Ca dependence, and binding on V-domain **S100A4**: *in vitro* RAGE binding in oligomeric and dimeric form, essentially needing Ca. Stimulates RAGE-mediated signaling for MMP13 activation in osteoarthritic cartilage **S100A5**: Interacts with RAGE at a micromolar affinity, binding to the V-domain and needing strict Ca dependence Further studies are needed to ascertain the physiological relevance of this interaction **S100A6**: Exhibits *in vitro* binding with RAGE through V and C2 domains, SPR screened interaction with GST-RAGE occurs in a Ca-dependent regime Effects in SH-SY5Y human neuroblastoma cells are reported *via* C2 domain binding. Binding with RAGE is unaffected by RAGE Glycosylation **S100A7**: Exhibits a chemotactic response *via* RAGE binding through granulocytes, monocytes, and lymphocytes **S100A8/A9**: Interacts with RAGE studied *via* immunoprecipitation. Promotes cell growth *via* p38MAPK, p44/42 kinase activation in tumor cells, and mediates endotoxin-induced cardiomyocyte functioning *via* RAGE dependence. RAGE-mediated signaling is noticed in S100A8/A9 treated HUVEC cells on AGEs pretreatment. A CML-modified S100A8/A9 in inflamed intestines activated NF-κB, exhibiting a RAGE-mediated inflammation.

(Table 2) cont.....

Members of the S100 family of proteins	Chromosomal loci	Prominent clinical activities known
S100A1-S100A18	1q21	**S100A11**: Modulates osteoarthritis *via* RAGE interaction, S100A11-RAGE activation manifests hypertrophy in chondrocytes, further screening is needed. Human keratinocytes also exhibit RAGE-mediated signaling *via* S100A11. **S100A12**: Hexameric S100A12 complexes with tetrameric RAGE in the C1-RAGE domain, irrespective of Ca presence, apo-S100A12-RAGE interaction being ~1000 fold weaker than those with Ca. **SPR: Recent studies on** S100A12-RAGE interaction revealed a sub-micromolar tetrameric S100A12-domain affinity. **S100A13**: RAGE-interaction was suggested *via* RAGE-mediated nuclear to cytoplasmic translocation in endothelial cells, on extracellular S100A13 addition. No molecular evidence yet for RAGE interaction, in-depth, studies are needed
S100B	21q22	• Binds strictly to the V-domain, higher levels correlated with brain trauma, ischemia, neurodegenerative, inflammatory, and psychiatric diseases. • Significant role in epileptogenesis, myocardial functioning, spatial and fear memory. RAGE-actions trigger neurite growth (nanomolar extents) and apoptosis (micromolar) • Low dose protects rat hippocampal cells from N-methyl-D-aspartate *via* NF-κB activation and RAGE engagement. Similar actions for LAN-5 neuroblastomas against Aβ peptide • **In human SH-Sy5Y cells**: 5 μM S100B promoted cell survival *via* PI3K/Akt/NF-κB pathway in a RAGE-dependent, ROS-generating regimen • **In Neuro2a cells**: 5 μM S100B activated mitogenic signaling *via* RAGE and p42/44 MAPK pathway • **In dorsal root ganglia neuron**: 500 nM S100B activated PI3/Akt kinase and NADPH oxidase *via* RAGE aggravated ROS generation • **Astrocytes**: Highest expression, release TNF-α, IL-6 in a RAGE-stimulated regime • **In monocytes**: RAGE dependent COX-2 activation *via* p38/ERK/NF-κB activation Generates superoxide radicals *via* NADPH oxidase activation Increased IP-10 mRNA and protein extents in a RAGE-driven manner • **In human aortic endothelial cells**: Induces monocyte chemoattractant protein a (MCP-1) expression*via*RAGE involvement • **In vascular smooth muscle cells**: Stimulated angiotensin II activated JAK2 phosphorylation and cell proliferation *via* RAGE involvement

Members of the S100 family of proteins	Chromosomal loci	Prominent clinical activities known
S100G	Xp22	Not a full member of the S100 family The major function is the cytosolic Ca^{+2} buffering, modulating Ca^{+2} absorption No extracellular function yet reported
S100P	4p16	A hallmark of ovarian, pancreatic, gastric, colorectal, breast and prostate cancers Forms a heterodimer with S100A1 under *in vitro* and *in vivo* conditions Mediates tumor growth, drug resistance and metastasis *via* RAGE binding Activates NF-κB *via* MAPK pathway in a RAGE driven regime in BxPC3, SW480 cells
S100Z	5q14	No extracellular function is yet reported

Significant diversity prevails pertaining to the large sub-members in S100A proteins, major reasons for this functional assortment are as follow:

- Broad diversification of the constituent family members (25 in humans).
- Distinct metal ion-binding properties of the individual S100 proteins.
- Spatial distribution in specific intracellular compartments or extracellular space, and
- The ability to form non-covalent homo- and hetero-dimers, allowing a dynamic exchange of the S100 subunits.

An emerging prospect of S100 proteins pertains to their null intrinsic catalytic activity. These RAGE ligands are generally believed to prevail like calmodulin and troponin C, exhibiting conformational modifications with modulated bioactivities *via* Ca^{2+}-binding. Upon Ca^{2+}-binding, the helices of S100 proteins rearrange to develop a cleft that forms the target protein binding site. In addition, some S100 members have been reported to bind Zn^{2+} and Cu^{2+}, suggesting their cofactor-like inter-dependence (rather than expected Ca^{+2}).

A befitting case of Ca^{2+} independent but Zn^{2+} assisted functioning of S100 proteins is the S100B interaction with tau protein (a prominent hallmark of AD, inhibition of tau protein phosphorylation by protein kinase II). Besides, Cu^{2+} binding to S100B is speculated to exhibit a neuroprotective impact. The Ca^{2+} and Zn^{2+} binding with S100A9, and S100A8 proteins is replete in myeloid cells, exhibiting multiple significant functions in both Ca^{+2} signaling as well as Zn^{+2} homeostasis although neither of these function as Ca^{2+} or Zn^{2+} buffering proteins in granulocytes.

Structure of S100 proteins

Most S100 proteins consist of two identical polypeptides (homodimers), held together by non-covalent linkages. The assembly is quite similar to calmodulins with the prominent distinction being the cell-specific expression pattern *i.e.* they are expressed in certain specific cells. The typical expression depends on environmental factors. However, calmodulin is ubiquitous, and a universal intracellular Ca^{2+} receptor is widely expressed in multiple cells.

Functions of S100 proteins

S100 proteins are normally present in cells derived from the neural crest (Schwann cells, and melanocytes), chondrocytes, adipocytes, myoepithelial cells, macrophages, Langerhans cells, dendritic cells, and keratinocytes. They may prevail in certain breast epithelial cells too.

S100 proteins have been screened as playing an important role in multiple intracellular and extracellular tasks [81] such as sustenance of protein phosphorylation, transcription factors, Ca^{2+} homeostasis, the dynamics of cytoskeletal components, enzymatic activities, cell growth and differentiation, and the inflammatory response. S100A7 (psoriasin) and S100A15 are known to function as cytokines in inflammation, particularly in autoimmune dermal pathologies, such as psoriasis [82].

Pathology Related to S100 Proteins

Several members of the S100 protein family have emerged as useful markers for certain tumors and epidermal differentiation. They have been reported to exist in melanomas, 100% schwannomas, 100% neurofibromas (weaker than schwannomas), 50% malignant peripheral nerve sheath tumors (may be weak or focal), paraganglioma stromal cells, histiocytoma, and clear-cell sarcomas.

Fig. (**5**) depicts the typical immunostained morphology of S100 distribution in paraganglioma stromal cells [83]. It is worth noting that S100 proteins are reliable markers for inflammatory diseases and are widely used as antimicrobials besides the optimization of cautionary measures about anatomic pathology.

LYSOPHOSPHATIDIC ACID

Lysophosphatidic acid (1- or 2-acyl-sn-glycerol 3-phosphate/radyl-glycer-1-phosphate, LPA) is a bioactive phospholipid having actions similar to growth factors. It is generated during the synthesis of cell membranes and is a well-known, robust extracellular signaling molecule in all eukaryotic tissues and blood plasma (serum). The information described here about the LPA fundamental molecular biology is presented after taking inputs from a 2021 review article by *Balijepalli and associates*. Readers are requested to refer to the cross-references in reference number [83].

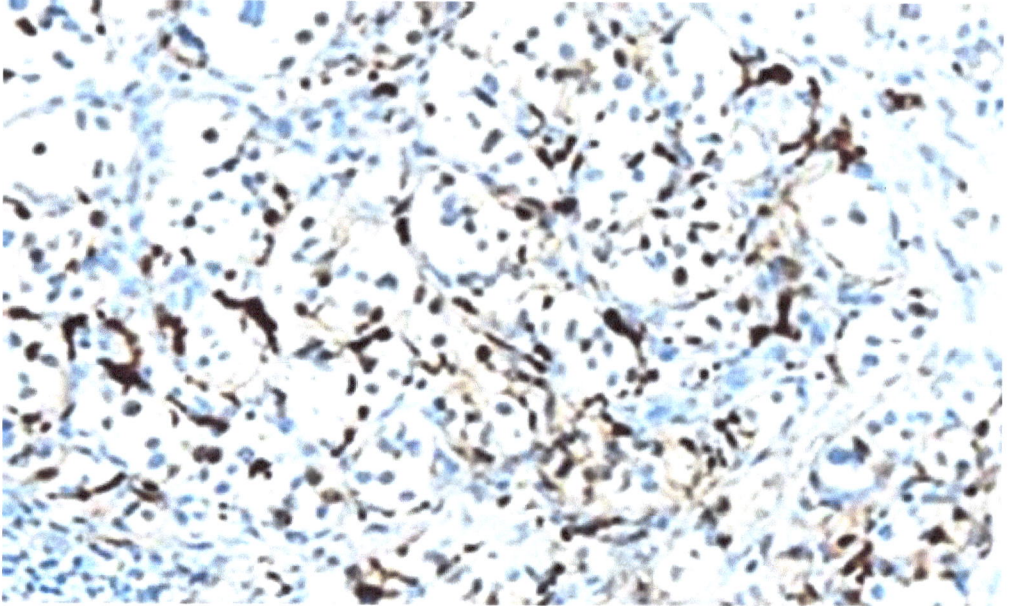

Fig. (5). S100 immunostaining marking the sustentacular cells in a paraganglioma.

- LPA includes all the small phospholipids having a single fatty acid substituent with an acyl or alkyl linkage to the glycerol backbone. An abundant molecule in mammals, often used in cell culture, is 18:1 (oleoyl)-LPA. LPA stimulates the proliferation of a wide variety of cells and is one of the significant growth factors in the sera, used for cell culture.

- Most LPA functions are the outcomes of the G protein-coupled receptors (GPCRs) agonist essence, also known as LPA receptors (LPARs), initially demonstrated as binding sites in 1994, prior to eventual cloning in 1996.

- LPARs are amongst the many GPCRs being studied as potential therapeutic targets for multiple human disorders, including cancer. Currently, the LPAR family includes six members, designated LPAR1-LPAR6. The LPAR nomenclature (LPAR 1-6) was established by the HUGO Gene Nomenclature Committee (https://www.genenames.org/data/genegroup/#!/group/205; accessed 10 August 2021).

- LPA acts through heterotrimeric G proteins G_s, $G_{i/o}$, $G_{q/11}$, and $G_{12/13}$, depending on the cell surface receptors.

- LPA stimulates cell proliferation, migration, and survival.

- In addition, LPA induces cellular morphology variations, enhances endothelial permeability, and inhibits gap-junction communication between adjacent cells. LPA also promotes wound healing and suppresses intestinal damage *via* irradiation.

- LPA receptors couple to multiple signaling pathways which are being clarified. These pathways include those initiated by the small GTPases RAS, RHO, and RAC, with RAS controlled cell-cycle progression and RHO/RAC signaling exhibiting a key role in cell migration and invasion.

- Recent evidence reveals LPA generation from the extracellular environment *via* lysophosphatidylcholine by 'autotaxin' (ATX/lysoPLD). ATX/lysoPLD is a ubiquitous exo-phosphodiesterase, originally identified as an autocrine motility factor for melanoma cells and is implicated in tumor progression. Through localized bioactive LPA generation, ATX/lysoPLD might support an invasive microenvironment for tumor cells, contributing to metastatic cascade.

- Both LPA receptors and ATX/lysoPLD are aberrantly expressed in several cancers.

- Targeted drugs against LPA receptors and/or ATX/lysoPLD have been demonstrated as effective towards reducing the tumor metastasis.

- LPA induces lung and mammary tumor proliferation, invasion and tumorigenesis *via* RAGE binding.

OLIGOMERIC FORMS OF AMYLOID-B PEPTIDE

- Amyloid β peptide comprises of 42 aa, derived from its precursor protein, amyloid β precursor protein (APP). The APP is a transmembrane glycoprotein that spans the membrane once. The gene for APP resides on chromosome 21.

- Amyloid β-oligomers are small assemblies or aggregates of this so-called amyloid β-protein, typically associated with the onset of AD.

- The amyloidogenic pathway is the process of Aβ biogenesis: APP is firstly cleaved by β-secretase, generating soluble β-APP fragments (sAPPβ) and C-terminal β fragments (CTFβ, C99). These C99 fragments are further cleaved by γ-secretase, generating APP intracellular domain (AICD) and Aβ.

- APP is initially cleaved by α-secretase (non-amyloidogenic pathway) or β-secretase (amyloidogenic pathway), generating membrane-tethered α- or β-C terminal fragments (CTFs). The cleavage of APP by α-secretase releases sAPPα from the cell surface, leaving an 83 AA C-terminal APP fragment (C83).

- The generation of sAPPα aggravates in response to electrical actions and the activation of muscarinic acetylcholine receptors, suggesting a neuronal activity aggravated α-secretase cleavage of APP. Further processing comprises an intramembrane cleavage of α- and β-CTFs by γ-secretase, generating P3 (3 kDa) and Aβ (4 kDa) peptides, respectively.

- Thus, the amyloidogenic processing of APP involves the sequential cleavage by β- and γ-secretase at the N and C termini of Aβ, respectively. The 99-amino-acid

C-terminal fragment of APP (C99) produced *via* β-secretase cleavage can be internalized and further processed by γ-secretase at multiple sites. This forms fragments of 43, 45, 46, 48, 49, and 51 AA cleaved to the main Aβ forms, the 40 AA Aβ40 and the 42 AA, Aβ42, in endocytic compartments.

- The γ-secretase cleavage of C99 generates an APP intracellular domain (AICD) capable of nuclear translocation, for regulating the gene expression, comprising apoptotic gene induction. The cleavage of APP/C99 by caspases forms a neurotoxic peptide (C31). The β-site APP cleaving enzyme is neuron-abundant and may accelerate the amyloidogenic processing in the brain to impair neuronal survival. The three-dimensional, human γ-secretase structure was first elucidated by single-particle cryo-electron microscopy in 2014 [84].

- The γ-secretase complex harbors a horseshoe-like transmembrane domain, comprising 19 transmembrane segments (TMs), and a large extracellular domain (ECD) from the nicastrin subunit, that localizes immediately above the hollow space formed by TM horseshoe. This structure plays a key role in understanding the γ-secretase functions.

- Typically, the γ-secretase complex consists of presenilin, nicastrin, presenilin enhancer 2, and anterior pharynx-defective 1 proteins. Presenilin is activated *via* auto-processing to generate N- and C-terminal cleavage products which contain aspartyl protease sites required for mature γ-secretase activity. Nicastrin, presenilin enhancer 2, and anterior pharynx-defective 1 are critical γ-secretase components and may modulate enzyme activity in response to physiological stimuli. This unique cleavage of APP provides therapeutic targets for AD [85].

- Human APP proteolysis occurs *via* non-amyloidogenic and amyloidogenic pathways. Non-amyloidogenic processing involves sequential APP treatment by membrane-bound αsecretases, cleaving within the Aβ domain to form the membrane-tethered α-C terminal fragment CTFα (C83) and the N-terminal APPα fragments. The CTFα is then cleaved by γ-secretases to form extracellular P3 and APP intracellular domain (AICD). Amyloidogenic processing of APP involves the sequential action of membrane-bound β and γsecretases, wherein the former cleaves APP in membrane-tethered C-terminal fragments β (CTFβ or C99) and N-terminal sAPPβ. CTFβ is subsequently cleaved by γ-secretases into the extracellular Aβ and APP intracellular domain (AICD).

Aβ Oligomers

- The amyloid fibrils are larger, insoluble entities, assembling as amyloid plaques to form histological lesions, the noted hallmarks of AD. The Aβ oligomers are soluble and may spread throughout the brain, exhibiting significant heterogeneity. There is a broad consensus for the preferential accumulation of a soluble high-molecular-weight species of ~100-200 kDa under relatively similar physiological conditions [86, 87].

- Aβ monomers can form higher-order assemblies ranging from low-molecula--weight oligomers, including dimers, trimers, tetramers, and pentamers, to moderate ones, (hexamers, nonamers, and dodecamers), forming protofibrils and fibrils.
- Distinct RAGE domains are involved in the pathogenesis of Aβ-induced cellular and neuronal toxicity *vis-à-vis* the aggregated state. Studies in this direction suggest a blockage of receptor sites as a potential strategy to attenuate neuronal death [88]. Further analysis using molecular modeling assays revealed RAGE's putative IgG V-like domains, in a dimeric conformation, encouraging thermodynamically favorable salt-bridge interactions with dimeric Aβ.
- Analysis revealed dimeric RAGE as a bi-valve structural shell, generating a flexible central cavity while the lateral walls are characterized by anti-parallel β-sheets of V-like domains. The study also suggested that as a cationic cluster, RAGE could serve as an ionic trap into which the ε-carboxymethylated lysyl residues of AGEs, may dock. The two Cys residues on each RAGE molecule are covalently bound to form an intramolecular disulfide bridge in the same manner as the conserved Cys residues in Ig-related molecules [89].
- Amyloid fibrils usually consist of 1-8 protofilaments (corresponding to a fibril), each being 2-7 nm in diameter, interacting laterally as flat ribbons to maintain a height of 2-7 nm (that of a single protofilament) and a width of 30 nm, more often as protofilaments.

ISLET AMYLOID POLYPEPTIDE

- Islet amyloid polypeptide (IAPP) is a 37-residue peptide secreted by islet β-cells which are stored with insulin inside the secretory granules and are released along with insulin. Studies on IAPP molecular significance report its modulation of insulin activity in skeletal muscles, influencing homeostasis, blood glucose levels, adiposity, and body weight [90].
- IAPP is processed from an 89-residue coding sequence, the proislet amyloid polypeptide (proIAPP, proamylin, proislet protein) formed in the pancreatic β-cells as a 67 AA, 7404 Dalton pro-peptide. This sequence undergoes post-translational modification including proteolytic cleavage to produce amylin.
- Islet amyloid polypeptide (IAPP or amylin) is one of the major secretory products of β-cells belonging to the pancreatic islets of Langerhans. It is a regulatory peptide with putative functions locally, (inhibiting insulin and glucagon secretion) and at distant targets.
- RAGE-IAPP binding is associated with manifold toxic responses, suggesting the IAPP/RAGE axis as a potential therapeutic target to mitigate the β cell impairment in metabolic diseases [33].

THE PHYSIOLOGICAL ROLE OF RECEPTOR FOR ADVANCED GLYCATION END PRODUCTS

Ligand engagement of RAGE stimulates multiple transcription factors such as NF-κB, STAT3, and signaling molecules including ERK, JNK, and AKT. Furthermore, RAGE and its multiple ligand interactions upregulate RAGE *via* positive feedback loops [91]. After RAGE activation, these ligands are expressed in various cells, including endothelial cells, vascular smooth muscle cells, lymphocytes, neurons, monocytes/macrophages, and podocytes [92 - 94]. A typical summary of such RAGE-ligand interaction outcomes is discussed in Table **2**.

Moreover, ligand-RAGE interactions are involved in the pathogenesis of diabetes mellitus, chronic renal failure, rheumatoid arthritis, atherosclerosis, neurodegenerative diseases, cancers, and immune/inflammatory responses [27, 31, 34, 41, 42, 75, 76]. Fig. (**6**) depicts the RAGE-ligand interactions mediated prominent physiological activities, such as bone formation, neurite differentiation, lung homeostasis, and immune regulation.

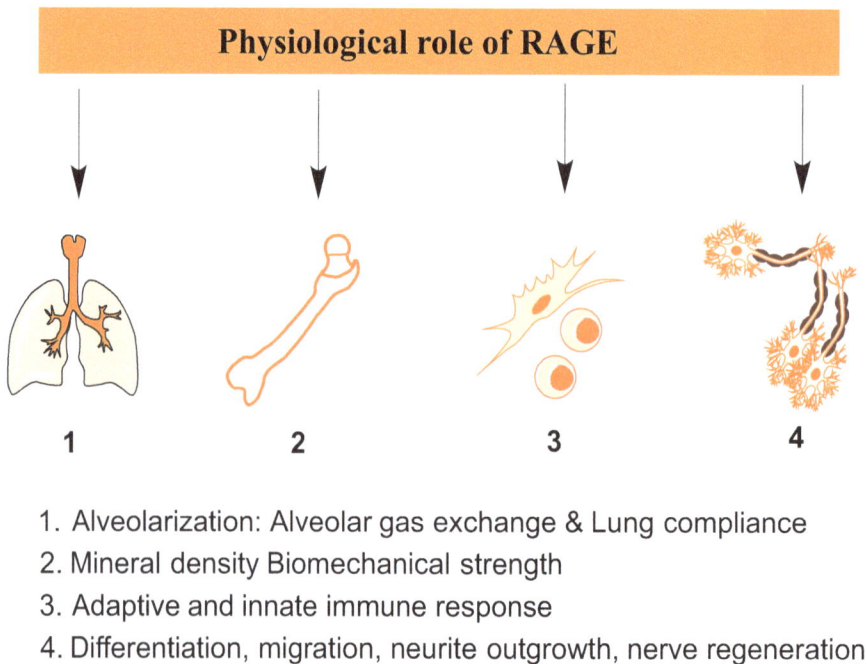

Physiological role of RAGE

1　　**2**　　**3**　　**4**

1. Alveolarization: Alveolar gas exchange & Lung compliance
2. Mineral density Biomechanical strength
3. Adaptive and innate immune response
4. Differentiation, migration, neurite outgrowth, nerve regeneration

Fig. (6). Prominent physiological actions of RAGE, depicting its functions in (a) lungs, (b) bone formation, (c) immune response regulation, and (d) neuron functioning.

Studies also decipher RAGE involvement in embryogenesis. In adult tissues, RAGE is highly expressed in the lungs, seconded by immunological regulation. Multiple studies describe the physiological role of RAGE in the coordinated nervous system functions and bone metabolism.

Thus, the physiological role of RAGE is discussed in the following sections:

Role of RAGE in Embryogenesis

Role of RAGE in Lung Homeostasis

Role of RAGE in Bone Morphogenesis

Role of RAGE in Neurite Outgrowth

Role of RAGE in Innate Immunity

Role of Receptor for Advanced Glycation End Products in Embryogenesis

- During rabbit embryogenesis, RAGE mRNA is expressed as early as in day 4 embryo, in the cell membrane of blastocysts, however, the exact role of RAGE in embryogenesis is yet to be understood. Studies on rats by *Reynolds and team* detected RAGE in the endothelial cells of embryonic vessels and alveolar capillaries [95].
- Erstwhile effort by *Hori and the group* noticed RAGE mRNA and protein expression in the embryonic (E17) cortical neurons with a regulatory influence for neurite outgrowth [25].
- However, most data dealing with RAGE expression during embryogenesis suggest its decisive role in lung organogenesis. RAGE expression in the fetal lung is associated with lung development. Despite its abundant expression in the lungs, RAGE depletion neither causes serious pulmonary changes nor reduces the lifespan. Therefore, the exact role of RAGE in embryogenic lung development remains unknown.

Role of Receptor for Advanced Glycation End Products in Lung Homeostasis

- Studies devoted to lung expression of RAGE demonstrate its histological actions in pulmonary endothelium, bronchial, and vascular smooth muscles besides the alveolar macrophages, lymphocytes, and on the visceral pleural surface in bovine tissues. Though it is well-reported that RAGE is considerably expressive in the alveolar epithelium cells there are conflicting results about the exact differentiation of these cells. Several studies have confirmed the RAGE prevalence in the basolateral membrane of lung alveolar type I (AT-1) epithelial cells. However, mice with impaired RAGE expression revealed no abnormalities in the alveolus. Noteworthy efforts by *Al-Robaiy and colleagues* elucidated the significance of high RAGE expression in sustaining respiratory motilities. Herein, analysis of RAGE knockout (RAGE$^{-/-}$) mice exhibited substantially dynamic lung compliance and suppressed expiratory airflow [96].

- The first-ever evidence demonstrating a preferential RAGE prevalence at intercellular contacts, was provided by *Bartling and associates* [97]. Based on these observations, RAGE was anticipated to have an adhesive role. Subsequently, it was noticed that RAGE indeed enhanced cellular adherence with the matrix proteins and, more excitingly, exhibited an ability to induce cell spreading. The relative extent of adherence could be moderated *via* blocking RAGE actions. These studies together established a significant role of RAGE in modulating alveolar epithelial cells-basement membrane adhesion. Besides the interaction of RAGE with the extracellular matrix (ECM) alongside the *trans* interaction of RAGE can promote tissue stiffness.

Role of Receptor for Advanced Glycation End Products in Bone Morphogenesis

- Recent investigations present experimental evidence for the role of RAGE in bone remodeling, demonstrating its prevalence in both osteoclasts and osteoblasts. For instance, in primary cultured bone marrow macrophages (BMM), RAGE is upregulated amidst the osteoclast differentiation. The BMMs differentiated from RAGE$^{-/-}$ bone marrow cultures exhibit a smaller size than the cells from WT cultures, suggesting a morphologic flaw in the *in vitro* differentiated RAGE$^{-/-}$ osteoclasts. Studies by *Zhou and colleagues* noticed reduced integrins-driven cell adhesion and signaling in BMMs [98].
- Erstwhile studies in RAGE knockout mice revealed enhanced bone mineral density and biomechanical strength, further being correlated with fewer osteoclasts. Studies by *Ding and associates* also revealed that lacking RAGE was the major source of diminished osteoclast function in RAGE knockout mice although it may not have a bearing on physiologically native osteoblast actions [99]. To a surprise, *Philip and team* observed attenuated gene expression related to osteoblast differentiation in RAGE-devoid mice [100]. Several other studies describe the role of RAGE in osteoclast maturation and function, elucidating its significance in the maturation of monocytes/macrophages representing immune system cell lineages.

Role of Receptor for Advanced Glycation End Products in Neurite Outgrowth

- The expression of RAGE in cell bodies and axonal actions was first demonstrated in the bovine nervous system motor neurons and certain populations of cortical neurons. Of note, the major RAGE ligands functional in the central nervous system are HMGB1, S100B, AGEs, and amyloid-β fibrils. The interaction of RAGE with AGEs or amyloid-β fibrils inevitably aggravates the inflammatory and oxidative stress, fueling the pathogenesis of certain neurological disorders *i.e.*, Parkinson's disease, AD, *etc*. On the other hand, RAGE ligands such as HMGB1 and S100B have demonstrated multiple neuronal functions, including neurite outgrowth, migration, differentiation, and survival. The expressive extent of RAGE-ligands also prevails as a decisive factor in predicting the effect of RAGE on cellular homeostasis.

- In general, low extents of RAGE-ligands proximity have been linked with beneficial events (neurite outgrowth) than the deleterious outcomes for higher extents (such as inflammatory signaling). Apart from the concentration, the cell type is another critical determinant for the outcome of a particular RAGE-ligand interaction. For example, in the neurons, HMGB1 binding of RAGE induces neurite outgrowth, while in microglia, it induces considerably deadly inflammatory signaling. The multiple effects of RAGE decipher its role in nerve regeneration *via* allocation of inflammation (in part) mediated *via* macrophage activation and axonal outgrowth. Blocking of RAGE impairs peripheral nerve regeneration, reducing the functional recovery of nerves besides decreasing the extent of infiltrating phagocytes.

Role of Receptor for Advanced Glycation End Products in Innate Immunity

- Enhanced RAGE activities have been viciously reported for inflammatory lesions of various diseases (rheumatoid arthritis, inflammatory kidney disease, arteriosclerosis, inflammatory bowel disease), mediated *via* RAGE's ability to bind multiple pro-inflammatory ligands (amyloid-β fibrils, S100 proteins, and HMGB1). These findings collectively establish RAGE as a prominent regulator of immune/inflammatory responses. Prevailing as a formidable link for multiple components of the innate immune system, *the* encoding gene for RAGE (AGER) is localized within the major histocompatibility (MHC) class III.

- RAGE is known to prevail on numerous immune cells including neutrophils, monocytes/macrophages, lymphocytes, and dendrites. In a significant attempt, *Liliensiek and teammates* demonstrated enhanced protection of RAGE-devoid mice from septic shock lethal consequences, exhibiting reliance on the innate immune response. Analysis revealed that RAGE-devoid mice exhibit a suppressed local inflammation with attenuated NF-κB activation in septic shock target organs [101]. RAGE impairment is characterized by a retarded inflammatory response *via* reduced central inflammation signaling pathways. However, RAGE stimulation on endothelium, mononuclear phagocytes, and lymphocytes has been elucidated as critical for cellular activation, releasing multiple pro-inflammatory mediators.

- Another useful study by *Chavaki and associates* revealed RAGE as a leukocyte integrin Mac-1 counter-receptor, exhibiting its direct involvement in leukocyte differentiation. The interaction of RAGE with leukocytes regulates the NF-κB dependent gene expression [28]. Recent studies have demonstrated the RAGE-TLR9 complex formation, the major DNA-screening receptor. This complex has been exhaustively used to quantify the pathogen DNA in multiple diseased conditions. Moving along the same lines, *Sirosis and associates* demonstrated RAGE moderated immunological recognition threshold for TLR9 activation, linking it with DNA uptake *via* endosomes [102].

- A surprising observation about RAGE-mediated inflammatory responses was made in the *in vivo* studies on RAGE-lacking mice, wherein no harsh inflammatory gathering was noticed in lung tissues. Studies probing RAGE's role in adaptive immune response were limited to innate immunity. Some recent observations offer crucial predictions in these matters, wherein sRAGE administration to naïve mice revealed a lymphocyte to spleen influx with a reduced B-cell count in the bone marrow. Apart from this, sRAGE also supported the B-cell activation and maturation, enabling a concomitant enhanced IgM and IgG secretion for T-cell activation. Separate studies established RAGE involvement in T-cell differentiation, adjacent to a Th1 phenotype with more abundant RAGE-mRNA in Th1 than Th2 cells [103].

THE PATHOPHYSIOLOGICAL ROLE OF RECEPTOR FOR ADVANCED GLYCATION END PRODUCTS

- Multiple studies decipher the role of high-level RAGE expression in mediating the diverse pathophysiological conditions. The major disorders aggravated and worsened *via* increased RAGE expression are asthma, fibrosis, COPD, acute lung injury (ALI), acute respiratory distress syndrome (ARDS), chronic bronchitis (CB), and several others of pulmonary origin. Those aggravated from the non-respiratory background are type 2 diabetic vascular complications, cardiovascular impediments, AD, inflammatory bowel disease, and other non-communicable ones. Except for lungs (a high native RAGE expression), the elevated RAGE expression has been reported as a critical hallmark for most cancers (namely, ovarian, pancreatic, colon, and hepatocellular). Readers are suggested to refer to other chapters of this book to know about RAGE's involvement in complicating other diseases.

- On the RAGE-mediated complication of various diseases; one study from our group reported the RAGE-TNFα interaction as the aggravating source of ROS and eventually, enhanced oxidative stress. Further trouble mounted by the RAGE expression in these conditions, mediated *via* NADPH oxidase and mitochondria [104]. Another significant finding associating RAGE with diseased conditions establishes its involvement in aggravating inflammatory stress *via* activation of the pro-inflammatory transcription factor, NF-κB [105].

- Apart from NF-κB, early growth response (Egr) gene-1, a distinct Zn-finger harboring transcription factor has been reported as being activated *via* tobacco smoke-triggered RAGE expression [95]. A related study on smoking-induced mice by *Reynolds and accomplices* revealed the RAGE-Ras-NF-κB axis as the aggravator for inflammation in general and specifically, for multiple lung disorders [106]. Similarly, the Ras-raf-MAPK pathway has been reported as critical for cell-cycle regulation [107, 108]. Apart from NF-κB and Egr-1, the transcription factor Nrf-2 has been demonstrated to mediate RAGE invasion of epithelial cells, wherein impaired RAGE actions resulted in moderated cigarette smoke-induced lung epithelial cell damage *vis-à-vis* Nrf2/DAMP signaling [109]. In this reference, an earlier study by *Schmidt and associates* observed RAGE-induced expression of vascular cell adhesion molecule 1 (VCAM-1), a cell surface pro-inflammatory and pro-oxidative adhesion molecule expressed on endothelial cells implicated in vasculopathy [110].

• Studies also established RAGE involvement in trauma-induced leukocyte adhesion for cremaster muscle venules, whereby, altered ligand-RAGE axis, sRAGE, and ICAM-1 expressions instigate the neutrophilic inflammation [111, 112]. Subsequent attempts by *Sukkar and colleagues* confirmed these findings, correlating the neutrophilic airway inflammation in asthma and COPD with reduced sRAGE (enhanced RAGE) expression [113]. Therefore, it is essential to understand the sRAGE expression for COPD occurrence and other diseases since sRAGE acts as a decoy receptor.

CONCLUSION

Altogether in this chapter, we have discussed various roles of RAGE in health and physiology. This cell surface immunoglobulin class of molecule was 1st discovered in the pulmonary tissues, where it is abundantly expressed. The chapter begins with a description of the salient aspects of the RAGE gene along with its various splice variants. Thereafter, the RAGE protein is characterized with a discussion on the solubilized forms of the RAGE gene *i.e.* sRAGE and esRAGE. Further moving on to the physiological relevance of RAGE, its diversified biochemical activities have been discussed through its ligand-specific interactions. The major biochemical aspects of RAGE-ligand biochemistry have been highlighted for AGEs, HMGB1, S100/Calgranulin family, lysophosphatidic acid, oligomeric forms of amyloid-β peptide, and islet amyloid-β peptide. The multiligand RAGE interactions decipher it as a prominent cell-signaling molecule exhibiting distinct responses in varied biochemical environments. Thereafter, the physiological significance of RAGE has been illustrated by its multifunctional activities towards the sustenance of embryogenesis, lung homeostasis, bone morphogenesis, neurite outgrowth, and innate immunity. While data about embryogenesis is not adequate except for some inceptive clues in rat and mouse models, bone morphogenesis does involve RAGE in the functional responses of osteoblasts and osteoclasts. The role of RAGE in neurite outgrowth is exhibited *via* its interactions with HMGB1 and S100, sustaining the outgrowth, migration, differentiation, and survival. For innate immunity, the involvement of RAGE *via* interactions with the proinflammatory and prooxidative transcription factors NF-κB, and Nrf-2 has been discussed. The most curious and turning point observation has been the role of RAGE in regulating normal lung physiology as per the analysis in rat and mouse models. Finally, a very brief outline of the role on various pathophysiological conditions is mentioned. Nevertheless, improved precision in biophysical characterization continues to enrich the interest in screening diverse RAGE expressions as decisive events in physiological and pathological activities. As the number of splice variants reported for humans remains low, much is anticipated from the diverse biochemistry of RAGE, encouraged by its multiligand interactions.

REFERENCES

[1]   Schmidt AM, Vianna M, Gerlach M, *et al.* Isolation and characterization of two binding proteins for advanced glycosylation end products from bovine lung which are present on the endothelial cell surface. J Biol Chem 1992; 267(21): 14987-97.
        [http://dx.doi.org/10.1016/S0021-9258(18)42137-0] [PMID: 1321822]

[2]   Hudson BI, Carter AM, Harja E, *et al.* Identification, classification, and expression of *RAGE* gene splice variants. FASEB J 2008; 22(5): 1572-80.
        [http://dx.doi.org/10.1096/fj.07-9909com] [PMID: 18089847]

[3]   Zhang L, Bukulin M, Kojro E, *et al.* Receptor for advanced glycation end products is subjected to protein ectodomain shedding by metalloproteinases. J Biol Chem 2008; 283(51): 35507-16.
        [http://dx.doi.org/10.1074/jbc.M806948200] [PMID: 18952609]

[4]   Raucci A, Cugusi S, Antonelli A, *et al.* A soluble form of the receptor for advanced glycation endproducts (RAGE) is produced by proteolytic cleavage of the membrane-bound form by the sheddase a disintegrin and metalloprotease 10 (ADAM10). FASEB J 2008; 22(10): 3716-27.
        [http://dx.doi.org/10.1096/fj.08-109033] [PMID: 18603587]

[5]   Brett J, Schmidt AM, Yan SD, *et al.* Survey of the distribution of a newly characterized receptor for advanced glycation end products in tissues. Am J Pathol 1993; 143(6): 1699-712.
        [PMID: 8256857]

[6]   Winden DR, Ferguson NT, Bukey BR, *et al.* Conditional over-expression of RAGE by embryonic alveolar epithelium compromises the respiratory membrane and impairs endothelial cell differentiation. Respir Res 2013; 14(1): 108.
        [http://dx.doi.org/10.1186/1465-9921-14-108] [PMID: 24134692]

[7]   Demling N, Ehrhardt C, Kasper M, Laue M, Knels L, Rieber EP. Promotion of cell adherence and spreading: a novel function of RAGE, the highly selective differentiation marker of human alveolar epithelial type I cells. Cell Tissue Res 2006; 323(3): 475-88.
        [http://dx.doi.org/10.1007/s00441-005-0069-0] [PMID: 16315007]

[8]   Wolf L, Herr C, Niederstraßer J, Beisswenger C, Bals R. Receptor for advanced glycation endproducts (RAGE) maintains pulmonary structure and regulates the response to cigarette smoke. PLoS One 2017; 12(7): e0180092.
        [http://dx.doi.org/10.1371/journal.pone.0180092] [PMID: 28678851]

[9]   Mukherjee TK, Mukhopadhyay S, Hoidal JR. Implication of receptor for advanced glycation end product (RAGE) in pulmonary health and pathophysiology. Respir Physiol Neurobiol 2008; 162(3): 210-5.
        [http://dx.doi.org/10.1016/j.resp.2008.07.001] [PMID: 18674642]

[10] Khaket TP, Kang SC, Mukherjee TK. The potential of receptor for advanced glycation end products (RAGE) as a therapeutic target for lung associated diseases. Curr Drug Targets 2019; 20(6): 679-89.
[http://dx.doi.org/10.2174/1389450120666181120102159] [PMID: 30457049]

[11] Vissing H, Aagaard L, Tommerup N, Boel E. Localization of the human gene for advanced glycosylation end product-specific receptor (AGER) to chromosome 6p21.3. Genomics 1994; 24(3): 606-8.
[http://dx.doi.org/10.1006/geno.1994.1676] [PMID: 7713518]

[12] Sugaya K, Fukagawa T, Matsumoto K, *et al.* Three genes in the human MHC class III region near the junction with the class II: gene for receptor of advanced glycosylation end products, PBX2 homeobox gene and a notch homolog, human counterpart of mouse mammary tumor gene int-3. Genomics 1994; 23(2): 408-19.
[http://dx.doi.org/10.1006/geno.1994.1517] [PMID: 7835890]

[13] Ensemble Genome Browser 91, 2018.

[14] Sterenczak KA, Willenbrock S, Barann M, *et al.* Cloning, characterisation, and comparative quantitative expression analyses of receptor for advanced glycation end products (RAGE) transcript forms. Gene 2009; 434(1-2): 35-42.
[http://dx.doi.org/10.1016/j.gene.2008.10.027] [PMID: 19061941]

[15] Yonekura H, Yamamoto Y, Sakurai S, *et al.* Novel splice variants of the receptor for advanced glycation end-products expressed in human vascular endothelial cells and pericytes, and their putative roles in diabetes-induced vascular injury. Biochem J 2003; 370(3): 1097-109.
[http://dx.doi.org/10.1042/bj20021371] [PMID: 12495433]

[16] Prasad K. Low levels of serum soluble receptors for advanced glycation end products, biomarkers for disease state: myth or reality. Int J Angiol 2014; 23(1): 011-6.
[http://dx.doi.org/10.1055/s-0033-1363423] [PMID: 24627612]

[17] Tam XHL, Shiu SWM, Leng L, Bucala R, Betteridge DJ, Tan KCB. Enhanced expression of receptor for advanced glycation end-products is associated with low circulating soluble isoforms of the receptor in Type 2 diabetes. Clin Sci (Lond) 2011; 120(2): 81-9.
[http://dx.doi.org/10.1042/CS20100256] [PMID: 20726839]

[18] Vazzana N, Santilli F, Cuccurullo C, Davì G. Soluble forms of RAGE in internal medicine. Intern Emerg Med 2009; 4(5): 389-401.
[http://dx.doi.org/10.1007/s11739-009-0300-1] [PMID: 19727582]

[19] Jules J, Maiguel D, Hudson BI. Alternative splicing of the RAGE cytoplasmic domain regulates cell signaling and function. PLoS One 2013; 8(11): e78267.
[http://dx.doi.org/10.1371/journal.pone.0078267] [PMID: 24260107]

[20] Soriano JB, Abajobir AA, Abate KH, *et al.* Global, regional, and national deaths, prevalence, disability-adjusted life years, and years lived with disability for chronic obstructive pulmonary disease and asthma, 1990–2015: a systematic analysis for the Global Burden of Disease Study 2015. Lancet Respir Med 2017; 5(9): 691-706.
[http://dx.doi.org/10.1016/S2213-2600(17)30293-X] [PMID: 28822787]

[21] Repapi E, Sayers I, Wain LV, *et al.* Genome-wide association study identifies five loci associated with lung function. Nat Genet 2010; 42(1): 36-44.
[http://dx.doi.org/10.1038/ng.501] [PMID: 20010834]

[22] Miller S, Henry AP, Hodge E, *et al.* The Ser82 RAGE variant affects lung function and serum RAGE in smokers and sRAGE production *in vitro*. PLoS One 2016; 11(10): e0164041.
[http://dx.doi.org/10.1371/journal.pone.0164041] [PMID: 27755550]

[23] Serveaux-Dancer M, Jabaudon M, Creveaux I, *et al.* Pathological implications of receptor for advanced glycation end-product (AGER) gene polymorphism. Dis Markers 2019; 2019: 1-17.
[http://dx.doi.org/10.1155/2019/2067353] [PMID: 30863465]

[24] Neeper M, Schmidt AM, Brett J, *et al.* Cloning and expression of a cell surface receptor for advanced glycosylation end products of proteins. J Biol Chem 1992; 267(21): 14998-5004.
[http://dx.doi.org/10.1016/S0021-9258(18)42138-2] [PMID: 1378843]

[25] Hori O, Brett J, Slattery T, *et al.* The receptor for advanced glycation end products (RAGE) is a cellular binding site for amphoterin. Mediation of neurite outgrowth and co-expression of rage and amphoterin in the developing nervous system. J Biol Chem 1995; 270(43): 25752-61.
[http://dx.doi.org/10.1074/jbc.270.43.25752] [PMID: 7592757]

[26] Hofmann MA, Drury S, Fu C, *et al.* RAGE mediates a novel proinflammatory axis: a central cell surface receptor for S100/calgranulin polypeptides. Cell 1999; 97(7): 889-901.
[http://dx.doi.org/10.1016/S0092-8674(00)80801-6] [PMID: 10399917]

[27] Yan SD, Chen X, Fu J, *et al.* RAGE and amyloid-β peptide neurotoxicity in Alzheimer's disease. Nature 1996; 382(6593): 685-91.
[http://dx.doi.org/10.1038/382685a0] [PMID: 8751438]

[28] Chavakis T, Bierhaus A, Al-Fakhri N, *et al.* The pattern recognition receptor (RAGE) is a counterreceptor for leukocyte integrins: a novel pathway for inflammatory cell recruitment. J Exp Med 2003; 198(10): 1507-15.
[http://dx.doi.org/10.1084/jem.20030800] [PMID: 14623906]

[29] Kim HJ, Jeong MS, Jang SB. Molecular characteristics of RAGE and advances in small-molecule inhibitors. Int J Mol Sci 2021; 22(13): 6904.
[http://dx.doi.org/10.3390/ijms22136904] [PMID: 34199060]

[30] Yamakawa N, Uchida T, Matthay MA, Makita K. Proteolytic release of the receptor for advanced glycation end products from *in vitro* and *in situ* alveolar epithelial cells. Am J Physiol Lung Cell Mol Physiol 2011; 300(4): L516-25.
[http://dx.doi.org/10.1152/ajplung.00118.2010] [PMID: 21257730]

[31] Yamamoto H, Watanabe T, Yamamoto Y, *et al.* RAGE in diabetic nephropathy. Curr Mol Med 2007; 7(8): 752-7.
[http://dx.doi.org/10.2174/156652407783220769] [PMID: 18331233]

[32] Taguchi A, Blood DC, del Toro G, *et al.* Blockade of RAGE–amphoterin signalling suppresses tumour growth and metastases. Nature 2000; 405(6784): 354-60.
[http://dx.doi.org/10.1038/35012626] [PMID: 10830965]

[33] Abedini A, Cao P, Plesner A, *et al.* RAGE binds preamyloid IAPP intermediates and mediates pancreatic β cell proteotoxicity. J Clin Invest 2018; 128(2): 682-98.
[http://dx.doi.org/10.1172/JCI85210] [PMID: 29337308]

[34] Rai V, Touré F, Chitayat S, *et al.* Lysophosphatidic acid targets vascular and oncogenic pathways *via* RAGE signaling. J Exp Med 2012; 209(13): 2339-50.
[http://dx.doi.org/10.1084/jem.20120873] [PMID: 23209312]

[35] Xue J, Manigrasso M, Scalabrin M, *et al.* Change in the molecular dimension of a RAGE-ligand complex triggers RAGE signaling. Structure 2016; 24(9): 1509-22.
[http://dx.doi.org/10.1016/j.str.2016.06.021] [PMID: 27524199]

[36] Park H, Boyington JC, Boyington JC. The 1.5 Å crystal structure of human receptor for advanced glycation endproducts (RAGE) ectodomains reveals unique features determining ligand binding. J Biol Chem 2010; 285(52): 40762-70.
[http://dx.doi.org/10.1074/jbc.M110.169276] [PMID: 20943659]

[37] Koch M, Chitayat S, Dattilo BM, *et al.* Structural basis for ligand recognition and activation of RAGE. Structure 2010; 18(10): 1342-52.
[http://dx.doi.org/10.1016/j.str.2010.05.017] [PMID: 20947022]

[38] Leclerc E, Fritz G, Vetter SW, Heizmann CW. Binding of S100 proteins to RAGE: An update. Biochim Biophys Acta Mol Cell Res 2009; 1793(6): 993-1007.

[http://dx.doi.org/10.1016/j.bbamcr.2008.11.016] [PMID: 19121341]

[39]   Perrone A, Giovino A, Benny J, Martinelli F. Advanced Glycation End Products (AGEs): biochemistry, signaling, analytical methods, and epigenetic effects. Oxid Med Cell Longev 2020; 2020: 1-18.
[http://dx.doi.org/10.1155/2020/3818196] [PMID: 32256950]

[40]   Kwak MS, Kim HS, Lee B, Kim YH, Son M, Shin JS. Immunological significance of HMGB1 post-translational modification and redox biology. Front Immunol 2020; 11: 1189.
[http://dx.doi.org/10.3389/fimmu.2020.01189] [PMID: 32587593]

[41]   Wang H, Bloom O, Zhang M, *et al.* HMG-1 as a late mediator of endotoxin lethality in mice. Science 1999; 285(5425): 248-51.
[http://dx.doi.org/10.1126/science.285.5425.248] [PMID: 10398600]

[42]   Andersson U, Wang H, Palmblad K, *et al.* High mobility group 1 protein (HMG-1) stimulates proinflammatory cytokine synthesis in human monocytes. J Exp Med 2000; 192(4): 565-70.
[http://dx.doi.org/10.1084/jem.192.4.565] [PMID: 10952726]

[43]   Blott EJ, Griffiths GM. Secretory lysosomes. Nat Rev Mol Cell Biol 2002; 3(2): 122-31.
[http://dx.doi.org/10.1038/nrm732] [PMID: 11836514]

[44]   Gardella S, Andrei C, Ferrera D, *et al.* The nuclear protein HMGB1 is secreted by monocytes *via* a non-classical, vesicle-mediated secretory pathway. EMBO Rep 2002; 3(10): 995-1001.
[http://dx.doi.org/10.1093/embo-reports/kvf198] [PMID: 12231511]

[45]   Rausch MP, Taraszka Hastings K. GILT modulates CD4+ T-cell tolerance to the melanocyte differentiation antigen tyrosinase-related protein 1. J Invest Dermatol 2012; 132(1): 154-62.
[http://dx.doi.org/10.1038/jid.2011.236] [PMID: 21833020]

[46]   Lackman RL, Cresswell P. Exposure of the promonocytic cell line THP-1 to *Escherichia coli* induces IFN-gamma-inducible lysosomal thiol reductase expression by inflammatory cytokines. J Immunol 2006; 177(7): 4833-40.
[http://dx.doi.org/10.4049/jimmunol.177.7.4833] [PMID: 16982925]

[47]   Lackman RL, Jamieson AM, Griffith JM, Geuze H, Cresswell P. Innate immune recognition triggers secretion of lysosomal enzymes by macrophages. Traffic 2007; 8(9): 1179-89.
[http://dx.doi.org/10.1111/j.1600-0854.2007.00600.x] [PMID: 17555533]

[48]   Chiang HS, Maric M. Lysosomal thiol reductase negatively regulates autophagy by altering glutathione synthesis and oxidation. Free Radic Biol Med 2011; 51(3): 688-99.
[http://dx.doi.org/10.1016/j.freeradbiomed.2011.05.015] [PMID: 21640818]

[49]   Tang D, Kang R, Livesey KM, *et al.* High-mobility group box 1 is essential for mitochondrial quality control. Cell Metab 2011; 13(6): 701-11.
[http://dx.doi.org/10.1016/j.cmet.2011.04.008] [PMID: 21641551]

[50]   Qi L, Sun X, Li FE, *et al.* HMGB1 promotes mitochondrial dysfunction-triggered striatal neurodegeneration *via* autophagy and apoptosis activation. PLoS One 2015; 10(11): e0142901.
[http://dx.doi.org/10.1371/journal.pone.0142901] [PMID: 26565401]

[51]   Huang CY, Chiang SF, Chen WTL, *et al.* HMGB1 promotes ERK-mediated mitochondrial Drp1 phosphorylation for chemoresistance through RAGE in colorectal cancer. Cell Death Dis 2018; 9(10): 1004.
[http://dx.doi.org/10.1038/s41419-018-1019-6] [PMID: 30258050]

[52]   Yang L, Ye F, Zeng L, Li Y, Chai W. Knockdown of HMGB1 suppresses hypoxia-induced mitochondrial biogenesis in pancreatic cancer cells. OncoTargets Ther 2020; 13: 1187-98.
[http://dx.doi.org/10.2147/OTT.S234530] [PMID: 32103987]

[53]   Bhat SM, Massey N, Shrestha D, *et al.* Transcriptomic and ultrastructural evidence indicate that anti-HMGB1 antibodies rescue organic dust-induced mitochondrial dysfunction. Cell Tissue Res 2022; 388(2): 373-98.

[http://dx.doi.org/10.1007/s00441-022-03602-3] [PMID: 35244775]

[54]   Rouhiainen A, Imai S, Rauvala H, Parkkinen J. Occurrence of amphoterin (HMG1) as an endogenous protein of human platelets that is exported to the cell surface upon platelet activation. Thromb Haemost 2000; 84(6): 1087-94.
[PMID: 11154118]

[55]   Passalacqua M, Zicca A, Sparatore B, Patrone M, Melloni E, Pontremoli S. Secretion and binding of HMG1 protein to the external surface of the membrane are required for murine erythroleukemia cell differentiation. FEBS Lett 1997; 400(3): 275-9.
[http://dx.doi.org/10.1016/S0014-5793(96)01402-0] [PMID: 9009213]

[56]   Todorova J, Pasheva E. High mobility group B1 protein interacts with its receptor RAGE in tumor cells but not in normal tissues. Oncol Lett 2012; 3(1): 214-8.
[http://dx.doi.org/10.3892/ol.2011.459] [PMID: 22740883]

[57]   Merenmies J, Pihlaskari R, Laitinen J, Wartiovaara J, Rauvala H. 30-kDa heparin-binding protein of brain (amphoterin) involved in neurite outgrowth. Amino acid sequence and localization in the filopodia of the advancing plasma membrane. J Biol Chem 1991; 266(25): 16722-9.
[http://dx.doi.org/10.1016/S0021-9258(18)55361-8] [PMID: 1885601]

[58]   Erlandsson Harris H, Andersson U. Mini-review: The nuclear protein HMGB1 as a proinflammatory mediator. Eur J Immunol 2004; 34(6): 1503-12.
[http://dx.doi.org/10.1002/eji.200424916] [PMID: 15162419]

[59]   Mistry AR, Falciola L, Monaco L, *et al.* Recombinant HMG1 protein produced in Pichia pastoris: a nonviral gene delivery agent. Biotechniques 1997; 22(4): 718-29.
[http://dx.doi.org/10.2144/97224rr01] [PMID: 9105624]

[60]   Cui X, Yao A, Jia L. Starvation insult induces the translocation of high mobility group box 1 to cytosolic compartments in glioma. Oncol Rep 2023; 50(6): 216.
[http://dx.doi.org/10.3892/or.2023.8653]

[61]   Morciano G, Marchi S, Morganti C, *et al.* Role of mitochondria-associated ER membranes in calcium regulation in cancer-specific settings. Neoplasia 2018; 20(5): 510-23.
[http://dx.doi.org/10.1016/j.neo.2018.03.005] [PMID: 29626751]

[62]   Murley A, Sarsam RD, Toulmay A, Yamada J, Prinz WA, Nunnari J. Ltc1 is an ER-localized sterol transporter and a component of ER–mitochondria and ER–vacuole contacts. J Cell Biol 2015; 209(4): 539-48.
[http://dx.doi.org/10.1083/jcb.201502033] [PMID: 25987606]

[63]   Bonaldi T, Talamo F, Scaffidi P, *et al.* Monocytic cells hyperacetylate chromatin protein HMGB1 to redirect it towards secretion. EMBO J 2003; 22(20): 5551-60.
[http://dx.doi.org/10.1093/emboj/cdg516] [PMID: 14532127]

[64]   Zhu S, Li W, Ward M, Sama A, Wang H. High mobility group box 1 protein as a potential drug target for infection- and injury-elicited inflammation. Inflamm Allergy Drug Targets 2010; 9(1): 60-72.
[http://dx.doi.org/10.2174/187152810791292872] [PMID: 19906009]

[65]   Xie J, Reverdatto S, Frolov A, Hoffmann R, Burz DS, Shekhtman A. Structural basis for pattern recognition by the receptor for advanced glycation end products (RAGE). J Biol Chem 2008; 283(40): 27255-69.
[http://dx.doi.org/10.1074/jbc.M801622200] [PMID: 18667420]

[66]   Yang H, Antoine DJ, Andersson U, Tracey KJ. The many faces of HMGB1: molecular structure-functional activity in inflammation, apoptosis, and chemotaxis. J Leukoc Biol 2013; 93(6): 865-73.
[http://dx.doi.org/10.1189/jlb.1212662] [PMID: 23446148]

[67]   Sims GP, Rowe DC, Rietdijk ST, Herbst R, Coyle AJ. HMGB1 and RAGE in inflammation and cancer. Annu Rev Immunol 2010; 28(1): 367-88.
[http://dx.doi.org/10.1146/annurev.immunol.021908.132603] [PMID: 20192808]

[68]   Kang R, Chen R, Zhang Q, *et al.* HMGB1 in health and disease. Mol Aspects Med 2014; 40: 1-116.
[http://dx.doi.org/10.1016/j.mam.2014.05.001] [PMID: 25010388]

[69]   Andersson U, Ottestad W, Tracey KJ. Extracellular HMGB1: a therapeutic target in severe pulmonary inflammation including COVID-19? Mol Med 2020; 26(1): 42.
[http://dx.doi.org/10.1186/s10020-020-00172-4] [PMID: 32380958]

[70]   Xu J, Jiang Y, Wang J, *et al.* Macrophage endocytosis of high-mobility group box 1 triggers pyroptosis. Cell Death Differ 2014; 21(8): 1229-39.
[http://dx.doi.org/10.1038/cdd.2014.40] [PMID: 24769733]

[71]   Chen L, Zhao Y, Lai D, *et al.* Neutrophil extracellular traps promote macrophage pyroptosis in sepsis. Cell Death Dis 2018; 9(6): 597.
[http://dx.doi.org/10.1038/s41419-018-0538-5] [PMID: 29789550]

[72]   Lu J, Yue Y, Xiong S. Extracellular HMGB1 augments macrophage inflammation by facilitating the endosomal accumulation of ALD-DNA *via* TLR2/4-mediated endocytosis. Biochim Biophys Acta Mol Basis Dis 2021; 1867(10): 166184.
[http://dx.doi.org/10.1016/j.bbadis.2021.166184] [PMID: 34087422]

[73]   Moore BW. A soluble protein characteristic of the nervous system. Biochem Biophys Res Commun 1965; 19(6): 739-44.
[http://dx.doi.org/10.1016/0006-291X(65)90320-7] [PMID: 4953930]

[74]   Marenholz I, Heizmann CW, Fritz G. S100 proteins in mouse and man: from evolution to function and pathology (including an update of the nomenclature). Biochem Biophys Res Commun 2004; 322(4): 1111-22.
[http://dx.doi.org/10.1016/j.bbrc.2004.07.096] [PMID: 15336958]

[75]   Newton RA, Hogg N. The human S100 protein MRP-14 is a novel activator of the β 2 integrin Mac-1 on neutrophils. J Immunol 1998; 160(3): 1427-35.
[http://dx.doi.org/10.4049/jimmunol.160.3.1427] [PMID: 9570563]

[76]   Huttunen HJ, Kuja-Panula J, Sorci G, Agnelletti AL, Donato R, Rauvala H. Coregulation of neurite outgrowth and cell survival by amphoterin and S100 proteins through receptor for advanced glycation end products (RAGE) activation. J Biol Chem 2000; 275(51): 40096-105.
[http://dx.doi.org/10.1074/jbc.M006993200] [PMID: 11007787]

[77]   Korfias S, Stranjalis G, Papadimitriou A, *et al.* Serum S-100B protein as a biochemical marker of brain injury: a review of current concepts. Curr Med Chem 2006; 13(30): 3719-31.
[http://dx.doi.org/10.2174/092986706779026129] [PMID: 17168733]

[78]   Ambartsumian N, Klingelhöfer J, Grigorian M, *et al.* The metastasis-associated Mts1(S100A4) protein could act as an angiogenic factor. Oncogene 2001; 20(34): 4685-95.
[http://dx.doi.org/10.1038/sj.onc.1204636] [PMID: 11498791]

[79]   Santamaria-Kisiel L, Rintala-Dempsey AC, Shaw GS. Calcium-dependent and -independent interactions of the S100 protein family. Biochem J 2006; 396(2): 201-14.
[http://dx.doi.org/10.1042/BJ20060195] [PMID: 16683912]

[80]   Donato R. Intracellular and extracellular roles of S100 proteins. Microsc Res Tech 2003; 60(6): 540-51.
[http://dx.doi.org/10.1002/jemt.10296] [PMID: 12645002]

[81]   Wolf R, Howard OMZ, Dong HF, *et al.* Chemotactic activity of S100A7 (Psoriasin) is mediated by the receptor for advanced glycation end products and potentiates inflammation with highly homologous but functionally distinct S100A15. J Immunol 2008; 181(2): 1499-506.
[http://dx.doi.org/10.4049/jimmunol.181.2.1499] [PMID: 18606705]

[82]   Nonaka D, Chiriboga L, Rubin BP. Differential expression of S100 protein subtypes in malignant melanoma, and benign and malignant peripheral nerve sheath tumors. J Cutan Pathol 2008; 35(11): 1014-9.

[http://dx.doi.org/10.1111/j.1600-0560.2007.00953.x] [PMID: 18547346]

[83]　Balijepalli P, Sitton CC, Meier KE. Lysophosphatidic acid signaling in cancer cells: what makes LPA so special? Cells 2021; 10(8): 2059.
[http://dx.doi.org/10.3390/cells10082059] [PMID: 34440828]

[84]　Lu P, Bai X, Ma D, *et al.* Three-dimensional structure of human γ-secretase. Nature 2014; 512(7513): 166-70.
[http://dx.doi.org/10.1038/nature13567] [PMID: 25043039]

[85]　Zhang X, Li Y, Xu H, Zhang Y. The Î³-secretase complex: from structure to function. Front Cell Neurosci 2014; 8: 427.
[http://dx.doi.org/10.3389/fncel.2014.00427] [PMID: 25565961]

[86]　Chen G, Xu T, Yan Y, *et al.* Amyloid beta: structure, biology and structure-based therapeutic development. Acta Pharmacol Sin 2017; 38(9): 1205-35.
[http://dx.doi.org/10.1038/aps.2017.28] [PMID: 28713158]

[87]　Hampel H, Hardy J, Blennow K, *et al.* The Amyloid-β pathway in Alzheimer's disease. Mol Psychiatry 2021; 26(10): 5481-503.
[http://dx.doi.org/10.1038/s41380-021-01249-0] [PMID: 34456336]

[88]　Sturchler E, Galichet A, Weibel M, Leclerc E, Heizmann CW. Site-specific blockade of RAGE-Vd prevents amyloid-beta oligomer neurotoxicity. J Neurosci 2008; 28(20): 5149-58.
[http://dx.doi.org/10.1523/JNEUROSCI.4878-07.2008] [PMID: 18480271]

[89]　Chaney MO, Stine WB, Kokjohn TA, *et al.* RAGE and amyloid beta interactions: Atomic force microscopy and molecular modeling. Biochim Biophys Acta Mol Basis Dis 2005; 1741(1-2): 199-205.
[http://dx.doi.org/10.1016/j.bbadis.2005.03.014] [PMID: 15882940]

[90]　Akter R, Cao P, Noor H, *et al.* Islet amyloid polypeptide: structure, function, and pathophysiology. J Diabetes Res 2016; 2016: 1-18.
[http://dx.doi.org/10.1155/2016/2798269] [PMID: 26649319]

[91]　Han SH, Kim YH, Mook-Jung I. RAGE: the beneficial and deleterious effects by diverse mechanisms of actions. Mol Cells 2011; 31(2): 91-8.
[http://dx.doi.org/10.1007/s10059-011-0030-x] [PMID: 21347704]

[92]　Goldin A, Beckman JA, Schmidt AM, Creager MA. Advanced glycation end products: sparking the development of diabetic vascular injury. Circulation 2006; 114(6): 597-605.
[http://dx.doi.org/10.1161/CIRCULATIONAHA.106.621854] [PMID: 16894049]

[93]　Harja E, Bu D, Hudson BI, *et al.* Vascular and inflammatory stresses mediate atherosclerosis *via* RAGE and its ligands in apoE–/– mice. J Clin Invest 2008; 118(1): 183-94.
[http://dx.doi.org/10.1172/JCI32703] [PMID: 18079965]

[94]　Vincent AM, Perrone L, Sullivan KA, *et al.* Receptor for advanced glycation end products activation injures primary sensory neurons *via* oxidative stress. Endocrinology 2007; 148(2): 548-58.
[http://dx.doi.org/10.1210/en.2006-0073] [PMID: 17095586]

[95]　Reynolds PR, Kasteler SD, Cosio MG, Sturrock A, Huecksteadt T, Hoidal JR. RAGE: developmental expression and positive feedback regulation by Egr-1 during cigarette smoke exposure in pulmonary epithelial cells. Am J Physiol Lung Cell Mol Physiol 2008; 294(6): L1094-101.
[http://dx.doi.org/10.1152/ajplung.00318.2007] [PMID: 18390831]

[96]　Al-Robaiy S, Weber B, Simm A, *et al.* The receptor for advanced glycation end-products supports lung tissue biomechanics. Am J Physiol Lung Cell Mol Physiol 2013; 305(7): L491-500.
[http://dx.doi.org/10.1152/ajplung.00090.2013] [PMID: 23997170]

[97]　Bartling B, Hofmann HS, Weigle B, Silber RE, Simm A. Down-regulation of the receptor for advanced glycation end-products (RAGE) supports non-small cell lung carcinoma. Carcinogenesis 2004; 26(2): 293-301.
[http://dx.doi.org/10.1093/carcin/bgh333] [PMID: 15539404]

[98] Zhou Z, Immel D, Xi CX, *et al.* Regulation of osteoclast function and bone mass by RAGE. J Exp Med 2006; 203(4): 1067-80.
[http://dx.doi.org/10.1084/jem.20051947] [PMID: 16606672]

[99] Ding KH, Wang ZZ, Hamrick MW, *et al.* Disordered osteoclast formation in RAGE-deficient mouse establishes an essential role for RAGE in diabetes related bone loss. Biochem Biophys Res Commun 2006; 340(4): 1091-7.
[http://dx.doi.org/10.1016/j.bbrc.2005.12.107] [PMID: 16403440]

[100] Philip BK, Childress PJ, Robling AG, *et al.* RAGE supports parathyroid hormone-induced gains in femoral trabecular bone. Am J Physiol Endocrinol Metab 2010; 298(3): E714-25.
[http://dx.doi.org/10.1152/ajpendo.00564.2009] [PMID: 20028966]

[101] Liliensiek B, Weigand MA, Bierhaus A, *et al.* Receptor for advanced glycation end products (RAGE) regulates sepsis but not the adaptive immune response. J Clin Invest 2004; 113(11): 1641-50.
[http://dx.doi.org/10.1172/JCI200418704] [PMID: 15173891]

[102] Sirois CM, Jin T, Miller AL, *et al.* RAGE is a nucleic acid receptor that promotes inflammatory responses to DNA. J Exp Med 2013; 210(11): 2447-63.
[http://dx.doi.org/10.1084/jem.20120201] [PMID: 24081950]

[103] Dong H, Zhang Y, Huang Y, Deng H. Pathophysiology of RAGE in inflammatory diseases. Front Immunol 2022; 13: 931473.
[http://dx.doi.org/10.3389/fimmu.2022.931473] [PMID: 35967420]

[104] Mukherjee TK, Mukhopadhyay S, Hoidal JR. The role of reactive oxygen species in TNFα-dependent expression of the receptor for advanced glycation end products in human umbilical vein endothelial cells. Biochim Biophys Acta Mol Cell Res 2005; 1744(2): 213-23.
[http://dx.doi.org/10.1016/j.bbamcr.2005.03.007] [PMID: 15893388]

[105] Mukhopadhyay S, Mukherjee TK. Bridging advanced glycation end product, receptor for advanced glycation end product and nitric oxide with hormonal replacement/estrogen therapy in healthy *versus* diabetic postmenopausal women: A perspective. Biochim Biophys Acta Mol Cell Res 2005; 1745(2): 145-55.
[http://dx.doi.org/10.1016/j.bbamcr.2005.03.010] [PMID: 15890418]

[106] Reynolds PR, Stogsdill JA, Stogsdill MP, Heimann NB. Up-regulation of receptors for advanced glycation end-products by alveolar epithelium influences cytodifferentiation and causes severe lung hypoplasia. Am J Respir Cell Mol Biol 2011; 45(6): 1195-202.
[http://dx.doi.org/10.1165/rcmb.2011-0170OC] [PMID: 21685154]

[107] Downward J. Cell cycle: Routine role for Ras. Curr Biol 1997; 7(4): R258-60.
[http://dx.doi.org/10.1016/S0960-9822(06)00116-3] [PMID: 9162506]

[108] Winston JT, Coats SR Sr, Wang YZ, Pledger WJ. Regulation of the cell cycle machinery by oncogenic ras. Oncogene 1996; 12(1): 127-34.
[PMID: 8552383]

[109] Lee H, Lee J, Hong SH, Rahman I, Yang SR. Inhibition of RAGE attenuates cigarette Smoke-Induced lung epithelial cell damage *via* RAGE-Mediated Nrf2/DAMP signaling. Front Pharmacol 2018; 9: 684.
[http://dx.doi.org/10.3389/fphar.2018.00684] [PMID: 30013476]

[110] Schmidt AM, Hori O, Chen JX, *et al.* Advanced glycation endproducts interacting with their endothelial receptor induce expression of vascular cell adhesion molecule-1 (VCAM-1) in cultured human endothelial cells and in mice. A potential mechanism for the accelerated vasculopathy of diabetes. J Clin Invest 1995; 96(3): 1395-403.
[http://dx.doi.org/10.1172/JCI118175] [PMID: 7544803]

[111] Frommhold D, Kamphues A, Hepper I, *et al.* RAGE and ICAM-1 cooperate in mediating leukocyte recruitment during acute inflammation *in vivo*. Blood 2010; 116(5): 841-9.

[http://dx.doi.org/10.1182/blood-2009-09-244293] [PMID: 20407037]

[112] Wang Y, Wang H, Piper MG, *et al.* sRAGE induces human monocyte survival and differentiation. J Immunol 2010; 185(3): 1822-35.
[http://dx.doi.org/10.4049/jimmunol.0903398] [PMID: 20574008]

[113] Sukkar MB, Wood LG, Tooze M, *et al.* Soluble RAGE is deficient in neutrophilic asthma and COPD. Eur Respir J 2012; 39(3): 721-9.
[http://dx.doi.org/10.1183/09031936.00022011] [PMID: 21920897]

CHAPTER 6

Receptor for Advanced Glycation End Products as a Mediator of Inflammation and Oxidative Stress

Ruma Rani[1,†], **Parth Malik**[2,3,†] and **Tapan Kumar Mukherjee**[4,*]

[1] *IICAR-National Research Centre on Equines, Hisar-125001, Haryana, India*

[2] *School of Chemical Sciences, Central University of Gujarat, Gandhinagar, Gujarat-382030, India*

[3] *Swarrnim Startup & Innovation University, Bhoyan-Rathod, Gandhinagar-Gujarat, India*

[4] *Amity Institute of Biotechnology, Amity University, New Town, Kolkata, West Bengal 700156, India*

Abstract: The receptor for advanced glycation end products (RAGEs) is a cell surface immunoglobulin class of molecules. RAGE prevails as a multiligand receptor capable of interacting with various ligands, the prominent amongst which is "advanced glycation end products (AGE)". The ligand-RAGE axis leads to an aggravated extent of inflammation and oxidative stress, activating various pro-inflammatory and pro-oxidative transcription factors such as nuclear factor kappa B (NF-κB). The binding of NF-κB to the promoter region of the RAGE gene activates its transcription. Once expressed, RAGE interacts further with its multiple ligands including AGE, HMGB1, S100, *etc.*, culminating in aggravated inflammatory and oxidative stress. Thus, RAGE which is a product of an increased level of inflammation and oxidative stress, once produced perpetuates a brutal cycle of self-propagation through sustained interaction with various ligands and subsequent inflammation and oxidation stress. Several levels of crosstalk possibilities prevail between pro-inflammatory and prooxidative reactive molecules. Sustaining a high level of pro-inflammatory and prooxidative reactions is the basic requirement to complicate various non-communicable disease conditions including diabetes-associated vascular complications, cardiovascular disorders (CVDs), pulmonary diseases, cancers, and others. This chapter describes the basic mechanism through which RAGE fuels the inflammatory and oxidative stress on a cellular front.

Keywords: Advanced Glycation End products (AGE), Inflammation, Oxidative Stress, Proinflammatory and Prooxidative Transcription Factors (NF-κB and SP1), Reactive Oxygen Species (ROS), Receptor for Advanced Glycation End Products (RAGE).

---

* **Corresponding author Tapan Kumar Mukherjee:** Amity Institute of Biotechnology, Amity University, New Town, Kolkata, West Bengal 700156, India; E-mail: tapan400@gmail.com
† These authors contributed equally to this work.

Tapan Kumar Mukherjee, Parth Malik & Ruma Rani (Eds.)
All rights reserved-© 2025 Bentham Science Publishers

INTRODUCTION

RAGE resides as a single-spanning transmembrane receptor protein belonging to the immunoglobulin class of proteins. The molecule RAGE is now, well-perceived as a pattern-recognition receptor (PRR) of the innate immune system. Being a multiligand receptor, RAGE interaction with various ligands confers a substantial significance to the three-dimensional RAGE structural configuration rather than its specific amino acid sequence. The well-characterized ligand molecules that can interact with the RAGE molecule are advanced glycated end products (AGE), S100, high mobility group box protein 1 (HMGB1, also known as amphoterin), amyloid β-peptide, and others [1, 2]. AGE is by far the, most well-characterized RAGE ligand. The substantial diversity of RAGE ligands could be described *by* exploring a common functional link for all ligands, such as persistent anionic surfaces, complementary to excessively alkaline and electropositive topography of RAGE ligand-binding domains. Such a sturdy electrostatic interaction induces the formation of a tight receptor-ligand complex, culminating in prolonged stimulation of downstream signaling pathways [3].

Initial studies by *Schmidt* and *associates* illustrated that the RAGE interaction with manifold ligands disseminates and propagates inflammatory conditions. While controlled, low inflammation (acute) is a homeostatic mechanism that eliminates infectious pathogens and mends damaged, stressed cells and tissues, excessive inflammation (chronic) is detrimental to the human body. Chronic inflammation complicates various non-communicable disease conditions in the human body. The involvement of RAGE in inflammatory stress has been confirmed by manifold experimental evidence. First, RAGE is aggravatedly expressed in all inflammatory lesions, including lung disorders [*e.g.* asthma, chronic obstructive pulmonary diseases (COPD), acute lung injury (ALI), acute respiratory disease syndrome (ARDS), cystic fibrosis (CF), chronic bronchitis (CB), sepsis and pulmonary hypertension, diabetes-associated vascular complications, cardiovascular complications (CVDs), specifically atherosclerosis, rheumatoid arthritis (RA), inflammatory kidney disease(glomerulonephritis), inflammatory bowel disease (IBD), various kinds of cancers (such as breast, ovary, pancreas, colon/ colorectal) and others [4 - 6]. Second, the RAGE promoter (the RNA polymerase binding site of the RAGE gene) houses the binding sequence of various pro-inflammatory and prooxidative transcription factors such as nuclear factor kappa B (NF-κB) and specificity protein 1 (SP1). Ligands-RAGE interaction activates these pro-inflammatory and pro-oxidative transcription factors. The third soluble RAGE (sRAGE), a condensed structure of the receptor spanning the extracellular ligand-binding domain (that subsequently competes with cell membrane-anchored RAGE for ligand binding), reduced inflammatory outcomes in all models screened to date, including delayed-type hypersensitivity (DTH), colitis and periodontitis. Studies using $F(ab)_2$ fragments mediated hindered ligand-RAGE binding complemented such observations, in particular for the highest concentration. Fourth, RAGE not only activates various adhesion molecules (*e.g.* vascular cell adhesion molecule-1 (VCAM-1), intercellular adhesion molecule-1 (ICAM-1), but a few studies also recognized RAGE as an integrin molecule, capable of bi-directional signaling (inside out and outside in). Of note, adhesion molecules are recognized as a cell-surface immunoglobulin class of molecules [7]. In addition to aggravating inflammation, RAGE also instigates the production of reactive oxygen species (ROS) and consequent oxidative stress, on a cellular front. Superoxide ($O_2^{\cdot-}$), hydroxyl (OH^-) and relatively less reactive hydrogen peroxides (H_2O_2) are the major cellular level ROS [8, 9].

In normal physiological conditions, the human cell cytoplasm maintains a robust mechanism to control the redox status of the cells. Of note, a reduced environment is hereby maintained *via a* pH of 7.2. This reducing environment of human cell cytoplasm is regulated by several endogenous antioxidants, like reduced glutathione (GSH) and thioredoxin. The *in vivo* cellular system produces vitamins C and E as exogenous food-derived antioxidants along with multiple enzymatic antioxidants such as peroxidases (*e.g.* glutathione peroxidase) and superoxide dismutase (SODs) [10]. Under normal physiological conditions, ROS is predominantly generated by mitochondria *via* carrier-mediated electron transfer such as flavin adenine dinucleotide (FAD), Nicotinamide Adenine Dinucleotide (NAD), coenzyme Q, and cytochrome C. Of note, the electron carriers are transferred from complex I to complex IV, *via* complex III and complex II. During this electron transfer, 1-2% electron leaks (single electron reduction) from complex III and to some extent from complex I occur, leading to ROS generation. However, under pathophysiological conditions, the ROS generation from the mammalian cell mitochondria considerably enhances, culminating in oxidative stress. Additionally, under complete physiological conditions, ROS is formed by several cellular oxidoreductases, like NADPH oxidase, xanthine oxidase, *etc.* This low level of ROS acts as a secondary messenger and therefore does not cause any adverse cellular effects [11 - 13]. High levels of ROS generation lead to oxidative stress which complicates various diseased conditions. Several lines of evidence indicate that enhanced oxidative stress induces the RAGE expression. Once RAGE is expressed, the ROS generation is further enhanced *via* interacting with its multiple ligands such as AGE. Therefore, RAGE perpetuates a vicious self-generation cycle *via* aggravated ROS generation [8].

Further, several lines of evidence indicate that oxidative stress and inflammation are closely interlinked*via*multiple crosstalk regulations. Chronic inflammation and oxidative stress complicate various diseased conditions, making it a prominent factor for the ligand-RAGE axis-dependent worsened diseased conditions [14]. This chapter provides a chronological description of the role of ligands-RAGE interaction in the generation of enhanced levels of inflammation and oxidative stress.

RECEPTOR FOR ADVANCED GLYCATION END PRODUCTS AS MEDIATORS OF INFLAMMATION

- Of note, RAGE is a single-spanning transmembrane receptor protein, prevailing amongst the immunoglobulin class of proteins. Although RAGE is often expressed in low levels, the receptor exhibits upregulation wherever its ligands accumulate. Further, RAGE is recognized as a multi-ligand receptor with advanced glycation end products (AGEs), high mobility group box protein 1 (HMGB-1), S100/Calgranulin, and β-amyloid, as its most well-illustrated receptors. RAGE shows similarity with pattern recognition receptors (PRRs) and is mainly related to innate immunity (Fig. **1**).
- The significance of PRRs concerning innate immunity is aided by their diversified allocation, diversity, and their dynamic essence as "sentinels", relying on a specific location, both at the cell surface as well as in intracellular compartments, as demonstrated by innumerable experimental findings. The observations herewith revealed enhanced infection susceptibility in humans, and mice exhibiting loss of function due to mutations affecting these receptors or associated signaling pathways.

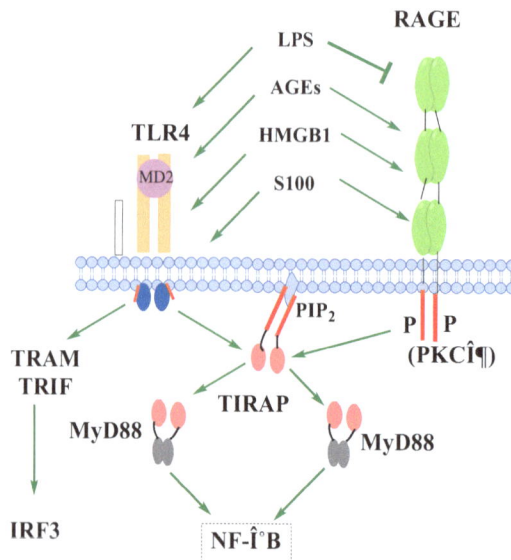

Fig. (1). Possible involvement of Receptor for Advanced Glycation End (RAGE) products in innate immune responses *via* ligand-TLR4 mediated crosstalk activities.

- There are four distinct PRR families, spanning the Toll-like receptors (TLRs), the C-type lectin receptors, retinoic acid-inducible gene-like receptors, and the nucleotide-binding oligomerization domain-like receptors. TLRs are the most important PRRs, being extensively dispersed in multiple cells [15]. While several crosstalk platforms exist between TLRs and RAGE, the exact classification of RAGE within a specific class of PRRs is yet unknown. Additionally, RAGE is demonstrated to support the activation of cell surface immunoglobulin class of adhesion molecules, like vascular cell adhesion molecule-1 (VCAM-1), intercellular adhesion molecule-1 (ICAM-1), and others. RAGE is also recognized as an integrin counter receptor, a sub-type of cell surface adhesion molecules equipped for bi-directional signaling.

- One of the major mechanisms through which the host's innate immunity works is by targeting damage-associated molecular patterns (DAMPs). DAMPs are expressed and released *via* pathogen-infected host cells after the PRRs-PAMPs interaction.

- The DAMPs comprise intracellular molecules, such as heat-shock proteins or HMGB-1(the well-studied RAGE ligands), and materials generated from the extracellular matrix (ECM) after tissue insult, such as hyaluronan fragments. Notable non-protein DAMPs comprise ATP, uric acid, heparin sulfate, and DNA. The masterpieces of pathogen-assisted host cell damage are the so-called release of danger signals or alarmins.

- Subsequently, the danger signals/alarmins released by the host cells, accumulate the innate immunological cells such as monocytes, macrophages, dendritic cells, and neutrophils in the vicinity of infected, damaged tissues and release cytokines (such as IL-6, TNFα), chemokines (*e.g.* macrophage chemotactic protein 1), interferons (*e.g.* IFNγ), reactive nitrogen species (*e.g.* peroxinitrite), reactive oxygen species (ROS, *e.g.* superoxide), acute phase proteins (*e.g.* C-reactive proteins), histamines, and other molecules leading to the inflammatory reactions [16].

- Such reactions are characterized by altered vasomotion, vascular damage, and leakage besides the accumulation of blood and body fluid (edema) in the affected tissues. Ideally, inflammation should promote the clearance of pathogens or damaged tissues after which, the body gradually heals itself to a normal state. Few studies also indicate the RAGE-mediated activation of acquired immunological cells such as T cells.

- However, in practicality, *while the limited and controlled level of inflammatory reactions (acute inflammation) serves as a homeostatic mechanism against invading pathogens (intended to injure the body cells and tissues), exaggerated and long-lasting inflammatory reactions (chronic inflammation) could be self-destructive to the human cells/ tissues and may even destroy various organs. A high level of inflammation is familiarly associated with various pathophysiological conditions and aggravates the diseased conditions* [17 - 19].

- Several lines of evidence confirm the RAGE involvement in the propagation of major non-communicable human diseases including lung disorders such as asthma, chronic bronchitis (CB), chronic obstructive pulmonary diseases (COPD), acute lung injury (ALI), and their worsened forms *i.e.* acute respiratory distress syndrome (ARDS), cystic fibrosis (CF), pulmonary hypertension and other pulmonary malfunctions. Cardiovascular complications (CVDs) and diabetes-associated vascular complications have also been demonstrated as being complicated *via* RAGE-ligand interaction. Various types of cancers that are complicated by increased RAGE-ligand interaction are breast, ovarian, uterine, pancreatic, hepatic, renal, colon, colorectal, and those of the nervous system. Some other prominent non-communicable sicknesses driven *via* aggravated RAGE-ligand interactions comprise inflammatory bowel disease, Alzheimer's disease, Parkinsonism, and so on. In all, these conditions are the outcomes of vigorous RAGE-ligand interactions, including aggravated inflammation and oxidative stress that collectively intensify a pathological status, intensifying the propagation of diseased conditions.

This section is subdivided into the following headings with a brief discussion:

- RAGE as an innate immunity regulator
- RAGE as an adaptive immunity regulator
- Leukocyte recruitment and inflammation as mediators of RAGE-dependent action
- Transcription factors mediated RAGE-associated inflammation.

Here is a brief discussion about them:

RAGE AS A REGULATOR OF INNATE IMMUNITY

Innate immunity is the one that is inherited from every preceding to successive generation, imparting non-specific protection to the living cells from invading pathogens. The mode of action of this primitive immunity is non-specific, whereby a generalized response is developed against various pathogens and differential extents of cellular injury. The immunological cells implicated in innate immunity do not exhibit any memory against invading pathogens. Several investigations claimed RAGE as a prominent mediator of innate immunity.

Under conventional *in vivo* conditions, while RAGE is moderately expressed by most body tissues, adult pulmonary tissues express it at a relatively higher extent. RAGE is involved in the development of embryonic and adult lungs, mediated *via the proliferation* and spreading of alveolar type 1 (AT-1) epithelial cells. However, RAGE is not needed to maintain normal lung physiology since RAGE-devoid mice live healthily without any noticeable deficiency. Studies on mice models lacking RAGE revealed a normal macroscopic pathology and histopathology. RAGE-lacking mice displayed normal initiation and inflammatory perpetuation. Also, RAGE-defunct mice remained protected from septic shock's deleterious effects, extensively driven *via* innate immunity. These findings demonstrated a sturdy correlation between RAGE and innate immunity [20, 21].

Several investigations reported considerable similarities between RAGE and pattern recognition receptor (PRR) molecules. Based on these observations, RAGE is now considered a PRR molecule. Several PRRs have been identified on multiple cellular surfaces, including innate immunological cells. Most prominent PRRs comprise the various toll-like receptors (TLRs), nucleotide-binding oligomerization domain (NOD)-Leucin Rich Repeats (LRR) containing receptors (NLR), the retinoic acid-inducible gene 1 (RIG-1) and retinoic acid like receptors (RLR; RIG-1-like helicases-RLH), the C-type lectin receptors (CLRs) and mannose receptors [22]. Amidst evolution, higher organisms build up numerous mechanisms to confront the fighting pathogens (viruses, bacteria, fungi, and protozoa). In the very first host defense, the pathogen-associated molecular pattern (PAMPs) expressed on pathogenic surfaces interacts with manifold PRRs, such as TLRs, Nod-like receptors and RAGE expressed mainly on disrupted host tissue or stressed cellular surfaces. Based on the results of a few studies, RAGE expressed on multiple cell surfaces was recognized as PRR [23]. Moreover, RAGE and TLR4 synergistic cooperation responded as an aggravator for countering inflammation in endothelial cells or murine models [24, 25].

From a structural viewpoint, the multiplicity of RAGE ligands and the recognition of AGEs as ligands form the genesis of the hypothesis considering RAGE as a PRR. The notable effort by *Xie and colleagues* used cell fluorescence resonance energy transfer to elucidate RAGE prevalence as a constitutive oligomer on the plasma membrane surface. The investigators further demonstrated constitutive RAGE oligomerization as formidably responsible for identifying varied configurations of AGE-modified proteins, having below 100 nM affinities [26]. The notion of pattern recognition implied to illustrate innate immunity receptors, such as the Toll-like and mannose receptors, that screen conserved molecular arrangements common to a large group of pathogens. Such unchanged patterns may vary from the specific carbohydrate moiety prevailing either on the pathogen cell wall or with the double-stranded viral RNA. Of note, the mannose receptor as a prominently studied and understood PRR serves as a paradigm of this class and exhibits several traits in commonality to all PRRs, including 1) a multidomain structure, 2) structurally identical repeating subunits, 3) the capacity to identify varied classes of ligands (proteins, carbohydrate chains, *etc.*), localized to specialized regions, and 4) specifically for polysaccharide binding, the affinity of monomeric units is quite low but increases with increment in the polymer chain length. Notably, RAGE exhibits the first and third characteristics of significance, but it does not possess repeating assortments of structurally similar domains, which could facilitate the patterned ligands recognition by the PRRs. Unlike the innate system PRRs which exclusively bind with exogenous ligands, almost all RAGE ligands are endogenous. The binding affinity of RAGE for AGEs stoutly varies with the extent of ligand glycation. Ironically, free glycated amino acids, carboxymethyl lysine (CML) and carboxyethyl lysine (CEL), do not bind with RAGE, unlike several AGE-modified proteins having a nanomolar affinity. However, in a manner distinct from other PRRs, RAGE can recognize multiple ligands, including AGEs and various DAMPs. This multiplicity could be reasoned *via a* common trait of all ligands, having a substantial anionic surface receptivity, supporting the richly alkaline and electropositive surface of RAGE ligand-binding domains (V, C1, and C2), existing on the cellular extracellular matrix. This strong electrostatic interaction results in the formation of a tight receptor-ligand complex, accounting for sustained downstream stimulation of proinflammatory and prooxidative signaling pathways [3].

RAGE AS A REGULATOR OF ADAPTIVE IMMUNITY

Although the findings inferring the RAGE axis involvement in manifesting adaptive immunity are of sufficiently lesser extension than those of innate immunity, the observations exhibit significance for molecular distinction. For instance, RAGE-devoid T cells exhibited a non-functional proliferation *in vitro* to nominal and alloantigen, analogous to a reduced IFN-γ and IL-2 generation [27]. Experimental data from RAGE-lacking mice amidst monitoring cell responses involved in autoimmunity and allograft rejection demonstrated that RAGE expression on T cells is implicated in early events associated with Th1 cell differentiation [28]. An erstwhile attempt revealed a constitutive RAGE expression on the diabetic patient's T cells. RAGE+ T cells from such patients exhibited a skewed phenotype, demonstrating the effect of environmental RAGE ligands on adaptive immune responses [29]. However, an astounding finding based on novel animal models having tissue-explicit RAGE expression inferred no significant RAGE involvement in the adaptive immune response in these animal models. The results further established that RAGE deletion protects from the septic shock lethal effects emanating from caecal ligation and puncture. These protective outcomes were reversed by the RAGE reconstitution in endothelial and hematopoietic cells. Their results established that the innate immune response is regulated by PRRs not merely in the initial stages but also at the perpetuation phase [30]. These contrasting results indicate room for further research to understand the exact RAGE involvement in adaptive immunity.

LEUKOCYTE RECRUITMENT AND INFLAMMATION AS MEDIATOR OF RAGE-DEPENDENT ACTION

- Integrins are a unique category of transmembrane adhesion molecules that mediate bi-directional signaling from "inside out" and "outside-in" of the cells.
- Human genome sequencing revealed as many as 18 α and 8 β integrin subunits, amongst which 24 functionally distinct ones are presently, believed to prevail in humans.
- With reference to the ligand binding explicit nature, mammalian integrins are broadly grouped into laminin-binding (α1β1, α2β1, α3β1, α6β1, α7β1, and α6β4), collagen-binding (α1β1, α2β1, α3β1, α10β1, and α11β1), leukocyte binding (αLβ2, αMβ2, αXβ2, and αDβ2) and RGD-recognizing (α5β1, αVβ1, αVβ3) entities.
- Integrins control cellular growth, proliferation, migration, signaling, cytokine activation, and release, exhibiting significant importance in regulating apoptosis and tissue repair, besides the events related to inflammation, chronic infections, and angiogenesis.
- RAGE functioning in adequacy is accompanied by adequate leukocyte recruitment *in vivo* , noticed in thioglycolate-induced acute peritonitis models and the *in vitro* investigations inferring its ability to do so *via interaction* with Mac-1. Of note, integrin Mac-1 prevails as a receptor on the neutrophils and monocyte/macrophage surface, supporting their adhesion to ECM proteins and multiple cell surface counter-receptors apart from sustaining phagocytosis, migration, aggregation, and degranulation.
- Multiple studies aimed towards understanding the relationship between RAGE and distinct adhesion molecules namely, vascular cell adhesion molecule-1 (VCAM-1) and intercellular adhesion molecule 1 (ICAM-1).In one such attempt, *Frommhold* and *associates* studied RAGE's biological relevance as a potential Mac-1 ligand, as elucidated in RAGE and ICAM-1 devoid mice wherein leukocyte adhesion was nearly non-functional in an acute surgically induced inflammation model, about the cremaster muscle *in vivo* . The observations of this study explained an intricate RAGE and ICAM-1 cooperation for sustaining a definitive leukocyte adhesion, with critical involvement of RAGE-Mac-1 and ICAM-1-Lymphocyte function-associated antigen-1 (LFA-1) interactions. Of note, LFA-1 is an integrin prevailing in lymphocytes and erstwhile leukocytes, exhibiting multiple important functions in emigration, the process by which leukocytes depart the bloodstream to access the tissues. LFA-1 also ensures a firm arrest of leukocyte activation. The investigation by *Frommhold and associates* further revealed that under a scenario of induced inflammation *via localized* treatment with pro-inflammatory and pro-oxidative cytokines TNF-α along with surgical trauma, leukocyte adhesion was only slightly moderated for missing RAGE and ICAM-1action.

The findings affirmed the alternative and still unknown integrin ligands, which prevail *via* close RAGE and ICAM-1 cooperation owing to stimulus dependence of β_2-integrin ligands [30]. Another attempt by the same group revealed distinctly regulated leukocyte localization by RAGE and ICAM-1 during stimulus-dependent acute inflammation [31]. Erstwhile studies noticed a prominent RAGE-driven leukocyte recruitment in diabetic mice alongside the upregulated RAGE expression. While RAGE merely contributed in part to the leukocyte enrolment in *vivo* in control mice, the leukocyte recruitment in diabetes-affected mice was surprisingly opposed by sRAGE (soluble RAGE). Furthermore, aggravated neutrophil extravasation in diabetes-affected wild-type mice was not observed in RAGE-defunct mice. These results complement a sharper role of RAGE-controlled leukocyte conscription in an aggravated RAGE expression at the inflammation site.

- It is now established as a consensus that under specific conditions (*e.g.* proteolytic cleavage/alternative splicing), the -NH_2 terminus region of single spanning transmembrane protein RAGE is released and enters circulation, being subsequently termed as soluble RAGE (sRAGE) or endogenous soluble RAGE (esRAGE). Since the truncated/released amino terminus region (sRAGE/esRAGE) through its various domains (V-C1-C2) binds with manifold ligands, it obstructs the RAGE ligand binding with membrane-bound RAGE. Therefore, sRAGE/esRAGE acts as a decoy receptor, emerging experimentally significant to screen the various RAGE-dependent actions. Thus, sRAGE/esRAGE acts as a potential opposing factor for leukocyte recruitment. Such interference of sRAGE with a prominent interaction sustaining the immune response stoutly supplements its broad-spectrum functioning.

- Furthermore, studies on RAGE-devoid mice revealed a substantial decrease (>50%) in the macrophages migrating to the peritoneum contrary to the wild-type mice, illustrating the significance of RAGE in the microangiopathies related to chronic inflammatory events. While RAGE is expressed to a small extent in normal tissues and the vasculature, the expression assumes a greater proportion in the regions where its ligands accumulate. Thus, the mechanism for leukocyte recruitment may be specifically suited to pathophysiological conditions implicated in high RAGE expressions, such as diabetes, chronic immune responses, atherosclerosis, or cancers. Simultaneously, enhanced RAGE expression could stimulate RAGE-driven cellular activation and dysfunction. Ligand interception of RAGE, (such as AGE), aggravates the vascular endothelium permeability, elevating the functions of adhesion molecules, such as VCAM-1, pro-coagulant tissue factor, or pro-inflammatory interleukin-6*via*RAGE-driven NF-κB activation. Thus, RAGE-controlled cellular activation probably leads to an aggravated localized transport of inflammatory cells. Apart from RAGE-mediated cellular activation, RAGE can also adjust leukocyte recruitment *via* functioning as a leucocyte-engaging endothelial cell adhesive receptor [32]. In 2007, *Chavakis and associates* demonstrated that apart from RAGE, HMGB-1 (a RAGE ligand) also induces the Endothelial Progenitor Cells (EPCs) chemotactic migration in a RAGE- and integrin-dependent regime. Additionally, HMGB1 swiftly mounts the RAGE- and integrin-dependent EPCs adhesion to mature endothelial cells, ICAM-1, and fibronectin. Another important HMGB1 action mediates β_1 and β_2 integrin activation*via*enhanced affinity, inducing the β_2 integrin's lateral motility on the EPC surface. Finally, EPC prestimulation with HMGB1 enhances the *in vivo* adhesion and EPCs homing at the sites of active angiogenesis and ischemia. Thus, this study provides insights into regulating the integrin activity in EPCs, unravelling a new function of HMGB1 *via* regulating the integrin-dependent EPCs homing [33].

- Finally, provoking RAGE-β2-integrins interaction could be a novel therapeutic anti-inflammatory approach in diabetes or chronic immune responses correlated with increased RAGE expression. The experimental results of *Chavakis group* studies deciphered that the occurrence feasibility of RAGE-β2-integrin Mac-1and p150,95, direct interactions illustrate RAGE as a new adhesion receptor, providing evidence for a novel leukocyte recruitment pathway which could be substantial in pathophysiological conditions like diabetes [32].

TRANSCRIPTION FACTORS AS MEDIATORS OF RAGE-DEPENDENT INFLAMMATION

To understand the transcription factors controlling RAGE expression, *Li and colleagues* cloned the 5'-flanking RAGE gene region and screened the prominent regulatory motifs. Screening the putative promoter locations divulged three potential NF-κB-like sites and two specificity protein 1 (SP1) binding sites [34]. Henceforth, in an erstwhile attempt, *Li and his group* observed manifold recognized binding elements within the RAGE gene-5'-flanking region, encoding for SP1. The molecular technological assays such as electrophoretic mobility shift (EMSA) and DNase I foot printing assays illustrated several functional SP1-binding sites within the region 2245 to 2240 bp of the RAGE gene promoter [35]. Thus, the RAGE gene promoter has NF-κB and SP1 transcription factor binding sites. Several subsequent investigations confirmed the NF-kB and SP1 binding sites on the RAGE gene promoter [8].

NF-κB consists of a family of transcription factors playing critical roles in inflammation, immunity, cell proliferation, differentiation, and survival. Inducible NF-κB stimulation relies on phosphorylation-induced proteasomal degradation of NF-κB proteins (IκBs) inhibitor, retaining the inactive NF-κB dimers in the unstimulated cells into the cytoplasm. Most of the assorted signaling pathways involved in NF-κB stimulation congregate on the IκB kinase (IKK) complex, typically modulating the IκB phosphorylation and is essential for NF-κB mediated signal transduction.

SP1 is associated with the SP/KLF family of transcription factors, the functional protein being 785 amino acids long, with 81 kDa molecular weight (M_w). Initially, SP1 was considered a general Zn finger family of a transcription factor typically needed for transcription of manifold 'housekeeping genes', the terminology suggesting their implication in metabolism, cell proliferation/growth, and cell death. Of note, zinc finger proteins are one of the most prominent in eukaryotic genomes, with extraordinarily diverse functions, comprising DNA recognition, RNA packaging, transcriptional activation, apoptotic regulation, protein folding and assembly, and lipid binding. Studies to date, have beyond doubt established that the SP1 transcription factor is intricately associated with normal development events.

SP1 activates the transcription of several cellular genes, which comprise presumed CG-rich SP-binding sites in their promoters. SP1 is a well-understood member of the transcription factor family, also having SP2, SP3, and SP4, as involved in manifold biological processes, proven decisive for cell growth, differentiation, apoptosis, and carcinogenesis. Of note, SP1 and SP3 bind to identical, if not the same, DNA tracts and struggle for binding, exhibiting potential for modulating the gene expression.

Significant evidence prevails about the SP-family of proteins that regulate the expression of cell proliferation and metastasis-regulating genes in multiple tumors. Usually, multi-cancer patients exhibit enhanced extents of SP1 as a negative prognostic factor. Multiple molecules (drugs) are known to impede the SP1 and remaining SP trans-activating activities for gene expression. Several pathways are implicated in SP protein suppression, typically operating either *via* direct intrusion with the SP proteins binding to their putative DNA sites or *via* encouraging the SP protein transcription factor degradation. In general, the SP transcription factors, and SP1-regulated gene suppression are drug-dependent and vary from one to another cell type. Recent attempts acknowledge the safety of some SP inhibiting compounds as supportive for their steadfast clinical usage in SP1 overexpressing tumor patients.

RAGE AS A MEDIATOR OF OXIDATIVE STRESS

- Reactive oxygen species (ROS) is a combined terminology for the reacting molecules and free radicals (FRs) generated from molecular oxygen. Oxygen in its ground state has two unpaired electrons ($1s^2$ $2s^2$ $2p^4$), making it easier to accept electrons and form FRs.

- FRs are the molecules containing unpaired electrons in the outermost orbitals and are highly reactive within the body *via* oxidizing (abstracting one or more electrons) other atoms, or less often *via* reduction (giving their electron(s) to other atoms.
- From a biochemical viewpoint, FRs are a class of highly reactive intermediate entities having their reactivity derived from the unshared electron that is able for a free prevalence for an infinitesimal time interval (10^9-10^{12} sec).
- As discussed in the introductory section, within the mitochondria four distinct complexes are involved in electron carrier (FAD, NAD) driven transport from complex I to complex IV *via* complex II and complex III involvement. Amidst this electron transport, ~1-2% of electrons are leaked, and there is single electron reduction leading to the formation of superoxide molecules. The superoxide thus produced is a vulnerable molecule with a high reactivity and is considered FR. Almost instantly super superoxide is converted to a more stable molecule such as hydrogen peroxide, which further breaks down to water and oxygen by various enzymes such as peroxidases. Thus, under physiological conditions, ROS are produced by mitochondria.
- The phagocytic as well as non-phagocytic NADPH oxidases prevailing in the phagocytic (*e.g.* macrophages, neutrophils) and non-phagocytic (*e.g.* endothelial cells) cells, xanthine oxidase and other oxidoreductases also account for ROS generation in the physiological conditions.
- The most common ROS are superoxide anion ($O_2^{\cdot-}$), hydroxyl radical (HO•), singlet oxygen ($^1O_2$), and relatively stable hydrogen peroxide (H_2O_2), all of which are more reactive than oxygen (O_2) itself. Table **1** (ahead) summarizes their distinctive ROS-generating mechanisms *via* biochemical means. It must be noted here that ROS generation is an essential requirement to sustain the electron transport chain (ETC), but on reaching a proportion that corresponds to the oxidation of native tissues and stored food reserves, the ROS excess formation needs treatment following antioxidant intake.

Table 1. Distinguishing aspects of common reactive oxygen species generating oxidizing agents.

Oxidizing Agent	Molecular formula	Chemical reaction
Superoxide anion	$O_2^{\cdot-}$	$NADPH + 2O_2 \longleftrightarrow NADP^+ + 2O_2^- + H^+$ $2O_2^- + H^+ \longrightarrow O_2 + H_2O_2$
Hydrogen peroxide	H_2O_2	$Hypoxanthine + H_2O + O_2 \rightleftharpoons Xanthine + H_2O_2$ $Xanthine + H_2O + O_2 \rightleftharpoons Uric\ acid + H_2O_2$
Hydroxyl radical	$^{\cdot}OH$	$Fe^{+2} + H_2O_2 \longrightarrow Fe^{+3} + OH^- + \cdot OH$

(Table 1) cont.....

Oxidizing Agent	Molecular formula	Chemical reaction
Hypochlorous acid	HOCl	$H_2O_2 + Cl^- \longrightarrow HOCl + H_2O$
Peroxyl radical	ROO˙	$R^{\bullet} + O_2 \longrightarrow ROO^{\bullet}$
Hydroperoxyl radical	HOO⁻	$O_2^- + H_2O \rightleftharpoons HOO^- + OH^-$

Because superoxide is toxic at high concentrations, nearly all organisms needing oxygen for survival exhibit multiple superoxide dismutases (SODs). SOD efficiently catalyzes the superoxide disproportionation as per the following reaction.

$$2H^+ + 2O_2^- \longrightarrow O_2 + H_2O_2$$

$$O_2 + e^- \rightarrow {^{\bullet}O_2^-} + e^- + 2H^+$$

Dismutation of superoxide generates hydrogen peroxide (H_2O_2), as below.

$${^{\bullet}O_2^-} + e^- + 2H^+ \rightarrow H_2O_2 + e^- + 2H^+$$

Hydrogen peroxide in response, may be partially or completely reduced to hydroxyl radical (˙OH) or water, respectively.

$$H_2O_2 + e^- \longrightarrow HO^- + {^{\bullet}OH}$$

$$2H^+ + 2e^- + H_2O_2 \rightarrow 2H_2O$$

ROS is labelled as a secondary cell signaling messenger. The low amount of ROS formed by mammalian cells in physiological processes aids in cell survival, proliferation, and various signaling events, manifesting the oxidative defense mechanisms for getting rid of bacteria and other pathogens and physiological homeostasis. Fig. (**2**) depicts the sequential biochemical processes whereby excess oxygen is eliminated as water, alongside generating superoxide, hydrogen peroxide, and hydroxyl radicals.

Major oxygen metabolites produced by one electron reduction

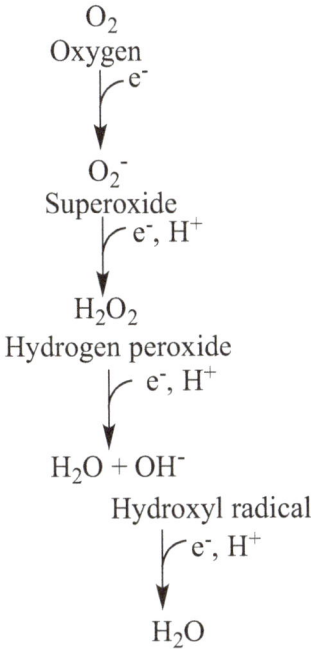

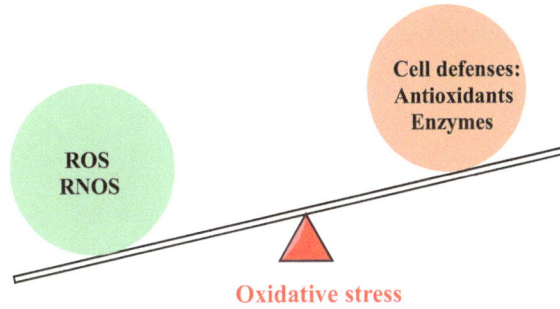

O_2
Oxygen
$\downarrow e^-$

O_2^-
Superoxide
$\downarrow e^-, H^+$

H_2O_2
Hydrogen peroxide
$\downarrow e^-, H^+$

$H_2O + OH^-$
Hydroxyl radical
$\downarrow e^-, H^+$

H_2O

ROS
RNOS

Cell defenses:
Antioxidants
Enzymes

Oxidative stress

Fig. (2). Reduction of oxygen by sequential one-electron steps. The four one-electron mediated O_2 reductions consecutively generate superoxide, hydrogen peroxide, hydroxyl radical, and water.

ROLE OF MITOCHONDRIA IN RAGE-DEPENDENT REACTIVE OXYGEN SPECIES GENERATION

Multiple studies describe the working mechanism of RAGE signaling-driven vascular disease induction. As mentioned earlier, the multi-ligand receptor RAGE interacts with several ligands including AGEs, HMGB1/amphoterin, s100/calgranulins, and others in distinct biochemical environments of specific redox status. The RAGE interaction with its ligands generates ROS which subsequently, propagates as pro-inflammatory and pro-oxidative events. Such events together result in a succession of diverse pathophysiological actions not limited to diabetes, but also significant in multiple erstwhile disorders, such as rheumatoid arthritis, atherosclerosis, inflammatory kidney disease, inflammatory bowel syndrome, sepsis, and Alzheimer's disease. Apart from these specific pathological outcomes, RAGE signaling also aggravates tumor growth and metastasis. The ROS formed *via* RAGE-ligand interactions intensely affects multiple pathologic events, as outlined in Fig. (**3**) [4, 5, 8, 9]. The NF-κB mediated signaling activities of RAGE culminate in multiple oxidative stress-sensitive pathologies.

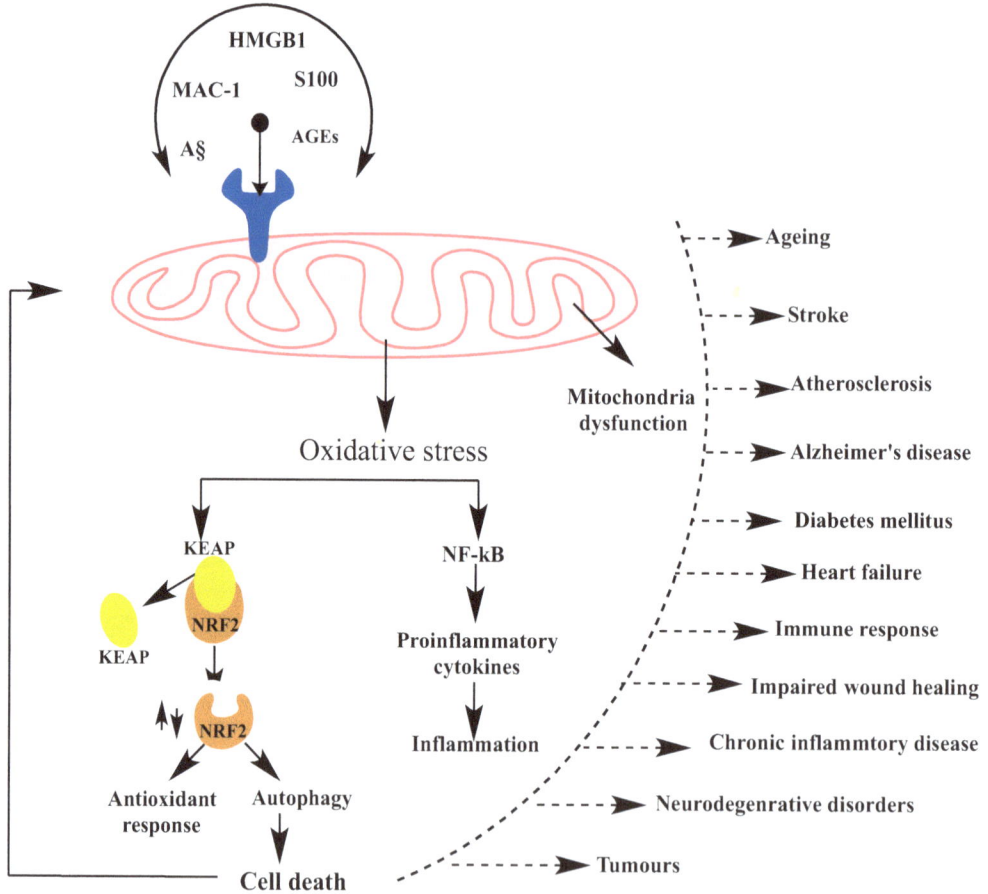

Fig. (3). Oxidative stress-mediated disorders caused by RAGE-ligand interactions, ligand diversity, and interaction duration may distinguish the disease severity.

In one of the studies reporting ROS generation from RAGE and its ligands interactions, *Mukherjee and colleagues* demonstrated the mechanism, through which pro-inflammatory cytokine TNF-α generates ROS from mitochondria and NADPH oxidase, which subsequently aggravates RAGE expression in the cultured human umbilical vein endothelial cells (HUVECs) (Fig. **4**). Major findings of this study noticed a TNFα-stimulated ROS production in the (HUVECs) through NADPH oxidase and mitochondrial electron transport chain.

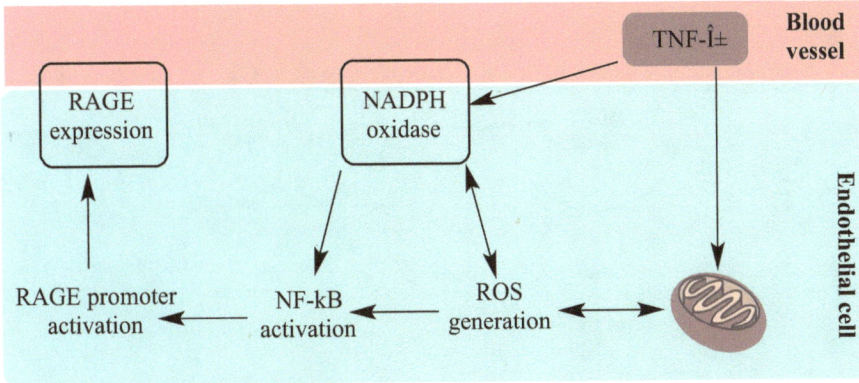

Fig. (4). Inflammation aggravated RAGE activities *via* the generation of reactive oxygen species on intercepting mitochondria and NADPH oxidases.

The ROS production, in turn, stimulates NF-κB actions as a downstream transcription factor that further fuels the RAGE expression. Fundamentally, ROS are the intermediate signaling moieties presenting a platform for the mitochondrial respiratory chain and NADPH oxidase interaction since opposed ROS generation by the mitochondrial respiratory chain affects its production from the NADPH oxidase and *vice-versa*. This study further asserts that RAGE expressed *via* enhanced ROS generation interacts with its various ligands such as AGE and HMGB1 to further intensify the ROS generation. Thus, at the cellular level, there is a vicious cycle of RAGE expression responsible for its maintenance and perpetuation [8].

The same research group henceforth treated cultured MCF-7 breast cancer cells with 17-α ethinyl estradiol (17α-EE), a synthetic estrogen, and examined the relationship of ROS generation with RAGE expression, proliferation, and cell survival. Analysis revealed a ROS generation following 17α-EE interaction with estrogen-related receptor gamma (ERRγ), whereby NF-κB activation leads to concomitant stimulated RAGE expression. The expressed RAGE further assists in ROS generation, with enhanced oxidative stress aggravated MCF-7 breast cancer cell number*via*modulating the cell proliferation (cyclin D1), cell survival (AKT), and anti-apoptotic proteins (Bcl$_2$). Fig. (**5**) depicts the stepwise mechanism by which 17α-EE mediated RAGE expression aggravates the MCF-7 breast cancer cell number. While RAGE-mediated ROS production acquires a decisive role in MCF-7 breast cancer cell survival and proliferation, erstwhile signaling pathways such as the RAGE-HMGB1 axis or hypoxia also contribute significantly to the inflammatory status of tumor cells. The investigators opined that possible future research based on these observations could improve the understanding of RAGE expression-mediated breast cancer cell proliferation and survival [36].

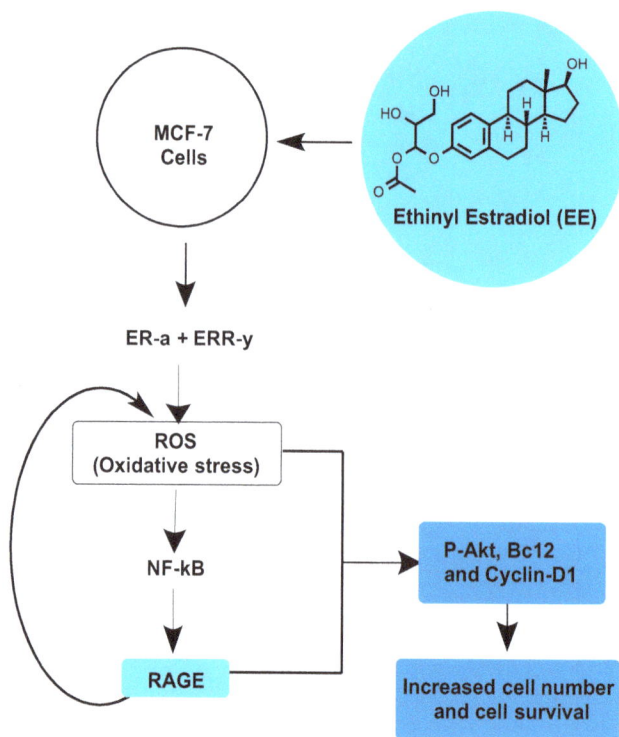

Fig. (5). Aggravation of the MCF-7 breast cancer cell generation*via*17α-EE stimulated RAGE expression.

Of note, the cell-organelle mitochondria (singular mitochondrion) are a significant source of intracellular ROS production. Multiple biological processes regulate the metabolism of food products, such as glycolysis (for glucose assimilation), Krebs cycle (generating intermediates for amino acids and fatty acids synthesis), and β-oxidation of fatty acids, generating electron carriers such as FAD (Flavin Adenine Dinucleotide) and NAD (Nicotinamide Adenine Dinucleotide). In comparison to the cytoplasmic occurrence of glycolysis, the Krebs cycle and fatty acids β-oxidation occur in mitochondria. FAD and NAD as the electron carriers, facilitate electron transfer from complex I to complex IV *via* intervening complexes II and III. On reaching electron carriers to complex IV, molecular oxygen is used and generates ATP by oxidative phosphorylation. In the mitochondria, a large proportion of consumed oxygen is transformed into water, although 1-2% remains as ROS due to electron-leakage from complex I (NADH Ubiquinone oxidoreductase) and complex III (Ubiquinol cytochrome C oxidoreductase) of mitochondrial electron transport chain (ETC) and single electron reduction [37].

Uncoupling proteins (UCPs) comprise a family of inner mitochondrial membrane proteins that assist the proton re-entry in the matrix without F_o-F_1 ATPase. The

UCPs disengage respiration from ATP generation, moderating the mitochondrial membrane potential ($\Delta\psi_m$), energy efficiency, and ROS generation [38, 39]. To date, studies have reported five UCPs (UCP1-5) with UCP1 being involved in thermogenesis whereas the function of the remaining proteins remains yet unknown.

Recent studies have presented indirect evidence for the possible involvement of AGEs and RAGE in COPD pathogenesis. In this context, a notable 2022 attempt by *Kwon and colleagues* observed the redox balance and mitochondrial changes associated with the mouse skeletal muscles, exposing inadequate AGE in RAGE knockout and wild-type, C57BL/6 models, made accessible to cigarette smoke (CS) till 8 months *via* immunoblotting, spectrophotometry, and high-resolution respirometry. Inspection revealed two-fold increased 4-HNE (4-Hydroxynonenal, a noted oxidative stress marker) extents with significantly suppressed contractile proteins, mitochondrial respiratory complexes, and uncoupling protein extents (p< 0.01). Amongst the noticeable functional changes after CS exposure was a higher dependence on complex-I assisted respiration (p< 0.01) with weakened respiration for fatty acids biosynthesis (p < 0.05). Interestingly, the RAGE suppression caused a 47% reduction in 4-HNE protein expression than CS-exposed control animals. These distinctions pointed out a RAGE-mediated oxidative stress induction following cigarette smoke inhalation that was observed as partly correlated with enhanced complex III protein expressions. It was also noted that RAGE suppression arrested the mitochondrial specific respiration (p< 0.05) irrespective of CS exposure, culminating in a proportionate enhancement of mitochondrial content (ascertained *via* citrate synthase activity, p < 0.001), revealing no change in muscle respiratory capacity. Collectively, the investigation revealed that RAGE knockdown guarded the skeletal muscle against the CS oxidative damage, till 8 months of exposure along with RAGE-regulated mitochondrial functioning, making RAGE a potential therapeutic target [40].

INVOLVEMENT OF NADPH OXIDASE IN RAGE-MEDIATED ROS GENERATION

• A diligent glance at the recent investigations renders it evident that apart from mitochondria, NADPH oxidase (NOX for normal cell NADPH oxidase, PHOX for phagocytic cell NADPH oxidase) driven ROS signaling performs an imperative role in cancer progression. The NOX family is a vital constituent of non-phagocytic cells, regulating cellular events such as cell proliferation, mitogenesis, and carcinogenesis *via* ROS generation. Fig. (**6**) outlines the possible enhanced inflammatory and concomitant oxidative stress conditions following NADPH oxidase and NF-κB modulated RAGE expressions across mitochondrial-ER crosstalk.

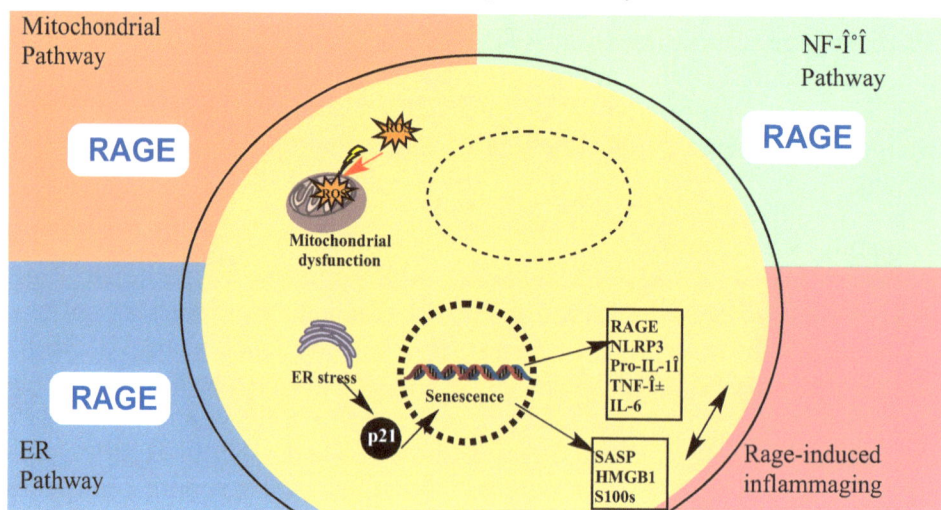

Fig. (6). Possible aggravated oxidative stress-mediated RAGE activities for inflammation and aging outcomes.

- Emerging experimental evidence reveals that NOX-generated ROS performs a decisive role in controlling breast cancer proliferation. The NOX signaling also contributes towards significant ROS generation inside the cells. Studies have suggested mitochondrial-NADPH oxidase crosstalk wherein mitochondrial redox signaling weakened control aggravates breast and ovarian tumorigenesis [41].
- The NOX-mitochondrial interactions have also been reported for endothelial cells, leading to ROS generation by TNFα [42]. A notable attempt from *Choi and colleagues*, herein revealed an upregulated NOX expression in a BLT2 (a receptor for leukotriene B4) reliant regime in breast cancer. Analysis revealed a BLT2 downregulation controlled by NADPH oxidase 1-derived ROS [43].
- Of note, BLT2 is highly upregulated in breast cancer cells, and its inhibition from an explicit antagonist, LY255283 or siBLT2 RNA interference, results in apoptosis in triple-negative breast cancer cells (MDA-MB-468 and MDA-M--453) [44]. Thereby, the results established a key role of BLT2 in breast cancer cell survival wherein estrogen receptors (ERs) are not necessarily involved. In the same study, NOX knockdown or treatment with an antioxidant molecule induced sudden apoptosis in the cancer cells [45].

A particular concentration of generated ROS exhibits multiple vital roles in different cancer processes. At high concentrations, ROS aggravates oxidative stress and apoptosis, leading to cell death whereas at low concentrations, ROS activates angiogenesis, paving the way for tumor cell invasion and metastasis

[46]. Generated ROS damages the tissues in various forms, such as oxidative stress-driven cardiovascular disease and neurodegeneration, culminating in carcinogenesis *via* damage to DNA, lipids, and cellular proteins. Thus, ROS serves as a noted inhibitor of proliferating cancer cells *via* apoptotic induction. Enhanced ROS generation in cancer cells promotes the onset of apoptosis or autophagy [47].

Nearly all cancer treatment therapies rely on ROS accumulation *via* concomitant suppression of antioxidant homeostasis in the cancer cells, aiming at the cancer stem cells (CSC) self-renewal, besides inhibiting epithelial-mesenchymal transition (EMT) and angiogenesis [48]. Several studies support the phenomenon of NADPH impasse or tumor cell inhibition that aids a tumor cell in evading the excessive ROS, which is ideally reversed as a tumor cell killing mechanism [46]. Thereby, ROS generation in a tumor cell regulates manifold important processes comprising apoptosis, and autophagy, enabling an understanding of their potential in the development of novel anti-cancer agents [49].

THE GENERATION OF OXIDATIVE STRESS BY HIGH LEVEL OF REACTIVE OXYGEN SPECIES

Like low ROS extents, moderately enhanced ROS generation may not be harmful to native physiology. This is so as slight enhancement in FR generation is nullified *via* *multiple* homeostatic antioxidant mechanisms prevailing in mammalian cells like reduced glutathione (GSH, most common reductive molecule in human cell cytoplasm), thioredoxin, catalase, multiple superoxide dismutase (SODs) and several others.

- The sensitive equilibration between ROS generation and ROS scavenging is affected by manifold stress inducers like UV radiations, bacterial LPS, cytokines like TNFα, *etc.*
- The imbalance due to ROS accumulation and its more than-required generation than being compensated *via* antioxidants, leads to a pathophysiological condition, termed oxidative stress. Oxidative stress can damage cellular proteins, lipids, and nucleic acids by interfering with their native redox status, causing various diseases.
- This oxidative stress complicates almost every non-communicable disease such as cancers, cardiovascular disorders (CVD), diabetes mellitus, chronic obstructive
- Experimentally, a high level of ROS can function as a pro-apoptotic trigger for various cancer cells, thereby aiding in apoptotic induction.
- Thereby, the detection/measurement of ROS generation does not merely aid in understanding the ROS-driven physiological cell signaling including MAPKs

and cell cycle activation, leading to cell division and eventual, oxidative stress-mediated complications such as CVD, COPD, and cancers. This enables manifold novel therapeutic avenues *via* targeting physiological ROS-generating molecules.

- Additionally, ROS and inflammatory molecules correlated with one another as several levels of crosstalk prevail between them, collectively complicating the diverse diseased conditions [14].

THE ANTIOXIDANT SYSTEM AND REGULATION OF OXIDATIVE STRESS

To mitigate oxidative/nitrosative stress, a cell's response to ROS/RNS *via* activating antioxidant defense is quite crucial. Antioxidant enzymes assumed a decisive role in moderating ROS/RNS extents, making transcription factor's redox regulation critical for ascertaining oxidative stress-sensitive gene expression and cellular response to stressful conditions. Several H_2O_2 sensors and pathways are activated, together regulating the manifold transcription factors including activating protein-1 (AP-1), nuclear factor erythroid 2-related factor 2 (Nrf2), cAMP-response element binding protein (CREB), heat shock factor 1 (HSF1), hypoxia-inducible factor (HIF-1), tumor protein 53 (TP53), NF-κB, Notch, specific protein 1 (SP1), which activate multiple genes, including those detoxifying the oxidizing molecules besides repairing and maintaining cellular homeostasis. Such actions are the backbone of regulatory events, involving multiple cellular functions like cell proliferation, differentiation, metabolism, and apoptosis. Furthermore, the family of FoxO-related transcription factors performs a critical role in sustaining the redox outcomes. Hydrogen peroxide-driven transcription factor regulation can happen at multiple levels: synthesis/degradation, cytoplasm-nuclear trafficking, DNA binding, and transactivation [50]. Despite the ROS generation, human cell cytoplasm amicably sustains a reduced environment. This is made feasible by a complex network of enzymatic (*e.g.* catalase, superoxide dismutase) and non-enzymatic (*e.g.* glutathione, thioredoxin) antioxidants, working together to scavenge excessive ROS. Besides, food-derived antioxidants (*e.g.* vitamins E, and C) also assist in maintaining a reduced intracellular environment.

ANTIOXIDANT ENZYMES AND THEIR WORKING MECHANISMS

- For regulating optimum cell signaling, it is desired that the diverse radical scavenging enzymes maintain a minimal ROS extent within the cell interior. More than the threshold ROS extent could deviate excessive signals inside the cell, besides the direct damage to the vital signaling pathways of intermediate species.
- Oxidative stress occurs in the event of an enhanced level of ROS formation and

therefore the inability to scavenge all the ROS by the anti-oxidative defense mechanism of the cells, leading to the accumulation of un-scavenged ROS. ROS can irreversibly tamper with the essential biological macromolecules including DNA, RNA proteins, and lipids, creating a likely scenario for carcinogenesis. Therefore, excessive ROS amounts must be promptly moderated *via manifold* defense mechanisms, comprising multiple antioxidants and detoxifying enzymes.

- On a molecular level, an antioxidant is defined as an entity capable of interfering with macromolecule oxidation. So, an antioxidant must be able to accommodate one or more of the electrons that are abnormally aggravating the chemical activity of free radicals (FRs) (the classical ROS analogies). The function of an antioxidant is to terminate these chain reactions by getting rid of FRs or inhibiting other oxidations and oxidizing themselves using the surplus oxygen available with FRs (supposedly ROS). In layman's terms, any molecule that can be robustly oxidized (withstand the addition of oxygen) can function as an antioxidant. In the redox sense, antioxidants such as polyphenols or thiols are classical reducing agents capable of extracting oxygen.

- Although oxidation reactions are vital for cells, they exhibit certain sensitivities. In general, plants and animals have various antioxidants such as vitamins C, E, and glutathione, as well as multiple enzymes capable of catalyzing the abstraction of oxygen such as catalase, superoxide dismutase (SOD), and glutathione peroxidases (GPx). Oxidative stress is biochemically the outcome of one or more impairments in the functioning of homeostatic antioxidant enzymes, that could otherwise alter the cellular permeability, leading to their lysis. The mechanistic operational steps followed by a defensive antioxidant are: 1) Jamming the FRs generation, 2) Scavenging the FRs using antioxidants, 3) Converting the toxic FRs into less toxic molecules, 4) Preventing the formation of secondary toxic metabolites and inflammatory mediators, 5) Impairing the chain propagation of secondary oxidants, 6) Repairing the damaged biomolecules, and 7) Functional restoration of the endogenous antioxidant defense system. These defense mechanisms act hand in hand to guard the native tissues from oxidative stress. Fig. (**7**) illustrates the integrated functioning of catalases, SODs, and GPx, whereby a physiological optimum redox balance is exercised by converting the deleterious hydrogen peroxide (H_2O_2) into H_2O and O_2.

The enzymatic antioxidants in the human body consist of three major classes, namely, SODs, catalases, and GPx, which unanimously exhibit key roles in sustaining cellular homeostasis (Table **2**). The initiation of the activities of these enzymes encompasses a precise response to a pollutant-mediated oxidative stress [51]. For instance, SOD scavenges superoxide radicals and converts them into H_2O_2, course catalyzing the superoxide anion (O^{2-}) breakdown into oxygen and hydrogen peroxide besides removing O_2 *via* catalyzing a dismutation reaction [52]. The reaction speed dramatically reduces in the absence of this enzyme. Fig. (**8**) depicts the $O_2^{\cdot}$ scavenging by SOD whereby the H_2O_2 is generated that is further converted into water and oxygen by catalases.

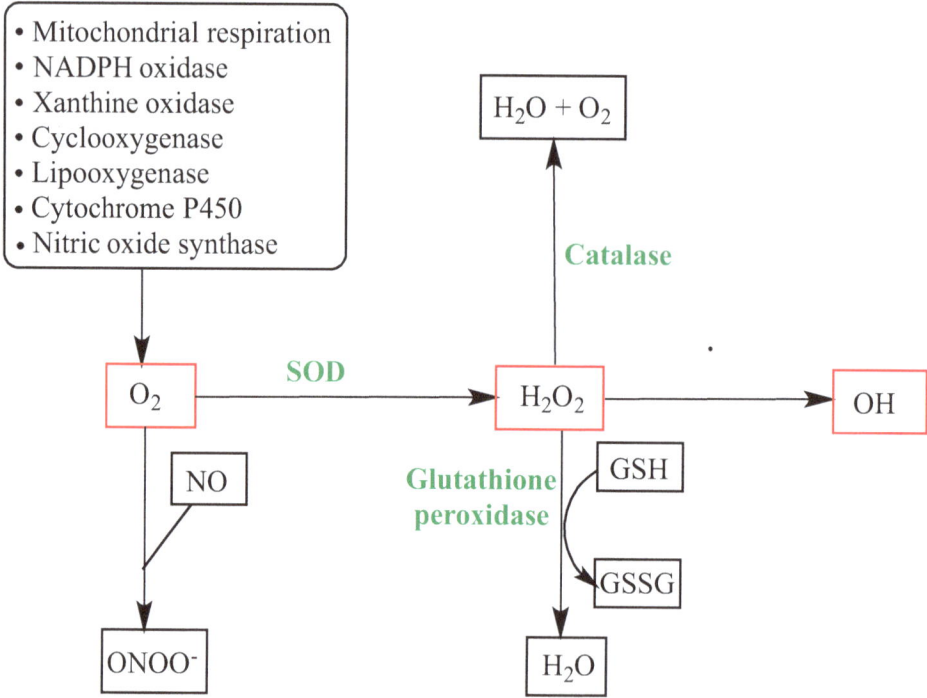

Fig. (7). Schematic representation of integrated functioning of enzymatic antioxidants, wherein superoxide dismutases, SODs function as the first line of defiance. The subsequent controls exercised by catalases and glutathione peroxidases are regulated as per the organ-specific requirement for H_2O_2 neutralization as H_2O and O_2. [Red boxes depict the reactive species].

Table 2. Working mechanisms of various enzymatic antioxidants, illustrating their diverse physiological significance and homeostasis-maintaining abilities.

Free radical scavenging agent	Ellipsis	Antioxidants Catalysed Reactions
Superoxide dismutase	SOD	$M^{(n+1)+}\text{-SOD} + O_2^- \longrightarrow M^{n+}\text{-SOD} + O_2$ $M^{n+}\text{-SOD} + O_2^- + 2H^+ \longrightarrow M^{(n+1)+}\text{-SOD} + H_2O_2$
Catalase	CAT	$2H_2O_2 \longrightarrow O_2 + 2H_2O$ $H_2O_2 + \text{Fe(III)-E} \longrightarrow H_2O + O = \text{Fe(IV)-E(.+)}$ $H_2O_2 + O = \text{Fe(IV)-E(.+)} \longrightarrow H_2O + \text{Fe(III)-E} + O_2$

(Table 2) cont.....

Free radical scavenging agent	Ellipsis	Antioxidants Catalysed Reactions
Glutathione peroxidase	GPx	$2GSH + H_2O_2 \longrightarrow GSSG + 2H_2O$ $2GSH + ROOH \longrightarrow GSSG + ROH + H_2O$
Thioredoxin	TRX	Adenosine monophosphate + sulfite + thioredoxin disulfide $=$ 5'-adenylyl sulfate + thioredoxin Adenosine-3',5'-bisphosphate + sulfite + thioredoxin disulfide $=$ 3'-phosphoadenylyl sulfate + thioredoxin
Peroxiredoxin	PRX	$2R'\text{-}SH + ROOH = R'\text{-}S\text{-}S\text{-}R' + H_2O + ROH$
Glutathione transferase	GST	$RX + GSH = HX + R\text{-}S\text{-}GSH$

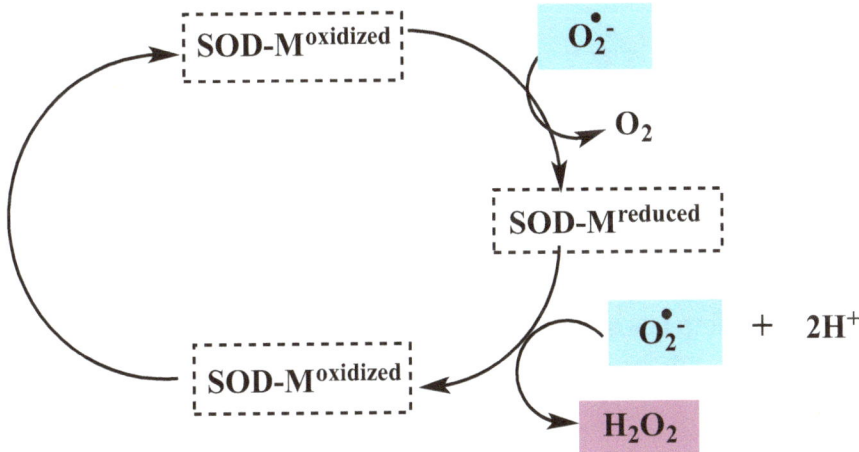

$M^{oxidized}/M^{reduced}:$ Cu^{2+}/Cu^{1+} for SOD1 and SOD3
MN^{3+}/Mn^{2+} for SOD2

Fig. (8). Mechanism of O^{2-} scavenging by superoxide dismutase, the enzymatic activity involves alternate reduction and re-oxidation of catalytic metal (*i.e.*, Cu or Mn) at the active site. This results in Cu or Mn as a key modulator of SOD activity of SOD1/SOD3 or SOD2.

Mammalian SODs prevail as cytoplasmic (Cu/ZnSOD), mitochondrial (MnSOD), and extracellular (Cu/ZnSOD), isoforms, each of which requires a metal catalyst as a cofactor for their optimum functioning. Emerging reports describe that in each sub-cellular location, SODs catalyze O^{2-} to H_2O_2 in exchange for their participation in cell signaling. Apart from this, SODs perform multiple decisive tasks in inhibiting the nitric oxide (NO) inactivation, concomitantly forbidding the peroxynitrate (NO_3-O^-) driven endothelial and mitochondrial impaired functioning.

The functioning of each SOD isoform is well-versed from the multiple investigations on genetically modified mice and viral vector-driven genetic transfer. The essential and much anticipated SOD actions in cardiovascular disorders have been a boost for most of the antioxidant therapies involving reinforcement of endogenous antioxidant safeguards towards thorough protection from oxidative stress. While the distinctive SOD features of their vascular tissue functioning are summarized in Table **3**, readers are advised to refer to more specific literature sources for morphological and functional distinctions of SOD isoforms as well as their multifaceted biological activities along with their pathological significance as FR scavengers and homeostatic regulator [53, 54].

Table 3. Diversity of Superoxide dismutase in vascular tissue functioning.

Isoform	Traits	Metal Cofactor	Physiological Residence
SOD1 (Cu/ZnSOD)	32 kDa, homodimer	Cu^{-2} (catalytic) and Zn^{+2} (stability)	Cytoplasm, mitochondria, nucleus, lysosomes, peroxisomes
SOD2 (MnSOD)	96 kDa, homotetramer	Mn^{+3} (catalytic)	Mitochondrial matrix
SOD3 (ecSOD)	135 kDa, homotetrameric secretory glycoprotein	Cu^{2+}(catalytic) and Zn^{-2} (stability)	Extracellular matrix, cell surface, extracellular fluids

- Catalase (H_2O_2 oxidoreductase) comprises 4 polypeptide chains, each having more than 500 amino acids besides four porphyrin heme (iron) groups, making it feasible for the enzyme's reaction with H_2O_2. Notable activity modes (mechanistic steps) of catalase for H_2O_2 degradation are (i) catalytic mode (converting H_2O_2 into H_2O and O_2), and (ii) peroxidic (activity oxidizing low M_w alcohols in low H_2O_2 concentrations) [55]. The catalase turnover rate is perhaps the highest among all the enzymatic antioxidants.
- The catalase-assisted H_2O_2 decomposition is a typical, first-order reaction with its rate being dependent on H_2O_2 concentration. Catalase as a cofactor is oxidized by one molecule of H_2O_2 after which, bound oxygen is transferred to the second substrate molecule. Catalase prevails in both eukaryotic and bacterial cells, often existing as oxidative particles in various mammalian cells except for the unaltered initial stage persistence in red blood cells (RBCs).
- As H_2O_2 serves as a substrate for the specific reaction to generate highly hydroxylated radicals, it is generally considered that the major antioxidant

defense action of catalase is to moderate the intracellular H_2O_2 accumulation. Multiple studies describe the catalase-mediated cells and tissues' oxidative protection exhaustively. Of note, catalase overexpression makes the cells defiant to H_2O_2 damage and concurrent injury [56].

- Studies on genetically engineered mice for overexpressed catalase revealed enhanced protection against myocardial infection on giving an adriamycin-rich diet and being treated with angiotensin for the exhibited high blood pressure [57, 58]. Similarly, catalase-deficient human subjects exhibited vulnerability to progressive oral gangrene owing to H_2O_2 elicited tissue damage, caused by peroxides-generating bacteria (streptococcus and pneumococcus) besides the bacterial phagocytic cells [59].

- The thioredoxin system comprises protein thioredoxin along with the enzyme thioredoxin reductase. Thioredoxin-dependent proteins exist in all living beings and are well-perceived ROS scavengers [60]. The physiological significance of thiol peroxidases (the functional domains of thioredoxins) is attributed to their role in catalase-driven H_2O_2 decomposition into H_2O and O_2, such as GPx and peroxiredoxins [61]. The typical process comprises H_2O_2 and organic hydroperoxides interconversion to water and analogous alcohols, respectively. Opposed to SODs and catalases, peroxidases use a thiol-based mechanism to neutralize the hydroperoxides after restoration of their reduced state by the GSH/GR or Trx/ TrxR system, on NADPH utilization [62].

- Peroxiredoxins exhibit a high affinity towards H_2O_2, inspired by their highly conserved active locus. The H_2O_2 reduction is typically facilitated by Thr/Cys/Arg arpeggio *via* stabilization of the transition state and further polarization of O-O peroxyl linkage, which engineers the electron transfer from catalytic cysteine [63]. The typical process enables a 10^5-10^8 M^{-1} s^{-1} reaction rate, several orders greater than H_2O_2 and protein thiol's reaction rate [64]. The sulfenic acid intermediate produced by the catalytic cysteine promptly reacts with two cysteine subunits in peroxiredoxins, releasing water *via* concomitant formation of water through a disulfide linkage (Fig. **9**). The cyclization terminates as the free thiol state is compensated by thioredoxin or GSH equivalent reductants. In the specific event of 1-cysteine peroxidase, distinguishing cysteine is missing and the generated sulfenic acid (on cysteine reaction with H_2O_2), is reduced by GSH along with glutathione-S-transferase (GST).

- GPx is an oxidoreductase (Enzyme Commission No.: 1.11.1.9), an erythrocytic enzyme that safeguards hemoglobin from oxidative damage. GPxs family of enzymes exist as homologous to selenocysteine (Sec) having mammalian GPx-1 which employs GSH as a necessitate co-substrate for H_2O_2 reduction to water. All GPxs (varying as per homology) do not necessarily need GSH due to Sec prevalence at the active site. Some of these, however, are indeed functionally recognized as thioredoxin-dependent peroxidases comprising redox-active cysteine rather than Sec. The GPx-1 exists as the readily prevalent member of GPx group of enzymes comprising erstwhile epithelial-specific enzyme widely prevalent in the intestine (GPx-2), a secreted subtype (GPx-3), and the rather widely expressed GPx-4 which differs from the remaining family members in terms of its substrate specificity.

- Functionally, the peroxide detoxification by mammalian GPx-1 occurs as a bi-substrate reaction wherein no saturation kinetics are yet reported. The typical peroxide reduction by enzymes is characterized *via* intermediate stable changes to the Sec active site [65, 66]. After the peroxide reaction, the selenic acid residue, Se-OH in GPx-1 is transformed into a selenic (Se-H) active site, wherein the reduction of selenenic acid from one GSH molecule forms glutathionylated selenol (Se-SG) intermediate (Fig. **10A**) [65, 66].

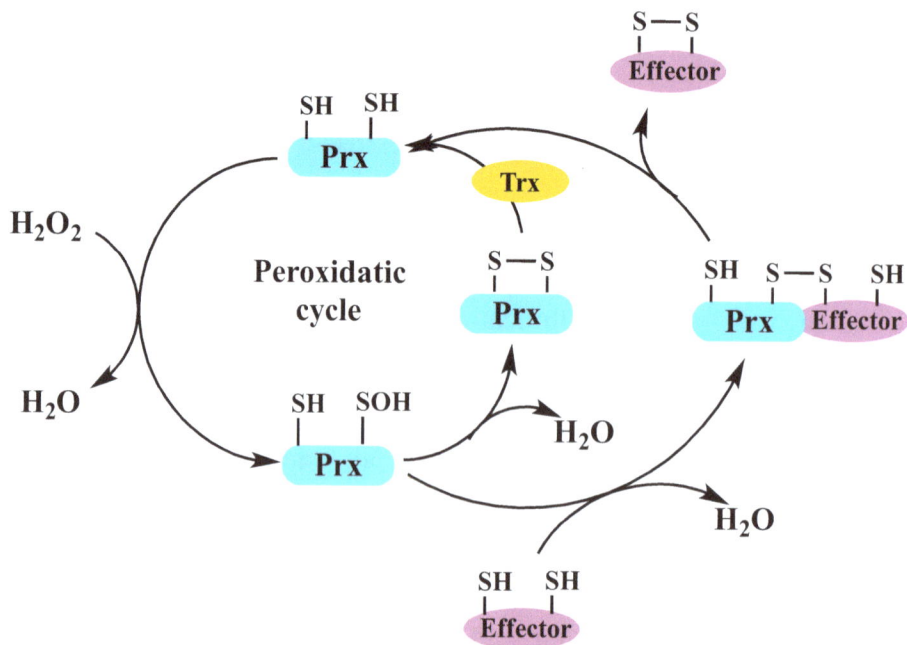

Fig. (9). Typical symbolification of peroxiredoxin functioning as peroxidase and H_2O_2-signal transducer. The reaction of the catalytic cysteine leads to the formation of a sulfenic acid intermediate. In the peroxidatic cycle (left), the sulfenic acid forms a disulfide with the resolving cysteine that is reduced by thioredoxin or GSH. In the Prx-redox relay (right), the sulfenic acid intermediate forms an intermolecular disulfide with the effector protein thiol group.

Thereafter, a subsequent GSH reduces the Se-SG bond, restoring the active site and concomitantly generating oxidizing glutathione (GSSG).

- Henceforth, the GSSG resolution is facilitated *via* NADPH-driven glutathione reductase activity, with NADP+/NADPH recycling mediated GSH pathway linkage to glucose-6-phosphate dehydrogenase and pentose-phosphate shunting (Fig. **10B**). So, regulation of cellular oxidative status by GPx-1 is mediated *via* enzymatic detoxification of non-radical hydroperoxides, directly by subsiding the hydroperoxides and GSH oxidation, notably the low M_w cellular thiols.

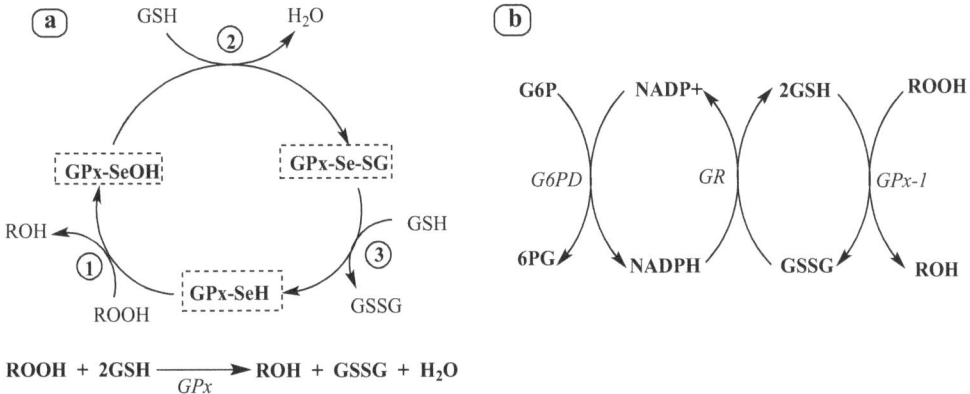

$$ROOH + 2GSH \xrightarrow{GPx} ROH + GSSG + H_2O$$

Fig. (10). (a) GPx-1 mediated H_2O_2 reduction. Enzyme-assisted peroxide inactivation by GPx-1 comprises the formation of multiple stable intermediates which manifest gradual modifications to the active site selenocysteine, (b) Redox pathways involved in sustaining the cofactors essential forGPx-1 activity. The GPx-1 reductively inactivates H_2O_2 and lipid hydroperoxides at the cost of GSH, typically oxidized to GSSG. The GSSG to GSH recycling is facilitated by glutathione reductase *via* provisioning reducing equivalents.

- The selenium-dependent GSH-peroxidase functioning was first reported by Rotruck, being optimized *via* glucose-6-phosphate dehydrogenase-GS--reductase driven GSH recycling to prevent the erythrocytes oxidation [67, 68]. Inspection revealed no protective effect of glucose-reduced GSH on the oxidation in erythrocytes from Se-devoid rats. Perhaps, analysis of H_2O_2 metabolism in cultured cells from GPx-1 devoid, wild mice exhibited a markedly enhanced oxidized GSH expression while GSSG did not very much in GPx-1 devoid cells [69]. Together, these observations provided an understanding of the GPx-1 enzymatic activity interlinkage with intracellular GSH/GSSG redox couple, mediating the cellular redox stress shift to facilitate the GSSG storage than reactive protein thiols. Ultimately, GPx-1 is best established as being linked with NADP+/NADPH redox couple, simultaneously being implicated in restoring normal GSH to GSSG stoichiometry (Fig. **11**). By modulating the NADP+ to NADPH balance, the GPx-1 mediated GSH oxidation could also interfere with the native pentose-phosphate pathway [70].
- The antioxidant functioning of GPx is enabled *via* hydrogen peroxide, lipid hydroperoxides, and other organic hydroperoxides chemical reduction [71]. The glutathione-S-transferases (GST) encompass a prominent group of detoxifying

enzymes, prevailing as a family of multifunctional proteins implicated in the cellular detoxification of cytotoxic and genotoxic compounds apart from guarding body tissues against oxidative degeneration (Table **4**).

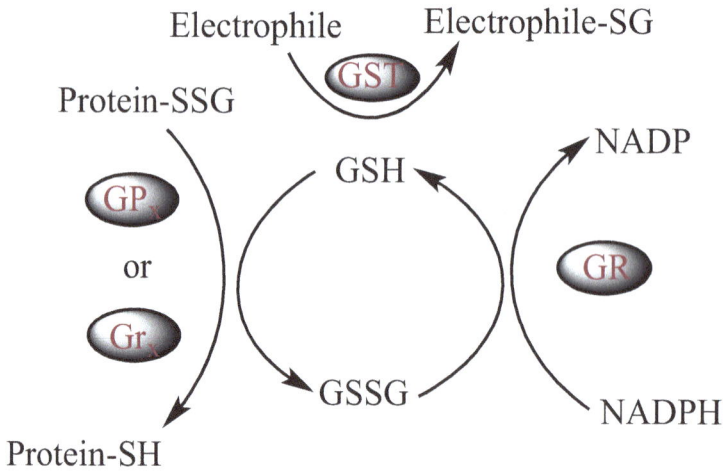

GP$_x$: Glutathione peroxidase   GST: Glutathione-*S*-transferase
Gr$_x$: Glutaredoxin   GSH: Reduced glutathione
GR : Glutathione reducatse   GSSG: Oxidized glutathione

Fig. (11). Sustenance of mammalian glutathione functioning, GSSG to GSH stoichiometry is the major confirmation of native oxidative regulation where enzymatic regulation *via redox* intonation exhibits a decisive role.

- Besides definitive functions in endogenous metabolism, these enzymes assist in xenobiotics detoxification including drugs, carcinogens, and environmental pollutants in humans and animals, and pesticides, and herbicide resistance in insects and plants [72].
- Some studies explain the GPx and GST inactivation by hydroperoxides, exerting their toxicity directly *via oxidizing* protein -SH groups or indirectly *via*-OH radical generation. Numerous GST isoenzymes also demonstrate GPx activity, which catalyzes the organic hydroperoxides reduction to corresponding alcohols [71].

Table 4. Functional diversity of glutathione-S-transferases.

GST type	Functions and protein targets
GSTA	• Detoxification, steroid isomerization, glutathione peroxidase functioning, lipid peroxidase reductase (4-HNE)
GSTM	• Detoxification, lipid peroxidation, prostaglandin isomerase functioning, catalysis of forward PSSG reaction • Interacts with ASK1, suppresses the ASK1, cJUN and p38 actions to inhibit cytokine and stress-driven apoptosis.
GSTO	• Detoxification, altered ryanodine receptor expression in ER, regulates dehydroascorbate and monomethyl arsenate reductase activity, inhibits ATP driven IL-1β processing, forward and reverse PSSG sustenance. • Interacts with ERK1/2, AKT, and JNK to stimulate their dephosphorylation, alters the cell-survival, apoptotic pathways
GSTP	• Detoxification and catalysis of forward PSSG reaction • Intercepts JNK to inhibit apoptosis, modulates TRAF2 expression for inhibited JNK and p38 activations, reduced cell death and stress induction
GSTT	• Detoxification, sustaining lipid peroxide reductase and sulfatase actions
GSTZ	• Detoxification, phenylalanine, and tyrosine (degradation) catabolism, maleylacetoacetate isomerize actions, interconversion of dichloroacetic acid to glyoxylic acid (β-oxidation), and metabolism of α-halo acids

ASK-1: apoptosis signal-regulated kinases, PSSG: protein S-glutathionylation, ERK: extracellular signal-regulated kinases, 4-HNE: 4-hydroxy-2-nonenal, TRAF2: tumor necrosis factor receptor-associated factor 2, JNK: c-Jun N-terminal kinase, IL-1β.

NON-ENZYMATIC ANTIOXIDANTS AND THEIR ACTION MECHANISMS

• Protein-bound and free thiols function as cellular reducing and guarding agents, conferring protection from deleterious inorganic pollutants, *via* their -SH functionality [73]. Thus, thiol sensitivity is prominently the first line of defense to confront oxidative stress. The thiol extent in a compound can be enhanced by an adaptive method to moderate oxidative stress *via* enhanced synthesis. However, conditions about oxidative stress may decrease thiol extents, causing inadequacy of adaptive mechanisms.

• Glutathione is a cellular antioxidant that performs a decisive role in attaining the native redox cellular state [74].

• Ascorbic acid is an antioxidant that prevails in plants and animals. However, its inability to generate in the human body mandates its intake from the diet. It can reduce and neutralize ROS [75]. Vitamin E is documented to arrest plasma membrane oxidation *by* getting rid of FR intermediates and reacting with the lipid radicals [76]. β-carotene has potent antioxidant properties and works *by* eliminating singlet oxygen to protect against an FR attack. Its richest food sources include egg yolk, milk, butter, spinach, carrots, tomato, and grains [77].

Table **5** distinguishes some important non-enzymatic antioxidants from their structural and food-grade significance viewpoints.

NATURAL ANTIOXIDANTS

- Of late, the protective effects of natural antioxidants against FR-induced toxicities have gathered significant interest. Flavonoids play important roles in guarding from oxidative stress-related complications, especially in the case of cancer. The richest and most common sources of flavonoids are vegetables, red wine, fruit, cocoa, and tea although these also prevail richly in beverages and foods exhibiting manifold bioactivities, of which antioxidation remains the most well-explored [78]. Natural antioxidants augment the ROS defiance of endogenous antioxidants, restoring an optimal balance of scavenging the reactive species. The antioxidant attributes of phenolics are manifested *via* distinctive mechanisms, such as FR-scavenging, hydrogen donation, singlet oxygen quenching, metal ion chelating, and substrate analog functioning towards superoxide and hydroxyl radicals [79].
- A significant 2021 study in this reference screened the parasitological efficacy of *Allium sativum* (AS) and *Curcuma longa* (CL) extracts in *Schistosoma mansoni*-infected mice models. The AS and Cl extracts were administered for 2 weeks post-seventh week of infection and the corresponding effects of Schistosoma worms were screened in terms of recovered worms, tissue egg count, and oogram morphology. Inspection revealed a substantial decline in recovered worms and egg count alongwith a noticeable variation in program morphology in all the mice treated with AS and CL compared to their untreated counterparts. The AS and CL significantly protected from hematological and biochemical disorders besides improving the antioxidant response. Further analysis revealed increased DNA fragmentation in the liver of schistosomiasis-infected mice compared with those of the untreated population. Based on these distinctions, the investigators concluded AS and CL as efficient anti-schistosomal drugs with an anticipated positivity to evaluate their combinatorial delivery for a better response subject to optimum treatment duration. The natural essence of AS and CL with such significant anti-schistosomal activity and enhanced physiological antioxidant actions, equated these to praziquantel (PZQ) for the elimination of young worms and eggs. It could be possible that AS and Cl in combination could provide enhanced efficacy with a moderated PZQ toxicity [80].
- Natural antioxidants are distinguished *via* manifold biochemical responses like suppressed ROS generation besides scavenging of FRs. In this regard, results of a recent study are worth mentioning, wherein anti-inflammatory attributes of *Camellia sinensis* Kuntze, *Acacia catechu* (catechin), *Curcuma longa* (curcumin), *Allium sativum* (garlic), *Punica granatum* (pomegranate) and

Azadirachata indica (neem) extracts were evaluated for their treatment efficacy of acute and chronic wound fibroblasts (AWFs and CWFs) [81]. Analysis of the biological efficacy for the phytoextracts (using Flow Cytometry) revealed no cytotoxicity at < 100 μg•mL^{-1} for normal human dermal fibroblasts. The garlic extract exhibited maximum cell viability, succeeded by catechin, epicatechin, curcumin, pomegranate peel, and neem, as per their respective IC$_{50}$ extents. About their anti-inflammatory essence, extracts of garlic, catechin, and epicatechin exhibited a substantial reduction in TGF-β and TNF-α activities in treated AWFs and CWFs compared to the non-treated AWFs and CWFs, nearly attaining a normal status. Erstwhile studies correlated these benefits to the benzoquinones, flavonoids, flavanol glycosides, alkaloids, carotenoids, catechols, glycosides, steroid glycosides, terpenoids, glycoalkaloids, mono, di- and triterpenes, saponins, and sterols in studied plants and herbs.

Table 5. Summary of salient non-enzymatic antioxidants enabling homeostasis maintenance in physiological conditions.

Name of the non-enzymatic scavenger	IUPAC/Chemical name	Chemical structure
β-carotene (vitamin A)	1,1′-[(1E,3E,5E,7E,9E,11E, 13E,15E,17E)-3,7,12,16-Tetramethyloctadeca-1,3,5,7,9,11,13,15,17-nonaene-1,18-diyl] bis(2,6, 6-trimethylcyclohex-1-ene)	
Ascorbic acid (vitamin C)	(2R)-2-[(1S)-1,2-dihydroxyethyl]-3,4-dihydroxy-2H-furan-5-one	
α-Tocopherol (vitamin E)	(2R)-2,5,7,8-tetramethyl-2-[(4R,8R)-4,8,12-trimethyltridecyl]-3,4-dihydrochromen-6-ol	
Glutathione (antioxidant)	γ-Glutamylcysteinylglycine	

RELATIONSHIP OF PRO-INFLAMMATORY REACTIONS AND OXIDATIVE STRESS

- Intricate relationship of oxidative stress and inflammation is discussed by multiple studies *via* ROS-modulated cell signaling actions [14]. Since both conditions amount to stress, the outcomes are inevitably caused due to excessive or undesired responses.
- It is well-known that almost all inflammatory responses comprise the generation of substantial ROS and reactive nitrogen (RNS) and chloride species amounts *via* activated phagocytic cells (neutrophils and macrophages). The prominent among these ROS/RNS producing intermediates are superoxide, hydrogen

peroxide, hydroxyl ion, nitric oxide, peroxynitrite and hypochlorous acid which unanimously aim at aggravating the oxidation, eventually resulting in physical or chemical stress. In all pathological inflammatory responses, there is a significant possibility of excessive ROS generation besides an accidental diffusion of generated ROS from the phagocytic cells, concomitantly aggravating the localized oxidative stress and surrounding tissue damage [82]. Quite interestingly, other than the usual generation from phagocytic cells, ROS could be generated from non-phagocytic cells too *via* immunologically guarding actions of pro-inflammatory cytokines [83, 84]. Studies reporting such possibilities have revealed synergistic participation of IFN-γ and pro-inflammatory components of bacterial cell wall lipopolysaccharide (LPS) for ROS formation in human pancreatic cancer cell lines *via* TLR4-NF-κB driven dual oxidase 2 (Duox2) expression, a vital member of NOX family [83].

- Recent investigations concurred with this possibility, wherein TLR costimulation aggravated the oxidative stress with simultaneous excessive pro-inflammatory cytokine generation [85]. A rather convincing trend has demonstrated ROS generation from inflammatory cytokine, IL-6 *via* enhanced NOX4 activity in non-small cell lung cancer (NSCLCs) [84]. Over-expressed NOX4 further resulted in enhanced IL-6 generation, creating a two-way feedback loop between IL-6 and NOX4, the respective inflammation and oxidative stress drivers. It seems quite prudent thereby, that oxidative and inflammatory stresses are intricately woven but only an excess inflammatory response mediates the trouble.

- Involvement of ROS (as H_2O_2) in the induction of inflammation occurs *via* NF-κB activation. Herein, oxidative stress is a decisive player implicated in the stimulation of NOD-like receptor protein 3 (NLRP3) inflammasome, an oligomeric complex that onset the innate immune responses *via* IL-1β and IL-18 maturation [86, 87]. Multiple mechanisms of ROS-driven NLRP3 inflammasome activation have been demonstrated of late, with the impaired mitochondrial functioning released ROS being involved in NLRP3 inflammasome activation, ultimately causing IL-1β secretion with enhanced localized inflammation [88].

- Evidence also deciphers the role of oxidized mitochondrial DNA in the NLRP3 inflammasome activation amidst apoptosis [86]. Another source of ROS-induced inflammation is the mutated DNA caused by ROS-elicited base modification. The base excision pair of oxidatively modified DNA base (7,8-dihydro-8-oxoguanine) by 8-oxoguanine-DNA glyoxalase-1 manifests a signaling cascade, onsetting NF-κB activation and subsequent enhanced inflammatory responses [89]. Apart from this, the oxidative stress-driven loss of extracellular redox potential of plasma cysteine as well as its disulfide form, activates monocyte adhesion near the vascular endothelial cells. This in turn, activates

NF-κB and subsequent IL-1β (a pro-inflammatory cytokine) expression [90, 91]. Multiple erstwhile cell-to-cell crosstalk pathways (like c-Jun N terminal kinase, p38 MAP kinase, and AP-1 transcription factor) together mediate continuous inflammatory sustenance through regulating oxidative stress.

- Fig. (12) explains the intricacy between oxidative and inflammatory stress conditions, whereby it seems quite evident that in case of oxidative stress manifested trouble in any of the organs, it is eventually likely to culminate in aggravated inflammation that further aggravates the oxidative stress due to localized accumulated response of neutralizing immune proteins. Alternatively, in case of inflammation happening early on, the hyperactivities of cytokines and chemokines inevitably result in exaggerated inflammation.

- For an idea about the RAGE signaling mediated altered inflammation and oxidative status, readers are advised to refer to erstwhile specific literature and section C (Fig. 6) [21, 92, 93].

- Owing to the complex redox control and mutual interplay of multiple molecular events, it becomes rather cumbersome to screen the first occurrence between inflammation and oxidative stress. Efforts to screen their occurrence concerning renal complications in spontaneously hypertensive rats (SHR) revealed a prior prevalence of oxidative stress and inflammation than the hypertensive manifestations deciphered their likely mutual relationship [94, 95]. Screening this relationship in 2- and 3-week-aged prehypertensive SHR and similar-aged WKY rats, the analysis revealed an evident increment in inflammatory and oxidative stress in SHR at the 3-week stage. Interestingly though, inspection of 2-week-old mice revealed enhanced expression of redox status exhibiting hallmarks while those for inflammation, exhibited a striking decrease. This indication of a prior modulation in oxidative status was further counter-screened by one-week treatment of 2-week aged SHR mice with antioxidants, whereby oxidative stress in kidneys was moderated with a substantial arrest of tubule-interstitial-renal cortex macrophage infiltration [96]. These observations affirmed the realization that oxidative stress is the primary renal abnormality in SHR. The beneficial effects of antioxidant therapy were indeed reciprocated for blood pressure and renal inflammation in hypertension-suffering animal models, but as of now, no concrete replication has been demonstrated in human subjects.

OXIDATIVE STRESS AND INFLAMMATION AS MEDIATORS OF VARIOUS RAGE-DEPENDENT DISEASE COMPLICATIONS

- The discussion in earlier sections has convincingly established that oxidative stress remains the major link for a manifestation of RAGE pathological effects. Multiple studies on *in vitro* cultured cells expressing RAGE herein have demonstrated that the ROS generation by RAGE is majorly attributed to its ligand interactions and subsequent altered gene expression. Apart from the mitochondrial amplification, activation of NADPH oxidase is the major mechanism by which RAGE contributes to oxidative stress. Recent studies on distinct *in vivo* studies on animal disease models decipher a binding of RAGE cytoplasmic domain with forming Dia1, whereby the activation of rac1 and NADPH oxidase in S100B (a putative RAGE ligand) treated primary murine aortic smooth muscle cells. Besides the cardiovascular complications, evidence on human subjects has correlated the RAGE ligands expression and soluble RAGE generation with oxidative stress-driven complications such as doxorubicin and acetaminophen toxicity, neurodegeneration, hyperlipidemia, diabetes, preeclampsia, rheumatoid arthritis, and pulmonary fibrosis [97].

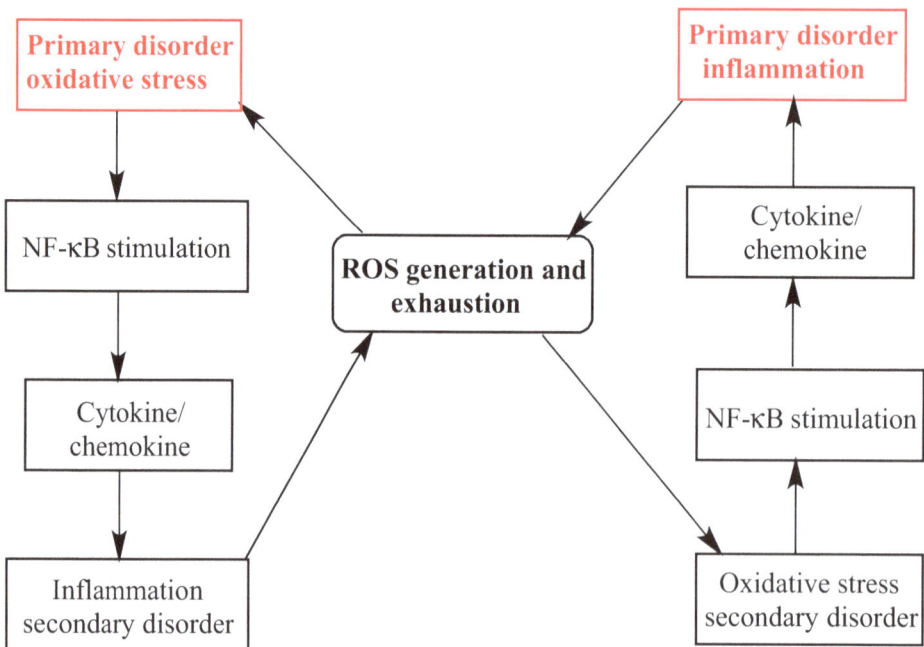

Fig. (12). Complex interplay between inflammatory and oxidative stress.

• RAGE and its soluble as well as endogenous secretory forms (sRAGE and en-RAGE) derive themselves from the TLRs superfamily, performing manifold critical roles in inflammation and autoimmunity, either directly or *via* binding with AGEs and advanced oxidation protein products (AOPPs). A noted attempt by *Riehl and colleagues* though established that RAGE is not direly essential to commence the inflammation but it does indeed sustain a regulated enhancement in the activity of manifold transcription factors such as Rb (retinoblastoma, a tumor suppressor protein)-E2F pathway, as per the observations of prominent staining in wild type mice contrary to the RAGE devoid mice on being stimulated with tetradecanoyl phorbol acetate (TPA) [98]. An erstwhile effort by *Leibold and associates* revealed that RAGE deletion on keratinocytes in mice models onsets a prompt resolution process after an initial inflammatory cascade, majorly due to the RAGE involvement in the TNF generation by keratinocytes [99].

• Separate studies by *Wolf and colleagues* on mice models revealed that in TPA-induced skin psoriasis lesions, pro-inflammatory cytokines such as MIP-2 (macrophage inflammatory protein-2), IL-1, and TNFα were considerably lower in the RAGE-devoid animals. Contrary to this, RAGE-devoid mice exhibited a higher extent of IL-1Ra (interleukin 1 receptor antagonist), as an anti-inflammatory effect in the *in vitro* conditions [100]. Another study by *Szczepanski and groupmates* reported the ear expression of RAGE and demonstrated the correlation between the HMGB1-RAGE pathway and cholesteatoma pathogenesis. The investigators noticed the sequence of events to aggravate the RAGE expression on skin cells, a rather consistent generation of IL-8, and a substantially better apoptotic safeguarding of keratinocytes *via* MEK (meiotic chromosome axis associated kinase) 1/2, MAPK (mitogen-activated protein kinase), and transcription factors like STAT3 and NF-κB. Interestingly, all these effects faded once the culture medium was furnished with RAGE-binding antibodies, which neutralizes the effects of RAGE [101].

• Significant evidence prevails to zero down the RAGE involvement in atherosclerotic lesions *via* ligand-dependent and independent signaling events. Studies on mice herein demonstrated an aggravated expression of vascular RAGE after diabetes induction and that RAGE knockdown attenuates the diabetes-triggered aggravation in atherosclerotic plaque region [102, 103]. Furthermore, studies on RAGE lacking diabetic mice revealed subdued macrophage recruitment, screened*via*CD68 staining, and further revealed a suppressed activity of prominent pro-inflammatory markers, such as immunoglobulin class of cell adhesion molecule vascular cell adhesion molecule 1 (VCAM-1) and transcription factor NF-κB [103]. More convincing efforts by *Park and teammates* revealed a scavenging of circulating AGEs on administering a soluble form, sRAGE, accounting for to suppressed progress of atherosclerosis in diabetic mice, irrespective of glucose and lipid intakes [104]. These observations beyond doubt establish the RAGE involvement in the pathogenesis of multiple disorders due to its aggravated ligand communication. The correlations of oxidative and inflammatory stress have been proven as reliable hallmarks for the diagnosis of cancers and related hyper-inflammatory disorders.

• Substantial positivity has been reported between the AOPPs, triglycerides aggravation, and total cholesterol extents. Exposure of umbilical vein endothelial cells to AOPPs has been reported as associated with nonphagocytic

NADPX oxidase (NOX) activation, enhanced superoxide generation besides NF-κB translocation, and VCAM-1, ICAM-1 upregulation [105]. A suppression of RAGE evaded inflammatory aggravations of AOPPs. Together these observations illustrate that RAGE signaling-driven post-translational modifications accounting for enhanced oxidative and inflammatory stress are critically involved in atherosclerotic plaque development. Erstwhile efforts also demonstrated a correlation between RAGE expression and NOX-mediated ROS generation, with RAGE-devoid mice exhibiting a reduction of diabetes-infuriated NOX subunits [102]. Cumulative findings from these studies demonstrate a prominent correlation and reciprocal linkage between NOX and multiple inflammatory pathways implicated in diabetes-aggravated atherosclerosis. The diversity of NOX with its multiple isoforms and cell-specific expressions is a major reason for an aggravated RAGE communication with its ligands under varying redox environments.

- Independent studies by *Thomas and the group also* demonstrated the excess sRAGE and AGEs generation as being correlated to a high risk of adverse renal outcomes with aggravated mortality. These observations made the investigators conclude that the AGE-RAGE axis might be a reliable therapeutic target for the therapeutic management of diabetes-associated kidney dysfunction (DKD).

- Finally, chronic hyperglycaemia was demonstrated as a manifesting source for hemodynamic variations characterized by agitating tension and aggressive frictional forces, leading to glomeruli-triggered renal injury *via* enhanced cytokines, pro-inflammatory markers, and growth factor activities which increase oxidative stress [106]. The observations made the investigators conclude that RAGE-mediated oxidative stress and increased inflammation, together contribute significantly to identifying novel therapeutic targets for treating renal complications. Thus, together all the discussed studies elucidate RAGE-mediated oxidative and inflammatory stress as being aggravated in multiple diseased conditions *via* enhanced RAGE communication with its ligands.

CONCLUSION

Under perfectly normal physiological conditions, eukaryotic cell organelle mitochondria and various oxidoreductases such as phagocytic and non-phagocytic NADPH oxidases and xanthine oxidases generate ROS. A low level of ROS acts as a secondary messenger and is used in several cell-signalling pathways. Apart from this, abundant non-enzymatic antioxidants such as reduced glutathione (GSH) and thioredoxin neutralize the excess ROS produced in normal physiological conditions. Vitamins C and E ingested from food intake are the other non-enzymatic antioxidants that maintain cellular redox homeostasis. Besides non-enzymatic antioxidants, excess activities of enzymatic antioxidants such as various superoxide dismutases (SODs) and peroxidases within the mammalian cells, also assist in neutralizing the excessive ROS generated in normal physiology. Thereby, under normal physiology, the abundance of enzymatic and non-enzymatic antioxidants in mammalian cell cytoplasm ensures a healthy physiology *via* maintaining a reduced intracellular environment. Analogous to ROS generation, in completely normal physiological conditions, the immunological cells of the body guard from infection and injury *via* generating inflammatory responses. Several lines of evidence indicate that both inflammatory and oxidative stresses are closely related *via cell-to-cell* crosstalk. In contrast to normal physiology, under various pathophysiological conditions, a high extent of ROS generation leads to oxidative stress. Similarly, in pathophysiological deviations, a substantial extent of pro-inflammatory reactions takes place. This aggravated inflammatory and oxidative stress propagates various non-communicable disease conditions such as cancers, diabetic vascular complications, and others.

Increased level of oxidation and inflammation aggravates the RAGE expression, the promoter sequence of which has the binding sites for pro-inflammatory transcription factors such as NF-κB. Under inflammatory and oxidative stress conditions, these pro-inflammatory transcription factors translocate from the cytoplasm to the nucleus and bind to the RAGE promoter to induce its expression. RAGE once expressed, interacts with several of its ligands including AGE and HMGB1. Multiple investigations provide experimental proof of RAGE and its ligand interaction mediated enhanced ROS generation as well as aggravated pro-inflammatory reactions, together stimulating the pro-inflammatory transcription factors such as NF-kβ. Thus, the cellular extent of RAGE is regulated by a nasty event of self-propagation. RAGE is implicated in the complication and propagation of several non-communicable diseases. This chapter describes the role of RAGE and its various ligands in regulating oxidative stress and pro-inflammatory reactions, which are the basis of the propagation of different disease conditions complicated by RAGE and RAGE-ligand interactions.

REFERENCES

[1]   Yan SF, Ramasamy R, Schmidt AM. Receptor for AGE (RAGE) and its ligands—cast into leading roles in diabetes and the inflammatory response. J Mol Med (Berl) 2009; 87(3): 235-47.
        [http://dx.doi.org/10.1007/s00109-009-0439-2] [PMID: 19189073]

[2]   Schmidt AM, Yan SD, Yan SF, Stern DM. The biology of the receptor for advanced glycation end products and its ligands. Biochim Biophys Acta Mol Cell Res 2000; 1498(2-3): 99-111.
        [http://dx.doi.org/10.1016/S0167-4889(00)00087-2] [PMID: 11108954]

[3]   Kierdorf K, Fritz G. RAGE regulation and signaling in inflammation and beyond. J Leukoc Biol 2013; 94(1): 55-68.
        [http://dx.doi.org/10.1189/jlb.1012519] [PMID: 23543766]

[4]   Mukherjee TK, Mukhopadhyay S, Hoidal JR. Implication of receptor for advanced glycation end

product (RAGE) in pulmonary health and pathophysiology. Respir Physiol Neurobiol 2008; 162(3): 210-5.
[http://dx.doi.org/10.1016/j.resp.2008.07.001] [PMID: 18674642]

[5]   Khaket TP, Kang SC, Mukherjee TK. The potential of receptor for advanced glycation end products (RAGE) as a therapeutic target for lung associated diseases. Curr Drug Targets 2019; 20(6): 679-89.
[http://dx.doi.org/10.2174/1389450120666181120102159] [PMID: 30457049]

[6]   Malik P, Chaudhry N, Mittal R, Mukherjee TK. Role of receptor for advanced glycation end products in the complication and progression of various types of cancers. Biochim Biophys Acta, Gen Subj 2015; 1850(9): 1898-904.
[http://dx.doi.org/10.1016/j.bbagen.2015.05.020] [PMID: 26028296]

[7]   Schmidt AM, Vianna M, Gerlach M, *et al.* Isolation and characterization of two binding proteins for advanced glycosylation end products from bovine lung which are present on the endothelial cell surface. J Biol Chem 1992; 267(21): 14987-97.
[http://dx.doi.org/10.1016/S0021-9258(18)42137-0] [PMID: 1321822]

[8]   Mukherjee TK, Mukhopadhyay S, Hoidal JR. The role of reactive oxygen species in TNFα-dependent expression of the receptor for advanced glycation end products in human umbilical vein endothelial cells. Biochim Biophys Acta Mol Cell Res 2005; 1744(2): 213-23.
[http://dx.doi.org/10.1016/j.bbamcr.2005.03.007] [PMID: 15893388]

[9]   Mukhopadhyay S, Mukherjee TK. Bridging advanced glycation end product, receptor for advanced glycation end product and nitric oxide with hormonal replacement/estrogen therapy in healthy versus diabetic postmenopausal women: A perspective. BBA-Mol Cell Res 2005; 1745: 145-55.
[http://dx.doi.org/10.1016/j.bbamcr.2005.03.010]

[10]   Kurutas EB. The importance of antioxidants which play the role in cellular response against oxidative/nitrosative stress: current state. Nutr J 2015; 15(1): 71.
[http://dx.doi.org/10.1186/s12937-016-0186-5] [PMID: 27456681]

[11]   Aon MA, Stanley BA, Sivakumaran V, *et al.* Glutathione/thioredoxin systems modulate mitochondrial H_2O_2 emission: An experimental-computational study. J Gen Physiol 2012; 139(6): 479-91.
[http://dx.doi.org/10.1085/jgp.201210772] [PMID: 22585969]

[12]   Ren X, Zou L, Zhang X, *et al.* Redox signaling mediated by thioredoxin and glutathione systems in the central nervous system. Antioxid Redox Signal 2017; 27(13): 989-1010.
[http://dx.doi.org/10.1089/ars.2016.6925] [PMID: 28443683]

[13]   Vermot A, Petit-Härtlein I, Smith SME, Fieschi F. NADPH oxidases (NOX): An overview from discovery, molecular mechanisms to physiology and pathology. Antioxidants 2021; 10(6): 890.
[http://dx.doi.org/10.3390/antiox10060890] [PMID: 34205998]

[14]   Camps J, García-Heredia A. Introduction: Oxidation and Inflammation, A molecular link between non-communicable diseases. oxidative stress and inflammation in non-communicable diseases - Molecular mechanisms and perspectives in therapeutics. Part of the Advances in Experimental Medicine and Biology book series (AEMB). 2014; 824: pp. 1-4.

[15]   Vijay K. Toll-like receptors in immunity and inflammatory diseases: Past, present, and future. Int Immunopharmacol 2018; 59: 391-412.
[http://dx.doi.org/10.1016/j.intimp.2018.03.002] [PMID: 29730580]

[16]   Bianchi ME. DAMPs, PAMPs and alarmins: all we need to know about danger. J Leukoc Biol 2007; 81(1): 1-5.
[http://dx.doi.org/10.1189/jlb.0306164] [PMID: 17032697]

[17]   Geto Z, Molla MD, Challa F, Belay Y, Getahun T. Mitochondrial dynamic dysfunction as a main triggering factor for inflammation associated chronic non-communicable diseases. J Inflamm Res 2020; 13: 97-107.
[http://dx.doi.org/10.2147/JIR.S232009] [PMID: 32110085]

[18] Furman D, Campisi J, Verdin E, *et al.* Chronic inflammation in the etiology of disease across the life span. Nat Med 2019; 25(12): 1822-32.
[http://dx.doi.org/10.1038/s41591-019-0675-0] [PMID: 31806905]

[19] Cirino G, Racagni G, Visioli F. Inflammation is at the root of all non-communicable diseases. Editorial, 4-5,
[http://dx.doi.org/10.36118/pharmadvances.2022.39]

[20] Lutterloh EC, Opal SM, Pittman DD, *et al.* Inhibition of the RAGE products increases survival in experimental models of severe sepsis and systemic infection. Crit Care 2007; 11(6): R122.
[http://dx.doi.org/10.1186/cc6184] [PMID: 18042296]

[21] Liliensiek B, Weigand MA, Bierhaus A, *et al.* Receptor for advanced glycation end products (RAGE) regulates sepsis but not the adaptive immune response. J Clin Invest 2004; 113(11): 1641-50.
[http://dx.doi.org/10.1172/JCI200418704] [PMID: 15173891]

[22] Amarante-Mendes GP, Adjemian S, Branco LM, Zanetti LC, Weinlich R, Bortoluci KR. Pattern recognition receptors and the host cell death molecular machinery. Front Immunol 2018; 9: 2379.
[http://dx.doi.org/10.3389/fimmu.2018.02379] [PMID: 30459758]

[23] Teissier T, Boulanger É. The receptor for advanced glycation end-products (RAGE) is an important pattern recognition receptor (PRR) for inflammaging. Biogerontology 2019; 20(3): 279-301.
[http://dx.doi.org/10.1007/s10522-019-09808-3] [PMID: 30968282]

[24] Yamamoto Y, Harashima A, Saito H, *et al.* Septic shock is associated with receptor for advanced glycation end products ligation of LPS. J Immunol 2011; 186(5): 3248-57.
[http://dx.doi.org/10.4049/jimmunol.1002253] [PMID: 21270403]

[25] Wang H, Tang Y, Fan Z, Lv B, Xiao X, Chen F. High-mobility group box 1 protein induces tissue factor expression in vascular endothelial cells *via* activation of NF-κB and Egr-1. Thromb Haemost 2009; 102(8): 352-9.
[http://dx.doi.org/10.1160/TH08-11-0759] [PMID: 19652887]

[26] Xie J, Reverdatto S, Frolov A, Hoffmann R, Burz DS, Shekhtman A. Structural basis for pattern recognition by the receptor for advanced glycation end products (RAGE). J Biol Chem 2008; 283(40): 27255-69.
[http://dx.doi.org/10.1074/jbc.M801622200] [PMID: 18667420]

[27] Moser B, Desai DD, Downie MP, *et al.* Receptor for advanced glycation end products expression on T cells contributes to antigen-specific cellular expansion *in vivo.* J Immunol 2007; 179(12): 8051-8.
[http://dx.doi.org/10.4049/jimmunol.179.12.8051] [PMID: 18056345]

[28] Chen Y, Akirav EM, Chen W, *et al.* RAGE ligation affects T cell activation and controls T cell differentiation. J Immunol 2008; 181(6): 4272-8.
[http://dx.doi.org/10.4049/jimmunol.181.6.4272] [PMID: 18768885]

[29] Akirav EM, Preston-Hurlburt P, Garyu J, *et al.* RAGE expression in human T cells: a link between environmental factors and adaptive immune responses. PLoS One 2012; 7(4): e34698.
[http://dx.doi.org/10.1371/journal.pone.0034698] [PMID: 22509345]

[30] Frommhold D, Kamphues A, Hepper I, *et al.* RAGE and ICAM-1 cooperate in mediating leukocyte recruitment during acute inflammation *in vivo.* Blood 2010; 116(5): 841-9.
[http://dx.doi.org/10.1182/blood-2009-09-244293] [PMID: 20407037]

[31] Frommhold D, Kamphues A, Dannenberg S, *et al.* RAGE and ICAM-1 differentially control leukocyte recruitment during acute inflammation in a stimulus-dependent manner. BMC Immunol 2011; 12(1): 56.
[http://dx.doi.org/10.1186/1471-2172-12-56] [PMID: 21970746]

[32] Chavakis T, Bierhaus A, Al-Fakhri N, *et al.* The pattern recognition receptor (RAGE) is a counterreceptor for leukocyte integrins: a novel pathway for inflammatory cell recruitment. J Exp Med 2003; 198(10): 1507-15.

[http://dx.doi.org/10.1084/jem.20030800] [PMID: 14623906]

[33]　Chavakis E, Hain A, Vinci M, *et al*. High-mobility group box 1 activates integrin-dependent homing of endothelial progenitor cells. Circ Res 2007; 100(2): 204-12.
[http://dx.doi.org/10.1161/01.RES.0000257774.55970.f4] [PMID: 17218606]

[34]　Li J, Schmidt AM. Characterization and functional analysis of the promoter of RAGE, the receptor for advanced glycation end products. J Biol Chem 1997; 272(26): 16498-506.
[http://dx.doi.org/10.1074/jbc.272.26.16498] [PMID: 9195959]

[35]　Li J, Qu X, Schmidt AM. Sp1-binding elements in the promoter of RAGE are essential for amphoterin-mediated gene expression in cultured neuroblastoma cells. J Biol Chem 1998; 273(47): 30870-8.
[http://dx.doi.org/10.1074/jbc.273.47.30870] [PMID: 9812979]

[36]　Lata K, Mukherjee TK. Knockdown of receptor for advanced glycation end products attenuate 17α-ethinyl-estradiol dependent proliferation and survival of MCF-7 breast cancer cells. Biochim Biophys Acta, Gen Subj 2014; 1840(3): 1083-91.
[http://dx.doi.org/10.1016/j.bbagen.2013.11.014] [PMID: 24252278]

[37]　Murphy MP. How mitochondria produce reactive oxygen species. Biochem J 2009; 417(1): 1-13.
[http://dx.doi.org/10.1042/BJ20081386] [PMID: 19061483]

[38]　Nègre-Salvayre A, Hirtz C, Carrera G, *et al*. A role for uncoupling protein-2 as a regulator of mitochondrial hydrogen peroxide generation. FASEB J 1997; 11(10): 809-15.
[http://dx.doi.org/10.1096/fasebj.11.10.9271366] [PMID: 9271366]

[39]　Cadenas S. Mitochondrial uncoupling, ROS generation and cardioprotection. Biochim Biophys Acta Bioenerg 2018; 1859(9): 940-50.
[http://dx.doi.org/10.1016/j.bbabio.2018.05.019] [PMID: 29859845]

[40]　Kwon OS, Decker ST, Zhao J, *et al*. The receptor for advanced glycation end products (RAGE) is involved in mitochondrial function and cigarette smoke-induced oxidative stress. Free Radic Biol Med 2023; 195: 261-9.
[http://dx.doi.org/10.1016/j.freeradbiomed.2022.12.089] [PMID: 36586455]

[41]　Desouki MM, Kulawiec M, Bansal S, Das GC, Singh KK. Cross talk between mitochondria and superoxide generating NADPH oxidase in breast and ovarian tumors. Cancer Biol Ther 2005; 4(12): 1367-73.
[http://dx.doi.org/10.4161/cbt.4.12.2233] [PMID: 16294028]

[42]　Brown DI, Griendling KK. Nox proteins in signal transduction. Free Radic Biol Med 2009; 47(9): 1239-53.
[http://dx.doi.org/10.1016/j.freeradbiomed.2009.07.023] [PMID: 19628035]

[43]　Choi JA, Lee JW, Kim H, *et al*. Pro-survival of estrogen receptor-negative breast cancer cells is regulated by a BLT2–reactive oxygen species-linked signaling pathway. Carcinogenesis 2010; 31(4): 543-51.
[http://dx.doi.org/10.1093/carcin/bgp203] [PMID: 19748928]

[44]　Park J, Jang JH, Park GS, Chung Y, You HJ, Kim JH. BLT2, a leukotriene B4 receptor 2, as a novel prognostic biomarker of triple-negative breast cancer. BMB Rep 2018; 51(8): 373-7.
[http://dx.doi.org/10.5483/BMBRep.2018.51.8.127] [PMID: 29898809]

[45]　Johar R, Sharma R, Kaur A, Mukherjee TK. Role of reactive oxygen species in estrogen dependant breast cancer complication. Anticancer Agents Med Chem 2015; 16(2): 190-9.
[http://dx.doi.org/10.2174/1871520615666150518092315] [PMID: 25980816]

[46]　Liou GY, Storz P. Reactive oxygen species in cancer. Free Radic Res 2010; 44(5): 479-96.
[http://dx.doi.org/10.3109/10715761003667554] [PMID: 20370557]

[47]　Redza-Dutordoir M, Averill-Bates DA. Activation of apoptosis signalling pathways by reactive oxygen species. Biochim Biophys Acta Mol Cell Res 2016; 1863(12): 2977-92.

[http://dx.doi.org/10.1016/j.bbamcr.2016.09.012] [PMID: 27646922]

[48] Perillo B, Di Donato M, Pezone A, *et al.* ROS in cancer therapy: the bright side of the moon. Exp Mol Med 2020; 52(2): 192-203.
[http://dx.doi.org/10.1038/s12276-020-0384-2] [PMID: 32060354]

[49] Li L, Ishdorj G, Gibson SB. Reactive oxygen species regulation of autophagy in cancer: Implications for cancer treatment. Free Radic Biol Med 2012; 53(7): 1399-410.
[http://dx.doi.org/10.1016/j.freeradbiomed.2012.07.011] [PMID: 22820461]

[50] Marinho HS, Real C, Cyrne L, Soares H, Antunes F. Hydrogen peroxide sensing, signaling and transcription factors regulation. Redox Biol 2014; 2: 535-62.
[http://dx.doi.org/10.1016/j.redox.2014.02.006] [PMID: 24634836]

[51] Sunitha J, Jeeva JS, Ananthalakshmi R, Rajkumari S, Ramesh M, Krishnan R. Enzymatic antioxidants and its role in oral diseases. J Pharm Bioallied Sci 2015; 7(6) (Suppl. 2): 331.
[http://dx.doi.org/10.4103/0975-7406.163438] [PMID: 26538872]

[52] Fukai T, Ushio-Fukai M. Superoxide dismutases: role in redox signaling, vascular function, and diseases. Antioxid Redox Signal 2011; 15(6): 1583-606.
[http://dx.doi.org/10.1089/ars.2011.3999] [PMID: 21473702]

[53] Miao L, St Clair DK. Regulation of superoxide dismutase genes: implications in disease. Free Radic Biol Med 2009; 47(4): 344-56.
[http://dx.doi.org/10.1016/j.freeradbiomed.2009.05.018] [PMID: 19477268]

[54] Zorov DB, Juhaszova M, Sollott SJ. Mitochondrial ROS-induced ROS release: An update and review. Biochim Biophys Acta Bioenerg 2006; 1757(5-6): 509-17.
[http://dx.doi.org/10.1016/j.bbabio.2006.04.029] [PMID: 16829228]

[55] Heck DE, Shakarjian M, Kim HD, Laskin JD, Vetrano AM. Mechanisms of oxidant generation by catalase. Ann N Y Acad Sci 2010; 1203(1): 120-5.
[http://dx.doi.org/10.1111/j.1749-6632.2010.05603.x] [PMID: 20716293]

[56] Nandi A, Yan LJ, Jana CK, Das N. Role of catalase in oxidative stress-and age-associated degenerative diseases. Oxid Med Cell Longev 2019; 2019: 1-19.
[http://dx.doi.org/10.1155/2019/9613090] [PMID: 31827713]

[57] Loperena R, Harrison DG. Oxidative stress and hypersensitive diseases. Med Clin North Am 2017; 101(1): 169-93.
[http://dx.doi.org/10.1016/j.mcna.2016.08.004] [PMID: 27884227]

[58] Münzel T, Camici GG, Maack C, Bonetti NR, Fuster V, Kovacic JC. Impact of oxidative stress on the heart and vasculature Part 2 of a 3-Part Series. J Am Coll Cardiol 2017; 70(2): 212-29.
[http://dx.doi.org/10.1016/j.jacc.2017.05.035] [PMID: 28683969]

[59] Takahara S. Progressive oral gangrene probably due to lack of catalase in the blood (acatalasaemia); report of nine cases. Lancet 1952; 260(6745): 1101-4.
[http://dx.doi.org/10.1016/S0140-6736(52)90939-2] [PMID: 12991731]

[60] Ahsan MK, Lekli I, Ray D, Yodoi J, Das DK. Redox regulation of cell survival by the thioredoxin superfamily: an implication of redox gene therapy in the heart. Antioxid Redox Signal 2009; 11(11): 2741-58.
[http://dx.doi.org/10.1089/ars.2009.2683] [PMID: 19583492]

[61] Flohé L, Toppo S, Cozza G, Ursini F. A comparison of thiol peroxidase mechanisms. Antioxid Redox Signal 2011; 15(3): 763-80.
[http://dx.doi.org/10.1089/ars.2010.3397] [PMID: 20649470]

[62] Maiorino FM, Brigelius-Flohé R, Aumann KD, *et al.* Diversity of glutathione peroxidases. Methods Enzymol 1995; 252: 38-53.
[http://dx.doi.org/10.1016/0076-6879(95)52007-4] [PMID: 7476373]

[63] Hall A, Parsonage D, Poole LB, Karplus PA. Structural evidence that peroxiredoxin catalytic power is based on transition-state stabilization. J Mol Biol 2010; 402(1): 194-209.
[http://dx.doi.org/10.1016/j.jmb.2010.07.022] [PMID: 20643143]

[64] Winterbourn CC, Hampton MB. Thiol chemistry and specificity in redox signaling. Free Radic Biol Med 2008; 45(5): 549-61.
[http://dx.doi.org/10.1016/j.freeradbiomed.2008.05.004] [PMID: 18544350]

[65] Flohe L, Günzler WA, Schock HH. Glutathione peroxidase: A selenoenzyme. FEBS Lett 1973; 32(1): 132-4.
[http://dx.doi.org/10.1016/0014-5793(73)80755-0] [PMID: 4736708]

[66] Kraus RJ, Prohaska JR, Ganther HE. Oxidized forms of ovine erythrocyte glutathione peroxidase cyanide inhibition of a 4-glutathione:4-selenoenzyme. Biochimica et Biophysica Acta (BBA) - Enzymology 1980; 615(1): 19-26.
[http://dx.doi.org/10.1016/0005-2744(80)90004-2] [PMID: 7426660]

[67] Rotruck JT, Pope AL, Ganther HE, Hoekstra WG. Prevention of oxidative damage to rat erythrocytes by dietary selenium. J Nutr 1972; 102(5): 689-96.
[http://dx.doi.org/10.1093/jn/102.5.689] [PMID: 5022203]

[68] Rotruck JT, Pope AL, Ganther HE, Swanson AB, Hafeman DG, Hoekstra WG. Selenium: biochemical role as a component of glutathione peroxidase. Science 1973; 179(4073): 588-90.
[http://dx.doi.org/10.1126/science.179.4073.588] [PMID: 4686466]

[69] Liddell JR, Hoepken HH, Crack PJ, Robinson SR, Dringen R. Glutathione peroxidase 1 and glutathione are required to protect mouse astrocytes from iron-mediated hydrogen peroxide toxicity. J Neurosci Res 2006; 84(3): 578-86.
[http://dx.doi.org/10.1002/jnr.20957] [PMID: 16721761]

[70] Fabregat I, Vitorica J, Satrustegui J, Machado A. The pentose phosphate cycle is regulated by NADPHNADP ratio in rat liver. Arch Biochem Biophys 1985; 236(1): 110-8.
[http://dx.doi.org/10.1016/0003-9861(85)90610-1] [PMID: 3966788]

[71] Lubos E, Loscalzo J, Handy DE. Glutathione peroxidase-1 in health and disease: from molecular mechanisms to therapeutic opportunities. Antioxid Redox Signal 2011; 15(7): 1957-97.
[http://dx.doi.org/10.1089/ars.2010.3586] [PMID: 21087145]

[72] Sheehan D, Meade G, Foley VM, Dowd CA. Structure, function and evolution of glutathione transferases: implications for classification of non-mammalian members of an ancient enzyme superfamily. Biochem J 2001; 360(1): 1-16.
[http://dx.doi.org/10.1042/bj3600001] [PMID: 11695986]

[73] Ulrich K, Jakob U. The role of thiols in antioxidant systems. Free Radic Biol Med 2019; 140: 14-27.
[http://dx.doi.org/10.1016/j.freeradbiomed.2019.05.035] [PMID: 31201851]

[74] Glutathione PJ. Integr Med (Encinitas) 2014; 13: 8-12.

[75] Njus D, Kelley PM, Tu YJ, Schlegel HB. Ascorbic acid: The chemistry underlying its antioxidant properties. Free Radic Biol Med 2020; 159: 37-43.
[http://dx.doi.org/10.1016/j.freeradbiomed.2020.07.013] [PMID: 32738399]

[76] Traber MG, Atkinson J. Vitamin E, antioxidant and nothing more 2007; 43: 4-15.
[http://dx.doi.org/10.1016/j.freeradbiomed.2007.03.024]

[77] Mueller L, Boehm V. Antioxidant activity of β-carotene compounds in different *in vitro* assays. Molecules 2011; 16(2): 1055-69.
[http://dx.doi.org/10.3390/molecules16021055] [PMID: 21350393]

[78] Waheed Janabi AH, Kamboh AA, Saeed M, *et al.* Flavonoid-rich foods (FRF): A promising nutraceutical approach against lifespan-shortening diseases. Iran J Basic Med Sci 2020; 23(2): 140-53.
[http://dx.doi.org/10.22038/IJBMS.2019.35125.8353] [PMID: 32405356]

[79]   Tsao R. Chemistry and biochemistry of dietary polyphenols. Nutrients 2010; 2(12): 1231-46.
        [http://dx.doi.org/10.3390/nu2121231] [PMID: 22254006]

[80]   Abu Almaaty AH, Rashed HAE, Soliman MFM, Fayad E, Althobaiti F, El-Shenawy NS.
        Parasitological and biochemical efficacy of the active ingredients of *Allium sativum* and *Curcuma longa* in *Schistosoma mansoni* infected mice. Molecules 2021; 26(15): 4542.
        [http://dx.doi.org/10.3390/molecules26154542] [PMID: 34361695]

[81]   Monika P, Chandraprabha MN, Murthy KNC. Catechin, epicatechin, curcumin, garlic, pomegranate peel and neem extracts of Indian origin showed enhanced anti-inflammatory potential in human primary acute and chronic wound derived fibroblasts by decreasing TGF-β and TNF-α expression. BMC Complement Med Ther 2023; 23(1): 181.
        [http://dx.doi.org/10.1186/s12906-023-03993-y] [PMID: 37268940]

[82]   Fialkow L, Wang Y, Downey GP. Reactive oxygen and nitrogen species as signaling molecules regulating neutrophil function. Free Radic Biol Med 2007; 42(2): 153-64.
        [http://dx.doi.org/10.1016/j.freeradbiomed.2006.09.030] [PMID: 17189821]

[83]   Wu Y, Lu J, Antony S, *et al.* Activation of TLR4 is required for the synergistic induction of dual oxidase 2 and dual oxidase A2 by IFN-γ and lipopolysaccharide in human pancreatic cancer cell lines. J Immunol 2013; 190(4): 1859-72.
        [http://dx.doi.org/10.4049/jimmunol.1201725] [PMID: 23296709]

[84]   Li J, Lan T, Zhang C, *et al.* Reciprocal activation between IL-6/STAT3 and NOX4/Akt signalings promotes proliferation and survival of non-small cell lung cancer cells. Oncotarget 2015; 6(2): 1031-48.
        [http://dx.doi.org/10.18632/oncotarget.2671] [PMID: 25504436]

[85]   Lavieri R, Piccioli P, Carta S, Delfino L, Castellani P, Rubartelli A. TLR costimulation causes oxidative stress with unbalance of proinflammatory and anti-inflammatory cytokine production. J Immunol 2014; 192(11): 5373-81.
        [http://dx.doi.org/10.4049/jimmunol.1303480] [PMID: 24771848]

[86]   Shimada K, Crother TR, Karlin J, *et al.* Oxidized mitochondrial DNA activates the NLRP3 inflammasome during apoptosis. Immunity 2012; 36(3): 401-14.
        [http://dx.doi.org/10.1016/j.immuni.2012.01.009] [PMID: 22342844]

[87]   Zhou R, Tardivel A, Thorens B, Choi I, Tschopp J. Thioredoxin-interacting protein links oxidative stress to inflammasome activation. Nat Immunol 2010; 11(2): 136-40.
        [http://dx.doi.org/10.1038/ni.1831] [PMID: 20023662]

[88]   Zhou R, Yazdi AS, Menu P, Tschopp J. A role for mitochondria in NLRP3 inflammasome activation. Nature 2011; 469(7329): 221-5.
        [http://dx.doi.org/10.1038/nature09663] [PMID: 21124315]

[89]   Aguilera-Aguirre L, Bacsi A, Radak Z, *et al.* Innate inflammation induced by the 8-oxoguanine DNA glycosylase-1-KRAS-NF-κB pathway. J Immunol 2014; 193(9): 4643-53.
        [http://dx.doi.org/10.4049/jimmunol.1401625] [PMID: 25267977]

[90]   Iyer SS, Accardi CJ, Ziegler TR, *et al.* Cysteine redox potential determines pro-inflammatory IL-1beta levels. PLoS One 2009; 4(3): e5017.
        [http://dx.doi.org/10.1371/journal.pone.0005017] [PMID: 19325908]

[91]   Go YM, Jones DP. Intracellular proatherogenic events and cell adhesion modulated by extracellular thiol/disulfide redox state. Circulation 2005; 111(22): 2973-80.
        [http://dx.doi.org/10.1161/CIRCULATIONAHA.104.515155] [PMID: 15927968]

[92]   Chen M, Wang T, Shen Y, *et al.* Knockout of RAGE ameliorates mainstream cigarette smoke-induced airway inflammation in mice. Int Immunopharmacol 2017; 50: 230-5.
        [http://dx.doi.org/10.1016/j.intimp.2017.06.018] [PMID: 28704797]

[93]   Malik P, Kumar Mukherjee T. Immunological methods for the determination of AGE-RAGE axis

generated glutathionylated and carbonylated proteins as oxidative stress markers. Methods 2022; 203: 354-63.
[http://dx.doi.org/10.1016/j.ymeth.2022.01.011] [PMID: 35114402]

[94]   Chabrashvili T, Tojo A, Onozato ML, *et al.* Expression and cellular localization of classic NADPH oxidase subunits in the spontaneously hypertensive rat kidney. Hypertension 2002; 39(2): 269-74.
[http://dx.doi.org/10.1161/hy0202.103264] [PMID: 11847196]

[95]   Wilcox CS. Oxidative stress and nitric oxide deficiency in the kidney: a critical link to hypertension? Am J Physiol Regul Integr Comp Physiol 2005; 289(4): R913-35.
[http://dx.doi.org/10.1152/ajpregu.00250.2005] [PMID: 16183628]

[96]   Biswas SK, Lopes De Faria JB, Biswas SK, Lopes De Faria JB. Which comes first: Renal inflammation or oxidative stress in spontaneously hypertensive rats? Free Radic Res 2007; 41(2): 216-24.
[http://dx.doi.org/10.1080/10715760601059672] [PMID: 17364948]

[97]   Daffu G, Del Pozo C, O'Shea K, Ananthakrishnan R, Ramasamy R, Schmidt A. Radical roles for RAGE in the pathogenesis of oxidative stress in cardiovascular diseases and beyond. Int J Mol Sci 2013; 14(10): 19891-910.
[http://dx.doi.org/10.3390/ijms141019891] [PMID: 24084731]

[98]   Riehl A, Bauer T, Brors B, *et al.* Identification of the Rage-dependent gene regulatory network in a mouse model of skin inflammation. BMC Genomics 2010; 11(1): 537.
[http://dx.doi.org/10.1186/1471-2164-11-537] [PMID: 20923549]

[99]   Leibold JS, Riehl A, Hettinger J, Durben M, Hess J, Angel P. Keratinocyte-specific deletion of the receptor RAGE modulates the kinetics of skin inflammation *in vivo.* J Invest Dermatol 2013; 133(10): 2400-6.
[http://dx.doi.org/10.1038/jid.2013.185] [PMID: 23594597]

[100]   Wolf R, Mascia F, Dharamsi A, *et al.* Gene from a psoriasis susceptibility locus primes the skin for inflammation. Sci Transl Med 2010; 2(61): 61ra90.
[http://dx.doi.org/10.1126/scitranslmed.3001108] [PMID: 21148126]

[101]   Szczepanski MJ, Luczak M, Olszewska E, *et al.* Molecular signaling of the HMGB1/RAGE axis contributes to cholesteatoma pathogenesis. J Mol Med (Berl) 2015; 93(3): 305-14.
[http://dx.doi.org/10.1007/s00109-014-1217-3] [PMID: 25385222]

[102]   Soro-Paavonen A, Watson AMD, Li J, *et al.* Receptor for advanced glycation end products (RAGE) deficiency attenuates the development of atherosclerosis in diabetes. Diabetes 2008; 57(9): 2461-9.
[http://dx.doi.org/10.2337/db07-1808] [PMID: 18511846]

[103]   Koulis C, Kanellakis P, Pickering RJ, *et al.* Role of bone-marrow- and non-bone-marrow-derived receptor for advanced glycation end-products (RAGE) in a mouse model of diabetes-associated atherosclerosis. Clin Sci (Lond) 2014; 127(7): 485-97.
[http://dx.doi.org/10.1042/CS20140045] [PMID: 24724734]

[104]   Park L, Raman KG, Lee KJ, *et al.* Suppression of accelerated diabetic atherosclerosis by the soluble receptor for advanced glycation endproducts. Nat Med 1998; 4(9): 1025-31.
[http://dx.doi.org/10.1038/2012] [PMID: 9734395]

[105]   Guo ZJ, Niu HX, Hou FF, *et al.* Advanced oxidation protein products activate vascular endothelial cells *via* a RAGE-mediated signaling pathway. Antioxid Redox Signal 2008; 10(10): 1699-712.
[http://dx.doi.org/10.1089/ars.2007.1999] [PMID: 18576917]

[106]   National Kidney F. KDOQI Clinical practice guideline for diabetes and CKD: 2012 Update. Am J Kidney Dis 2012; 60(5): 850-86.
[http://dx.doi.org/10.1053/j.ajkd.2012.07.005] [PMID: 23067652]

Receptor for Advanced Glycation End Products in Pulmonary Diseases

Parth Malik[1,2,†], **Ruma Rani**[3,†] and **Tapan Kumar Mukherjee**[4,*]

¹ School of Chemical Sciences, Central University of Gujarat, Gandhinagar, Gujarat-382030, India

² Swarrnim Startup & Innovation University, Bhoyan-Rathod, Gandhinagar-Gujarat, India

³ ICAR-National Research Centre on Equines, Hisar-125001, Haryana, India

⁴ Amity Institute of Biotechnology, Amity University, New Town, Kolkata, West Bengal 700156, India

Abstract: The receptor for advanced glycation end products (RAGE) is characterized as a multi-ligand pattern recognition receptor molecule exhibiting physiologically profuse expression in the lung alveolar type 1 (AT-1) epithelial cell's basolateral region. Advanced glycation end products (AGEs) are the most prominent among the multiple ligands of RAGE in lung tissues. Other major RAGE ligands comprise high mobility group box protein 1 (HMGB-1) and S100/calgranulin. In various pathophysiological conditions, lung tissues express the supraphysiological level of RAGE and its multiple ligands. In physiological conditions, the interaction of RAGE with its ligands assists in the maturity, spreading, and maintenance-enabled homeostasis of lung epithelial cells. Thus, physiologically abundant expression of RAGE in the lung AT-1 cells maintains their morphology and specific architecture. In physiological conditions, high basal level expression of RAGE in the lung tissues guards against the development of non-small cell lung cancers (NSCLCs), wherein decreased RAGE extents are correlated with non-small cell lung cancer (NSCLC) complications. However, in the lung tissues under pathophysiological conditions, supraphysiological expression of RAGE and its various ligands stimulates inflammation and oxidative stress-related cell signaling molecules. This aggravated extent of inflammation and oxidative stress in the lung tissues leads to the propagation of manifold lung diseases namely, asthma, chronic obstructive pulmonary disease (COPD), idiopathic pulmonary fibrosis (IPF), and cystic fibrosis (CF), acute lung injury (ALI) and acute respiratory distress syndrome (ARDS), pneumonia, sepsis, bronchopulmonary dysplasia, and pulmonary hypertension. This chapter describes the physiological and pathophysiological role of RAGE in the lungs and the anti-RAGE therapy against various lung diseases.

* **Corresponding author Tapan Kumar Mukherjee:** Amity Institute of Biotechnology, Amity University, New Town, Kolkata, West Bengal 700156, India; E-mail: tapan400@gmail.com
† These authors contributed equally to this work.

Keywords: Advanced glycation end products (AGE), Inflammatory stress, Lung diseases, Non-small cell lung cancers (NSCLCs), Oxidative stress, Receptor for advanced glycation end products (RAGE), RAGE-ligand interactions, RAGE inhibiting pharmacological agents.

INTRODUCTION

The receptor for advanced glycation end products (RAGE) was first identified in the bovine lung endothelium as a target protein for advanced glycosylated end products (AGEs). RAGE prevails as a single-chain cell surface immunoglobulin (Ig) class of protein molecule and is thus recognized as a membrane-bound protein (mRAGE). RAGE is also available as a soluble RAGE (sRAGE) or endogenous soluble RAGE (esRAGE), generated *via* proteolytic cleavage or alternative mRNA splicing. In adult lung tissues, RAGE is highly expressed in alveolar type-1(AT-1) epithelial cells. Besides AT-1 cells, RAGE is also expressed in erstwhile lung cells, including endothelial and smooth muscle cells. The mRAGE is an extensively glycosylated, single-spanning transmembrane protein containing a ligand-binding extracellular structure [one V-type and two C-type domains], a transmembrane region, and a C-terminal cytosolic domain involved in signal transduction. Missing transmembrane and cytoplasmic domains is a noted feature of sRAGE although it does exhibit extracellular domains [1].

RAGE recognizes a substantial diverse group of ligands, including advanced glycation end products (AGEs), amyloid β-peptides, high mobility group box protein 1 (HMGB1), S100/calgranulin, advanced oxidation protein products (AOPPs), oxidized low-density lipoprotein receptor- 1 (Lox-1), fasciclin EGF-like, laminin-type EGF-like link domain-containing scavenger receptor-1/2 (FEEL1/2) and CD36. Thus, RAGE is considered a multi-ligand receptor molecule. The lung tissues expressed all major RAGE ligands, *viz*. AGEs, HMGB-1, and S100/calgranulin [2].

Conventional lung physiology is characterized by RAGE involvement in the adherence, spreading, and homeostasis maintenance of various cells including AT1 cells. RAGE is also involved in the overall homeostatic regulation of multiple lung functions. Evidence comes from the RAGE knockout (RAGE$^{-/-}$) mice, which exhibited considerable aggravation in dynamic lung compliance with a suppressed maximal expiratory airflow [3, 4]. Nonetheless, high-level expression of RAGE or its ligands (exclusively, AGEs and HMGB1, S100, and subsequent aggravated RAGE-ligand interactions) leads to a complicated state of multiple lung diseases, *viz*. asthma, chronic bronchitis (CB), acute lung injury (ALI) and more severe regimes such as acute respiratory distress syndrome ARDS), chronic obstructive pulmonary disorders (COPD), sepsis, cystic fibrosis

(CF), and idiopathic pulmonary fibrosis (IPF). The inevitable outcome of enhanced RAGE-ligand interactions is the aggravated extent of inflammatory and oxidative stress, leading to an aggravated state of various pulmonary diseases. Differentially altered immune responses are primarily caused by endogenous antigens, inhaled allergens, respiratory pathogens, and environmental pollutants. Airway mucosal surface-associated pattern recognition receptor (PRR) molecules identify these antigens as RAGE, and these antigens are ultimately eliminated by inducing immune responses. Since immunological cells such as monocytes, macrophages, neutrophils, and leukocytes secrete most of the RAGE ligands, these cells are essential for the initiation and propagation of RAGE-dependent inflammatory responses and oxidative stress. Inactivating RAGE or RAGE-ligand interaction may moderate the inflammatory and oxidative stress, thus easing the complications of various lung diseases. This chapter describes the physiological and pathophysiological aspects of RAGE in lung tissues [5, 6].

CELL-SPECIFIC EXPRESSION OF RAGE IN THE LUNG TISSUES

The cell surface protein RAGE was first identified in bovine lungs. RAGE is most copiously expressed in the lung tissues compared to other tissues in the human body. *Neeper and* colleagues for the 1st time detected RAGE expression in the bovine lung tissues. Several subsequent investigations examined cell-explicit RAGE expression in the lungs. While initial studies by *Katsuoka and colleagues* localized RAGE in the pulmonary epithelial type 2(AT2) cells, studies of *Demling, Shirasawa, and Fehrenbach* groups demonstrated the exclusive RAGE expression on the alveolar type 1 (ATI) epithelial cells. Of note, these studies did not identify the RAGE expression in AT2 epithelial or capillary endothelial cells. *Fehrenbach and associates* used quantitative immunoelectron microscopy in rat and human lung specimens, establishing RAGE localization in the basal cell membrane of AT1 cells. *Shirasawa and colleagues* using an identical approach also identified RAGE expression on the ATI epithelial cell's basolateral membrane. Likewise, *Demling and teammates* also localized RAGE at the human lung AT1 cell basolateral membrane. Thereby, RAGE prevalence in the normal lungs is majorly within the AT1 cells' basolateral membrane, now a consensual viewpoint. Apart from the AT1 cells, RAGE expression has been demonstrated in the lung endothelium, bronchial, and vascular smooth muscle, alveolar macrophages, leiomyocytes, and on the visceral pleural surface in bovine tissues. Multiple subsequent attempts demonstrated synergy with plentiful RAGE expression in the endothelial cells exhibiting vascular origin [1, 6 - 12].

THE PHYSIOLOGICAL ROLE OF RAGE IN THE LUNG TISSUES

As already mentioned, the highest RAGE expression is noticed in the lung-AT1 cells. Of note, a typical mammalian lung consists of a tree-like airway and a honeycomb-like gas exchange compartment. The AT1 and AT2 cells surround the gaseous exchange-sensitive alveolar surface. These cells are in close contact with underlying capillaries and fibroblasts [13 - 15]. However, AT1 cells are flat and cover >95% of the gaseous exchange surface, whereas AT2 cells are cuboidal and are recognized as AT1 cell progenitors, attributing to surfactant generation and homeostasis [16, 17]. During development, a population of bi-potential progenitors expressing markers of both AT1 and AT2 cells, differentiate into mature AT1 or AT2 cells by enhancing further signatures for corresponding cell fate(s), eliminating the markers for alternative cell fate [18, 19]. The studies of *Bartling and associates* were the first crucial evidence regarding preferential RAGE localization at the AT-1 cell's basolateral surface. The investigators demonstrated that RAGE suppression stimulates lung cancer [20]. Localization of RAGE expression at the cell-to-cell contact sites inferred a possible adhesion molecule resembling the functioning of RAGE. Later it was found that RAGE enhances the cell's adherence with the matrix proteins and, most remarkably, has an outstanding ability to induce cell spreading. This was verified by counter-screened, suppressed cell-matrix adherence on impairing either the RAGE expression or *via* forbidden RAGE-ligand interactions. These studies altogether provided vital experimental evidence for a possible role of RAGE in modulating the alveolar epithelial cells-basement membrane adhesion [21]. Other than the RAGE-extracellular matrix (ECM) interactions, the trans interaction of RAGE can aggravate tissue stiffness. Aimed at understanding the physiological role of RAGE in pulmonary tissues, *Demling and colleagues* studied full-length human RAGE cDNA-transfected in the human HEK-293 cells (kidney epithelial cell lines). Analysis revealed RAGE-dependent cell adherence and spreading of the transfected cells, exhibiting a reduction upon inactivation of RAGE. These results suggested the possible role of RAGE in the adherence and spreading of AT-1 cells [9]

Subsequent experiments by *Lizotte and teammates* on fetal rat lungs illustrated enhanced RAGE expression with fetal growth. Of note, in these studies, hyperoxia-suppressed RAGE expression in the fetal lungs, based on which, investigators concluded that RAGE had a decisive role in pulmonary homeostasis [3]. In a similar type of study, *Kindermann and accomplices* also showed RAGE-modulated lung development and lung sensitivity to hyperoxic injury in newborn mice [22].

Further, *Marinakis and groupmates* observed ceasing polarization and pulmonary AT1 cell differentiation on partial to complete RAGE elimination. These observations made the investigators conclude that lung epithelial cells-basal lamina adhesion is regulated by RAGE-basal lamina and heparin sulfate chains electrostatic interactions [21]. Parallel attempts by *Al-Robaiy and colleagues* demonstrated the significance of high RAGE expression in supporting the respiratory mechanics of the lung. Lungs from RAGE knockout (RAGE$^{-/-}$) mice exhibit a substantial increase in dynamic compliance and a decrease in maximal expiratory airflow. A deficiency of RAGE culminated in lost lung structure, increased compliance, decreased elasticity of the respiratory tissues, and increased concentrations of serum protein albumin in bronchoalveolar lavage (BAL) fluids [23]. To conclude, the lung functions of RAGE could vary from adherence, spread, and maintenance homeostasis of various cells including AT1 cells.

THE PATHOPHYSIOLOGICAL ROLE OF RAGE IN THE COMPLICATION OF LUNG-ASSOCIATED DISEASES

RAGE and many of its ligands such as AGE and HMGB1 are viciously involved in manifesting inflammatory and oxidative stress complications. The pro-inflammatory and pro-oxidative actions of RAGE ligands are well demonstrated, both in RAGE-bound and free states. RAGE is recognized as a cause and consequence of the inflammatory and oxidative stresses associated with lung diseases. Under aggravated inflammatory and oxidative stress conditions, varied RAGE or its ligands such as AGE, HMGB1, S100, *etc.* expression, play a key role in complicating lung-associated diseases. Of note, RAGE is predominantly localized at the tissue injury sites, where RAGE-ligand axis aggression generates inflammation and oxidative stress and complicates lung-associated diseases. The major lung diseases complicated by RAGE are asthma, CB, COPD, ALI, ARDS, IPF, CF, and Sepsis. It is also claimed that RAGE is implicated in lung transplantation rejection [5, 24 - 28].

The following paragraphs summarize the important studies implicating the potential involvement of RAGE in complicating various lung-associated diseases.

ROLE OF RAGE IN THE COMPLICATION OF ASTHMA

The typical heterogeneity of asthma is characterized by chronic inflammation, reversible broncho-constriction, and mucus hypersecretion. The immune system response is characteristic of allergic and non-allergic eosinophilic or neutrophilic asthma. Thus, eosinophils or neutrophils are responsible for asthmatic chronic inflammation. However, in severe asthma, airway inflammatory response usually comprises neutrophil accumulation because of ongoing neutrophil influx, uncontrolled activation, and impaired neutrophil clearance [6, 28 - 30].

Ahead paragraphs briefly describe the role of RAGE in the complication of asthma.

CONSEQUENCES OF RAGE EXPRESSION IN ASTHMA COMPLICATIONS

Multiple investigations *via* cell culture experiments, animal models (using rats/mice), and asthma patients have demonstrated a RAGE-dependent enhanced extent of inflammation as responsible for worsening the pathological outcomes of asthma. In one such study, RAGE was observed as instrumental in instigating airway inflammation *via* activating the histone deacetylase 1 (HDAC1) pathway in a toluene di-isocyanate-induced murine asthma model. During asthmatic airway inflammation (AAI), RAGE is known to stimulate IL-33-mediated activation of type 2 innate lymphoid cells (ILC2 cells), which consequently promotes the IL-5 and IL-13-mediated pathophysiological consequences of asthma. Another important study observed a positive correlation of rs2070600 (cause G82S) substitution in the RAGE ligand-binding domain and forced expiratory volume in 1 second (FEV1) that impairs asthma. Separate investigations on ovalbumin-induced airway asthma also demonstrated RAGE involvement in T-cell activation. To support these observations, RAGE knockdown in rats was observed as the moderating link for house dust mite (HDM) or ovalbumin-induced airway remodeling, eosinophilic inflammation, and airway hypersensitivity. Thus, RAGE plays a crucial role in airway allergens-dependent immune response stimulation and its direct involvement in asthmatic pathogenesis [31 - 33].

LIGAND-RAGE INTERACTIONS INVOLVED IN ASTHMATIC COMPLICATIONS

Recent research has demonstrated the importance of HMGB1-RAGE interaction-based signaling for allergy-induced airway sensitization and inflammation (caused by ovalbumin, cockroaches, and HDM). By directly activating TLR2, TLR4, and RAGE-NF-κB signaling in CD4+ naive T cells or indirectly in dendritic cells, HMGB1 can induce Th2 and Th17 cell differentiation. In an elegant effort, *Hou and colleagues* observed a positive association between upregulated HMGB1 level and disease severity in asthmatic and COPD subjects. Analysis of the sputum of asthmatic patients showed a direct correlation between HMGB1, RAGE levels, and the neutrophil percentage, with their concurrent extents being associated with impaired FEV1. Treatment, however, decreased the expressions of HMGBl and RAGE, thereby highlighting the importance of both HMGB1 and RAGE in inflammation and the pathophysiology of asthma. In addition to HMGB1, macrophages, and neutrophils have higher concentrations of

S100A8/A9 RAGE ligands and members of the Ca^{2+} binding S100 protein family. There have been reports linking these ligands to asthmatic complications like remodeling and inflammation of the airways. Additionally, the protein level of S100A12 increased in asthmatic patients, supporting IgE-mediated lung functions and RAGE-mediated mast cell degranulation [34 - 38].

ROLE OF SOLUBLE RAGE IN THE COMPLICATION OF ASTHMA

In children's sputum, a corroborative connection between asthmatic severity and lower soluble RAGE (sRAGE) level was observed. In contrast to healthy controls, a lower plasma sRAGE level was seen in individuals with neutrophilic asthma. Similarly, children with asthma who did not respond well to the first therapy had a significantly larger sRAGE extent, which was directly correlated with the number of asthma exacerbations that occurred in the previous six months. Besides, sRAGE expression has been reported as positively associated with FEV1 and inversely correlated with eosinophilic count and total IgE count. A noted attempt herein demonstrated reduced sRAGE generation following allergens contact while intratracheal sRAGE administration culminated in suppressed HMGB1 signaling. Results established a simultaneous regulation of HMGB1 expression, neutrophilic inflammation, and Th17-type responses *via* sRAGE expression in neutrophilic asthma. This inverse relationship of sRAGE with disease progression might be due to its decoy receptor functioning. Thus, sRAGE may be used to arrest asthma [31, 39 - 42].

ROLE OF RAGE IN THE CHRONIC OBSTRUCTIVE PULMONARY DISEASE

Progression of airflow obstruction, airway inflammation, elevated neutrophil count, CD8+ T cells, mast cells, macrophages, and persistent inflammation in alveolar tissues are the hallmarks of chronic obstructive pulmonary disease (COPD)/emphysema. In COPD, air-space enlargement is caused by several factors, including oxidative stress, inflammation, protease/anti-protease imbalance, and apoptosis. Notably, 80-90% of COPD patients have chronic inflammation and oxidative stress as a result of tobacco smoke exposure [43].

CONSEQUENCES OF RAGE EXPRESSION IN AGGRAVATED CHRONIC OBSTRUCTIVE PULMONARY DISEASE COMPLICATIONS

RAGE over-expression is a prominent hallmark of smoking-habituated COPD sufferers. In an experiment, exposure to cigarette smoke extract (CSE) decreases the distribution of RAGE and produces pro-inflammatory cytokines by triggering redox-sensitive DAMP (Damage-associated molecular patterns) signaling*via*Nrf2 (nuclear factor erythroid 2-related factor 2). Excessive expression of RAGE could aggravate the production of nitric oxide (NO), which in turn may promote the expression and activation of pro-inflammatory transcription factors such as NF-κB and cytokines, which in turn may exacerbate the complications associated with COPD. These complications can be reversed on treatment with RAGE inhibitor, FPS-ZM1.

LIGANDS-RAGE INTERACTION IN THE COMPLICATION OF CHRONIC OBSTRUCTIVE PULMONARY DISEASE

Both experimental and clinical studies concur with RAGE and its ligands (prominently, HMGB1 and S100) involvement in COPD complexity. RAGE knockout has been demonstrated as a key strategy to ameliorate cigarette smoke-induced airway inflammation in mice, possibly by impairing HMGB1, S100A8, S100A9, and S100A12 expressions. Furthermore, clinical administration of enalapril (EPRL, a drug used to treat high blood pressure and congestive heart failure) also attenuated cigarette smoke-induced pulmonary and renal endothelial cell injury *via* impairing the AGE-RAGE pathway. Fig. (**1**) depicts the EPRL chemical structure, highlighting its dominant hydrophobic terminal sensitivity with a vulnerable molecular geometry caused by in and out of the plane orientations in the middle domain.

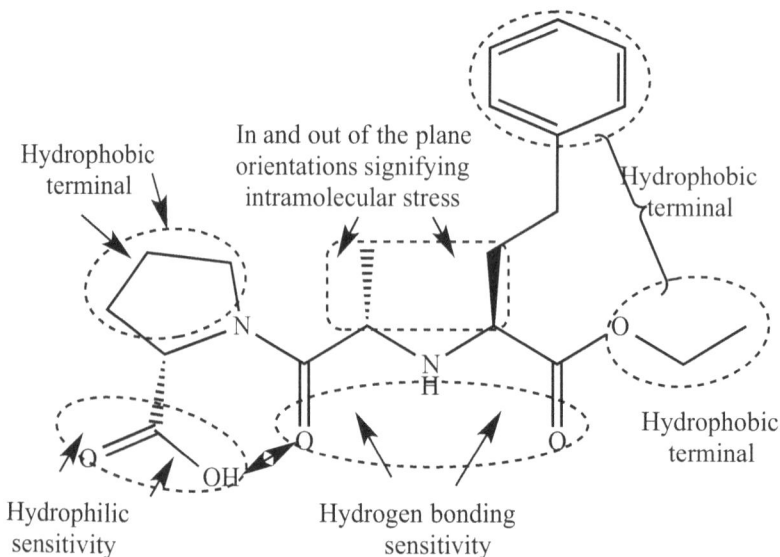

Fig. (1). Chemical structure of enalapril, depicting philicphobic sensitivity and in-out planar asymmetry attributing to an intramolecular stressed state.

An aggressive hydrophobicity at accessible terminal locations infers a low bioavailability of this drug through its unaided delivery. Furthermore, its complex structure with considerable intramolecular stress likely seems to aggravate the toxicity concerns for its oral intake. These constraints suggest a higher suitability for this drug's efficacy in being delivered through nanocarriers. EPRL is sparingly soluble in water and ethanol, whereas it is readily soluble in methanol. It is delivered as a pro-drug to counter its native structural complexity. After its oral intake, this drug is bio-activated by ethyl ester hydrolysis to enalapril *via* active angiotensin-modifying enzyme inhibitor [44 - 46].

Apart from the above-stated aggravated inflammatory outcomes of RAGE (via post-cigarette smoke exposure driven ligand interactions), RAGE also plays a key role in regulating pulmonary structural integrity, while cigarette smoke exposure aggravates inflammation-driven lung damage. RAGE deficiency has been shown as a key link to the loss of lung structure characterized by enhanced mean chord length, upregulated extents of bronchoalveolar lavage (BAL) fluid proteins, and decreased respiratory system elastance. RAGE targeting should be executed in strictly controlled conditions. A surprising observation leading to moderated HMGB1 and sRAGE expression was demonstrated by *Zhang and associates*, whereby acute exacerbation to the convalescence phase transition in COPD caused a considerable suppression of plasma HMGB1 and sRAGE (RAGE decoy receptor) [47].

ROLE OF SOLUBLE RAGE IN THE COMPLICATION OF CHRONIC OBSTRUCTIVE PULMONARY DISEASE

Recent investigations have elucidated an association of AGER locus (rs2070600) of the RAGE gene polymorphisms with the sRAGE extents, having been revealed as inversely related to emphysema (a pathological situation affecting the air spaces distal to the terminal bronchiole) severity. Studies also reported rs2070600T (Ser82) AGER alleles as related to higher FEV1, FEV1/FVC, and lower serum sRAGE levels in UK smokers. Another research attempt confirmed the serum and bronchoalveolar lavage (BAL) fluid sRAGE suppression in elastase-induced experimental COPD mice and COPD patients. Thus, RAGE inhibition can be protective in COPD, wherein no TLR4 signaling impairment could guard against airway neutrophilia and hyper-responsiveness (AHR) in COPD [47 - 52].

ROLE OF RAGE IN THE COMPLICATION OF ACUTE LUNG INJURY AND ACUTE RESPIRATORY DISTRESS SYNDROME

The conditions known as acute respiratory failure, acute lung injury (ALI), and their severe variant, acute respiratory distress syndrome (ARDS), are identified by neutrophil-derived inflammation, non-cardiogenic edema, severe systemic hypoxemia, and diffused alveolar damage. ALI/ARDS can be brought on by several different clinical conditions, including sepsis, acute pancreatitis, trauma, and direct lung injury, which includes pneumonia, gastric aspiration, and toxic inhalation. The American-European Consensus Conference (AECC) recognized ALI and ARDS by the ratio of arterial oxygen to the fraction of inspired and below 200 mm Hg for ARDS.

LUNG DAMAGE ASSOCIATED WITH RAGE EXPRESSION

Recent studies have identified the significance of RAGE in the development of ALI/ARDS. Given that AT-I cells comprise over 95% of the internal lung surface area, impairment to these cells is a strong marker of lung injury, particularly in ARDS and ALI. Since RAGE is primarily linked to the basolateral membrane of AT-I cells, its release in the pulmonary edema fluid of ALI/ARDS patients suggests damage to the membranes of AT-I cells based on the severity of the disease. In a noteworthy effort, *Calfee and colleagues* [53] provided evidence in support of these findings, specifically demonstrating a higher baseline plasma RAGE expression in enhanced severity of lung injury and aggravated mortality rates. Enhanced RAGE extent could also be noticed in BAL amidst lung damage on intratracheal instillation of hydrochloric acid (HCl), bacterial lipopolysaccharide (LPS), or *Escherichia coli,* a Gram-negative bacterium and hyperoxia (higher partial pressure of oxygen). On the other hand, RAGE suppression can also effectively reduce TNF-α and IL-1β extents in LPS-treated AT-1 cells. Of further significance is the RAGE suppression in protecting against hyperoxia-induced ALI, accompanied by a marked reduction in protein leakage and secretion of pro-inflammatory cytokines.

RAGE-LIGAND INTERACTION IN THE COMPLICATION OF ACUTE LUNG INJURY AND ACUTE RESPIRATORY DISTRESS SYNDROME

RAGE inhibition, particularly targeting the HMGB1-RAGE axis, can attenuate LPS-induced ALI *via* peroxisome proliferator-activated receptor-gamma (PPARγ) mediated heme oxygenase 1 (HMOX-1) activation. RAGE also exhibits the potential to trigger fibro-proliferative phase progression of ARDS since RAGE−/− mice are guarded against bleomycin (BMC) induced lung fibrosis. One study observed that lidocaine hydrochloride (LCD) (Fig. **2**) could attenuate cecal ligation and puncture (CLP, a common protocol for *in vivo* sepsis modeling) induced upregulated HMGB1 and RAGE expressions. The strategy successfully downregulated the expressions of pro-inflammatory transcription factor NF-κB and MAPK signaling pathways. The onset of RAGE-HMGB1 interaction has also been demonstrated to trigger neutrophil accumulation, the inception of lung edema, and elevated pulmonary levels of cytokines such as IL-1, TNFα, and MIP-2 (macrophage inflammatory protein-2). Of note, LCD is an anesthetic agent commonly applied locally or in topical anesthesia.

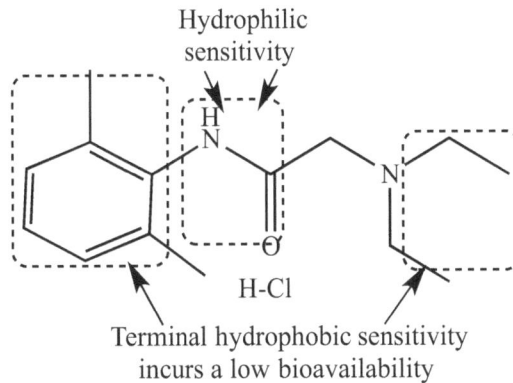

Fig. (2). Chemical structure of lidocaine hydrochloride (LCD), working *via* blocking the signals at the dermal nerve endings. Terminal hydrophobicity signifies a weaker bioavailability with hydrogen chloride supplementation indicating its pH-sensitive action.

This drug has also been examined for its antiarrhythmic and analgesic attributes and can be used as an accessory to tracheal intubation. Unlike the unconsciousness exhibited by common anesthetics for surgery, LCD works by interrupting the signals at dermal nerve terminals and rarely exhibits any unconsciousness. Also demarcated in its chemical structure, the terminal hydrophobicity of LCD impairs its water solubility, exhibiting instability on being exposed to air. Such attributes argue well for the nanocarrier-mediated administration of LCD, protecting its structure for a longer duration. The normal prescription of LCD comprises its HCl solution (also referred to as a viscous LCD) administered to accomplish numbness of mucous membranes of the mouth, gums, and throat. In general, LCD base is a weaker compound that readily combines with acids to form salts. Due to its unstable nature at a pH of 7.9, LCD is administered as its acidic preparation to modulate its solubility and prolong its shelf-life. To augment the duration of its anesthetic effect, LCD is delivered *via* supplementing with epinephrine, which moderates its toxicity *via* prompt homeostasis. Pain and uneasiness during sub-cutaneous injection of LCD can be reduced by diluting it with normal saline in 1:10 proportion.

In humans with severe trauma, plasma HMGB1 levels correlate positively with the severity of injury, gradually advancing to ALI. Targeting HMGB1 by glycyrrhizin (GHZ) ameliorates LPS-induced, IL-33 expression-mediated lung injury. Of note, GHZ prevails as a major constituent of licorice root, used in traditional medicine to treat bronchitis, gastritis, and jaundice. The drug exhibits a complex structure with multiple non-planar molecular orientations conferring significant stress to the native molecule (Fig. **3**).

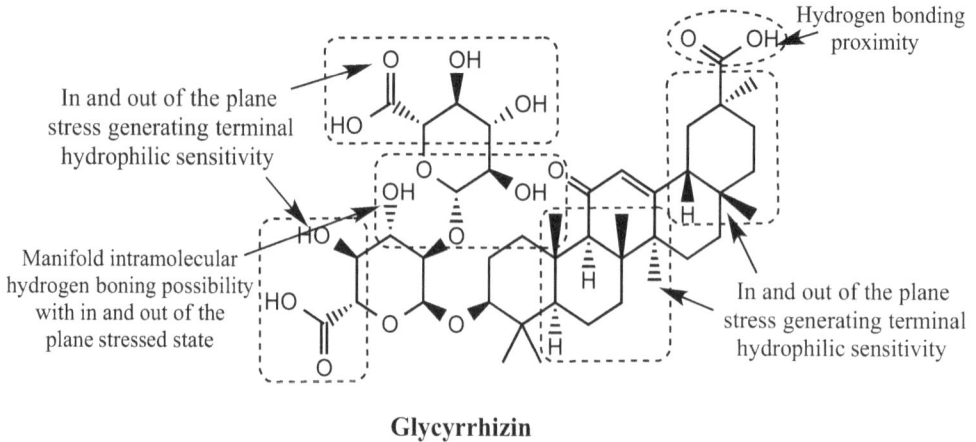

Glycyrrhizin

Fig. (3). Chemical structure of glycyrrhizin, depicting its characteristic hydrophilic and hydrophobic sensitivity. Multiple O and –OH substituents confer a hydrophilic sensitivity but in and out of the plane orientations generate a residual intramolecular stress.

Studies on GHZ intake demonstrate that including too much licorice (a molecule 50 times sweeter than sugar) could deplete the potassium that extends from the body, in turn increasing the vulnerability to cardiac problems, majorly resulting in increased blood pressure. Recommendations warn that eating more than 5 g of licorice for several weeks could worsen the side effects (including cardiac arrest). The compound does not suffer from a low bioavailability with a 3.5 h half-life and (1-10) mg•ml^{-1} aq solubility at 20°C, although concentration moderation could improve the efficacy [54, 55].

The HMGB1-RAGE signaling axis promotes progression to a severe, acute exudative phase of ARDS. Additionally, the profibrotic (conditions promoting thickening and stiffness of the tissues) activity of RAGE is due to HMGB1-RAGE signaling leading to enhanced generation of profibrotic growth factors and epithelial-mesenchymal transition (EST). Another RAGE ligand, S100A12 in higher concentrations in pulmonary and BAL, has been reported as a criticality in ARDS patients.

ROLE OF SOLUBLE RAGE IN INACUTE LUNG INJURY AND ACUTE RESPIRATORY DISTRESS SYNDROME

Attenuation of LPS-induced lung inflammation and lung injury in mice, following sRAGE administration, is well versed. Besides, in corrosively injured mice, the rate of alveolar fluid clearance (AFC) has an inverse correlation with plasma sRAGE and BAL levels. Thus, the RAGE level could be a crucial indicator of cell damage and disease severity, especially in ALI/ARDS.

ROLE OF RAGE IN THE COMPLICATION OF IDIOPATHIC PULMONARY FIBROSIS

Idiopathic pulmonary fibrosis (IPF) is a parenchymal lung disease associated with myofibroblast proliferation. IPF starts with damage to the alveolar epithelium, which then leads to the creation of fibroblastic foci, destroying lung parenchymal architecture. The alveolar epithelial cells (AEC) produce myofibroblasts by epithelial-to-mesenchymal transition (EMT). Transforming growth factor beta (TGF-β) prompts the alveolar EMT in human lung epithelial cells *via* the SMAD2- or SMAD3-subordinate pathway. Of note, TGF-β is a cytokine that controls proliferation, cellular differentiation, and other functions in most cells, and SMADs (Smad protein family consists of eight members and regulates the transcriptional effects of TGF-β signaling in vertebrates) are the major signal transducers for receptors.

LIGAND-RAGE INTERACTIONS IN IDIOPATHIC PULMONARY FIBROSIS

RAGE could be applied with prognostic and diagnostic biomarkers in the IPF pathogenesis. The AGE-RAGE interaction can attenuate TGF-β-dependent EMT through Smad7 expression. In a study, *Kyung and colleagues* observed increased levels of AGE synthesis and higher RAGE expression in IPF patients [56]. Therefore, elevated basal RAGE expression in the lungs supported its protective functioning in pulmonary fibrosis by inhibiting the TGF-β-induced EMT. Additionally, AGE treatment in rats enhances type IV collagen and fibronectin expression with a consequent increase in renal TGF-β1 and connective tissue growth factor (CTGF) expression.

Fig. (**4a** and **b**) depicts the BMC and aminoguanidine chemical structures, wherein a complex structure of the former makes its unaltered administration critical. Although the presence of manifold OH, O, and –NH_2 functionalities at the accessible terminal locations decipher a hydrophilic site, their adjacent placements create a vulnerability about the desired extent of medium (the desired physiological locations) interactions. Such features intensify the prevalence of self-interactions with manifold possibilities of intramolecular hydrogen bonding. Apart from this, the adjacent in and out of the plane orientations of substituents at nearby loci confer a resultant stress on the molecule. Therefore, controlling oscillations in these molecules is a challenge in their delivery. Opposed to this, a much simpler structure of aminoguanidine with no congested substituents placements and a hydrophilic sensitivity of NH and –NH_2 functionalities convey a correlation with structural traits of AGE inhibiting agents. It seems that the hydrophilic essence of aminoguanidine engages AGEs, which prohibits its interaction with RAGE. Since circulating AGE levels were significantly higher in IPF patients exhibiting diabetic complications, certain anti-diabetic agents also attenuated lung fibrosis up to an extent. However, increased expression of pro-fibrotic cytokines such as TGF-β by AGEs further complicates its regulatory mechanism.

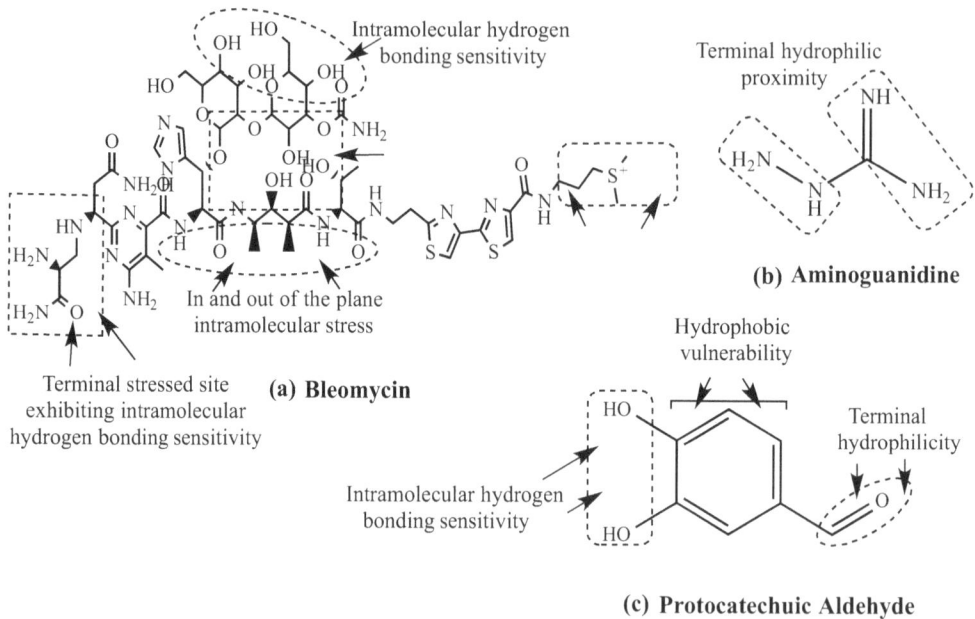

Fig. (4). Chemical structures of (a) Bleomycin (BMC), (b) aminoguanidine, and (c) protocatechuic aldehyde. While BMC is an antibiotic targeting rapidly dividing cells, aminoguanidine is a noted inhibitor of advanced glycation end products. Similarly, the protocatechuic aldehyde is an HMGB1-RAGE pathway modulator administered as a protective agent for pulmonary fibrosis.

The above observations thereby authenticate AGE's involvement in the pathogenesis of fibrosis. AGE accumulation was observed as directly correlated with the progression of BMC-induced pulmonary fibrosis (BIPF, BMC is an antibiotic which kills the profusely dividing cells), and utilization of an AGE inhibitor, aminoguanidine resulted in BIPF attenuation.

On the other hand, one study demonstrated that protocatechuic aldehyde (PA, a regulator of G-protein coupled estrogen receptor-1, exhibits protective effects in endothelial dysfunction and atherosclerosis) prevents experimental pulmonary fibrosis by modulating the HMGB1-RAGE pathway [57]. Fig. (**4c**) depicts the chemical structure of PA, wherein accessible terminals exhibit a hydrophilic proximity *vis-à-vis* correlation with its function of inhibiting HMGB1 activity. However, the central benzene ring in the structure confers a hydrophobic spacing of the hydrophilic ends, making the complete structure amphipathic. Adjacently placed –OH groups at the left terminal create a considerable possibility of intramolecular hydrogen bonding, and the sole carbonyl functionality at the right terminal imparts an electrophilic sensitivity to the molecule. Possible modulation of structural activities could be encompassed by replacing H of the left terminal –OH groups with varied substituents.

PA consumption also ameliorated BMC-induced pulmonary fibrosis *via* suppressed levels of vimentin, fibroblast growth factor 2 (FGF-2), HMGBI, human lung fibroblasts (HLF-1) proliferation, platelet-derived growth factor (PDGF) and TGF-β1-mediated EMT along with a consequent E-cadherin and RAGE enhancement. Thus, HMGB1 reduction probably holds the key for PA treatment, restoring EMT and pulmonary fibrosis.

Evidence of RAGE-HMGB1 signaling involvement in BMC-induced lung fibrosis was also provided by *He and colleagues*, wherein the specific involvement of RAGE in the acute inflammation and chronic fibrotic phases of lung injury induced by intratracheal instillation of BMC in mice models, was screened. BMC-induced lung fibrosis was initially screened in wild-type and RAGE-deficient mice models wherein administration to wild-type mice caused an initial pneumonitis which gradually evolved into fibrosis. Quite curiously, RAGE-deficient mice exhibited an early inflammatory response owing to which the mice were guarded off from the late fibrotic effects of BMC. It was a striking coincidence to note that this RAGE deficiency manifested protection was accompanied by suppressed pulmonary extents of pro-active RAGE-inducible fibrotic cytokines, TGF-β, and PDGF. Yet another notable observation was the stimulated activation of HMGB1 on BMC administration, a putative RAGE ligand secreted by the inflammatory cells that accumulated in the adjacent airspace.

The confirmatory screening was made using the fact that co-culture with HMGB1 manifested EMT in alveolar type II epithelial cells from wild-type mice. This characteristic trait was complemented by the null response of RAGE-deficient mice alveolar type II epithelial cells to the HMGB1 treatment. The findings zeroed in on an explicit involvement of RAGE-HMGB1 axis orientation in the EMT induction. Besides, BMC intake also activated profibrotic cytokines, TGF-β, and PDGF only in the wild-type mouse lungs. Collectively, the findings made the investigators conclude that RAGE expression aggravates BMC-induced lung fibrosis *via* EMT and pro-fibrotic cytokine generation. Therefore, targeting RAGE signaling could be a potent therapeutic strategy for pulmonary fibrosis treatment [58].

The critical involvement of RAGE in the manifestation of idiopathic pulmonary fibrosis (IPF) was also highlighted by *Kumar and groupmates*, wherein findings of manifold yester attempts were discussed significantly [59]. The authors took eminent note of a 2008 Letter to Editor by *Bargagli and the team*. The recalled work assessed the extent of RAGE in the broncho-alveolar lavage (BAL) of IPF sufferers, Desquamative Interstitial Pneumonia (DIP) besides comparative RAGE expressions in healthy smokers and non-smokers. Investigators therein demonstrated that a reduced RAGE expression could aggravate the complications of IPF sufferers, which was emphasized *via* similar findings of multiple attempts [60, 61] for the generalization that RAGE could be a decisive guarding agent in pulmonary fibrosis. However, despite several pieces of evidence, the view of the authors meets in contrast with a wider consensus about RAGE-aggravated fibrosis in multiple organs [58, 62, 63].

In light of such contradictions for RAGE involvement in IPF, *Kumar and the group* elaborated on some critical molecular aspects of RAGE involvement in IPF complications. First and foremost, the reported polymorphisms in the advanced glycation end product receptor (AGER) gene have been viciously examined for aggravated diabetic complications, such as microalbuminuria, nephropathy, retinopathy, and cardiovascular complications *via* modulating the AGE-RAGE interaction. Thereby, the authors raised the possibility of explicit RAGE polymorphism as the contributory factor for contrasting results on IPF. The second aspect pointed out by authors relates to manifold RAGE splice variants, generated as outcomes of alternate mRNA splicing or N-terminal truncation. In light of this, authors have raised the involvement of a specific splice variant in disease complications while others have not. This verification could clarify the conceptions regarding RAGE involvement in IPF pathophysiology.

The subsequent aspect discussed by the authors takes note of low CD40 expression in the lung fibroblasts under normal conditions. Noting that CD40 signaling in lung fibroblasts is significantly implicated in proliferation, cytokine generation, and extracellular matrix secretion. It is pertinent to note that under inflammatory conditions, CD40 expression increases which could result in IPF complications [64]. These findings collectively strengthen a need to ascertain the role of RAGE in CD40 expression and its impact on IPF. The fourth aspect pointed out by *Kumar and group* has talked about the prospect that ligand-bound sRAGE prevalent in BAL has a similar potential of binding with RAGE antibody that is used to ascertain the extent of RAGE *via* ELISA as discussed by *Bargagli and colleagues* [65]. Results of three recent studies are discussed ahead which implicate RAGE-aggravated IPF manifestations, a consolidation for future studies.

- The first study herein is a 2016 research attempt by *Machhua and the group* whereby abnormal epithelial-mesenchymal transition was noticed as being critically involved in IPF for the fastened aging process, a prominent hallmark of modulated wound healing. Taking note of AGE generation *via* lipid-protein non-enzymatic reactions with multiple oxidants, the investigators screened RAGE participation in lung fibrosis and alveolar homeostasis. Screening the AGE-RAGE aging-mediated complications in IPF, the investigators examined 16 IPF and 9 control lung samples obtained *via* surgical lung biopsy. The distinctive AGE levels and RAGE expression of IPF patients and control subjects were determined *via* RT-PCR, Western blot, and immunohistochemical studies whereby an effect of AGEs on cell viability of primary lung fibrotic fibroblasts and alveolar epithelial cells was studied. Cell transformation of fibrotic fibroblasts cultured in the glycated matrices was ascertained under varied experimental conditions. Analysis of AGE levels and RAGE expression in the IPF and control cells revealed that enhanced AGE level accompanied by deceased RAGE expression is responsible for the IPF complications. Immunohistochemical screening revealed higher AGEs staining, associated with ECM proteins at the apical surface of alveolar epithelial cells (AECs) residing in the vicinity of fibroblast foci for fibrosis-infected lungs. Opposed to this, RAGEs prevailed more exhaustively at the AECs membrane within the control lungs. In the fibrosis-affected tissues, RAGE expression was undetectable along the AEC membrane. Subsequently, the effect of AGE generation on cell viability was observed as distinct for AECs and fibrosis-affected fibroblast cells. The presence of AGEs even at low concentrations decreased the viability in AECs contrary to a moderate impact on fibroblast viability. The enhanced ECM glycation in fibrotic cells was confirmed by increased fibroblast expression. The observed distinctions suggested a role of enhanced AGE to RAGE stoichiometry in IPF, with a rapid aging-driven tissue reaction and abnormal lung wound healing in IPF patients [66].
- The subsequent study is a 2021 research attempt by *Baek and colleagues* whereby their findings suggest that reduced RAGE was associated with increased fibrotic genes in bleomycin (BLM)-induced mice and patients with IPF. Therefore, RAGE could be useful as a biomarker for prognosis and diagnosis in IPF pathogenesis [67].
- Finally, the last PubMed-recognized attempt is a 2022 study by *Piao and colleagues* wherein the efficacy of a RAGE antagonist peptide (RAP), optimized *via* the RAGE binding site of HMGB1 protein was screened in a BLM-induced mice model suffering from IPF. The BLM was administered intratracheally after which RAP was injected twice *via* intratracheal instillation, the first and third

day after manifested BLM response. To screen the further mechanistic details, mice administered with BLM were sacrificed on the 7th day and the lungs were preserved. Compared to the control group, the RAP-administered mice showed a reduced pulmonary hydroxyproline level. Related protein expressions that were altered included those of tumor growth factor-β (TGF-β), α-smooth muscle actin (α-SMA), and collagen, all exhibiting a dose-dependent moderation. To screen the long-term effects of RAGE inhibition, the mice challenged with BLM were treated with RAP (*via* tracheal route) every 7 days till 28 days. Inspection of retrieved and harvested lung samples revealed that on repeated RAP administration, the functional extents of hydroxyproline, TGF-β, α-SMA, and collagen were substantially suppressed. The observations collectively established the therapeutic significance of RAP-inhibited RAGE expression as a mechanism of IPF treatment [68].

ROLE OF SOLUBLE RAGE IN IDIOPATHIC PULMONARY FIBROSIS

A few studies demonstrated a positive association between decreased esRAGE levels in BAL and blood with poor prognosis in IPF patients. Analysis revealed esRAGE as a vital linking agent for the IPF pathophysiology, elucidating serum esRAGE as a potential IPF prognostic marker. *Manichaiku and accomplices* witnessed a non-significant association between AGER loci (rs2070600) of RAGE gene variations with IPF. Their study further showed that lower plasma sRAGE levels may be a biological measure of disease severity in IPF. Variation at the rs2070600 single-nucleotide polymorphism was not associated with IPF risk [69]. Therefore, more studies are required to explain the specific pathophysiological conditions under which RAGE may aggravate or attenuate IPF.

ROLE OF RAGE IN THE COMPLICATION OF CYSTIC FIBROSIS

Cystic fibrosis (CF) is an autosomal recessive disorder that affects multiple organs, although a decline in respiratory function represents the major source of morbidity and mortality. In CF, lung injury *via* sustained inflammation is the predominant reason for morbidity and mortality. The underlying chronic inflammatory processes are troublesome for the diagnosis of acute infectious exacerbations.

LIGAND-RAGE INTERACTIONS IN CYSTIC FIBROSIS COMPLICATION

Very scarce information is available about the significance of RAGE in CF diagnosis. However, some studies claimed that the chronic inflammatory state in the airways of CF patients is related to the enhanced RAGE expression. *Foell and colleagues* showed how RAGE-dependent inflammation contributes to the declining health of individuals with cystic fibrosis. While sRAGE is not detectable in CF fluids, RAGE expression is upregulated in CF airway neutrophils and plasma [70]. In general, neutrophils offer the first line of defense against pathogens; however, *Makam and colleagues* showed that neutrophils appear to prefer persistent infections in the airway lumen of patients with cystic fibrosis (CF); this phenomenon may be attributed to neutrophilic RAGE [71]. Along with RAGE, patients' sputum from acute infectious exacerbations also considerably increases RAGE ligands, particularly the S100A12 level, which shows a strong association with the neutrophil count. The decrease in the S100A12 level on intravenous antibiotic treatment strongly supported its implications for CF prognosis. Studies also claimed that elevated HMGB-1extents and epithelial sodium channel hyperactivity (ENaC) are hallmarks of lung CF. HMGB-1-RAGE signaling can perpetuate pro-inflammatory cytokine signaling by activating ENaC, aiding the continued HMGB-1 nucleus to the cytoplasm translocation for extracellular release. This is so as HMGB-1-RAGE signaling to ENaC hyperactivity contributes to both pro-inflammatory responses and airway inflammation, making it an intriguing signaling pathway as a target of a drug in the CF treatment. In CF airways, neutrophils promote the mammalian target of the rapamycin (mTOR) pathway, cAMP response element-binding protein (CREB), and phosphorylation of the eukaryotic initiation factor 4E that consequently increases RAGE expression.

Localized within the nucleus and acting as a transcription factor, CREB binds to the cAMP response element of the target gene promoters after getting phosphorylated at Ser133 loci by the multiple receptors stimulated protein kinases (such as protein kinase A, calmodulin-dependent protein kinase, MAPK, *etc.*). After activation, CREB binding protein is recruited to commence the transcription [72]. CREB genes in humans and mice comprise 11 exons and 3 isoforms, distinguished as a, b, and 1: generated *via* alternative splicing [73]. Known for expression in most tissues, these isoforms resemble each other in most functions. Primary structural investigations decipher the full-length CREB sequence being comprised of 4 N to C-terminus functional domains, namely: Q1 basal transcriptional activity domain, kinase inducible domain, glutamine-rich Q2 domain (for consecutive activation), and a leucine zipper domain (bZIP), forming a homodimer and regulating the DNA binding [74]. The Q1 region of functional CREB localizes at the N-terminus, interacting with TATA-binding protein to promote gene transcription [75]. The kinase inducible domain is located in the mid-region, central to which is Ser133, the phosphorylation of which by manifold protein kinases is essential for CREB activation [76]. Fig. (**5**) illustrates the CREB signaling network, commencing *via* adenylate cyclase activation on the G-protein coupled receptors stimulation by neurotransmitters whereby cAMP extents are increased, readying the protein kinase A (PKA). Apart from PKA nuclear translocation and concomitant phosphorylation of CREB at Ser133 residue, activation of PI3K/Akt/GSK3b, Ras/Raf/MEK/ERK/p90/RSK and ERK/MSK1 signaling pathways together enhance the CREB phosphorylation at multiple locations [77].

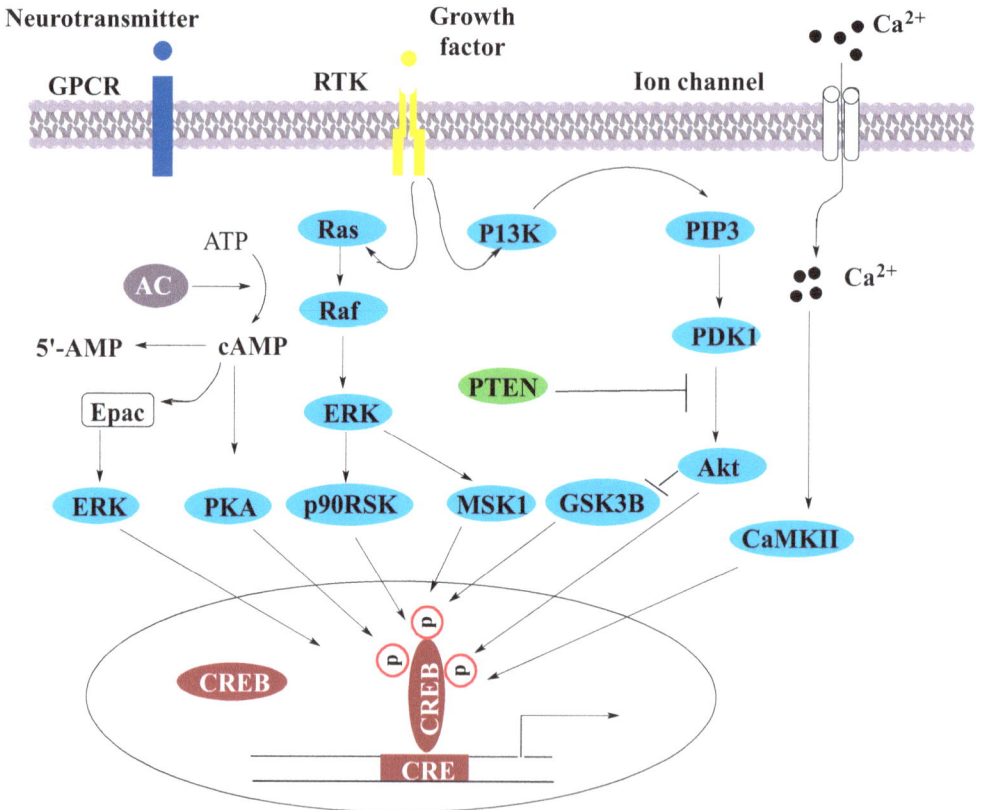

Fig. (5). Representation of signaling cascade of CREB. Upon activation of cellular G-protein-coupled receptors (GPCRs) by neurotransmitters, adenylate cyclase (AC) is stimulated, leading to increased cAMP levels. Elevated cAMP levels activate protein kinase A (PKA). The catalytic subunits of PKA then move to the nucleus, where they phosphorylate CREB at Ser133. In addition, the binding of growth factors to receptor tyrosine kinases (RTKs) triggers the activation of multiple signaling pathways, including PI3K/Akt/GSK3β, Ras/Raf/MEK/ERK/p90/RSK, and ERK/MSK1. These pathways further enhance CREB phosphorylation at various sites. Furthermore, the activation of excitatory NMDA receptors increases CREB phosphorylation through Ca^{2+}/CaMK-dependent pathways.

The functions of CREB in neurons have been described as significant in multiple intracellular events, including proliferation, differentiation, survival, long-term synaptic potentiation, neurogenesis, and neuronal plasticity [78, 79]. Herein, coordinated activities of RAGE-CREB interactions assume significance with RAGE interception of amphoterin (one of its many known ligands), being known to encourage neurite outgrowth and cell migration. A notable 2002 study from *Huttunen and the team* revealed that RAGE-mediated alterations in cell morphology are accompanied by impaired proliferation and modulated gene expression in the neuroblastoma cells. Further, amphoterin binding with RAGE results in prompt phosphorylation and nuclear translocation of CREB, the chief regulator of chromogranin expression [80].

Of note, chromogranin B (a constituent of secretory vesicles in neurons) is known to be upregulated *via* RAGE signaling amidst neuroblastoma cell differentiation alongside chromogranin A and secretogranin (the noted members of the chromogranin family). An interesting aspect of RAGE-CREB coordinated activities is that the inhibition of ERK1/2-Rsk2-dependent CREB phosphorylation impaired the chromogranin gene expression upregulation on RAGE activation. Thereby, RAGE signaling could mediate neuronal differentiation *via* REB activation and inducing chromogranin gene expression. The effects were counter-screened *via* amphoterin-stimulated RAGE-dependent neuronal differentiation in embryonic stem cells, accompanied by upregulated neuronal markers: light neurofilament protein, III-tubulin, CREB activation, and enhanced chromogranin A and B expressions. The observations were absent in signaling deficient RAGE-amphoterin mutant stimulated embryonic stem cells.

ROLE OF RAGE IN THE COMPLICATION OF BRONCHIOLITIS

Bronchiolitis is a common lung infection that occurs in young children and infants. The disease represents the most common outcome of concomitant inflammation and congestion in the small airways. It is generally caused by a respiratory syncytial virus (RSV) (70-80% cases) and occurs most commonly in infants. Bronchiolitis severity is primarily manifested *via* clinical signatures such as coughing, breath shortness and wheezing, traditional inflammatory biomarkers are inadequate to ascertain the extent of underlying inflammation and lung damage.

RELATIONSHIP OF RAGE AND SRAGE WITH BRONCHIOLITIS

Over the last several years, some studies explored the relationship between bronchiolitis and sRAGE. One of these attempts noticed a considerably higher sRAGE extent in pediatric bronchiolitis sufferers than in healthy subjects. Patients with high sRAGE extents act as decoy receptors and exhibit protection by binding to the RAGE ligands, thus affecting the impaired lung RAGE axis. However, *Ergon and accomplices* observed suppressed sRAGE levels in acute bronchiolitis patients alongwith no correlation for disease severity [81]. Therefore, more clinical studies with large samples are urgently needed to understand the sRAGE functioning in bronchiolitis.

ROLE OF RAGE IN THE COMPLICATION OF SEPSIS

Sepsis is an impending, critical inflammatory condition that occurs when the body's response to an infection damages the native tissues. Bacterial and other microbial infections remain the most common cause of sepsis. Infection primarily occurs in the lungs, stomach, kidneys, or bladder. Under severe conditions, chronic extents of inflammation damage most of the native organs, leading to concomitant functional failure and eventually death. Despite extensive research, the precise and consensual inflammatory dynamics of sepsis remain unexplored.

LIGAND-RAGE INTERACTIONS IN SEPSIS COMPLICATION

Previously, it has been demonstrated that RAGE is a crucial component of the detrimental effects emanating from acute inflammatory disorders, like sepsis. It has been demonstrated that RAGE can stimulate various signaling pathways sustaining cellular inflammation, dysfunction, and tissue damage. Thereby, RAGE can be a promising target for sepsis regulation as well. Although the definite role of RAGE in sepsis is not known yet, studies on RAGE null mice exhibited improved survival with higher arterial oxygenation in calprotectin (CLP, a heterodimer formed by two S-100 calcium-binding proteins, S100A8 and S100A9) driven sepsis. In a noted effort, *Liliensiek and colleagues* revealed that RAGE knockdown impairs the inflammatory cells, *vis-à-vis* reduced septic response wherein as many as 80% of mice survived [4]. Another worthy effort by *Lutterloh and groupmates* demonstrated that treatment with an anti-RAGE antibody resulted in a protective outcome in the CLP model of polymicrobial-induced sepsis [82]. Surprisingly, both homozygous and heterozygous RAGE knockdown mice survived in better numbers than wild-type mice. In yet another substantial effort, *Hu and colleagues* evaluated whether the RAGE pathway participated in the actions of dexamethasone (DEX) on sepsis-stimulated ALI in rats. Their results demonstrated that DEX attenuates the aggravation of sepsis-stimulated ALI *via* downregulation of the RAGE pathway, which has potential implications in clinical therapy [83, 84].

The observations of the above-discussed studies unanimously establish RAGE as a decisive mediator in sepsis pathogenesis. In this context, a meta-analysis demonstrated that RAGE inhibition provides a significant advantage to the survival rates in polymicrobial infections followed by bacterial infections. The protective effect of RAGE inhibition in experimental sepsis is much anticipated, at least in part, due to the inhibition of one of its ligands, S100/calgranulins, and HMGB1. Studies on these possibilities have demonstrated that the expression of RAGE ligands, S100/calgranulins, and HMGB1 is aggravated in sepsis patients. Moreover, HMGB-1 levels were also enhanced amidst CLP-induced polymicrobial sepsis while anti-HMGB-1 antibody administration moderated the sepsis mortality. In a different study, the decoy RAGE receptor *i.e.*, sRAGE treatment was examined as potent enough to regulate infection-driven chronic inflammation, confirming the involvement of RAGE in infection-driven inflammation [85]. This RAGE suppression augments sepsis survival by up to 40% compared to control (17%). In a surprising turnover, *Matsumoto and colleagues* identified a positive correlation between increased serum sRAGE level and severity of sepsis, yet again demonstrating the complexity of sRAGE-based diagnosis [86]. Thus, RAGE has also been implicated in the inflammation-driven sepsis manifestation; invariably attributed to systemic response syndrome mediated multiple organ failures. However, the variation of sepsis epidemiology and the distinct structure-activity scenarios of RAGE ligands such as HMGB1 further complicate the rationale of RAGE targeting as a feasible sepsis treatment mechanism. To authenticate the exact role of RAGE involvement in sepsis pathogenesis and screen the efficacy of anti-RAGE therapy against sepsis, detailed studies exploring the workability of each mechanism are necessary.

ROLE OF RAGE IN THE COMPLICATION OF PULMONARY ARTERIAL HYPERTENSION

Pulmonary arterial hypertension (PAH) is a rare disease characterized by elevated pulmonary arterial pressure (PAP) due to increased vasoconstriction and pulmonary microvasculature remodeling, leading to right ventricular failure and ultimately death. In PAH, enhanced pulmonary artery smooth muscle cell (PASMC) proliferation and suppressed apoptosis result in enhanced pulmonary arterial pressure and vascular resistance. Untreated chronic thromboembolic pulmonary hypertension (CTEPH) is also recognized as a tragic disease. The development of PAH is characterized by complex interactions involving multiple effectors. The primary form of PAH is classified as idiopathic PAH, while the secondary diseased state, including congenital heart disease with systemic-t--pulmonary shunts, makes up pulmonary vascular disease resembling idiopathic PAH in histological variations. These possibilities suggest a substantial commonness in disease progression mechanisms although an updated classification of PAH, suggesting pathological conditions leading to PAH with distinct origin, has been reported recently [87, 88].

LIGAND-RAGE INTERACTIONS IN THE COMPLICATION OF PULMONARY ARTERIAL HYPERTENSION

In their proteomic analysis, *Wilkins and team* demonstrated RAGE as one of the most upregulated proteins in PAH lung tissues compared to control lung tissues, leading to the hypothesis that RAGE is implicated in PAH etiology [89]. RAGE is also described as a modulator of vasoreactivity and proliferative events in response to pulmonary circulation to chronichypoxia. Plasma levels of sRAGE, but not HMGB1, might be a novel marker illustrating the pathological conditions of PAH patients [90, 91]. RAGE is claimed to be a novel, potent PAH therapeutic target.

ROLE OF RAGE IN TRAUMATIC BRAIN INJURY MEDIATED PULMONARY DYSFUNCTION AND LUNG TRANSPLANTATION REJECTION

Traumatic brain injury (TBI) results in systemic inflammatory responses that affect the lungs. In the case of lung transplantation, the donor's lungs are usually procured from brain-dead donors, wherein ~ 40-70% have TBI, leaving a meager 15% of lungs as suitable for transplantation. TBI results in systemic incendiary reactions that influence the lung, particularly during lung transplantation.

INVOLVEMENT OF LIGAND-RAGE INTERACTIONS IN TRANSPLANTATION REJECTION

Both in experimental and clinical studies, RAGE and its ligands, particularly HMGB1 have been demonstrated as being involved in lung transplant rejection. High-level RAGE expression in the donor's lungs is associated with primary graft dysfunction and possible transplant rejection. A noted study by *Weber and colleagues* provided evidence about TBI-induced systemic hypoxia, ALI, and pulmonary neutrophil attenuation *via* RAGE knockdown in mice models [92]. Indeed, lungs from RAGE-deficient, TBI donors are unable to develop ALI on transplantation, emphasizing the decisive role of RAGE in concurrent lung transplant rejection. During organ transplantation, early graft injury largely relies on ischemia-reperfusion injury (IRI), with aggravated cell damage and death. The cellular injury provoked HMGB1-RAGE interactions' dependence on endogenous pro-inflammatory mediators, resulting in leukocyte chemotaxis, vascular barrier disruption, and eventually, graft rejection. Moreover, elevated systemic HMGB1 in donors also correlated with impaired systemic lung oxygenation, both in pre- and post-transplantation stages, adequately supported by reversed hypoxia and improved lung compliances on neutralizing TBI-mediated HMGB1 actions. The role of HMGB-1 is defined by its isoform as the extracellular thiol can bind to RAGE and initiate leukocyte chemo-attraction while the disulfide isoform exhibits a high TLR-4 affinity, assisting in the release of pro-inflammatory cytokines and other mediators. In addition, the HMGB1 involvement in RAGE-mediated NF-κB activation made it conceivable that pulmonary dysfunction caused by TBI includes the HMGB1-mediated generation of pro-inflammatory cytokines. On the other hand, elevated plasma sRAGE levels were also screened as associated with primary graft dysfunction on lung transplantation. An early increment in the plasma sRAGE extent within 6 hours of lung transplantation is associated with longer mechanical ventilation, intensive care durations, and the development of chronic lung allograft dysfunction. Thus, HMGB1-RAGE interaction promotes cytokine-mediated inflammation and vascular permeability that subsequently culminates in pulmonary dysfunction followed by lung transplant rejection.

ANTI-RAGE THERAPY AGAINST VARIOUS PULMONARY DISEASES

Ever since the inception of RAGE-multiligand interactions, their correlation has been elucidated in several lung and pulmonary disorders. Of note, normal lung physiology is characterized by a threshold RAGE expression, which is irreversibly damaged in case of adequate RAGE actions. Within the lungs, RAGE is commonly expressed in bronchiolar epithelia, type II alveolar pneumocytes, alveolar macrophages, and the endothelium of larger arteries. Several studies have reckoned tumor suppressive RAGE functions in lung cancer [93]. The ironic prospect regarding RAGE implication in lung cancer pertains to its considerable presence in normal alveolar epithelium contrary to a reduced extent for lung carcinomas. This distinction has been established in lung cancer tissues and the tumor progression in lung cancer cell lines subcutaneously grafted mice models [94].

The modulated inflammatory responses triggered *via* aggressive RAGE-ligand interactions comprise the major source of abrupt pulmonary conditions, leading to a multitude of complications including those of oxidative and inflammatory stress, culminating in ALI, ARDS, asthma, COPD, and other pulmonary criticalities (Fig. **6**).

Herein, selected studies establishing the relevance of targeting RAGE signaling as a treatment strategy to counter pulmonary diseases are discussed.

• As early as in 2009, *Su and colleagues* [95] examined the feasibility of enhanced BAL-RAGE level, in experimental models of direct acute lung injury (ALI), typified *via* alveolar epithelial cell injury. The investigators performed ELISA screening for RAGE before examining direct or indirect ALI in mice models *via* retrieving BAL at definitive endpoints to monitor RAGE expression. Further screened correlations were the relative extents of BAL-RAGE with (i) the severity of lung injury in acid and hyperoxia-induced ALI, and (ii) the protective effect of novel mechanisms *via* mesenchymal stem cells (MSCs) in LPS-manifested ALI. Analysis revealed that in ALI modes of direct lung injury induced by intratracheal instillation of acid, LPS, or *E. coli*, the BAL RAGE exhibited 58-, 22- and 13-times elevated expressions. Opposed to this, BAL

RAGE could not be ascertained in direct ALI models induced by intraperitoneal injection of thiourea or *via* intravenous injection of MHC-I monoclonal antibody (MAb) that generates a mouse model of transfusion-associated ALI. Analysis revealed a correlation of BAL-RAGE with the severity of lung injury in acid and hyperoxia-manifested ALI. Further, for LPS-induced ALI, BAL-RAGE was slightly reduced on MSC treatment. Together the findings provided evidence for BAL-RAGE extent as a reliable ALI marker, exhibiting significance in distinguishing direct and indirect ALI models besides assessing the response to erstwhile therapies [95].

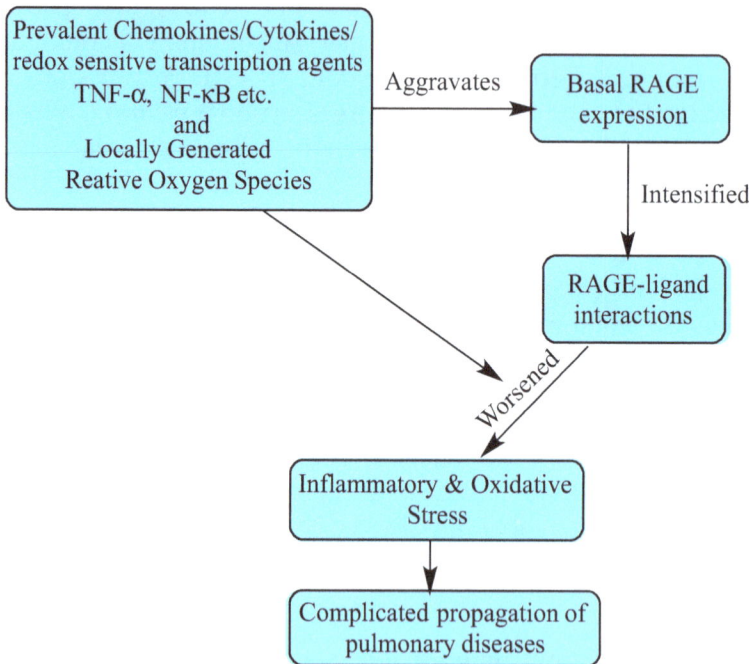

Fig. (6). Possible involvement of modulated native redox environment by aggressive interactions of RAGE with its ligands (HMGB1, AGEs, S100A, and others).

- Another noteworthy attempt to screen RAGE involvement in pulmonary disorders was a 2015 study by *Sambamurthy and colleagues* [96] wherein the involvement of RAGE in the pathogenesis of emphysema was screened in mice models. Analysis of RAGE lacking mice, exposed to cigarette smoke exhibited substantial protection from smoke-induced emphysema, inferred *via* airspace enlargement and no significant reduction in lung tissue elasticity rather than their air-exposed controls. Prevalence of emphysema screened as an outcome of aggravated inflammatory cell-mediated elastolysis. Apart from this, acute exposure to cigarette smoke in RAGE-lacking mice also signified an impaired and prompt localized migration of neutrophils, accounting for ~6-fold reduction than the wild-type mice. Thus, it was elucidated that moderated RAGE expression decreased the neutrophil localization owing to which even continued cigarette smoke exposure could not aggravate emphysema [96].

- The second 2015 evidence discussing the role of RAGE in pulmonary health is a comprehensive review article by *Yonchuk and accomplices* [97] wherein the significance of RAGE as are reliable marker to monitor COPD heterogeneity and disease severity has been demonstrated. Authors have taken a critical note of sRAGE which being a soluble regime of a prominent pro-inflammatory receptor, serves as a decoy for RAGE ligands and prohibits their binding with RAGE. This literature source is one of the foremost attempts to discuss the expression of sRAGE as a biomarker for emphysema prevalence and progression. The notable cohort studies, Treatment of Emphysema with a Selective Retinoid Agonist (TESRA) and Evaluation of COPD Longitudinally to Identify Predictive Surrogate Endpoints (ECLIPSE) demonstrated sRAGE expression levels as related to emphysema. In the first case (TESRA), sRAGE emerged as a marker of the diffusing capacity of the lung for carbon monoxide (CO) and lung density while for the ECLIPSE cohort, sRAGE extents exhibited a correlation with emphysema, GOLD (Global Initiative for Chronic Obtrusive Lung Disease) stage and COPD status, remaining significant on covariate adjustment. Both Cohorts revealed sRAGE, ICAM-1, and chemokine ligand 20 as the surrogate measures of emphysema with the most significant association for sRAGE-CT scan-assessed emphysema. Although the exact relationship of these parameters is still warranted, a likely consensus was the fact that low extents of sRAGE are the hallmarks not only for COPD but more exclusively for emphysema and its persistence.Some interesting inferences listed herewith underlined an association of sRAGE with emphysema in sufferers exhibiting a chronic airflow restriction, but it can't be exactly generalized for emphysema patients lacking airflow obstruction. Thereby, challenges like screening the sRAGE levels in COPD patients are the need of the hour and something that could be of real clinical significance [97].

- Relying on the repeated susceptibility of the RAGE gene as being implicated as COPD manifesting gene *via* manifold genome-wide association studies, the next study herein screened the specific mechanism by which RAGE aggravates COPD pathogenesis. Encouraged by the collaborative efforts from South Korea and the United States, the study by *Lee and colleagues* [98] examined the prowess of RAGE explicit antagonist, FPS-ZM1 in the *in vitro* and *in vivo* COPD models. The examined mice were intratracheally injected with elastase and FPS-ZM1 after which infiltrated inflammatory cells and cytokines were screened using ELISA. The RAGE cellular levels in the cells were ascertained in serum and BAL fluid of mice besides the lungs and serum of human volunteers and COP sufferers. Assessment of the damage-associated molecular pattern (DAMP) pathway was made at the *in vitro* and *in vivo* mode using immunofluorescence staining, western blot analysis, and ELISA. Analysis revealed a membrane RAGE expression-mediated inflammatory response and of sRAGE acting as a decoy, being correlated with the aggravated expression of DAMP-related signaling pathway *via* Nrf2 transcription factor activation. The RAGE involvement in COPD manifestation was concluded by the emphysema reversal in mice lungs on RAGE antagonist FPS-ZM1 administration. Additionally, the RAGE-inhibiting actions of FPS-ZM1 were accompanied by moderated lung inflammation only in Nrf2++ mice models. The study presented the first-ever experimental evidence of RAGE involvement in COPD aggravation accompanied by its corresponding therapeutic targeting as a reliable treatment option for COPD. Association of lung inflammation was additionally inferred as a critical COPD hallmark, signifying intensified RAGE-ligand interactions [98].

- Another significant effortfrom 2017, examined the mechanisms by which RAGE inhibition moderates the lung insult in experimental models. The French study by *Blondonnet and groupmates* [99] analyzed the effect of RAGE in anesthetized C57B/JRj mice, which were divided into 4 groups, three of which underwent orotracheal instillation of acid and were subsequently treated with anti-RAGE mAb or recombinant sRAGE (as a decoy receptor). The fourth group of mice served as control models. Screening of lung injury in the studied mice was accomplished by analyzing the blood gases, alveolar permeability, histology, and cytokine expression. Apart from these, lung expression and distribution of epithelial channel, ENaC, Na-K-ATPase, and aquaporin, were also screened. Analysis revealed a suppressed alveolar inflammation in acid-harmed mice with an anti-RAGE MAb or sRAGE moderated lung injury, arterial oxygenation ability, and suppressed alveolar inflammation. Administration of anti-RAGE therapies restored the alveolar fluid clearance (AFC) besides increasing aquaporin complex (AQP-5) lung expression in the alveolar cells. Besides these protective impacts, the therapeutic validation of RAGE targeting was also inferred from the suppressed alveolar type I epithelial cell insult in a translational mouse model having ARDS (inferred *via* restored AFC and lung AQP-5 expression). The observations collectively validated the significance of RAGE targeting as the potential strategy to moderate the ARDS health risks, although the modulated pathways affected by RAGE inhibition still need a detailed screening [99].

- The last major effort of 2017 attempted to screen for the role of RAGE in pulmonary complications and examined the effects of RAGE opposing antibodies in a crush injury model, taking clues from the inflammatory regulating and pattern recognition receptor ability of RAGE. A prominent hallmark of crush injury is the aggressive inflammation exhibited by the concerned patients, concomitantly resulting in multiple organ failure. The study by *Matsumoto and colleagues* [100] analyzed RAGE-mediated aggressive inflammation in rats, which were randomly categorized amongst sham (RAGE-Sh) group, crush (control models), or the anti-RAGE antibody-treated crush injury-positive animals. To screen the RAGE involvement in cumulative inflammation, both hindlimbs of rats were subjected to pressure for 6 h *via* lifting 3 kg blocks and a subsequent release. Sample retrieval was made on 3, 6, and 24 h after pressure release. The possible RAGE involvement in the aggravated symptoms was inferred by enhanced serum sRAGE expression in the systemic tissues at 6 h. While a possible lung injury was confirmed by histological screening post 6 and 24 h, the administration of MAbs against RAGE before releasing the compression, suppressed the aggravated RAGE expression in lung alveoli besides impairing the sRAGE and cytokine IL-6 (RAGE-associated mediators) expressions, reducing the lung injury and improving 7-day survival frequency. Thereby, reduced inflammatory adversities using anti-RAGE antibodies before compression release inferred a favorable prognosis to reduce the severity of crush injury-driven inflammation aggravation [100].

• Experimental proof of RAGE involvement in COPD pathogenesis was also revealed by a 2018 attempt wherein cigarette smoke-driven oxidative stress and apoptotic induction were examined. Knowing the profuse RAGE expression in alveolar epithelium, the investigators screened the role of RAGE actions in increased COPD vulnerability *via* probing the cigarette smoke-exposed human alveolar type II epithelial cell line. To probe such RAGE implications in COPD, cigarette smoke-exposed lung epithelial cells were treated with RAGE explicit antagonist, FPSZM1. Analysis revealed aggressive formation of reactive oxygen nitrile species (RONS) on exposure to cigarette smoke, besides altered pro-inflammatory cytokine levels, and suppressed physiological antioxidant defense system. Mechanistic screening inferred cigarette smoke modulated RAGE distribution through aggravated DAMP signaling *vis-à-vis* Nrf2 cellular activity. The observed findings were counter-verified *via* SB202190 (a p38 inhibitor) or Sp600125 (JNK inhibitor) pre-treatment, being unable to recover the altered RAGE distribution. Contrary to this, treatment with FPSZM1 considerably activated the anti-oxidative and anti-inflammatory effects, inhibiting the stimulation of redox-sensitive DAMP signaling *via* Nrf2 migration even on cigarette smoke exposure. Therefore, considering the findings altogether, the investigators concluded that RAGE and Nrf2 together played decisive roles in sustaining the alveolar epithelial tissue integrity and healthy status. These findings could be a breakthrough explanation for the desired RAGE expression in native lung tissues. Alternatively, it could be stated that cigarette smoke interferes with the native RAGE-Nrf2 coordination to affect the integrity of lung alveolar tissue [101].

- Moving to 2021, a potential effort by *Zhang and associates* [102] examined sRAGE-triggered neutrophilic airway inflammation as a causative factor of asthma. Screening the possible role of sRAGE in mucus hypersecretion, investigators treated mice suffering from neutrophilic asthma (NA) with adeno-associated virus (AAV) 9-sRAGE and inhibitors. Collagen deposition and goblet cell hyperplasia in the lungs were ascertained *via* periodic acid-Schiff (PAS) and Masson staining. Analysis revealed airway mucus hypersecretion in the neutrophilic asthma-positive mice. Observing an effective transfection of mice by AAV9-sRAGE *via* tail-vein injection and intranasal drip, scrutiny inferred enhanced sRAGE extents on AAV9-sRAGE transfection with concomitant inhibition of collagen gathering and the MUC5AC, MUC5B expressions. Prominent RAGE involvement triggered aggravated inflammation worsened NA pathogenesis, inferred *via* suppressed MUC5AC extents on administering HMGB1, RAGE, and PI3K inhibitors, in NA-positive mice and cultured HMGB1-induced human bronchial epithelial cells. Apart from the above, phosphor-extracellular signal-regulated kinase (ERK) protein was inhibited by sRAGE intervention. Collectively, the findings suggest a therapeutic significance of RAGE for treating mucus hypersecretion in neutrophilic asthma [102].

- Another noteworthy effort from 2022 correlated the RAGE-aggravated inflammatory conditions with the COVID-19 manifestation, deciphered from the accumulated prevalence of pro-inflammatory molecules such as AGEs, calprotectin, HMGB1, cytokines, angiotensin-converting enzyme 2 (ACE2) and others in the alveolar region of plasma and lungs. Although not an exclusive research study, the contribution by *Salehi and colleagues* [103] underlined the critical implication of RAGE *via* mitogen-activated protein kinase (MAPK), together worsening the severity of chronic inflammatory disorders, such as diabetes mellitus and ARDS. The dominant expression of the RAGE gene in the alveolar epithelial cells of the pulmonary system has been emphasized by the authors, with the clinical trials aiming to screen the feasible correlation between soluble RAGE isoforms (sRAGE and esRAGE) and the ARDS, ALI severity. The authors, thereafter, discussed multiple studies establishing the role of the RAGE-ligand axis and sRAGE/esRAGE extents in acute respiratory illness, especially focused on COVID-19-accompanied ARDS pathologies [103]. In their view with retrospective effect, screening of sRAGE/esRAGE in COVID-19-ARDS sufferers is a reliable hallmark for screening inceptive pulmonary involvement [104].

- A more recent study screened the possible involvement of RAGE in lung cancer pathogenesis by studying the staining pattern of anti-RAGE immunohistochemical markers in lung adenocarcinomas, specifically in a lepidic regimen of adenocarcinomas. The major reason to probe such RAGE actions was stated by the investigators as RAGE involvement in multiple pro-oxidative and pro-inflammatory conditions. To screen the staining pattern, 32 patients exhibiting positivity for lepidic pattern dominant lung adenocarcinoma were analyzed from 2009 to 2022, from the archives of the Medical Pathology Department at Selcuk University. Comparative analysis of staining regimen for the adenocarcinoma region and the adjoining normal lung tissue using immunohistochemical anti-RAGE staining revealed significant membranous staining in normal alveolar structures in the tumor vicinity. Interestingly, all patients exhibited immunohistochemical anti-RAGE antibodies besides substantial staining loss in all patterns of the adenocarcinoma region. It was noticed that anti-RAGE staining could predict the tumor size, pleural invasion, and the tumor spread within air spaces, jointly affecting the lung cancer stage and prognosis. Thereby, the study established anti-RAGE antibodies as auxiliary probes for lung adenocarcinoma diagnosis and screening of prognostic factors. The authors still concluded with a need for larger case studies to probe the role of anti-RAGE immunohistochemical staining as reactive changes post fibrosis or inflammation [105].

- Lately, a significant effort by *Xiong and colleagues* [106] studied the involvement of AECs autophagy and inflammatory stress *via* aggressive RAGE expression in ALI sufferers. Aiming to screen the involvement of RAGE activation signals in the impaired functioning of the alveolar epithelial barrier (*via* autophagic death), the investigators studied ALI-positive C57BL/6 and AGER gene defunct mice models. The study commenced *via* the treatment of A549 lung cancer cells and primary type II AECs with siRNA to suppress the AGER gene expression. Monitoring the autophagy inhibition *via* 3-methyladenine (3-MA), lung injury *via* histopathological scrutiny, cell viability using cell counting kit-8, interleukin (ILK)-6, 8 and sRAGE expressions (in serum and BAL) using ELISA, the investigators probed the autophagic and apoptotic involvement of RAGE signaling through western blot, immunohistochemistry, immunofluorescence and TEM screening. To recall herewith, ROS-mediated autophagic apoptosis for cancer treatment is well-known [107].

Analysis revealed aggravated RAGE expression *via* LPS, leading to autophagic activation squarely in mice lung tissues and A549 lung cancer cells as well as primary AT2 cells. A notable observation was the positive correlation of sRAGE expression with IL-6 and IL-8 extents. Besides, in comparison to wild-type mice, inflammation and apoptosis in lung tissues were noticed as much lower in the AGER-negative mice models. Specific involvement of autophagy in apoptotic induction was concluded from the reduced lung damage following 3-MA treatment. The explicit RAGE involvement was inferred from the inhibition of LPS-induced autophagy on AGER knockdown, with reduced lung injury. The findings complemented the *in vitro* observations of suppressed LPS activation on RAGE knockdown, inducing autophagy and apoptosis of A549 lung cancer and primary AT2 cells. Besides, RAGE activation also reduced the activity of the STAT3 signaling pathway. In all, the noticed findings established a prominent role of RAGE in the pathogenic manifestation of AT2 cell injury. Alternatively, it could be said that RAGE inhibition was relieved from the LPS-manifested lung insult *via* impaired autophagic apoptosis of AECs [107].

The above evidence establishes the role of RAGE expression and the concomitant aggressive ligand interactions as the hallmark of most aggravated inflammatory and oxidative stress conditions. Therapeutic targeting of RAGE with natural antioxidants seems a viable strategy herein with the depleted *in vivo* antioxidant functioning being tackled to some extent. Still, focused clinical and *in vivo* investigations are required to make any consensual statement although a reliable prospect emerging herein is the aggravated response of RAGE and its varied interactions with ligands.

CONCLUSION

RAGE is a cell surface immunoglobulin molecule class of molecules exhibiting structural and functional similarities with PRRs. Experimental evidence also prevails for a structure-function RAGE resemblance with adhesion molecules and cell surface immunoglobulin class of molecules. The initial RAGE discovery was made in bovine pulmonary endothelial cells. Subsequently, it was observed that AT-1 epithelial cells are the primary source of lung RAGE expression with AT-2 epithelial cells and pulmonary vascular smooth muscle cells being recognized as other noted sources. Lung tissues remain the vicious sites for multiple ligands of this multi-ligand receptor, the prominent of which are AGE, HMGB-1, and various S100. The adhesion and spreading of AT-1 epithelial cells besides the lung homeostasis maintenance are some prominent physiological functions of RAGE. RAGE is also implicated in the complication of various lung diseases such as asthma, CB, COPD, ALI, ARDS, IPF, CF, and sepsis. It is also claimed that RAGE is implicated in lung transplantation rejection. Enhanced inflammatory and oxidative stresses are the prerequisites of RAGE-ligand-mediated complications in different lung diseases. Inhibition or attenuation of RAGE-ligand interaction(s) could be a potential strategy to treat the sufferers of the above diseases.

Overall, this chapter addresses the whereabouts of manifold RAGE interactions and their biochemical significance based on the respective prevalence of different ligands. Even though studies have established an essentiality of threshold RAGE prevalence for the normal functioning of lungs (more commonly homeostasis of lung physiology), the RAGE-ligand interactions squarely reign as hallmarks of different pulmonary disorders. Spanning the diversity of COPD, asthma, emphysema, and multiple other respiratory conditions, RAGE-ligand interactions profusely mediate their essence *via* oxidative and inflammatory stress. Despite all this, clinical progress and commercial validation of anti-RAGE therapies is still a long way to go. Although manifold studies have reported their success in animal models (rats, mice, and human volunteers), elaborate guidelines for human validity continue to hamper their wholesome application and feasibility. Anticipating antioxidant and anti-inflammatory actions of anti-RAGE therapies, and their co-delivery with natural compounds (edible plant-derived polyphenols, in particular) presents a bright aspect for their moderated dosage and toxicity.

REFERENCES

[1]   Neeper M, Schmidt AM, Brett J, *et al.* Cloning and expression of a cell surface receptor for advanced glycosylation end products of proteins. J Biol Chem 1992; 267(21): 14998-5004.
        [http://dx.doi.org/10.1016/S0021-9258(18)42138-2] [PMID: 1378843]

[2]   Bucciarelli LG, Wendt T, Rong L, *et al.* RAGE is a multiligand receptor of the immunoglobulin superfamily: implications for homeostasis and chronic disease. Cell Mol Life Sci 2002; 59(7): 1117-28.
        [http://dx.doi.org/10.1007/s00018-002-8491-x] [PMID: 12222959]

[3]   Lizotte PP, Hanford LE, Enghild JJ, Nozik-Grayck E, Giles BL, Oury TD. Developmental expression of the receptor for advanced glycation end-products (RAGE) and its response to hyperoxia in the neonatal rat lung. BMC Dev Biol 2007; 7(1): 15.
        [http://dx.doi.org/10.1186/1471-213X-7-15] [PMID: 17343756]

[4]   Liliensiek B, Weigand MA, Bierhaus A, *et al.* Receptor for advanced glycation end products (RAGE) regulates sepsis but not the adaptive immune response. J Clin Invest 2004; 113(11): 1641-50.
        [http://dx.doi.org/10.1172/JCI200418704] [PMID: 15173891]

[5]   Khaket TP, Kang SC, Mukherjee TK. The Potential of Receptor for Advanced Glycation End Products (RAGE) as a therapeutic target for lung associated diseases. Curr Drug Targets 2019; 20(6): 679-89.
        [http://dx.doi.org/10.2174/1389450120666181120102159] [PMID: 30457049]

[6]   Mukherjee TK, Mukhopadhyay S, Hoidal JR. Implication of RAGE in pulmonary health and pathophysiology. Respir Physiol Neurobiol 2008; 162: 210-5.
        [http://dx.doi.org/10.1016/j.resp.2008.07.001] [PMID: 18674642]

[7]   Mukherjee TK, Mukhopadhyay S, Hoidal JR. The role of reactive oxygen species in TNFα-dependent expression of the receptor for advanced glycation end products in human umbilical vein endothelial cells. Biochim Biophys Acta Mol Cell Res 2005; 1744(2): 213-23.
        [http://dx.doi.org/10.1016/j.bbamcr.2005.03.007] [PMID: 15893388]

[8]   Mukherjee TK, Reynolds PR, Hoidal JR. Differential effect of estrogen receptor alpha and beta agonists on the receptor for advanced glycation end product expression in human microvascular endothelial cells. Biochim Biophys Acta Mol Cell Res 2005; 1745(3): 300-9.
        [http://dx.doi.org/10.1016/j.bbamcr.2005.03.012] [PMID: 15878629]

[9] Demling N, Ehrhardt C, Kasper M, Laue M, Knels L, Rieber EP. Promotion of cell adherence and spreading: a novel function of RAGE, the highly selective differentiation marker of human alveolar epithelial type I cells. Cell Tissue Res 2006; 323(3): 475-88.
[http://dx.doi.org/10.1007/s00441-005-0069-0] [PMID: 16315007]

[10] Shirasawa M, Fujiwara N, Hirabayashi S, *et al.* Receptor for advanced glycation end-products is a marker of type I lung alveolar cells. Genes Cells 2004; 9(2): 165-74.
[http://dx.doi.org/10.1111/j.1356-9597.2004.00712.x] [PMID: 15009093]

[11] Fehrenbach H, Kasper M, Tschernig T, Shearman MS, Schuh D, Müller M. Receptor for advanced glycation endproducts (RAGE) exhibits highly differential cellular and subcellular localisation in rat and human lung. Cell Mol Biol 1998; 44(7): 1147-57.
[PMID: 9846897]

[12] Katsuoka F, Kawakami Y, Arai T, *et al.* Type II alveolar epithelial cells in lung express receptor for advanced glycation end products (RAGE) gene. Biochem Biophys Res Commun 1997; 238(2): 512-6.
[http://dx.doi.org/10.1006/bbrc.1997.7263] [PMID: 9299542]

[13] Williams MC. Alveolar type I cells: molecular phenotype and development. Annu Rev Physiol 2003; 65(1): 669-95.
[http://dx.doi.org/10.1146/annurev.physiol.65.092101.142446] [PMID: 12428023]

[14] Herzog EL, Brody AR, Colby TV, Mason R, Williams MC. Knowns and unknowns of the alveolus. Proc Am Thorac Soc 2008; 5(7): 778-82.
[http://dx.doi.org/10.1513/pats.200803-028HR] [PMID: 18757317]

[15] Weibel ER. On the tricks alveolar epithelial cells play to make a good lung. Am J Respir Crit Care Med 2015; 191(5): 504-13.
[http://dx.doi.org/10.1164/rccm.201409-1663OE] [PMID: 25723823]

[16] Crapo JD, Barry BE, Gehr P, Bachofen M, Weibel ER. Cell number and cell characteristics of the normal human lung. Am Rev Respir Dis 1982; 126(2): 332-7.
[http://dx.doi.org/10.1164/arrd.1982.126.2.332] [PMID: 7103258]

[17] Yang J, Hernandez BJ, Alanis DM, *et al.* Development and plasticity of alveolar type 1 cells. Development 2015; 143(1): dev.130005.
[http://dx.doi.org/10.1242/dev.130005] [PMID: 26586225]

[18] Desai TJ, Brownfield DG, Krasnow MA. Alveolar progenitor and stem cells in lung development, renewal and cancer. Nature 2014; 507(7491): 190-4.
[http://dx.doi.org/10.1038/nature12930] [PMID: 24499815]

[19] Treutlein B, Brownfield DG, Wu AR, *et al.* Reconstructing lineage hierarchies of the distal lung epithelium using single-cell RNA-seq. Nature 2014; 509(7500): 371-5.
[http://dx.doi.org/10.1038/nature13173] [PMID: 24739965]

[20] Bartling B, Hofmann HS, Weigle B, Silber RE, Simm A. Down-regulation of the receptor for advanced glycation end-products (RAGE) supports non-small cell lung carcinoma. Carcinogenesis 2004; 26(2): 293-301.
[http://dx.doi.org/10.1093/carcin/bgh333] [PMID: 15539404]

[21] Marinakis E, Bagkos G, Piperi C, Roussou P, Diamanti-Kandarakis E. Critical role of RAGE in lung physiology and tumorigenesis: a potential target of therapeutic intervention? Clinical Chemistry and Laboratory Medicine (CCLM) 2014; 52(2): 189-200.
[http://dx.doi.org/10.1515/cclm-2013-0578] [PMID: 24108211]

[22] Kindermann A, Binder L, Baier J, *et al.* Severe but not moderate hyperoxia of newborn mice causes an emphysematous lung phenotype in adulthood without persisting oxidative stress and inflammation 2019; 19(1): 245.
[http://dx.doi.org/10.1186/s12890-019-0993-5]

[23] Al-Robaiy S, Weber B, Simm A, *et al.* The receptor for advanced glycation end-products supports

lung tissue biomechanics. Am J Physiol Lung Cell Mol Physiol 2013; 305(7): L491-500.
[http://dx.doi.org/10.1152/ajplung.00090.2013] [PMID: 23997170]

[24] Chuah YK, Basir R, Talib H, Tie TH, Nordin N. Receptor for advanced glycation end products and its involvement in inflammatory diseases. Int J Inflamm 2013; 2013: 1-15.
[http://dx.doi.org/10.1155/2013/403460] [PMID: 24102034]

[25] Oczypok EA, Perkins TN, Oury TD. All the "RAGE" in lung disease: The receptor for advanced glycation endproducts (RAGE) is a major mediator of pulmonary inflammatory responses. Paediatr Respir Rev 2017; 23: 40-9.
[http://dx.doi.org/10.1016/j.prrv.2017.03.012] [PMID: 28416135]

[26] Ramasamy R, Yan SF, Schmidt AM. RAGE: therapeutic target and biomarker of the inflammatory response—the evidence mounts. J Leukoc Biol 2009; 86(3): 505-12.
[http://dx.doi.org/10.1189/jlb.0409230] [PMID: 19477910]

[27] Wautier MP, Chappey O, Corda S, Stern DM, Schmidt AM, Wautier JL. Activation of NADPH oxidase by AGE links oxidant stress to altered gene expression *via* RAGE. Am J Physiol Endocrinol Metab 2001; 280(5): E685-94.
[http://dx.doi.org/10.1152/ajpendo.2001.280.5.E685] [PMID: 11287350]

[28] Sukkar MB, Ullah MA, Gan WJ, *et al.* RAGE: a new frontier in chronic airways disease. Br J Pharmacol 2012; 167(6): 1161-76.
[http://dx.doi.org/10.1111/j.1476-5381.2012.01984.x] [PMID: 22506507]

[29] Barnes PJ. Immunology of asthma and chronic obstructive pulmonary disease. Nat Rev Immunol 2008; 8(3): 183-92.
[http://dx.doi.org/10.1038/nri2254] [PMID: 18274560]

[30] Busse WW, Lemanske RF Jr. Asthma. N Engl J Med 2001; 344(5): 350-62.
[http://dx.doi.org/10.1056/NEJM200102013440507] [PMID: 11172168]

[31] Perkins TN, Donnell ML, Oury TD. The axis of the receptor for advanced glycation endproducts in asthma and allergic airway disease. Allergy 2021; 76(5): 1350-66.
[http://dx.doi.org/10.1111/all.14600] [PMID: 32976640]

[32] Oczypok EA, Milutinovic PS, Alcorn JF, *et al.* Pulmonary receptor for advanced glycation end-products promotes asthma pathogenesis through IL-33 and accumulation of group 2 innate lymphoid cells. J Allergy Clin Immunol 2015; 136(3): 747-756.e4.
[http://dx.doi.org/10.1016/j.jaci.2015.03.011] [PMID: 25930197]

[33] Akirav EM, Henegariu O, Preston-Hurlburt P, Schmidt AM, Clynes R, Herold KC. The receptor for advanced glycation end products (RAGE) affects T cell differentiation in OVA induced asthma. PLoS One 2014; 9(4): e95678.
[http://dx.doi.org/10.1371/journal.pone.0095678] [PMID: 24759895]

[34] Ullah MA, Loh Z, Gan WJ, *et al.* Receptor for advanced glycation end products and its ligand high-mobility group box-1 mediate allergic airway sensitization and airway inflammation. J Allergy Clin Immunol 2014; 134(2): 440-450.e3.
[http://dx.doi.org/10.1016/j.jaci.2013.12.1035] [PMID: 24506934]

[35] Li R, Wang J, Zhu F, *et al.* HMGB1 regulates T helper 2 and T helper17 cell differentiation both directly and indirectly in asthmatic mice. Mol Immunol 2018; 97: 45-55.
[http://dx.doi.org/10.1016/j.molimm.2018.02.014] [PMID: 29567318]

[36] Zhou Y, Jiang Y, Wang W, *et al.* HMGB1 and RAGE levels in induced sputum correlate with asthma severity and neutrophil percentage. Hum Immunol 2012; 73(11): 1171-4.
[http://dx.doi.org/10.1016/j.humimm.2012.08.016] [PMID: 22960399]

[37] Halayko AJ, Ghavami S. S100A8/A9: a mediator of severe asthma pathogenesis and morbidity?This article is one of a selection of papers published in a special issue celebrating the 125th anniversary of the Faculty of Medicine at the University of Manitoba. Can J Physiol Pharmacol 2009; 87(10): 743-55.

[http://dx.doi.org/10.1139/Y09-054] [PMID: 19898558]

[38]   Yang Z, Yan WX, Cai H, *et al.* S100A12 provokes mast cell activation: A potential amplification pathway in asthma and innate immunity. J Allergy Clin Immunol 2007; 119(1): 106-14.
[http://dx.doi.org/10.1016/j.jaci.2006.08.021] [PMID: 17208591]

[39]   El-Seify M, Fouda E, Nabih E. Serum level of soluble receptor for advanced glycation end products in asthmatic children and its correlation to severity and pulmonary functions. Clin Lab 2014; 60(06/2014): 957-62.
[http://dx.doi.org/10.7754/Clin.Lab.2013.130418] [PMID: 25016700]

[40]   Lyu Y, Zhao H, Ye Y, *et al.* Decreased soluble RAGE in neutrophilic asthma is correlated with disease severity and RAGE G82S variants. Mol Med Rep 2017; 17(3): 4131-7.
[http://dx.doi.org/10.3892/mmr.2017.8302] [PMID: 29257350]

[41]   Bediwy A, Hassan SM, El-Najjar MR. Receptor of advanced glycation end products in childhood asthma exacerbation. Eur Respir J 2016; 48 (Suppl. 60): OA4803.
[http://dx.doi.org/10.1183/13993003.congress-2016.OA4803]

[42]   Zhang F, Su X, Huang G, *et al.* sRAGE alleviates neutrophilic asthma by blocking HMGB1/RAGE signalling in airway dendritic cells. Sci Rep 2017; 7(1): 14268.
[http://dx.doi.org/10.1038/s41598-017-14667-4] [PMID: 29079726]

[43]   Gangemi S, Casciaro M, Trapani G, *et al.* Association between HMGB1 and COPD A systematic review. Mediators Inflamm 2015; 2015(1): 164913.
[http://dx.doi.org/10.1155/2015/164913] [PMID: 26798204]

[44]   Sweet CS, Gaul SL, Reitz PM, Blaine EH, Ribeiro LT. Mechanism of action of enalapril in experimental hypertension and acute left ventricular failure. J Hypertens Suppl 1983; 1(1): 53-63.
[PMID: 6100609]

[45]   Bout MR, Vromans H. Influence of commonly used excipients on the chemical degradation of enalapril maleate in its solid state: The role of condensed water. Eur J Pharm Sci 2022; 171: 106121.
[http://dx.doi.org/10.1016/j.ejps.2022.106121] [PMID: 35007714]

[46]   Qiao B, Du C, Dong R, *et al.* Experiment and computation of solubility and dissolution properties for enalpril maleate and its intermediate in pure solvents. J Chem Eng Data 2018; 63(12): acs.jced.8b00333.
[http://dx.doi.org/10.1021/acs.jced.8b00333]

[47]   Zhang Y, Li S, Wang G, *et al.* Changes of HMGB1 and sRAGE during the recovery of COPD exacerbation. J Thorac Dis 2014; 6(6): 734-41.
[http://dx.doi.org/10.3978/j.issn.2072-1439.2014.04.31] [PMID: 24976997]

[48]   Malik P, Hoidal JR, Mukherjee TK. Implication of RAGE polymorhic variants in COPD complication and Anti-COPD therapeutic potential of sRAGE. COPD 2021; 18(6): 737-48.
[http://dx.doi.org/10.1080/15412555.2021.1984417] [PMID: 34615424]

[49]   Sharma A, Kaur S, Sarkar M, Sarin BC, Changotra H. The AGE-RAGE Axis and RAGE genetics in chronic obstructive pulmonary disease. Clin Rev Allergy Immunol 2021; 60(2): 244-58.
[http://dx.doi.org/10.1007/s12016-020-08815-4] [PMID: 33170477]

[50]   Smith DJ, Yerkovich ST, Towers MA, Carroll ML, Thomas R, Upham JW. Reduced soluble receptor for advanced glycation end products in chronic obstructive pulmonary disease. Eur Respir J 2010; 37: 516-22.
[http://dx.doi.org/10.1183/09031936.00029310] [PMID: 20595148]

[51]   Kanazawa H, Tochino Y, Asai K, Ichimaru Y, Watanabe T, Hirata K. Validity of HMGB1 measurement in epithelial lining fluid in patients with COPD. Eur J Clin Invest 2012; 42(4): 419-26.
[http://dx.doi.org/10.1111/j.1365-2362.2011.02598.x] [PMID: 21950682]

[52]   Chen L, Wang T, Guo L, *et al.* Overexpression of RAGE contributes to cigarette smoke-induced nitric oxide generation in COPD. Lung 2014; 192(2): 267-75.

[http://dx.doi.org/10.1007/s00408-014-9561-1] [PMID: 24535058]

[53] Calfee CS, Ware LB, Eisner MD, *et al.* Plasma receptor for advanced glycation end products and clinical outcomes in acute lung injury. Thorax 2008; 63(12): 1083-9.
[http://dx.doi.org/10.1136/thx.2008.095588] [PMID: 18566109]

[54] Yamamura Y, Kawakami J, Santa T, *et al.* Pharmacokinetic profile of glycyrrhizin in healthy volunteers by a new high-performance liquid chromatographic method. J Pharm Sci 1992; 81(10): 1042-6.
[http://dx.doi.org/10.1002/jps.2600811018] [PMID: 1432618]

[55] Polyakov NE, Leshina TV. Physicochemical approaches to the study of the antioxidant activity of Glycyrrhizin. Russ J Phys Chem A Focus Chem 2023; 97(5): 828-35.
[http://dx.doi.org/10.1134/S0036024423050229]

[56] Kyung SY, Byun KH, Yoon JY, *et al.* Advanced glycation end-products and receptor for advanced glycation end-products expression in patients with idiopathic pulmonary fibrosis and NSIP. Int J Clin Exp Pathol 2013; 7(1): 221-8.
[PMID: 24427342]

[57] Zhang L, Ji Y, Kang Z, Lv C, Jiang W. Protocatechuic aldehyde ameliorates experimental pulmonary fibrosis by modulating HMGB1/RAGE pathway. Toxicol Appl Pharmacol 2015; 283(1): 50-6.
[http://dx.doi.org/10.1016/j.taap.2015.01.001] [PMID: 25582705]

[58] He M, Kubo H, Ishizawa K, *et al.* The role of the receptor for advanced glycation end-products in lung fibrosis. Am J Physiol Lung Cell Mol Physiol 2007; 293(6): L1427-36.
[http://dx.doi.org/10.1152/ajplung.00075.2007] [PMID: 17951314]

[59] Kumar S, Jain N, Mukherjee TK. Role of RAGE in idiopathic pulmonary fibrosis: Promising prospects. Respir Physiol Neurobiol 2009; 165(2-3): 121-2.
[http://dx.doi.org/10.1016/j.resp.2008.10.018]

[60] Englert JM, Hanford LE, Kaminski N, *et al.* A role for the receptor for advanced glycation end products in idiopathic pulmonary fibrosis. Am J Pathol 2008; 172(3): 583-91.
[http://dx.doi.org/10.2353/ajpath.2008.070569] [PMID: 18245812]

[61] Queisser MA, Kouri FM, Königshoff M, *et al.* Loss of RAGE in pulmonary fibrosis: molecular relations to functional changes in pulmonary cell types. Am J Respir Cell Mol Biol 2008; 39(3): 337-45.
[http://dx.doi.org/10.1165/rcmb.2007-0244OC] [PMID: 18421017]

[62] Li JH, Wang W, Huang XR, *et al.* Advanced glycation end products induce tubular epithelial-myofibroblast transition through the RAGE-ERK1/2 MAP kinase signaling pathway. Am J Pathol 2004; 164(4): 1389-97.
[http://dx.doi.org/10.1016/S0002-9440(10)63225-7] [PMID: 15039226]

[63] De Vriese AS, Tilton RG, Mortier S, Lameire NH. Myofibroblast transdifferentiation of mesothelial cells is mediated by RAGE and contributes to peritoneal fibrosis in uraemia. Nephrol Dial Transplant 2006; 21(9): 2549-55.
[http://dx.doi.org/10.1093/ndt/gfl271] [PMID: 16757496]

[64] Selman M, King TE Jr, Pardo A. Idiopathic pulmonary fibrosis: prevailing and evolving hypotheses about its pathogenesis and implications for therapy. Ann Intern Med 2001; 134(2): 136-51.
[http://dx.doi.org/10.7326/0003-4819-134-2-200101160-00015] [PMID: 11177318]

[65] Bargagli E, Penza F, Bianchi N, *et al.* Controversial role of RAGE in the pathogenesis of idiopathic pulmonary fibrosis. Respir Physiol Neurobiol 2008; •••
[http://dx.doi.org/10.1016/j.resp.2008.10.017] [PMID: 19026768]

[66] Machahua C, Montes-Worboys A, Llatjos R, *et al.* Increased AGE-RAGE ratio in idiopathic pulmonary fibrosis. Respir Res 2016; 17(1): 144.
[http://dx.doi.org/10.1186/s12931-016-0460-2] [PMID: 27816054]

[67]   Baek H, Jang S, Park J, *et al.* Reduced receptor for advanced glycation end products is associated with α-SMA expression in patients with idiopathic pulmonary fibrosis and mice. Lab Anim Res 2021; 37(1): 28.
[http://dx.doi.org/10.1186/s42826-021-00105-0] [PMID: 34600594]

[68]   Piao C, Zhuang C, Ko MK, Hwang DW, Lee M. Pulmonary delivery of a recombinant RAGE antagonist peptide derived from high-mobility group box-1 in a bleomycin-induced pulmonary fibrosis animal model. J Drug Target 2022; 30(7): 792-9.
[http://dx.doi.org/10.1080/1061186X.2022.2069781] [PMID: 35451894]

[69]   Manichaikul A, Sun L, Borczuk AC, *et al.* Plasma soluble receptor for advanced glycation end products in idiopathic pulmonary fibrosis. Ann Am Thorac Soc 2017; 14(5): 628-35.
[http://dx.doi.org/10.1513/AnnalsATS.201606-485OC] [PMID: 28248552]

[70]   Foell D, Seeliger S, Vogl T, *et al.* Expression of S100A12 (EN-RAGE) in cystic fibrosis. Thorax 2003; 58(7): 613-7.
[http://dx.doi.org/10.1136/thorax.58.7.613] [PMID: 12832680]

[71]   Makam M, Diaz D, Laval J, *et al.* Activation of critical, host-induced, metabolic and stress pathways marks neutrophil entry into cystic fibrosis lungs. Proc Natl Acad Sci USA 2009; 106(14): 5779-83.
[http://dx.doi.org/10.1073/pnas.0813410106] [PMID: 19293384]

[72]   Dyson HJ, Wright PE. Role of intrinsic protein disorder in the function and interactions of the transcriptional coactivators CREB-binding protein (CBP) and p300. J Biol Chem 2016; 291(13): 6714-22.
[http://dx.doi.org/10.1074/jbc.R115.692020] [PMID: 26851278]

[73]   Ichiki T. Role of cAMP response element binding protein in cardiovascular remodeling: good, bad, or both? Arterioscler Thromb Vasc Biol 2006; 26(3): 449-55.
[http://dx.doi.org/10.1161/01.ATV.0000196747.79349.d1] [PMID: 16293792]

[74]   Xu W, Kasper LH, Lerach S, Jeevan T, Brindle PK. Individual CREB-target genes dictate usage of distinct cAMP-responsive coactivation mechanisms. EMBO J 2007; 26(12): 2890-903.
[http://dx.doi.org/10.1038/sj.emboj.7601734] [PMID: 17525731]

[75]   Felinski EA, Quinn PG. The coactivator dTAF$_{II}$110/hTAF$_{II}$135 is sufficient to recruit a polymerase complex and activate basal transcription mediated by CREB. Proc Natl Acad Sci USA 2001; 98(23): 13078-83.
[http://dx.doi.org/10.1073/pnas.241337698] [PMID: 11687654]

[76]   Sun L, Zhao R, Zhang L, *et al.* Prevention of vascular smooth muscle cell proliferation and injury-induced neointimal hyperplasia by CREB-mediated p21 induction: An insight from a plant polyphenol. Biochem Pharmacol 2016; 103: 40-52.
[http://dx.doi.org/10.1016/j.bcp.2016.01.015] [PMID: 26807478]

[77]   Wang H, Xu J, Lazarovici P, Quirion R, Zheng W. cAMP response element binding protein (CREB): A possible signaling molecule link in the pathophysiology of schizophrenia. Front Mol Neurosci 2018; 11: 255.
[http://dx.doi.org/10.3389/fnmol.2018.00255] [PMID: 30214393]

[78]   Alberini CM. Transcription factors in long-term memory and synaptic plasticity. Physiol Rev 2009; 89(1): 121-45.
[http://dx.doi.org/10.1152/physrev.00017.2008] [PMID: 19126756]

[79]   Kitagawa H, Sugo N, Morimatsu M, Arai Y, Yanagida T, Yamamoto N. Activity-dependent dynamics of the transcription factor of cAMP response element binding protein in cortical neurons revealed by single molecule imaging. J Neurosci 2017; 37(1): 1-10.
[http://dx.doi.org/10.1523/JNEUROSCI.0943-16.2016] [PMID: 28053025]

[80]   Huttunen HJ, Kuja-Panula J, Rauvala H. Receptor for advanced glycation end products (RAGE) signaling induces CREB-dependent chromogranin expression during neuronal differentiation. J Biol

Chem 2002; 277(41): 38635-46.
[http://dx.doi.org/10.1074/jbc.M202515200] [PMID: 12167613]

[81]   Egron C, Roszyk L, Rochette E, *et al.* Serum soluble receptor for advanced glycation end-products during acute bronchiolitis in infant: Prospective study in 93 cases. Pediatr Pulmonol 2018; 53(10): 1429-35.
[http://dx.doi.org/10.1002/ppul.24141] [PMID: 30113140]

[82]   Lutterloh EC, Opal SM, Pittman DD, *et al.* Inhibition of the RAGE products increases survival in experimental models of severe sepsis and systemic infection. Crit Care 2007; 11(6): R122.
[http://dx.doi.org/10.1186/cc6184] [PMID: 18042296]

[83]   Hu H, Shi D, Hu C, Yuan X, Zhang J, Sun H. Dexmedetomidine mitigates CLP-stimulated acute lung injury *via* restraining the RAGE pathway. Am J Transl Res 2017; 9(12): 5245-58.
[PMID: 29312480]

[84]   Kong Q, Wu X, Qiu Z, Huang Q, Xia Z, Song X. Protective effect of Dexmedetomidine on acute lung injury *via* the upgradation of tumor necrosis factor-α induced protein-8-like 2 in septic mice. Inflammation 2020; 43(3): 833-46.
[http://dx.doi.org/10.1007/s10753-019-01169-w] [PMID: 31927655]

[85]   Ansar W, Ghosh S. Inflammation and inflammatory diseases, markers and mediators: Role of CRP in some inflammatory diseases. Biol. C Reactive Protein in Health & Dis 2016; pp. 67-107.
[http://dx.doi.org/10.1007/978-81-322-2680-2_4]

[86]   Matsumoto H, Matsumoto N, Ogura H, *et al.* The clinical significance of circulating soluble RAGE in patients with severe sepsis. J Trauma Acute Care Surg 2015; 78(6): 1086-94.
[http://dx.doi.org/10.1097/TA.0000000000000651] [PMID: 26002402]

[87]   Moser B, Megerle A, Bekos C, *et al.* Local and systemic RAGE axis changes in pulmonary hypertension: CTEPH and iPAH. PLoS One 2014; 9(9): e106440.
[http://dx.doi.org/10.1371/journal.pone.0106440] [PMID: 25188497]

[88]   Prasad K. AGE-RAGE stress in the pathophysiology of pulmonary hypertension and its treatment. Int J Angiol 2019; 28(2): 071-9.
[http://dx.doi.org/10.1055/s-0039-1687818] [PMID: 31384104]

[89]   Abdul-Salam VB, Wharton J, Cupitt J, Berryman M, Edwards RJ, Wilkins MR. Proteomic analysis of lung tissues from patients with pulmonary arterial hypertension. Circulation 2010; 122(20): 2058-67.
[http://dx.doi.org/10.1161/CIRCULATIONAHA.110.972745] [PMID: 21041689]

[90]   Diekmann F, Chouvarine P, Sallmon H, *et al.* Soluble Receptor for Advanced Glycation End Products (sRAGE) is a sensitive biomarker in human pulmonary arterial hypertension. Int J Mol Sci 2021; 22(16): 8591.
[http://dx.doi.org/10.3390/ijms22168591] [PMID: 34445297]

[91]   Suzuki S, Nakazato K, Sugimoto K, *et al.* Plasma Levels of Receptor for Advanced Glycation End-Products and High-Mobility Group Box 1 in Patients With Pulmonary Hypertension. Int Heart J 2016; 57(2): 234-40.
[http://dx.doi.org/10.1536/ihj.15-188] [PMID: 26973260]

[92]   Weber DJ, Gracon ASA, Ripsch MS, *et al.* The HMGB1-RAGE axis mediates traumatic brain injury–induced pulmonary dysfunction in lung transplantation. Sci Transl Med 2014; 6(252): 252ra124.
[http://dx.doi.org/10.1126/scitranslmed.3009443] [PMID: 25186179]

[93]   Mukherjee TK, Malik P, Hoidal JR. Receptor for Advanced Glycation End Products (RAGE) and its polymorphic variants as predictive diagnostic and prognostic markers of NSCLCs: a perspective. Curr Oncol Rep 2021; 23(1): 12.
[http://dx.doi.org/10.1007/s11912-020-00992-x] [PMID: 33399986]

[94]   Malik P, Chaudhry N, Mittal R, Mukherjee TK. Role of receptor for advanced glycation end products

in the complication and progression of various types of cancers. Biochim Biophys Acta, Gen Subj 2015; 1850(9): 1898-904.
[http://dx.doi.org/10.1016/j.bbagen.2015.05.020] [PMID: 26028296]

[95]   Su X, Looney MR, Gupta N, Matthay MA. Receptor for advanced glycation end-products (RAGE) is an indicator of direct lung injury in models of experimental lung injury. Am J Physiol Lung Cell Mol Physiol 2009; 297(1): L1-5.
[http://dx.doi.org/10.1152/ajplung.90546.2008] [PMID: 19411309]

[96]   Sambamurthy N, Leme AS, Oury TD, Shapiro SD. The receptor for advanced glycation end products (RAGE) contributes to the progression of emphysema in mice. PLoS One 2015; 10(3): e0118979.
[http://dx.doi.org/10.1371/journal.pone.0118979] [PMID: 25781626]

[97]   Yonchuk JG, Silverman EK, Bowler RP, *et al.* Circulating soluble receptor for advanced glycation end products (sRAGE) as a biomarker of emphysema and the RAGE axis in the lung. Am J Respir Crit Care Med 2015; 192(7): 785-92.
[http://dx.doi.org/10.1164/rccm.201501-0137PP] [PMID: 26132989]

[98]   Lee H, Park JR, Kim WJ, *et al.* Blockade of RAGE ameliorates elastase-induced emphysema development and progression *via* RAGE-DAMP signaling. FASEB J 2017; 31(5): 2076-89.
[http://dx.doi.org/10.1096/fj.201601155R] [PMID: 28148566]

[99]   Blondonnet R, Audard J, Belville C, *et al.* RAGE inhibition reduces acute lung injury in mice. Sci Rep 2017; 7(1): 7208.
[http://dx.doi.org/10.1038/s41598-017-07638-2] [PMID: 28775380]

[100]   Matsumoto H, Matsumoto N, Shimazaki J, *et al.* Therapeutic effectiveness of anti-RAGE antibody administration in a rat model of crush injury. Sci Rep 2017; 7(1): 12255.
[http://dx.doi.org/10.1038/s41598-017-12065-4] [PMID: 28947744]

[101]   Lee H, Lee J, Hong SH, Rahman I, Yang SR. Inhibition of RAGE attenuates cigarette smoke-induced lung epithelial cell damage *via* RAGE-mediated Nrf2/DAMP signaling. Front Pharmacol 2018; 9: 684.
[http://dx.doi.org/10.3389/fphar.2018.00684] [PMID: 30013476]

[102]   Zhang X, Xie J, Sun H, Wei Q, Nong G. sRAGE Inhibits the mucus hypersecretion in a mouse model with neutrophilic asthma. Immunol Invest 2022; 51(5): 1243-56.
[http://dx.doi.org/10.1080/08820139.2021.1928183] [PMID: 34018452]

[103]   Salehi M, Amiri S, Ilghari D, Hasham LFA, Piri H. The remarkable roles of the Receptor for Advanced Glycation End Products (RAGE) and its soluble isoforms in COVID-19: The importance of RAGE pathway in the lung injuries. Indian J Clin Biochem 2023; 38(2): 159-71.
[http://dx.doi.org/10.1007/s12291-022-01081-5] [PMID: 35999871]

[104]   Singh R, Malik P, Kumar M, Kumar R, Alam MS, Mukherjee TK. Secondary fungal infections in SARS-CoV-2 patients: pathological whereabouts, cautionary measures, and steadfast treatments. Pharmacol Rep 2023; 75(4): 817-37.
[http://dx.doi.org/10.1007/s43440-023-00506-z] [PMID: 37354313]

[105]   Celik M, Ates MC, Gencel E, *et al.* 2023, The role of immunohistochemical anti-RAGE antibody in lepidic pattern lung adenocarcinomas. Chron. Precis Med Res 2023; 4(1): 34-8.
[http://dx.doi.org/10.5281/zenodo.7716241]

[106]   Xiong X, Dou J, Shi J, *et al.* RAGE inhibition alleviates lipopolysaccharides-induced lung injury *via* directly suppressing autophagic apoptosis of type II alveolar epithelial cells. Respir Res 2023; 24(1): 24.
[http://dx.doi.org/10.1186/s12931-023-02332-6] [PMID: 36691012]

[107]   Mukherjee T, Kashyap D, Sharma A, *et al.* Reactive oxygen species (ROS): an activator of apoptosis and autophagy in cancer. J Biol Chem Sci 2016; 3(2): 256-64.
[http://dx.doi.org/10.1016/j.biopha.2021.112142]

Receptor for Advanced Glycation End Products in Cardiovascular and Diabetic Complication(s)

Ruma Rani[1,†], Parth Malik[2,3,†] and **Tapan Kumar Mukherjee[4,*]**

[1] *ICAR-National Research Centre on Equines, Hisar-125001, Haryana, India*

[2] *School of Chemical Sciences, Central University of Gujarat, Gandhinagar, Gujarat-382030, India*

[3] *Swarrnim Startup & Innovation University, Bhoyan-Rathod, Gandhinagar-Gujarat, India*

[4] *Amity Institute of Biotechnology, Amity University, New Town, Kolkata, West Bengal 700156, India*

Abstract: The Receptor for Advanced Glycation End Products (RAGE) has emerged as a pivotal player in the pathogenesis of cardiovascular and diabetic complications. An in-depth exploration of RAGE involvement in the disease processes, elucidating the molecular mechanisms, signaling pathways, and the associated pathological outcomes, is discussed. In diabetes, chronic hyperglycemia leads to the formation and accumulation of advanced glycation end products (AGEs), which activate RAGE and subsequently initiate a cascade of pro-inflammatory and pro-oxidative events. These processes contribute to the development and progression of diabetic vascular complications, including atherosclerosis, neuropathy, nephropathy, and retinopathy. In the cardiovascular system, RAGE activation promotes vascular inflammation, endothelial dysfunction, and vascular smooth muscle cell proliferation, all of which are critical in the pathogenesis of atherosclerosis and cardiovascular diseases. Furthermore, RAGE-mediated oxidative stress and inflammation have been implicated in the progression of heart failure and post-ischemic injury. Targeting RAGE signaling thereby emerges as a promising therapeutic approach to mitigate the detrimental effects of chronic hyperglycemia and vascular inflammation in diabetic and cardiovascular diseases. A comprehensive understanding of the multifaceted RAGE functions in cardiovascular complications such as atherosclerosis, peripheral arterial disease, atrial fibrillation, thrombotic disorder, myocardial infarction, vascular calcification, and the role of RAGE in diabetes-associated cardiac fibrosis, is discussed with a focus on therapeutic significance.

---

[*] **Corresponding author Tapan Kumar Mukherjee:** Amity Institute of Biotechnology, Amity University, New Town, Kolkata, West Bengal 700156, India; E-mail: tapan400@gmail.com

[†] These authors contributed equally to this work.

Keywords: Atherosclerosis, Atrial Fibrillation, Diabetic Nephropathy, Diabetic Retinopathy, Diabetic Neuropathy, Myocardial Infarction, Peripheral Arterial Disease, Receptor for Advanced Glycation End Products (RAGE), Vascular Calcification.

INTRODUCTION

In type-1 and 2 diabetes (T1D, T2D), cardiovascular disease (CVD) squarely remains a leading cause of morbidity and mortality [1 - 4]. The detrimental impacts of consistently high plasma glucose on various organs vary based on the kind of cells and tissues involved. Vascular endothelial cells (ECs), for example, display high quantities of glucose transporter 1 (GLUT 1), making them more susceptible to the damaging effects of hyperglycemia due to the inability to restrict intracellular glucose concentrations. The increased glucose levels in the circulation, skin, and other tissues cause glycation, a non-enzymatic process, resulting in plasma proteins and glucose covalent adducts formation. Many diabetic complications are primarily caused by protein glycation [5]. Some structural and functional proteins, including collagen and those from plasma, may become glycated due to persistently high glucose extents [6]. The alteration of drug binding in plasma, platelet activation and the generation of oxygen free radicals, impaired fibrinolysis, and immune system functioning are the prominent harmful effects emanating from non-enzymatic modification of plasma proteins, such as albumin, fibrinogen, and globulins (Fig. **1**) [5 - 7]. Conversely, structural collagen impairment modifies osteoblast differentiation, which results in skeletal fragility and bone remodeling [8, 9]. Increased extracellular matrix (ECM) formation and polyol pathway activation are the traits of diabetic phenotype that are acquired by renal mesangial cells over-expressing GLUT 1 [3]. The death rate for people with diabetes and myocardial infarction (MI) or stroke is almost twice as high, resulting in a projected 12-year reduction in life expectancy [10]. Important gaps in the treatment arsenal for diabetes and cardiovascular disease (CVD) remain, despite efforts to control blood pressure, cholesterol, and lifestyle factors. These hurdles highlight an urgent need for disease-modifying treatments and feasible cautions, *vis-à-vis* life threats.

One of the major problems of diabetes is accumulated glycation products [11, 12]. It has been discovered that the levels of intra- and extracellular AGEs in clinical and experimental models of diabetes are higher than in young, healthy controls [13, 14]. These correlative data infer a likely role of glycation in the pathophysiology of diabetes in disease advancement. All major biomolecules are affected by glycation; damage levels are predicted as 0.1% of basic phospholipids, 1 in 107 nucleotides on DNA, and 0.1-1% for lysine and arginine residues on proteins [15]. The extent of damage is likely to have detrimental effects. For example, the production and build-up of protein AGEs can lead to metabolic malfunction. With the progressive glycation of arginine and lysine residues, charge neutralization can alter the protein structure and the consequent functional integrity [16]. In contrast to the non-glycated protein, glycated collagen is stiff and nonelastic [17, 18].

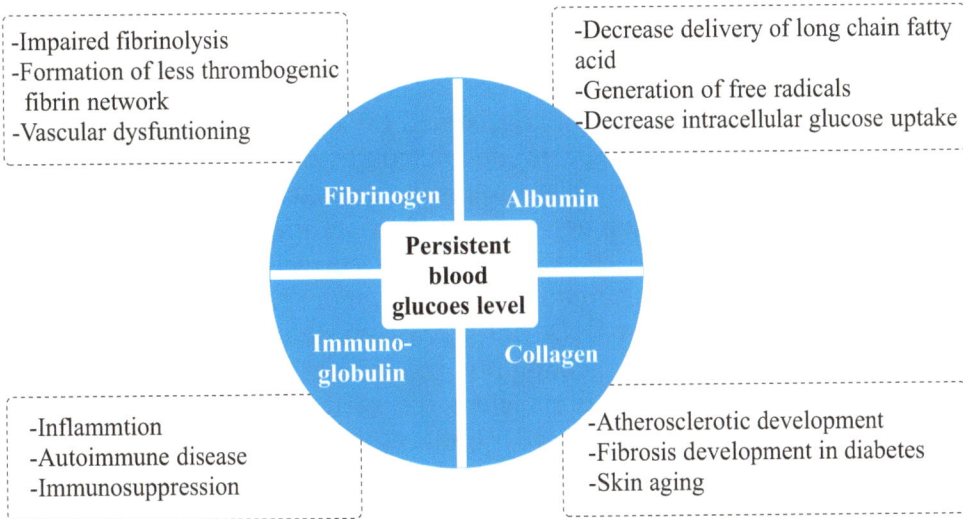

Fig. (1). Long-term diabetes with persistently high blood sugar causes structural and functional alterations in several bodily proteins, such as albumin, globulins, fibrinogen, and collagens. Glycation of these proteins is linked to the induction of harmful effects in the body.

The shift in charge distribution may also aggravate protein aggregation, such as lens crystalline-mediated cataract development in elderly with diabetes and other conditions [19, 20]. The function of a protein may also be affected by conformational changes, glycation of amino acid residues at substrate binding sites, and allosteric regulation of enzymes. For example, it has been demonstrated that glutamate dehydrogenase isolated from bovine liver exhibits decreased activity on methyl glyoxal-induced glycation of lys 126 and arg 463, affecting the enzyme's substrate binding capacity besides its allosteric activator (adenosine diphosphate, ADP), respectively [21].

Cell-ECM interactions are also impacted by the glycation of extracellular matrix (ECM) proteins. For example, EC detachment caused by methylglyoxal (MGO) reaction with arginine residues on the RGD and GFOGER motifs in the collagen integrin binding sites, interfered with cell-ECM interactions [7]. The formation of lipid glycation adducts, which increase membrane fluidity [22], may also have an impact on membrane interactions [23]. Glycation can occur in lipids with a free $-NH_2$ group, like phosphatidylethanolamine, but not in those not having a free

–NH$_2$ group, like phosphatidylcholine [24]. Lipid glycation is hypothesized to encourage lipid peroxidation, ultimately leading to oxidative damage [25 - 27].

For instance, glycation and lipid oxidation products were found as elevated when human low-density lipoprotein (LDL) was incubated with 200 mM glucose for 12 days *in vitro*. This finding seems to corroborate the theory that lipid glycation induces, or at least intensifies lipid peroxidation [25]. It is challenging to differentiate between lipid glycation and peroxidation as the primary initiating event and a secondary downstream event, especially *in vivo* , since both these may occur concurrently [25]. Similarly, strand breaks, double helix unwinding, mutations, and the creation of nucleotide-nucleotide alongside the DNA-protein cross-links are just a few effects that can result from the glycation of DNA [28 - 33]. The extent of modification appears varied as per the used glycation source. For instance, when DNA and DNA polymerase I derived from *Escherichia coli* were incubated *in vitro*, MGO produced ten times more DNA-protein cross-links than glyoxal [34]. These DNA-DNA polymerase crosslinks have the potential to impede replication, in turn encouraging the frameshift mutations [34]. By stopping the binding of transcription factors, the steric hindrance caused by DNA glycation adducts may potentially hinder transcription [35]. In other words, DNA glycation may change gene expression, besides affecting genome integrity. Aside from causing direct glycation damage to biomolecules, AGEs, particularly those from extracellular locations have the potential to activate the pro-inflammatory transcription factor NF-κB and downstream signaling molecules like p21, MAP kinases, and JNK *via* attachment with cell surface receptors like the receptor for AGEs (RAGE) [36, 37].

GLYCATED PROTEINS AS A MODEL OF TISSUE INJURY LINKED TO DIABETES MELLITUS AND AGING

Protein glycation is a non-enzymatic process that occurs when glucose attaches to proteins, lipids, or nucleic acids. This process can interfere with a protein's normal functions by altering its molecular structure, enzymatic activity, and receptor functioning. It is generally accepted that macromolecule exposure to varied glucose concentrations is an *in vitro* predictor of hyperglycemia, a suitable model for understanding functional degradation in diabetes mellitus and aging, specifically for low-turnover and extracellular structures [38]. For instance, structural changes in proteins provide new fluorophores [39] that are comparable to those linked with abnormal tissue changes in diabetes patients [40]. Studies on glycation explain the significance of structural changes therein as caused *via* covalent attachment of glucose to –NH$_2$ groups through the Amadori pathway, followed by rearrangement and dehydration reactions [38, 39, 41]. However, glucose, like other α-hydroxy aldehydes [42], is susceptible to oxidation by transition metals (*via* their enediol moiety), resulting in the generation of H_2O_2 and reactive intermediates like ·OH and ketoaldehydes [43]. The glucose oxidative chemistry may be a source of macromolecular changes during experimental glycation.

Oxidative stress may contribute to the tissue damage linked to diabetes [44] and aging [45], as per the view of numerous studies. Although the cause of this improper oxidation remains unknown, it may be related to a rise in the concentration of redox catalysts and/or materials that are easily peroxidized, generating free radicals. For instance, the concentration of Cu^{2+}, which easily catalyzes processes involving H_2O_2 and free radicals, rises in plasma with aging and the onset of diabetes [46, 47]. It is unclear whether biological Cu^{2+} can catalyze radical reactions [48]; even though Cu^{2+} levels in normal eye lenses are known to rise with age [49], as well as in idiopathic cataracts linked to aging or other risk factors like ocular inflammation [50]. According to *Wolff and teammates* [43], the transition metal-catalyzed enediol oxidation plays a key role in the tissue damage linked to aging and diabetes. This led to examining the role of free radicals and H_2O_2, generated by glucose enediol autoxidation, in the structural changes that occur when protein is exposed to glucose.

When a protein encounters glucose, it cleaves, changes its conformation, and produces fluorescent adducts known as "glycofluorophores." It is assumed that these alterations are the consequence of glucose's covalent bonding with protein–NH_2 groups. However, hydroxyl radicals generated by glucose auto-oxidation, or a closely related process, are necessary for the fragmentation and conformational changes. Additionally, antioxidants separate the structural damage caused by glucose exposure to protein from the protein's incorporation of monosaccharides. Additionally, it remains well-known that the synthesis of glycofluorophores is reliant on oxidative processes related to ketoaldehyde generation that are catalyzed by metals. These findings suggest that antioxidants may have a therapeutic impact if experimental glycation serves as a suitable model for the tissue damage noticed in diabetes mellitus [51].

In addition to being a major contributor to the formation and accumulation of AGEs, oxidative stress has been linked to the aggression of several diseases, including chronic conditions like diabetes, AD, and aging [52 - 54]. Since oxidative stress is linked to changes in biological substances and cellular activity that manifest a pathogenic environment, it is increasingly believed to have a fundamental mechanistic role in these conditions, specifically for oxidative damage to proteins. The formation of too many ROS molecules from glucose autoxidation and the non-enzymatic, covalent attachment of glucose molecules with circulating proteins, together culminating in AGE development, are the main sources of RAGE-mediated oxidative stress [55].

RELATIONSHIP OF CARDIOVASCULAR COMPLICATION AND DIABETES

A thin line of boundary separates diabetes and heart or cardiovascular disease (CVD). Diabetes increases the risk of CVD, which can result in heart attacks and strokes. Adults with diabetes have a 1.7 fold higher risk of dying from CVD in the US than people without the disease. This elevated risk is brought on by a higher chance of MI and stroke. Diabetes-related high blood sugar levels can harm blood vessels as well as the nerves that regulate the function of the heart and arteries. Heart disease may eventually result from these damages. Additional variables that may cause harm to blood vessels are smoking, high blood pressure, and high cholesterol. The diabetes sufferers typically experience heart disease at a younger age. Diabetes alone can result in heart failure even when other conventional cardiac risk factors are not present. The role of RAGE in various cardiovascular complications like atherosclerosis, vascular calcification, atrial fibrillation, thrombotic disorders, myocardial infarction, and vascular calcification is discussed ahead.

ROLE OF RAGE IN ATHEROSCLEROSIS

The most significant underlying pathological event for CVD is atherosclerosis. The important triggers for atherosclerosis are modified or oxidized LDL, oxidized proteins, diabetic vasculature, and AGEs. The process of atherosclerosis includes fatty streak formation, followed by the development of atheroma and plaque formation [56]. Several studies involving animal models and human subjects indicated the involvement of RAGE and its various ligands (AGEs) in atherosclerosis.

Initially, ApoE (apolipoprotein E) knockout mice were used to generate animal models of atherosclerosis and diabetes (Type I, TID). Elevated blood sugar was the source of accelerated atherosclerosis in these mice [57]. Daily intraperitoneal injections of recombinant sRAGE prevented the development of aggressive atherosclerosis in these mice, without affecting glucose or lipid levels. In an alternative study, sRAGE treatment prevented the advancement of atherosclerosis in diabetic mice lacking ApoE or LDLR (Low-Density Lipoprotein Receptor), even though the mice already had the disease [58]. Mice lacking ApoE, transgenic mice expressing RAGE [in ECs] with deleted cytoplasmic domain, or mice with a global genetic deletion of AGER (the RAGE gene) squarely revealed a significant attenuation in atherosclerosis. This was indeed true for the animals having diabetes, but it was especially true for those having hyperglycemia [59]. Studies employing transgenic mice in which the cytoplasmic domain of RAGE was removed in ECs have revealed significant roles of EC-RAGE in endothelial function and signal transduction. RAGE ligands specifically increased Vascular Cell Adhesion Molecule 1 (VCAM1) in wild-type aortic ECs, but not in ECs lacking the RAGE cytoplasmic domain [59]. Bu and colleagues studied the transcriptome of the aorta in the T1D condition in mice devoid of ApoE with or without simultaneous AGER deletion. These results demonstrate a significant RAGE-dependent alteration of the TGF-signaling pathways. Studies on ROCK1 (a serine/threonine kinase protein) modulation in smooth muscle cells suggest that the concurrent RAGE expression influenced diabetic atherosclerosis in mice using ROCK1 signaling [60].

Important roles for myeloid AGER (RAGE) in diabetic atherosclerosis were also found in studies based on bone marrow transplantation [61]. The RAGE ligand-RAGE interaction suppressed the cholesterol transporters (ABCA1, ABCG1), moderating cholesterol efflux to APOA1 and HDL in human THP1 cells and murine bone marrow-derived macrophages (BMDMs). This was caused, at least in part, by these transporters' PPAR-dependent regulation [62]. Of note, AGER deletion led to a significant decrease in vascular inflammation in diabetic mice suffering from atherosclerosis, along with genes regulating cholesterol efflux.

RAGE is expressed in both non-diabetic and diabetic atherosclerotic lesions in human subjects. A notable effort by *Egaña and the group* noticed elevated RAGE expression in diabetes, co-localizing in the lesion area with aggravated oxidative and inflammatory stress. In a study involving healthy control subjects and individuals with T1D, levels of sRAGE and esRAGE declined with age, irrespective of diabetes status, pubertal stage, or sex. The research was conducted at baseline (ages 8-18) and again after a 5-year interval. During the follow-up testing, baseline sRAGE in the diabetic subjects exhibited an inverse correlation with C reactive protein (CRP), but sRAGE and esRAGE levels had a positive correlation with carotid intima-media thickness (IMT) [63].

As per the view of many other studies, high extents of sRAGE may be protective, at least in the early stages of the disease or possibly when acute CVD events are worsened. Even when early metabolic susceptibility subsets in individuals without proven diabetes are identified, the levels of sRAGE may still be correlated with CVD risk markers. A 1-hour glucose tolerance test (GTT) revealed a high serum post 155 mg/dl glucose load level (GLL) in participants with metabolic illness who were not diabetic. Thereafter, the esRAGE levels of these subjects were examined and were found to lower with higher levels of the RAGE ligand, S100A12 compared to control patients, whose 1-h post-GLL was higher [1]. To fully test whether sRAGE levels, including both sRAGE and esRAGE, correlate with the risk of CVD development, long-term prospective studies on individuals with and without varying degrees of metabolic dysfunction are required.

ATHEROSCLEROSIS-RELATED STUDIES IN HUMAN SUBJECTS

A great deal of research has been performed on sRAGE levels in humans to examine the potential links between the RAGE pathway and diabetic CVDs. In research involving T1D patients and healthy control participants, sRAGE and esRAGE levels decreased with age, regardless of sex, diabetes status, or pubertal stage, both at baseline (age 8-18 years) and on five years of follow-up. At the follow-up testing, baseline sRAGE was negatively correlated with the CRP levels in the diabetic subject group, whereas sRAGE and esRAGE levels were positively correlated with carotid IMT [63]. The scientists concluded that although there was no protection against anomalies of arterial stiffness or wall thickness, high levels of baseline sRAGE might shield against inflammation five years later [63].

One research examined whether sRAGE levels in individuals with metabolic dysfunction who do not have diagnosed diabetes could serve as proxy markers for the early stages of atherosclerosis. On examining the levels of esRAGE in non-diabetic people with metabolic illness, 1-hour glucose tolerance testing (GTT) showed a high level (155 mg/dl) in the blood post-glucose load. Compared to control patients, whose 1-h post-glucose load level was less than 155 mg/dl, these individuals exhibited lower levels of esRAGE and higher extents of RAGE ligand S100A12, concurrently with increased pulse wave velocity (PWV) and carotid IMT [64]. As per the notification of investigations, the RAGE pathway may be involved in the variability of metabolic dysfunction observed in individuals with normal glucose tolerance.

In a different attempt, participants without a prior diabetic history were divided into three categories: pre-diabetes, controls, and newly diagnosed Type 2 Diabetes (T2D). Compared to controls, the pre-diabetic subjects had lower levels of esRAGE and higher levels of S100A12. Peripheral blood mononuclear cells (PBMCs) from the subjects with lower esRAGE also showed lower levels of the esRAGE splice variant, indicating that this splice variant's lower transcription may contribute to the lower esRAGE systemic levels [65]. According to statistical analysis, the primary factors influencing IMT were age, glycosylated haemoglobin, and esRAGE, while the main factors influencing PWV were blood pressure (systolic) and S100A12 levels [65].

In a 3-year longitudinal study of 1,002 subjects with CVD, 933 participants were tested for sRAGE levels, which were then segregated by quartiles. After 3 years of follow-up, 16% of individuals showed a new CVD event (MI, stroke, and CVD death). Even after accounting for variables related to CVD, the patients with the highest quartile of sRAGE showed the highest incidence of recurrent CVD [66]. When taken as a whole, these studies contribute to a substantial amount of literature about the connection between sRAGE-diabetes-CVD and offer the following insights: (1) High levels of sRAGE may be protective, at least during the early stages of the disease or possibly when acute CVD events are being actively exacerbated; and (2) even after early metabolic vulnerability subsets in individuals without diabetes confirmation, sRAGE levels may correspond with CVD risk markers. These factors highlight the need for long-term prospective investigations in individuals with and without different degrees of metabolic dysfunction to thoroughly investigate whether sRAGE and esRAGE levels are associated with first events and recurring CVD occurrences.

ATHEROSCLEROSIS-RELATED STUDIES IN ANIMAL MODELS

Mice lacking ApoE were treated with streptozotocin to mimic T1D in early *in vivo* experiments, consequently leading to accelerated atherosclerosis acquiring when the mice were in hyperglycemic conditions [67]. In diabetic mice lacking ApoE, daily intraperitoneal recombinant sRAGE injection therapy decreased the incidence of accelerated atherosclerosis without affecting glucose or lipid levels. In a separate investigation, sRAGE therapy of diabetic mice lacking ApoE but already having atherosclerosis stopped the disease progression [68]. No matter the level of diabetes, but especially in the mice suffering from hyperglycemia, there was a notable reduction in atherosclerosis in mice lacking ApoE or LDLR. Similar was the condition in transgenic mice expressing RAGE with deleted cytoplasmic domain [in endothelial functions, ECs] or in mice with a global genetic AGER deletion [59, 60, 68]. Studies employing transgenic mice having deleted RAGE cytoplasmic domain in ECs demonstrated the significant functions that EC-RAGE plays in signal transduction and endothelial function. In ECs from wild-type aortas, RAGE ligands elevated inflammatory markers, including VCAM1. However, this was not so in ECs devoid of the RAGE cytoplasmic domain [69]. In the T1D condition, *Bu and colleagues* examined the transcriptome of the aortas in mice lacking ApoE, either in conjunction with or independently of the simultaneous AGER deletion. These investigations revealed a considerable RAGE-dependent regulation of the ROCK1 branch of the TGF-β signaling pathway in SMCs, indicating a major role for SMC-RAGE in ROCK1 signaling-mediated diabetic atherosclerosis, in mice [70]. Key roles for myeloid AGER in diabetic atherosclerosis were also uncovered through bone marrow transplantation studies [72]. Ligand-RAGE interaction in macrophages impaired the functioning of cholesterol transporters, ABCA1 and ABCG1, at least partially due to their PPAR-γ-dependent regulation in human THP1 cells and murine BMDMs [62]. It also significantly attenuated cholesterol efflux to APOA1 and HDL. In diabetic atherosclerotic mice, deletion of AGER resulted in a considerable reduction of vascular inflammation, independent of genes controlling cholesterol efflux.

ROLE OF RAGE IN PERIPHERAL ARTERIAL DISEASE

Peripheral arterial disease (PAD) generally denotes a pathophysiological condition wherein reduced blood flow to limbs (typically legs) is caused by constricted arteries. Atherosclerosis, a gradual accumulation of fatty deposits in the arterial walls, is usually the cause of this narrowing. Because of this, the tissues in the limbs might not get enough blood or oxygen, resulting in symptoms like numbness, cramping, or discomfort in the legs, especially when exercising. Untreated PAD can result in major side effects such as ulcers, infections, and in extreme situations, amputation or tissue death (gangrene) [1]. Moreover, diabetes is linked to a higher risk of limb amputation and all-cause mortality in PAD patients; the 5-year mortality rate is 30% in diabetic patients with critical limb ischemia. Furthermore, PAD advances more quickly and is more severe in diabetic individuals [69, 70]. The discussion on PAD ahead focuses on the pathophysiological role along with culture and *in vivo* studies of RAGE ligands in PAD, including AGEs, high mobility group box 1 (HMGB1), and S100A proteins.

PERIPHERAL ARTERIAL DISEASE-RELATED STUDIES IN HUMAN SUBJECTS

There is growing evidence that the (RAGE) ligand-RAGE pathway has a role in the etiology of PAD [71]. Interestingly, the percentage of infrainguinal vein tissue stained for AGE, CML, RAGE, and S100A12 was similar in patients with and without diabetes when used for vascular grafting in this disorder. This suggests that factors unrelated to glucose may have also played a role in the recruitment of the RAGE pathway in PAD. Analyzed models of varied settings exhibited higher levels of the RAGE ligands S100A12 and carboxy-methyl lysine (CML)-interacting AGE proportions in their blood. A noteworthy attempt to monitor sRAGE levels in patients with CAD and PAD revealed that people with both disorders had lower total sRAGE levels than for a singular disorder alone [72]. This data raises the possibility that, at least in part, the causes of disease in each region may be different. In a population-based cohort study, the levels of S100, RAGE ligands, and esRAGE (together, the RAGE score) were assessed in 106 PAD patients, both with and without amputation. Analysis revealed that shorter amputation-free lives in T2D patients were associated with higher plasma S100A12 concentrations and an overall RAGE score, suggesting a role for the RAGE pathway in the PAD severity [1]. The baseline plasma levels of S100A12 and RAGE score, determined by averaging Z-standardized values of S100A12,

CML, and endogenous secreted RAGE form, were found as significantly associated with amputation or death in a 12-year prospective cohort study of 146 diabetic patients without PAD. The age-and-sex-adjusted hazard ratio for amputation or death was 1.29 per 100 ng/mL, increasing the S100A12 by 1.79 per 1 unit (standard deviation) increment in the RAGE score [73]. Additionally, compared to controls, PAD patients had greater plasma S100A12 levels, particularly those with diabetic positivity [74]. Another 3-year prospective cohort trial revealed individuals suffering from PAD and above 75% (8.6 ng/mL) S100A12 values to have a lower amputation-free survival rate than those with a < 75% S100A12 extent. In the latter group, the risk of death or amputation was three times higher than in the former. According to these findings, plasma S100A12 may be involved in the onset, course, and prognosis of PAD, a potentially fatal illness. In diabetic individuals suffering from PAD, HMGB1 levels in the vasculature of amputated feet were considerably greater than non-diabetic controls. These levels were positively linked with oxidative stress, inflammatory indicators, and the grade of arterial stenosis [75]. Plasma concentrations of HMGB1 were significantly higher in 24 patients with PAD than in the 10 (controls), having matched age-, sex-, and body weight. Curiously, the plasma fibrinogen was the sole correlate of HMGB1 in PAD patients [76]. As earlier noticed, it is still unknown whether HMGB1 promotes or inhibits collateral vessel development and limb ischemia in PAD patients. HMGB1 levels in the blood, however, might be a biomarker of PAD.

PERIPHERAL ARTERIAL DISEASE-RELATED STUDIES IN ANIMAL MODELS

Through unilateral hind limb ischemia, ischemic damage of the peripheral vascular system has been imposed on PAD animal models to examine these theories. Measurements of blood flow using laser Doppler imaging and angiogenesis responses represent significant findings from these studies. An elegant research effort herein noticed a considerable rise in RAGE expression for hindlimb ischemia sufferers when molecularly targeted nanoparticles were used for multimodal imaging in a mouse model [77]. A related study in this reference inferred enhanced use of an anti-RAGE antibody fragment in RAGE imaging for diabetic *versus* non-diabetic hind limb ischemia [78]. Anti-RAGE antibodies were observed to exhibit improved angiogenesis and blood flow recovery in diabetic mice, revealing good results in hind limb ischemia-inflicted diabetic and non-diabetic animals with mice having entirely empty AGER [1].

Significant distinctions between RAGE signaling in the peripheral vascular system and atherosclerosis were brought to light by a screened hind limb ischemia model. AGER-devoid mice had significantly fewer immune cells (macrophages and T cells) in atherosclerotic lesions than diabetic, ApoE null mice expressing AGER [60], but both diabetic and non-diabetic animals had significantly more macrophages during the peak of immune cell infiltration into the ischemic hind limb than the wild-type animals [84]. Simultaneously, the ischemic hind leg muscle tissue of AGER null mice exhibited considerably higher mRNA transcript levels for the CCL2 and EGR genes, being related to pro-inflammatory processes and the recruitment of inflammatory cells [79]. Ablation of AGER (and delivery of anti-RAGE antibodies in T1D mice subjected to hind limb ischemia) improved vascular disease in both cases, despite clear differences between atherosclerosis and hind limb ischemia while screening RAGE prevalence in immune cell composition. These findings suggest that RAGE ligands and responses in diverse forms of vascular injury are regulated by niche-specific cues, and they also demonstrate the complexity and adaptability of RAGE signaling in immune cells, having varied vascular depots and *in vivo* contexts.

ROLE OF RAGE IN ATRIAL FIBRILLATION

Heart arrhythmia conditions of the cardio-vasculature are sometimes accompanied by abnormalities of the heart's rhythm, such as atrial fibrillation (AF) [80]. The incidence of AF is higher in diabetic patients, especially in those with longer disease duration or poor glycemic control. It has been hypothesized that RAGE ligand AGEs, through their capacity to enhance stiffness, oxidative stress, and fibrosis, contribute to this phenomenon [81]. Cardiovascular diseases are often accompanied by abnormalities in cardiac rhythm, such as atrial fibrillation (AF). It has been proposed that AGEs and RAGE ligands enhance stiffness, oxidative stress, and fibrosis, which may contribute to the greater incidence of AF in diabetic patients, particularly those with poor glycemic control or prolonged illness [81]. In AF-affected humans, analysis of ligand-RAGE pathway markers revealed higher fluorescent AGE and sRAGE levels in the AF group over the control group when 38 AF sufferers were compared to 59 individuals with normal sinus rhythm, particularly for non-diabetic patients [75]. The study observed a correlation between left atrial dimensions and the AGEs, sRAGE indicators [82]. Plasma sRAGE and esRAGE levels were deciphered as higher in Caucasian individuals with persistent AF than paroxysmal AF in a different study [83]. Higher plasma levels of sRAGE were noticed as independently correlated with a lower rate of AF recurrence following catheter ablation in diabetic individuals in another study focused at monitoring the effects of therapeutic intervention [84]. On the other hand, 1068 individuals in the Atherosclerosis Risk in Communities (ARIC) study had established baseline sRAGE levels when they were enrolled. These participants exhibited multiple measures of inflammation. Analysis revealed a correlation of the sRAGE lower quartile with the higher baseline levels of inflammatory markers (hsCRP, white blood cell count, and fibrinogen) than the highest quartile of sRAGE. However, there was no correlation with sRAGE, on a perspective consideration of the data (6-year change in inflammatory markers). Furthermore, no discernible correlations between sRAGE levels and AF risk were observed [85]. All these investigations point to the possibility that sRAGE levels could be biomarkers for AF, at least in some populations and under some circumstances. To test these ideas definitively, detailed studies are required.

ROLE OF RAGE IN THROMBOTIC DISORDERS

Blood clots in blood arteries caused by thrombotic diseases (TD) can result in potentially fatal situations such as deep vein thrombosis, MI, and stroke. Although the precise function of RAGE in TD remains unclear, several studies have indicated that it may contribute to thrombosis through a variety of methods. One significant aspect is the role of RAGE in promoting endothelial dysfunction, a hallmark of TD. Increased arterial permeability, inflammation, and thrombosis are the results of endothelial dysfunction, characterized by compromised EC activity. When ECs encounter RAGE, they express adhesion molecules like ICAM-1 and VCAM-1, aiding in platelets and leukocytes adherence to the endothelium and recruiting them, initiating the thrombotic process. It has been demonstrated that RAGE directly interacts with thrombin, tissue factor (TF), plasminogen activator inhibitor-1 (PAI-1), and other elements of the coagulation cascade to promote a prothrombotic state. For example, the extrinsic coagulation pathway is initiated when RAGE interacts with TF to boost TF expression and activity. The inhibition of fibrinolysis by RAGE-mediated activation of PAI-1 improves thrombus stability to a substantial extent. Furthermore, RAGE activation promotes platelet aggregation and activation, which contributes to thrombus formation.

Numerous investigations employed animal models, and several clinical studies have furnished proof endorsing the RAGE involvement in TDs. It has been demonstrated that RAGE inhibition or genetic deletion reduces thrombosis in a variety of experimental models, suggesting RAGE as a promising target for TDs treatment.

THROMBOTIC DISORDERS RELATED STUDIES IN HUMAN SUBJECTS

Numerous investigations that have employed animal models and clinical observations have inferred RAGE involvement in thrombotic diseases. It has been demonstrated that inhibition or genetic deletion RAGE B2 glycoprotein 1 (also known as anti-b2-GP1) is one of the primary targets of the anti-phospholipid antibody in the prothrombotic condition known as anti-phospholipid syndrome (APS). Upregulation of RAGE and changed HMGB1 cellular localization were noticed when platelets and monocytes derived from healthy human subjects were incubated with anti-b2-GP1 [91]. In serum tests, patients with APS exhibited considerably greater sRAGE and HMGB1 levels over the controls, and there was a clear relationship between HMGB1 levels and the length of specific conditions [86]. Additional research revealed that HMGB1 binds to platelets and the activation of platelets upregulated the RAGE expression and that human coronary artery thrombi rich in platelets exhibit high levels of HMGB1 expression [87]. In erythrocytes isolated from human saphenous veins, exposure to AGEs boosted neutrophil adherence and ROS generation. Simvastatin therapy counteracted these prothrombotic stimuli and decreased RAGE expression [88]. Nevertheless, additional research investigated the potential correlation between the outcomes in symptomatic CAD and the levels of platelet HMGB1. Comparing patients with stable CAD, unstable CAD, non-ST segment elevation myocardial infarction (NSTEMI), and ST-segment elevation myocardial infarction (STEMI), the investigators found no variations in the HMGB1 expression of platelets [89]. Furthermore, among the participants who also experienced MI, there was no association between left ventricular ejection fraction (LVEF). Monocyte behavior was similarly impacted by HMGB1 generated from platelets. It was demonstrated that HMGB1 activated the MAPK pathway in these cells in a TLR4-dependent manner, inhibiting monocyte death, and stimulating monocyte migration *via* RAGE [90]. Therefore, interactions between platelet HMGB1-RAGE may influence different cell types, in turn raising the CVD risk and severity.

THROMBOTIC DISORDERS RELATED STUDIES IN ANIMAL MODELS

In animal models, it was discovered that neutrophil-derived S100A8/A9 enhanced thrombopoietin production, leading to reticulated thrombocytosis, especially in diabetic-positive cases. However, this effect was mitigated in mice by either lowering blood glucose levels with dapagliflozin or preventing S100A8/A9 binding to RAGE using laquinimod [91]. The investigators of this study compared their results with human diabetic patients, where reticulated thrombocytosis was linked with the levels of glycosylated haemoglobin and S100A8/A9 [91]. Disulfide HMGB1 was revealed to promote monocyte recruitment and, *via* RAGE, to stimulate the development of pro-thrombotic neutrophil extracellular traps (NETs) in other mouse investigations. This procedure exposed more HMGB1 to their extracellular DNA strands, spreading the HMGB1 and NETs prothrombotic effects [92]. Altogether, these investigations connect RAGE and its ligands to platelet disruption and to the over-expression of prothrombotic pathways, which are connected not only to the cardiovascular consequences of diabetes but also to diabetes itself.

ROLE OF RAGE IN MYOCARDIAL INFARCTION

Myocardial infarction (MI) is a critical complication of diabetes, which happens when a portion of the heart's blood supply is cut off for an extended period, causing damage or even heart muscle death. Numerous processes, including endothelial dysfunction, inflammation, oxidative stress, and fibrosis, are involved in the role of RAGE in MI. Endothelial dysfunction, which is a prelude to atherosclerosis and MI, is brought on by RAGE activation on ECs. Pro-inflammatory cytokines and adhesion molecules like VCAM-1 and ICAM-1 are expressed when RAGE binds to its ligands, including AGEs, S100 proteins, and HMGB1. These molecules play a crucial role in the development of coronary artery disease (CAD) and MI by facilitating the recruitment of leukocytes to the endothelium surface, promoting inflammation and atherogenesis. Additionally, the oxidative environment changes low-density lipoprotein (LDL) into oxidized LDL (oxLDL), increasing the risk of plaque rupture and eventual MI. Fibrosis is a result of increased fibroblast proliferation and collagen deposition brought on by RAGE activation. Heart failure can result from impaired cardiac function caused by excessive fibrosis. Moreover, the prolonged inflammatory response in MI is aggravated by RAGE-mediated signaling pathways, specifically the NF-κB pathway, exacerbating cardiac damage. Several studies have highlighted the critical role of RAGE in MI.

MYOCARDIAL INFARCTION-RELATED STUDIES IN HUMAN SUBJECTS

Studies on human subjects with MI and related CVD disorders have focused majorly on the varied outcomes of the ligand-RAGE axis. In one such noteworthy effort, 276 T2D Japanese citizens with no history of CVD were recruited. Their baseline biochemical and clinical information, including serum sRAGE levels, was investigated, followed by the screening of a potential correlation between these variables and CVD occurrence feasibility. With a median follow-up length of 5.6 years, 25 additional CVD incidents were reported. According to the tertile analysis, there was a positive correlation between serum sRAGE levels and an elevated CVD risk. Even after controlling traditional coronary risk variables, multivariate Cox proportional hazards regression analysis revealed an independent association between serum sRAGE levels and CVD. In conclusion, higher sRAGE levels were linked to higher CVD risk in T2D-positive individuals [93]. A subsequent three-year longitudinal cohort study of 1002 in-patients with angiographically proven CVD, sRAGE levels were measured in 933 patients using an ELISA kit at baseline. The combined endpoint included MI, stroke/TIA, and cardiovascular death. The sRAGE values were analyzed in quartiles, and adjusted hazard ratios (HR) were computed using various CVD risk factors. In all,

886 patients having completed the follow-up indicated a 16% overall incidence of the combined endpoint. The incidence of recurrent CVD was considerably higher in patients in the highest sRAGE quartile. Multivariate Cox regression confirmed the association between high sRAGE levels and new CVD events [66]. According to *Larsen and groupmates* [94], sRAGE has a dual phase-dependent association with residual CVD risk in an acute coronary event. The S100A12 and sRAGE were measured in 524 patients within 24 hours after an acute coronary syndrome (ACS), and in a sub-group of 114 patients, 6 weeks later. After a year, this sub-group also finished a follow-up echocardiogram. For recurrent major adverse cardiovascular events (MACE), defined as recurrent ACS or cardiovascular mortality, the median follow-up period was 25.7 ± 12.6 months. Based on Cox proportional hazard analysis, MACE risk exhibited a positive association with baseline S100A12 and sRAGE, even with null conventional cardiovascular risk factors. Additionally, declining left ventricular function and a higher post-discharge hospitalization for heart failure were linked to high sRAGE. Conversely, recurrent ACS risk was reduced in patients with increased sRAGE at 6 weeks than the baseline [94].

Another significant attempt by *Paradela-Dobarro and colleagues* [95] compared the behavior of AGEs and their soluble receptor (sRAGE) in two patient cohorts: those with heart failure (HF) and acute coronary syndrome (ACS). In a unicentric observational clinical trial, 102 patients squarely having chronic heart failure (HF) and ACS were matched according to age and gender. Quantitative plasma fluorescent spectroscopy was used to assess fluorescent AGEs at inclusion, whereas ELISA kits were used to measure total sRAGE and esRAGE levels. The primary objective of the study was to ascertain cardiac death, and the secondary endpoints were the incidence of non-fatal, MI, or HF readmission, with a 5-year follow-up. Patients with HF exhibited increased glycation characteristics, whereas there were no variations in sRAGE forms between the HF and ACS cohorts, except for cRAGE, which was higher in the HF group. The major distinction in HF and ACS patients was based on the AGE-RAGE axis. In HF, all the sRAGE forms showed a direct relationship with the glycation parameters. The independent value of the sRAGE forms on each CVD was supported by esRAGE, being an independent predictor of bad long-term prognosis only for HF.

The onset of cardiogenic shock is a serious side effect of acute MI. The effects on levels of MMP9 and Tissue Inhibitors of Metalloproteinase (TIMPs) were investigated because it was demonstrated that higher levels of monocyte RAGE and lower levels of plasma sRAGE were associated with higher mortality in cardiogenic shock. Of note, the MMPs have been studied earlier to contribute to sRAGE generation by cleaving its extracellular domains. Analysis revealed acute MI survivors with a higher MMP9 activity, whereas those acute MI individuals who experienced cardiogenic shock, exhibited a lower MMP9 activity [96]. Herein, elevated blood S100A12 concentration was independently linked to an increased hospitalization risk for HF in patients with T2D, but not an increased risk of MACE. The potential therapeutic use of S100A12 for HF prediction is still restricted compared to NT-proBNP. S100A12, however, might be a contender, *vis-à-vis* multi-marker strategy for HF risk evaluation in diabetic individuals.

MYOCARDIAL INFARCTION-RELATED STUDIES IN ANIMAL MODELS

Research on the isolated perfused heart and *in vivo* MI caused by occlusion/reperfusion of the left anterior coronary artery in rats and mice reveals that RAGE blockade protects diabetic and non-diabetic animals against ischemia/reperfusion (I/R) injury [97 - 99]. The suppression of RAGE was accomplished using sRAGE or genetically modified mice, *i.e.*, AGER null mice or transgenic mice expressing cytoplasmic domain-deleted RAGE (in macrophages or ECs). Blockade of the RAGE axis was associated in these experiments with decreased infarct size, decreased myocardial necrosis, enhanced cardiac function, and ATP recovery. Induction of hypoxia/re-oxygenation (H/R) in cultured cardiomyocytes promoted JNK MAP kinase activation and GSK-3b dephosphorylation in a RAGE-dependent manner. This was inhibited in cells lacking AGER or when sRAGE was administered to wild-type cardiomyocytes [100].

Several other studies also probed the S100/calgranulin-RAGE connections using a rat model of MI and cultured cardiomyocytes. In one such effort, it was noticed that S100B may contribute to cardiomyocyte apoptosis (*via* RAGE) by activating ERK1/2-MAPK and p53 signaling [101]. An erstwhile effort involved coronary artery ligation in wild-type and S100B-deleted T1D animals [102]. The S100B and RAGE expression in the heart increase independently in MI and diabetes. However, only in diabetic mice did S100B expression become less pronounced in the post-MI stages. Following MI, the diabetic S100B-deleted animals revealed a higher dilatation of the left ventricle than the diabetic wild-type mice. This was accompanied by a greater impairment of cardiac function, GLUT4 expression, and systemic levels of AGE [102]. Collectively all studies suggested that S100B expression might aid in modifying heart metabolism in diabetic sufferers with MI positivity. Separate investigations also linked ligand HMGB1 to experimental MI in both diabetic and non-diabetic mice, at least in part *via* RAGE [103]. Therefore, it is crucial to determine when the actions of RAGE ligands occur acutely or chronically to determine the optimal circumstances for RAGE antagonistic effects in MI. Lastly, research on the possible roles of RAGE cytoplasmic domain binding partner (forming DIAPH1) in myocardial I/R injury has commenced. In a study focusing on this, the DIAPH1 expression was elevated in wild-type animals on I/R induction. Similarly, H/R activation upregulated DIAPH1 expression in cultured H9C2 and AC16 cardiomyocytes [35]. Global deletion of Diaph1 on experimental MI reduced infarct size and preserved cardiac function relative to the Diaph1-expressing control mice, exhibiting consistency with mediating Diaph1 functions in myocardial damage. Silencing Diaph1 in H/R increased sarcoplasmic Ca-ATPase activity with a reduced Na-Ca exchanger expression in H9C2 cells.

ROLE OF RAGE IN VASCULAR CALCIFICATION

Vascular calcification (VC), a condition strongly associated with several CVDs, is mostly influenced by the RAGE, particularly in individuals with diabetes. When AGEs attach to RAGE, oxidative stress is generated, activating signaling pathways like NF-κB and mitogen-activated protein kinase (MAPK). Increased expression of inflammatory and pro-fibrotic factors results from this activation, encouraging arterial calcification and the development of atherosclerosis and other CVDs [104].

VASCULAR CALCIFICATION-RELATED STUDIES IN HUMAN SUBJECTS

Studies have probed the role of the ligand-RAGE axis in human individuals with VC. In 199 patients on hemodialysis for whom vascular calcium scores were obtained (49.2% of the subjects had diabetes), circulating sRAGE levels are negatively associated with Ca scores independent of the S100A12 level and inflammatory markers [105]. The degree of aortic calcification was adversely associated with esRAGE levels, considerably lower in non-diabetic participants receiving hemodialysis than in control subjects [106]. In SMCs isolated from the saphenous veins of patients undergoing coronary artery bypass grafting (CABG), exposure to high levels of glucose resulted in NADPH oxidase- and protein kinase C-dependent HMGB1 translocation of to the nucleus, which increased calcification through an NF-κB-dependent regulation of bone morphogenetic protein 2 (BMP2) [107]. Exposure of vascular SMCs to AGEs enhanced calcification, at least partially through the activation of p38 MAPK, is consistent with the roles of RAGE ligands in both processes [108, 109]. These arguments collectively infer that RAGE signaling plays a role in vascular calcification, in diabetic and non-diabetic environments. This is most likely due to the production of RAGE ligands, such as AGEs, and other pro-inflammatory/pro-oxidative ligands in diseases like advanced renal disease.

VASCULAR CALCIFICATION-RELATED STUDIES IN ANIMAL MODELS

Research conducted on mice models using several techniques to promote calcification highlighted the role of RAGE in these *in vivo* processes. A high phosphate diet was given to a subset of mice having sham surgery or chronic kidney disease (CKD) in the mice having both ApoE and AGER. The AGER mRNA was significantly higher in the CKD arteries compared to controls, and after 12 weeks of CKD, ApoE null mice's blood levels of RAGE ligands, AGEs, and S100/calgranulins increased. Enhanced expression of Runx, missing in mice lacking AGER, was accompanied by increased vascular calcification in the CKD-suffering, ApoE null animals [110]. *in vitro*, the stimulation of SMCs with RAGE ligand S100A12, strengthened the mineralization and osteoblast transformation, inhibited by AGER deletion [110]. In other investigations, transgenic mice over-expressing RAGE ligand S100A12 and subjected to CKD, revealed increased vascular calcification *via* NADPH oxidase-dependent mechanisms, illuminating direct mediating roles for S100/calgranulins in the pathogenesis of vascular calcification [111]. Of note, FGF23 exhibited a slight upregulation in the heart and vascular tissues of S100 transgenic mice with CKD, but not in CKD wild-type or CKD AGER null S100 transgenic mice. These observations suggest S100/RAGE involvement in the upregulation of FGF23 and pro-inflammatory factors contributing to vascular calcification in valvular calcification (mitral and aortic valve) associated with CKD [112]. In conclusion, vascular calcification is worsened by the buildup of RAGE ligands in CKD conditions other than diabetes, at least partially due to RAGE. The research on S100 transgenic mice highlights the multi-ligand contributions of the RAGE pathway in CVD and calcification, as does the seemingly strong effect of 100/calgranulins.

ROLE OF RAGE IN DIABETES-ASSOCIATED CARDIAC FIBROSIS

RAGE plays a critical role in the development of diabetes-associated cardiac fibrosis. Cardiac fibrosis is a hallmark of diabetic cardiomyopathy, characterized by the excessive accumulation of ECM proteins, leading to stiffening and functional impairment of the heart. Myocardial fibrosis is part of the remodeling process that occurs in HF. Although the exact process is still unknown, numerous studies have demonstrated the involvement of the RAGE and AGEs in the activation of NF-κB and MAPK pathways, being triggered by AGEs binding with RAGE. This inflammatory reaction encourages cardiac fibroblasts (CFs) to activate, proliferate, and change to myofibroblasts, releasing copious amounts of collagen and other matrix components, and exacerbating fibrosis [113, 114].

According to experimental research, the degree of cardiac fibrosis in diabetic patients can be decreased by eliminating RAGE or preventing its interaction with AGEs. Decreased endothelium-to-mesenchymal transition (EMT), a process wherein ECs acquire mesenchymal, fibrogenic features, amidst linkage to this reduction, playing a role in the fibrotic process. Moreover, pro-fibrotic and pro-inflammatory molecules such as connective tissue growth factor (CTGF) and TGF-β are expressed at lower levels when RAGE expression is reduced [115]. Another worthy effort by *Liang and colleagues* [116] elucidated the mechanism by which the AGEs-RAGE axis mediates the activation of cardiac fibroblasts (CFs) in heart failure. Transverse aortic constriction (TAC) was utilized to create a model of heart failure in C57BL/6J wild-type (WT) mice. Relevant ions were found following a six-week therapy. AGEs were higher in TAC-treated mice than in mice who were given a sham operation. On acquiring TAC positivity, mice with functional cardiac protection showed less hypertrophy and fibrosis when RAGE was inhibited. Notably, autophagy acted as a mediator in CF activation, which later developed into myofibroblasts and played a role in fibrosis. To confirm the *in vivo* results, CFs from newborn Sprague-Dawley rats were collected and treated *in vitro* with AGEs, bovine serum albumin, and short hairpin RNA (shRNA) for RAGE. These findings imply that CF activation brought on by autophagy plays a role in the AGEs-RAGE axis involvement in the pathophysiology of myocardial fibrosis in HF. By reducing CF activation, inhibition of the AGEs-RAGE axis reduced cardiac dysfunction and myocardial fibrosis in TAC-positive mice. This work demonstrates how autophagy in HF activates CFs, a process that contributes to the myocardial fibrosis pathophysiology. In HF, inhibiting the AGEs-RAGE axis reduces cardiac dysfunction and eases myocardial fibrosis, making RAGE a feasible target for therapy in HF and myocardial fibrosis.

ROLE OF RAGE IN OBESITY AND DIABETES

Obesity is linked to an increased risk of CVD and T2DM. There is a connection between RAGE, inflammation, obesity, T2DM, and CVD that goes beyond managing the risk factors for their development. Studies have demonstrated that RAGE expression affects insulin resistance and fat cell hypertrophy in mice and that it is upregulated in the adipose tissue of obese people [117, 118]. It was demonstrated by *Masayo Monden* and *colleagues* [119] that RAGE could control the process of fat cell hypertrophy directly *in vitro*. Significant reduced fat cell hypertrophy can be achieved by inhibiting RAGE *via* siRNA intervention. In addition, early suppression of TLR2 mRNA expression in adipose tissue is linked to RAGE deficiency. Consequently, it appears that fat cell hypertrophy and the control of TLR2 by RAGE are connected [120]. The body weight, epididymal fat weight, and big size of RAGE defective mice fed a high-fat diet are much lower than those of WT mice, suggesting that RAGE knockout mice exhibit apparent resistance to obesity caused by high-fat diets [119]. Of late, it has been suggested that HMGB1 is an adipokine expressed twice as much in fat cells of obese people and functions as an innate pro-inflammatory mediator. This suggests that HMGB1 may have a connection to metabolic syndrome and insulin resistance [120 - 122].

An imbalance between energy intake and consumption in the human body leads to several illnesses, including obesity. At this point, the body will trigger the recruitment of certain inflammatory cells to essential metabolic organs, mostly due to the accumulation of macrophages in visceral fat. The occurrence of inflammatory reactions is regulated by variations in macrophage M1 and M2 phenotypes, which are particularly noticeable in adipose tissue, the liver, and other regions when metabolic disorders like obesity and diabetes are present. Adipose tissue is an essential endocrine organ that secretes a variety of physiologically active chemicals, including adiponectin and leptin [123]. Adipose tissue macrophages (ATMs) are one of these that primarily aid in accomplishing adipose tissue homeostasis and healthy function by engulfing the dead fat cells. According to studies, metabolic inflammation in adipose tissue during obesity causes ATMs to change to a pro-inflammatory phenotype, primarily accompanied by an increase in the infiltration of pro-inflammatory macrophages. Once macrophages are activated, ATMs release inflammatory factors that interact with fat cells to intensify the inflammatory response [124]. Under normal conditions, ATMs are primarily M2 anti-inflammatory macrophages. M1 pro-inflammatory macrophages in adipose tissue steadily rise as obesity progresses, while M2 anti-inflammatory macrophages steadily decline, resulting in macrophage activation to restore normalcy.

From an M2 to an M1 pro-inflammatory type, the polarization state is shifted, binding to ligands in macrophages and adipocytes to stimulate NF-κβ signaling within white adipose tissue *via* chemokine and cytokine production, linked to insulin resistance [125] (Fig. **2**). Consequently, it is feasible to identify novel targets for the treatment of obesity in macrophages, which will lower the obesity-related metabolic syndrome. In a noted attempt, *Feng and colleagues* [126] postulated that adipose tissue insulin resistance is related to RAGE and its ligands and that inflammation linked to obesity is significantly influenced by the activation of the AGE-RAGE axis. The C57BL/6J mice (WT) and RAGE deficient (RAGE$^{-/-}$) mice were fed a high-fat diet (HFD) and subjected to glucose and insulin tolerance tests. Analysis revealed RAGE$^{-/-}$ mice exhibited a reduced body weight and epididymal adipose tissue (eAT) mass along with enhanced insulin sensitivity and glucose tolerance. In the adipose tissues of WT mice (given a regular chow diet), exogenous MGO inhibited insulin-stimulated AKT activation, but not in RAGE$^{-/-}$ animals. On the other hand, MGO therapy did not lessen AKT phosphorylation in WT-HFD mice, generated by insulin in obese mice. Additionally, it was discovered that the adipose tissue of RAGE$^{-/-}$-HFD mice had reduced insulin-induced AKT phosphorylation. Further, RAGE$^{-/-}$ mice exhibited signs of enhanced adipose tissue browning with better inflammatory profiles. This observation aligns with the discovery that RAGE$^{-/-}$ mice had lower plasma levels of FFA, glycerol, IL-6, and leptin than the WT mice. All observations related to the possibility that RAGE-mediated adipose tissue inflammation and insulin signaling are significant events, deciphering insulin-signaling and RAGE-mediated adipose tissue inflammation as significant processes in the emergence of obesity-associated insulin resistance.

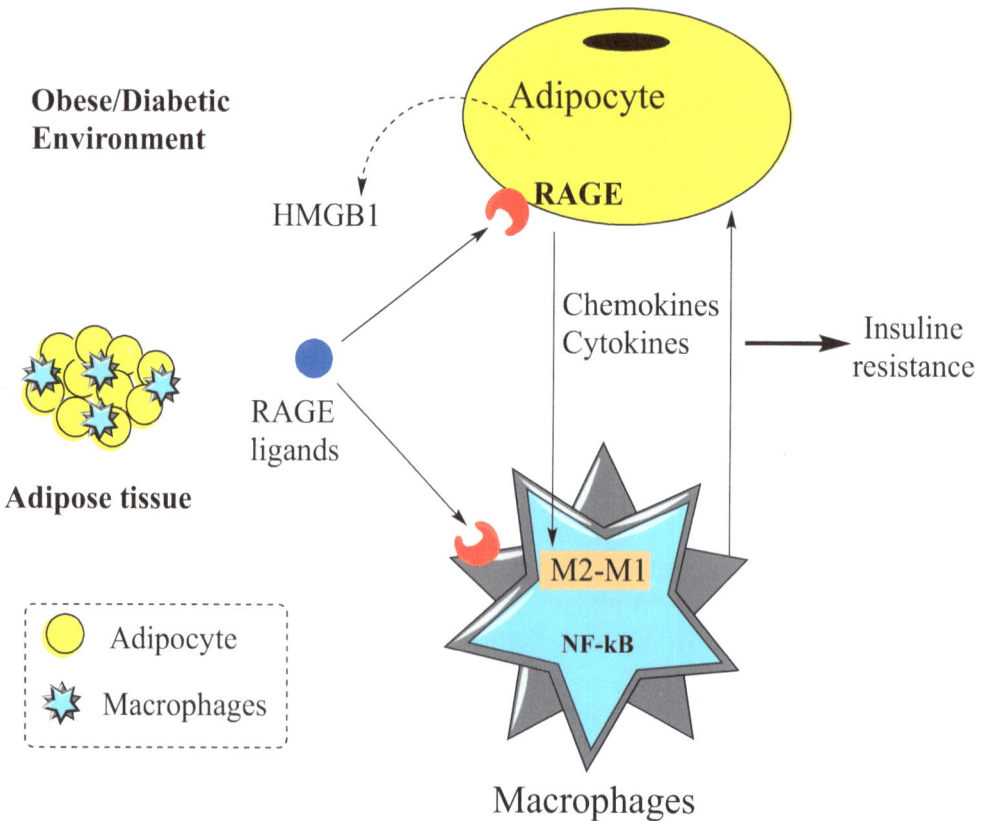

Fig. (2). RAGE signaling in adipose tissue macrophages.

ROLE OF RAGE IN DIABETIC NEUROPATHY

The early stages of diabetic neuropathy (DN) are characterized by positive symptoms such as pain, hypersensitivity, tingling, cramps, and cold feet, while the later consequences are primarily described by a loss of function such as delayed wound healing and loss of sensory perception. These symptoms vary depending on the stage of the disease. Lowering blood glucose levels is insufficient to prevent and reverse neuropathy in patients with T2D, and elevated blood glucose alone cannot account for the DN onset and progression [127]. Clinical research, however, has shown that bringing blood glucose levels back to normal does not stop neuropathic symptoms in T2D patients and only partially prevents DN onset in T1D patients [128, 129]. Furthermore, reduction of blood glucose does not moderate DN symptoms in either type 1 or 2 diabetic patients [128]. Pre-diabetic patient trials provide conclusive evidence that hyperglycemia on its own does not produce diabetic neuropathic pain during T2D. A noteworthy effort herein by *Bongaerts and associates* revealed peripheral neuropathy as more common in non-diabetic people with suppressed glucose tolerance and impaired fasting glucose (IFG) to a similar degree as in diabetic sufferers [130]. Peripheral neuropathy symptoms were revealed as more common in pre-diabetic patients than in people with normal glucose tolerance, according to a study by *Lu and groupmates* [131]. The neuropathic symptoms in pre-diabetic patients were revealed as independently correlated with reduced glucose tolerance in both studies. The findings thereby elucidate reduced glucose tolerance as a more reliable marker of diabetic complications than IFG or glycated hemoglobin (HbA1c), the clinical measure used to manage glycemic control in diabetes. Even on normalized individual glucose extents, diabetic individuals are more likely to have accumulated triose phosphates (fructose-1, 6-bisphosphate, glyceraldehyde-3-phosphate, and dihydroxyacetone phosphate) [132]. The concentration of triose phosphates in diabetes individuals is positively linked with the MGO concentration, a reactive metabolite formed when these intermediates are not phosphorylated.

MGO has been demonstrated to play a role in the acquisition of function by altering the DN of neuronal ion channels that are implicated in chemosensing and the generation of action potentials in nociceptive nerve terminals. A unifying mechanism for the DN onset and progression is provided by the direct and indirect effects of MGO on nerves or neuronal microvascular networks. In the sciatic nerve and neuronal tissues, the effect of increased metabolic flux and decreased detoxification by Glyoxalase I in combination leads to MGO accumulation during diabetes. The alteration of membrane ion channels, namely $Na_v1.8$ and TRPA1 results in increased neural excitability and hyperalgesic symptoms. Persistent hyperexcitability may eventually result in neuronal damage, excitotoxicity, and Ca^{+2} overload. The direct effect of MGO on mitochondria is the accumulation of ROS, which further promotes cytotoxicity. The pro-inflammatory immune response is triggered by the release of cytokines and damaged-associated molecular pattern (DAMPs) molecules such as S100 and HMGB1 as shown in (Fig. **3**).

The RAGE-enhanced pro-inflammatory response can cause delayed and dysregulated regeneration of the damaged nerve, axonal shrinkage, dwindling and demyelination and scarring, and ultimately complete loss of nerve fiber integrity and sensory perception [133]. This is especially true in diabetic subjects with RAGE expression. A noteworthy effort from *Juranek and associates* demonstrated that RAGE induced inflammation, preventing the recovery of sensory and motor function on sciatic nerve compression injury in diabetics [134]. Another pertinent attempt by *Thakur and colleagues* [135] demonstrated the amelioration of increased HMGB1, TLR4, and Nod-like receptor protein 3 (NLRP3) levels in dorsal root ganglia of diabetic animals with painful DN by treatment with the HMGB1 inhibitor.

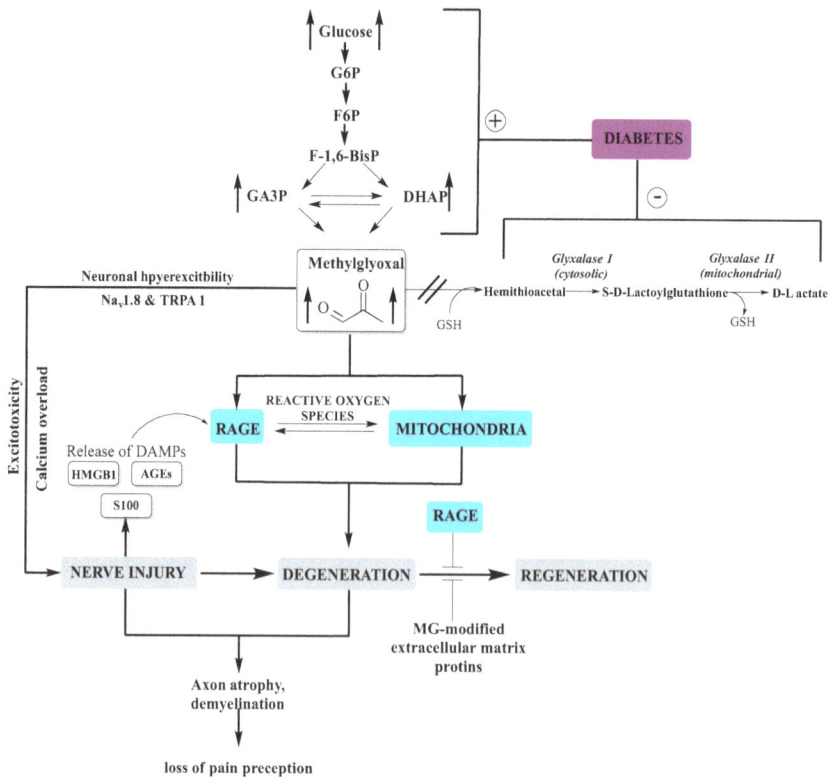

Fig. (3). Schematic representation of mechanism for the development and progression of Diabetic nephropathy. *Glucose- 6-phosphate (G6P), Fructose-1, 6-bisphosphate (F-1,6-biP), Glyceraldehyde-- -phosphate (GA3P), Dihydroxyacetone phosphate (DHAP), Ribose-5- phosphate (R5P), reduced glutathione (GSH).*

In a significant study, *Osonoi and accomplices* [136] reported that the pro-inflammatory macrophages infiltrated the sciatic nerve of T1D mice decreased retrograde axonal transport (RAT), atrophied the neurons in the dorsal root ganglia, and reduced insulin sensitivity. The RAGE-null mice restored the ganglion cells, ensuring intact insulin sensitivity with increased anti-inflammatory macrophage populations besides RAT. The peripheral nerve deficiencies were prevented when bone marrow from RAGE-null mice was transplanted into diabetic mice, indicating that RAGE plays a significant role in determining the polarity of macrophages in diabetic polyneuropathy [136]. However, new and more potent therapeutic methods for DN treatment may be optimized by focusing on the MGO buildup and forbidden RAGE interactions.

ROLE OF RAGE IN DIABETIC RETINOPATHY

Diabetic retinopathy (DR), a multifactorial condition largely associated with retinal inflammation and vascular abnormalities, is the primary cause of visual loss or impairment, in older and working-age adults worldwide [137]. RAGE is found in retinal pathological response cells, such as the pre-retinal and neovascular membranes, and is expressed in glial cells, Müller cells, nerve cells, vascular ECs, pericytes, and RPE cells [138]. Numerous ligands interact with RAGE to trigger VEGF release and activate NF-κB, leading to intracellular inflammation and oxidative stress which is involved in the DR pathophysiology of retinal vascular endothelial damage, promoting neo-vascularization [139, 140]. In patients with DR, retinal inflammation accelerates the development of neo-vascularization and vascular growth factors, chemokines, and cytokines [141]. Several clinical investigations have demonstrated elevated levels of inflammatory markers, such as CRP, IL-6, TNF-α, NF-κB, VCAM-1, and ICAM-1, in serum, vitreous, and retinal tissues of DR patients. Apart from AGEs, various additional ligands can also bind to RAGE, including HMGB-1, Mac-1, β-amyloid peptide, and S100 family members [142].

The S100 protein family binds to RAGE and triggers NF-κB, which aggravates the release of pro-inflammatory cytokines. These cytokines induce the translocation of neutrophils, monocytes, and macrophages, culminating in neuroinflammation and glial cell activation. In the retina, S100B expressed by Müller glia and astrocytes, interacts with RAGE to activate PI3K/AKT, NF-κB, p38MAPK, and JNK, amongst other inflammatory cytokines [143, 144]. The binding of RAGE with β2-integrin Mac-1 (CD11a/CD18) on leukocytes mediates leukocyte adhesion to ECs while that with S100-B enhances the RAGE binding with Mac-1. The elevated RAGE expression and the process of leukocyte recruitment by the RAGE-Mac-1 interaction are further relevant to pathophysiological diseases like diabetes, cancer, atherosclerosis, and chronic immunological responses (Fig. **4**). Therefore, inhibition of the S100 protein family binding to the RAGE and antagonizing the RAGE and β2-integrin Mac-1 interaction could be a potential novel target for delaying disease progression.

The stimulation of AGEs-RAGE signaling induces Müller cell activation, promoting VEGF expression that leads to retinal vascular endothelial damage and neovascularization [150]. In DN, PI3K/AKT signaling is essential for neovascularization. In addition to working with VEGF to support retinal ECs migration and proliferation, activation of PI3K/AKT signaling by AGEs in conjunction with RAGE may prolong endothelial cell survival and ultimately aid in neovascularization [145, 146]. Many downstream target proteins, including mTOR phosphorylation for cell growth, proliferation, and adverse expression suppression (resulting in apoptosis), can be mediated by activated AKT [147, 148]. According to *Li and groupmates* [149], Src kinase is activated by RAGE-AGE interaction wherein the downstream signal transduction pathways include PI3K, FAK, and MAPK. By increasing ERK phosphorylation in ECs, activated Src not only stimulates angiogenesis *via* downstream signaling pathways, but also conveys signals to moesin, VE-cadherin, and FAK, damaging the endothelial cell barrier and increasing vascular permeability [154]. The development of diabetic kidney disease is influenced by the toxicity of glycolaldehyde-derived angiogenesis (AGE2) and glycolaldehyde-derived AGE3 (likely to result in excessive angiogenesis). In a noteworthy effort, *Yamazaki and colleagues* discovered that AGE3 activated the mTOR signaling pathway in ECs by binding to scavenger receptors (CD163, CD36, and LOX-1). RAGE stimulates scavenger receptor expression, aiding the toxic AGEs for angiogenesis promotion [150].

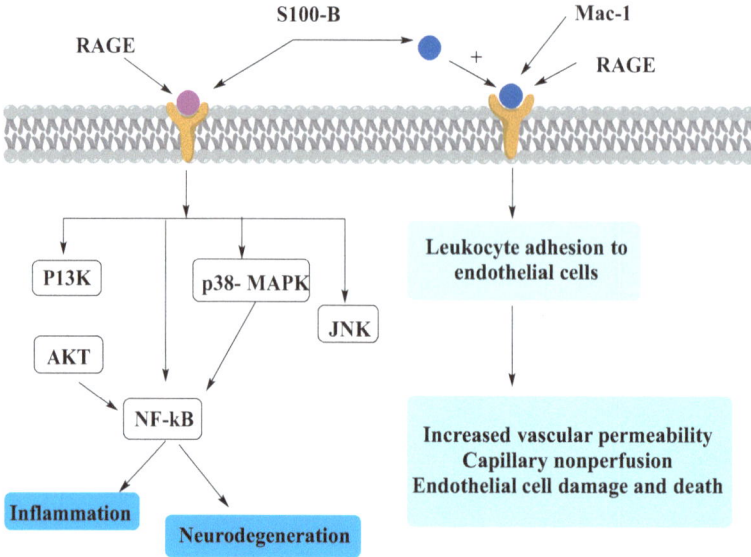

Fig. (4). Typical interaction scenario of the S100 protein family with RAGE which activates NF-κB. S100-B enhances the binding of RAGE to Mac-1 and further mediates leukocyte adhesion to endothelial cells (ECs), increasing vascular permeability, endothelial cell damage, and death.

HMGB1 functions as a pro-angiogenic factor and a pro-inflammatory cytokine. The first to show elevated levels of RAGE and its ligand, HMGB1 in the vitreous cavity and pre-retinal membranes of PDR patients was *Pachydaki and groupmates* [151]. Additionally, *Asrar and colleagues* [152] showed a strong correlation between the extent of neovascularization and the HMGB1 levels in the vitreous humor of PDR patients [152, 153]. In a subsequent effort, *Shen and associates* reported a prevalence of pro-inflammatory and angiogenic states in PDR, as demonstrated by higher serum and vitreous levels of HMGB-1, RAGE, VEGF, and IL-1β. HMGB-1 could be a novel treatment target to impede the PDR advancement, through its cytokine-bridging actions of inflammatory response and angiogenesis in DR [154]. Signaling cascades mediated by HMGB1/RAGE comprise p38 MAPK, NF-κB, and ERK1/2 pathways. The pro-inflammatory and pro-angiogenic effects of retinal ECs are facilitated by the activation of the HMGB1/RAGE axis and downstream pathways. This also activates NF-κB, up-regulating the expression of growth factors, chemokines, and pro-inflammatory cytokines besides increasing the ICAM-1 and VCAM-1 expression on the ECs surface. RAGE also contributes to the activation of Mac-1 on neutrophils. NF-κB is reactivated by Mac-1 interaction with ICAM-1 [155, 156]. According to *Orlova and associates*, HMGB1 stimulated NF-κB activation in neutrophils, dose-dependently increasing the Mac-1 and RAGE interaction [157]. To activate MAPK, (which includes ERK1/2, p38MAPK, JNK, and finally the transcription factor NF-κB), HMGB1 binds to RAGE. Through its interaction with TLR4, HMGB1 activates IKK-β and IKK-α, promotes activated NF-κB nuclear translocation, and increases the expression of pro-inflammatory and pro-angiogenic genes, causing the hematopoietic and ECs to release a variety of pro-inflammatory and pro-angiogenic cytokines. Activation of EC, macrophages, EPC, and mesangioblast fibroblastic factors is associated with new blood vessel formation [158 - 160] (Fig. **5**).

Thus, it appears that HMGB1/RAGE/TLR4 contributes to neovascularization and inflammatory damage in DR, suggesting a possible impairment of its effects as a new treatment option. RAGE inhibitors can block HMGB1-induced neovascularization *in vivo*, and this RAGE suppression reduces the inflammatory response in cells grown in high glucose [161]. Inhibiting RAGE has been shown to alleviate diabetes-associated vascular permeability damage using RAGE fusion proteins in streptozotocin-induced diabetic animals [162, 163]. Therefore, factors interrupting RAGE-ligand interactions and activating downstream signaling pathways provide novel insights for the prevention and treatment of diabetic retinopathy [164].

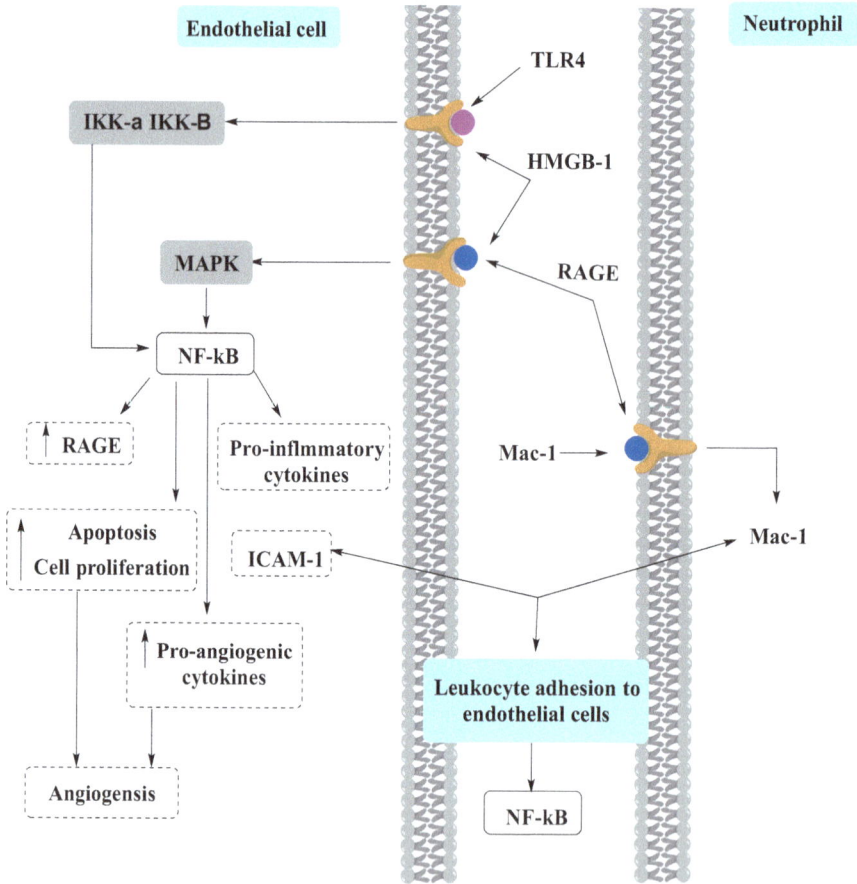

Fig. (5).   Typical interaction of RAGE with HMGB1 to promote inflammatory factors release and neovascularization. RAGE contributes to activating Mac-1 on neutrophils. Mac-1 and ICAM-1 combine to promote leukocyte adhesion with endothelial cells (ECs), further reactivating the NF-κB.

ROLE OF RAGE IN DIABETIC NEPHROPATHY

Nephropathy is the deterioration of kidney function. End-stage renal disease, or ESRD, is the term used for the latter stage of nephropathy. The most frequent cause of ESRD, according to the CDC, is diabetes. About 26 million Americans were estimated to have diabetes in 2011, and over 200,000 of these had ESRD and were either receiving chronic renal dialysis or had received a kidney transplant. Diabetic nephropathy can result from either T1D or T2D, albeit type 1 is more likely to cause ESRD. The hallmark of early-stage diabetic nephropathy is microalbuminuria, a typical signifier of glomerular filtration barrier damage rather than changes in glomerular pressure or filtration rate alone. This damage is caused by ultrastructural changes in podocytes and glomerular ECs. Within ten years, 20% of individuals experience end-stage renal disease, and almost 20% acquire nephropathy due to microalbuminuria [165]. On one hand, there is currently no treatment for T1DM and about 20% of patients get end-stage renal failure in less than ten years, and 75% of patients exhibit so, in less than twenty years. However, microalbuminuria and nephropathy are evident in T2DM patients shortly after diabetes mellitus diagnosis. Hyper aminoacidemia, a glomerular hyperfiltration promoter, and hyperglycemia are the metabolic modifications that change renal hemodynamics, facilitating fibrosis and inflammation in diabetes' initial stage [165, 166].

In a noted effort, *Wendt and colleagues* [167] discovered that diabetic nephropathy also involves elevated levels of RAGE ligands other than AGEs, such as S100/calgranulins, playing a key role in inflammatory cell infiltration. The regulation of inflammatory responses is also demonstrated by HMGB1 passive release from damaged cells or activated immune cells on infection, injury, and sterile inflammation [168, 169]. This release can also be mediated by inflammasomes, large caspase-1-activating protein complexes, such as NLRP3 which can also function as a cytokine [170, 171]. According to *Chen and associates* [172], there was a notable rise in the blood HMGB1 level of T2D patients than healthy individuals. This increase was positively correlated with TNF-α and IL-6 (inflammatory cytokines) concentrations. These results imply that the pathophysiology of diabetic nephropathy depends on the control of ligand-RAGE interactions.

In the past, high sRAGE concentrations have been observed in people with impaired renal function, despite the varied underlying reasons [173], in people with various forms of diabetes [174], T1D [175], T2D [176, 177], and in people without diabetes [178, 179]. According to *Wadén and colleagues* [180], individuals with DN had the highest baseline sRAGE concentrations than those with microalbuminuria or a normal albumin excretion rate (AER). In the competing risks analysis, baseline sRAGE expression was linked to the development of ESRD from macroalbuminuria.

Furthermore, renal failure and urine albumin excretion are caused by T1D induction in C57BL/6J mice using streptozotocin. Additionally, renal enlargement, elevated glomerular mesangial matrix, progressive glomerulosclerosis, and increased VEGF, and TGF-β expressions were observed. Surprisingly, RAGE knockout mice showed substantial inhibition of these alterations [181]. In the noted study of *Jensen and colleagues* [182], the treatment of RAGE-neutralizing antibodies to db./db. mice resulted in a decrease in urine albumin excretion, besides increasing mesangial regions and thickening of the basement membrane, all of which are indicative of early diabetic nephropathy. Likewise, another study by *Reinger and colleagues* [181] demonstrated that combining RAGE knockouts with OVE26 mice provided a spontaneous T1D animal. The changes herein enhanced glomerulosclerosis thickened the glomerular basement membrane, decreased podocyte foot process effacement and maintained renal function. Thus, RAGE plays a major role in the development of diabetic nephropathy, and inhibiting RAGE may help diabetic patients bypass kidney damage.

CONCLUSION

In summary, a key role for RAGE in the etiology of cardiovascular and diabetes problems may prevail. The AGEs are one of the many ligands that RAGE, a member of the immunoglobulin superfamily, interacts with. AGEs are common in hyperglycemic circumstances found in diabetes. An increased risk of diabetes and the incidence of CVDs can result from the AGE-RAGE axis' role as a trigger for insulin resistance and altered leptin secretion. Thus, it will be challenging to identify the exact mechanism of action that could inhibit these pathological responses.

Notably, research on human subjects emphasizes that measuring sRAGE and esRAGE levels may give rise to a biomarker to monitor the activity of the RAGE signaling. The levels of sRAGEs are deciphered as lower in chronic illness without exacerbation than in control persons. Blocking the AGEs-RAGE interaction, either using sRAGE or other inhibitors, has been suggested as a potential therapeutic strategy to prevent or reduce manifestations related to CVDs and diabetes. In the recent past, the development of RAGE-targeting agents, such as aptamers, small molecules, and RAGE antagonist peptides has made it possible to inhibit ligand-RAGE interaction, thereby reducing cardiovascular and diabetic complications.

REFERENCES

[1]   Egaña-Gorroño L, López-Díez R, Yepuri G, *et al.* Receptor for Advanced Glycation End Products (RAGE) and Mechanisms and Therapeutic Opportunities in Diabetes and Cardiovascular Disease: Insights From Human Subjects and Animal Models. Front Cardiovasc Med 2020; 7: 37.
       [http://dx.doi.org/10.3389/fcvm.2020.00037] [PMID: 32211423]

[2]   Miller RG, Costacou T, Orchard TJ. Risk factor modeling for cardiovascular disease in type 1 diabetes in the pittsburgh epidemiology of diabetes complications (EDC) study: a comparison with the diabetes control and complications trial/epidemiology of diabetes interventions and complications study (DCCT/EDIC). Diabetes 2019; 68(2): 409-19.
       [http://dx.doi.org/10.2337/db18-0515] [PMID: 30409781]

[3]   Paneni F, Beckman JA, Creager MA, Cosentino F. Diabetes and vascular disease: pathophysiology, clinical consequences, and medical therapy: part I. Eur Heart J 2013; 34(31): 2436-43.

[http://dx.doi.org/10.1093/eurheartj/eht149] [PMID: 23641007]

[4]   Livingstone SJ, Looker HC, Hothersall EJ, *et al.* Risk of cardiovascular disease and total mortality in adults with type 1 diabetes: Scottish registry linkage study. PLoS Med 2012; 9(10): e1001321.
[http://dx.doi.org/10.1371/journal.pmed.1001321] [PMID: 23055834]

[5]   Negre-Salvayre A, Salvayre R, Augé N, Pamplona R, Portero-Otín M. Hyperglycemia and glycation in diabetic complications. Antioxid Redox Signal 2009; 11(12): 3071-109.
[http://dx.doi.org/10.1089/ars.2009.2484] [PMID: 19489690]

[6]   Khan N, Bakshi KS, Jaggi AS, Singh N. Ameliorative potential of spironolactone in diabetes induced hyperalgesia in mice. Yakugaku Zasshi 2009; 129(5): 593-9.
[http://dx.doi.org/10.1248/yakushi.129.593] [PMID: 19420890]

[7]   Helou C, Marier D, Jacolot P, *et al.* Microorganisms and Maillard reaction products: a review of the literature and recent findings. Amino Acids 2014; 46(2): 267-77.
[http://dx.doi.org/10.1007/s00726-013-1496-y] [PMID: 23588491]

[8]   Alikhani M, Alikhani Z, Boyd C, *et al.* Advanced glycation end products stimulate osteoblast apoptosis *via* the MAP kinase and cytosolic apoptotic pathways. Bone 2007; 40(2): 345-53.
[http://dx.doi.org/10.1016/j.bone.2006.09.011] [PMID: 17064973]

[9]   Saito M, Fujii K, Mori Y, Marumo K. Role of collagen enzymatic and glycation induced cross-links as a determinant of bone quality in spontaneously diabetic WBN/Kob rats. Osteoporos Int 2006; 17(10): 1514-23.
[http://dx.doi.org/10.1007/s00198-006-0155-5] [PMID: 16770520]

[10] Pennells L, Kaptoge S, Wood A, *et al.* Equalization of four cardiovascular risk algorithms after systematic recalibration: individual-participant meta-analysis of 86 prospective studies. Eur Heart J 2019; 40(7): 621-31.
[http://dx.doi.org/10.1093/eurheartj/ehy653] [PMID: 30476079]

[11] Brownlee M. Biochemistry and molecular cell biology of diabetic complications. Nature 2001; 414(6865): 813-20.
[http://dx.doi.org/10.1038/414813a] [PMID: 11742414]

[12] Rabbani N, Thornalley PJ. Glyoxalase in diabetes, obesity and related disorders. Semin Cell Dev Biol 2011; 22(3): 309-17.
[http://dx.doi.org/10.1016/j.semcdb.2011.02.015] [PMID: 21335095]

[13] Dobler D, Ahmed N, Song L, Eboigbodin KE, Thornalley PJ. Increased dicarbonyl metabolism in endothelial cells in hyperglycemia induces anoikis and impairs angiogenesis by RGD and GFOGER motif modification. Diabetes 2006; 55(7): 1961-9.
[http://dx.doi.org/10.2337/db05-1634] [PMID: 16804064]

[14] Li H, Nakamura S, Miyazaki S, *et al.* N2-carboxyethyl-2′-deoxyguanosine, a DNA glycation marker, in kidneys and aortas of diabetic and uremic patients. Kidney Int 2006; 69(2): 388-92.
[http://dx.doi.org/10.1038/sj.ki.5000064] [PMID: 16408131]

[15] Thornalley PJ. The enzymatic defence against glycation in health, disease and therapeutics: a symposium to examine the concept. Biochem Soc Trans 2003; 31(6): 1341-2.
[http://dx.doi.org/10.1042/bst0311341] [PMID: 14641059]

[16] Luthra M, Balasubramanian D. Nonenzymatic glycation alters protein structure and stability. A study of two eye lens crystallins. J Biol Chem 1993; 268(24): 18119-27.
[http://dx.doi.org/10.1016/S0021-9258(17)46819-0] [PMID: 8349689]

[17] Bailey AJ, Paul RG, Knott L. Mechanisms of maturation and ageing of collagen. Mech Ageing Dev 1998; 106(1-2): 1-56.
[http://dx.doi.org/10.1016/S0047-6374(98)00119-5] [PMID: 9883973]

[18] Ulrich P, Cerami A. Protein glycation, diabetes, and aging. Recent Prog Horm Res 2001; 56(1): 1-22.
[http://dx.doi.org/10.1210/rp.56.1.1] [PMID: 11237208]

[19] Crabbe MJC, Cooper LR, Corne DW. Use of essential and molecular dynamics to study γB-crystallin unfolding after non-enzymic post-translational modifications. Comput Biol Chem 2003; 27(4-5): 507-10.
[http://dx.doi.org/10.1016/S1476-9271(03)00048-3] [PMID: 14642758]

[20] Sell DR, Monnier VM. Ornithine is a novel amino acid and a marker of arginine damage by oxoaldehydes in senescent proteins. Ann N Y Acad Sci 2005; 1043(1): 118-28.
[http://dx.doi.org/10.1196/annals.1333.015] [PMID: 16037230]

[21] Hamelin M, Mary J, Vostry M, Friguet B, Bakala H. Glycation damage targets glutamate dehydrogenase in the rat liver mitochondrial matrix during aging. FEBS J 2007; 274(22): 5949-61.
[http://dx.doi.org/10.1111/j.1742-4658.2007.06118.x] [PMID: 17949437]

[22] Ravandi A, Kuksis A, Marai L, *et al.* Isolation and identification of glycated aminophospholipids from red cells and plasma of diabetic blood. FEBS Lett 1996; 381(1-2): 77-81.
[http://dx.doi.org/10.1016/0014-5793(96)00064-6] [PMID: 8641444]

[23] Requena JR, Ahmed MU, Fountain CW, *et al.* Carboxymethylethanolamine, a biomarker of phospholipid modification during the maillard reaction *in vivo*. J Biol Chem 1997; 272(28): 17473-9.
[http://dx.doi.org/10.1074/jbc.272.28.17473] [PMID: 9211892]

[24] Pamplona R, Bellmunt MJ, Portero M, Riba D, Prat J. Chromatographic evidence for amadori product formation in rat liver aminophospholipids. Life Sci 1995; 57(9): 873-9.
[http://dx.doi.org/10.1016/0024-3205(95)02020-J] [PMID: 7630316]

[25] Bucala R, Makita Z, Koschinsky T, Cerami A, Vlassara H. Lipid advanced glycosylation: pathway for lipid oxidation *in vivo*. Proc Natl Acad Sci USA 1993; 90(14): 6434-8.
[http://dx.doi.org/10.1073/pnas.90.14.6434] [PMID: 8341651]

[26] Breitling-Utzmann CM, Unger A, Friedl DA, Lederer MO. Identification and quantification of phosphatidylethanolamine-derived glucosylamines and aminoketoses from human erythrocytes--influence of glycation products on lipid peroxidation. Arch Biochem Biophys 2001; 391(2): 245-54.
[http://dx.doi.org/10.1006/abbi.2001.2406] [PMID: 11437356]

[27] Oak JH, Nakagawa K, Miyazawa T. Synthetically prepared Amadori-glycated phosphatidylethanolamine can trigger lipid peroxidation *via* free radical reactions. FEBS Lett 2000; 481(1): 26-30.
[http://dx.doi.org/10.1016/S0014-5793(00)01966-9] [PMID: 10984609]

[28] Chen HJC, Chen YC. Analysis of glyoxal-induced DNA cross-links by capillary liquid chromatography nanospray ionization tandem mass spectrometry. Chem Res Toxicol 2009; 22(7): 1334-41.
[http://dx.doi.org/10.1021/tx900129e] [PMID: 19527002]

[29] Murata-Kamiya N, Kamiya H, Kaji H, Kasai H. Glyoxal, a major product of DNA oxidation, induces mutations at G:C sites on a shuttle vector plasmid replicated in mammalian cells. Nucleic Acids Res 1997; 25(10): 1897-902.
[http://dx.doi.org/10.1093/nar/25.10.1897] [PMID: 9115355]

[30] Kasai H, Iwamoto-Tanaka N, Fukada S. DNA modifications by the mutagen glyoxal: adduction to G and C, deamination of C and GC and GA cross-linking. Carcinogenesis 1998; 19(8): 1459-65.
[http://dx.doi.org/10.1093/carcin/19.8.1459] [PMID: 9744543]

[31] Pischetsrieder M, Seidel W, Münch G, Schinzel R. N(2)-(1-Carboxyethyl)deoxyguanosine, a nonenzymatic glycation adduct of DNA, induces single-strand breaks and increases mutation frequencies. Biochem Biophys Res Commun 1999; 264(2): 544-9.
[http://dx.doi.org/10.1006/bbrc.1999.1528] [PMID: 10529399]

[32] Wuenschell GE, Tamae D, Cercillieux A, Yamanaka R, Yu C, Termini J. Mutagenic potential of DNA glycation: miscoding by (R)- and (S)-N2-(1-carboxyethyl)-2'-deoxyguanosine. Biochemistry 2010; 49(9): 1814-21.

[http://dx.doi.org/10.1021/bi901924b] [PMID: 20143879]

[33]   Seidel W, Pischetsrieder M. DNA-glycation leads to depurination by the loss of N2-carboxyethylguanine *in vitro*. Cell Mol Biol 1998; 44(7): 1165-70.
[PMID: 9846899]

[34]   Murata-Kamiya N, Kamiya H. Methylglyoxal, an endogenous aldehyde, crosslinks DNA polymerase and the substrate DNA. Nucleic Acids Res 2001; 29(16): 3433-8.
[http://dx.doi.org/10.1093/nar/29.16.3433] [PMID: 11504881]

[35]   Breyer V, Frischmann M, Bidmon C, Schemm A, Schiebel K, Pischetsrieder M. Analysis and biological relevance of advanced glycation end-products of DNA in eukaryotic cells. FEBS J 2008; 275(5): 914-25.
[http://dx.doi.org/10.1111/j.1742-4658.2008.06255.x] [PMID: 18215162]

[36]   Yan SF, Ramasamy R, Naka Y, Schmidt AM. Glycation, Inflammation, and RAGE. Circ Res 2003; 93(12): 1159-69.
[http://dx.doi.org/10.1161/01.RES.0000103862.26506.3D] [PMID: 14670831]

[37]   Yan SF, Ramasamy R, Schmidt AM. Mechanisms of Disease: advanced glycation end-products and their receptor in inflammation and diabetes complications. Nat Clin Pract Endocrinol Metab 2008; 4(5): 285-93.
[http://dx.doi.org/10.1038/ncpendmet0786] [PMID: 18332897]

[38]   Cerami A. Aging of proteins and nucleic acids: what is the role of glucose? Trends Biochem Sci 1986; 11(8): 311-4.
[http://dx.doi.org/10.1016/0968-0004(86)90281-1]

[39]   Pongor S, Ulrich PC, Bencsath FA, Cerami A. Aging of proteins: isolation and identification of a fluorescent chromophore from the reaction of polypeptides with glucose. Proc Natl Acad Sci USA 1984; 81(9): 2684-8.
[http://dx.doi.org/10.1073/pnas.81.9.2684] [PMID: 6585821]

[40]   Monnier VM, Vishwanath V, Frank KE, Elmets CA, Dauchot P, Kohn RR. Relation between complications of type I diabetes mellitus and collagen-linked fluorescence. N Engl J Med 1986; 314(7): 403-8.
[http://dx.doi.org/10.1056/NEJM198602133140702] [PMID: 3945267]

[41]   Harding JJ. Nonenzymatic covalent posttranslational modification of proteins *in vivo*. Adv Protein Chem 1985; 37: 247-334.
[http://dx.doi.org/10.1016/S0065-3233(08)60066-2] [PMID: 3904349]

[42]   Wolff SP, Crabbe MJC, Thornalley PJ. The autoxidation of glyceraldehyde and other simple monosaccharides. Experientia 1984; 40(3): 244-6.
[http://dx.doi.org/10.1007/BF01947562]

[43]   Wolff SP, Dean RT. Glucose autoxidation and protein modification. The potential role of 'autoxidative glycosylation' in diabetes. Biochem J 1987; 245(1): 243-50.
[http://dx.doi.org/10.1042/bj2450243] [PMID: 3117042]

[44]   Wolff SP, Wang GM, Spector A. Pro-oxidant activation of ocular reductants. 1. Copper and riboflavin stimulate ascorbate oxidation causing lens epithelial cytotoxicity *in vitro*. Exp Eye Res 1987; 45(6): 777-89.
[http://dx.doi.org/10.1016/S0014-4835(87)80095-7] [PMID: 2828093]

[45]   Harman D. The aging process. Proc Natl Acad Sci USA 1981; 78(11): 7124-8.
[http://dx.doi.org/10.1073/pnas.78.11.7124] [PMID: 6947277]

[46]   Harman D. The free radical theory of aging: effect of age on serum copper levels. J Gerontol 1965; 20(2): 151-3.
[http://dx.doi.org/10.1093/geronj/20.2.151] [PMID: 14284786]

[47]   Noto R, Alicata R, Sfogliano L, Neri S, Bifarella M. A study of cupremia in a group of elderly

diabetics. Acta Diabetol Lat 1984; 21(1): 79-85.
[http://dx.doi.org/10.1007/BF02624767] [PMID: 6858545]

[48] Gutteridge JMC, Winyard PG, Blake DR, Lunec J, Brailsford S, Halliwell B. The behaviour of caeruloplasmin in stored human extracellular fluids in relation to ferroxidase II activity, lipid peroxidation and phenanthroline-detectable copper. Biochem J 1985; 230(2): 517-23.
[http://dx.doi.org/10.1042/bj2300517] [PMID: 4052055]

[49] Nath R, Srivastava SK, Singh K. Accumulation of copper and inhibition of lactate dehydrogenase activity in human senile cataractous lens. Indian J Exp Biol 1969; 7(1): 25-6.
[PMID: 5771164]

[50] McGahan MC, Bito LZ. The pathophysiology of the ocular microenvironment. I. Preliminary report on the possible involvement of copper in ocular inflammation. Curr Eye Res 1982-1983; 2(12): 883-5.
[http://dx.doi.org/10.3109/02713688209020026] [PMID: 7187644]

[51] Hunt JV, Dean RT, Wolff SP. Hydroxyl radical production and autoxidative glycosylation. Glucose autoxidation as the cause of protein damage in the experimental glycation model of diabetes mellitus and ageing. Biochem J 1988; 256(1): 205-12.
[http://dx.doi.org/10.1042/bj2560205] [PMID: 2851978]

[52] Brownlee M. Advanced protein glycosylation in diabetes and aging. Annu Rev Med 1995; 46: 223-34.
[http://dx.doi.org/10.1146/annurev.med.46.1.223] [PMID: 7598459]

[53] Baynes JW. Role of oxidative stress in development of complications in diabetes. Diabetes 1991; 40(4): 405-12.
[http://dx.doi.org/10.2337/diab.40.4.405] [PMID: 2010041]

[54] Yan SD, Yan SF, Chen X, *et al.* Non-enzymatically glycated tau in Alzheimer's disease induces neuronal oxidant stress resulting in cytokine gene expression and release of amyloid β-peptide. Nat Med 1995; 1(7): 693-9.
[http://dx.doi.org/10.1038/nm0795-693] [PMID: 7585153]

[55] Giacco F, Brownlee M. Oxidative stress and diabetic complications. Circ Res 2010; 107(9): 1058-70.
[http://dx.doi.org/10.1161/CIRCRESAHA.110.223545] [PMID: 21030723]

[56] Rafieian-Kopaei M, Setorki M, Doudi M, Baradaran A, Nasri H. Atherosclerosis: process, indicators, risk factors and new hopes. Int J Prev Med 2014; 5(8): 927-46.
[PMID: 25489440]

[57] Piedrahita JA, Zhang SH, Hagaman JR, Oliver PM, Maeda N. Generation of mice carrying a mutant apolipoprotein E gene inactivated by gene targeting in embryonic stem cells. Proc Natl Acad Sci USA 1992; 89(10): 4471-5.
[http://dx.doi.org/10.1073/pnas.89.10.4471] [PMID: 1584779]

[58] Bucciarelli LG, Wendt T, Qu W, *et al.* RAGE blockade stabilizes established atherosclerosis in diabetic apolipoprotein E-null mice. Circulation 2002; 106(22): 2827-35.
[http://dx.doi.org/10.1161/01.CIR.0000039325.03698.36] [PMID: 12451010]

[59] Harja E, Bu D, Hudson BI, *et al.* Vascular and inflammatory stresses mediate atherosclerosis *via* RAGE and its ligands in apoE–/– mice. J Clin Invest 2008; 118(1): 183-94.
[http://dx.doi.org/10.1172/JCI32703] [PMID: 18079965]

[60] Bu D, Rai V, Shen X, *et al.* Activation of the ROCK1 branch of the transforming growth factor-beta pathway contributes to RAGE-dependent acceleration of atherosclerosis in diabetic ApoE-null mice. Circ Res 2010; 106(6): 1040-51.
[http://dx.doi.org/10.1161/CIRCRESAHA.109.201103] [PMID: 20133903]

[61] Koulis C, Kanellakis P, Pickering RJ, *et al.* Role of bone-marrow- and non-bone-marrow-derived receptor for advanced glycation end-products (RAGE) in a mouse model of diabetes-associated atherosclerosis. Clin Sci (Lond) 2014; 127(7): 485-97.
[http://dx.doi.org/10.1042/CS20140045] [PMID: 24724734]

[62]   Daffu G, Shen X, Senatus L, *et al.* RAGE Suppresses ABCG1-mediated macrophage cholesterol efflux in diabetes. Diabetes 2015; 64(12): 4046-60.
[http://dx.doi.org/10.2337/db15-0575] [PMID: 26253613]

[63]   Heier M, Margeirsdottir HD, Gaarder M, *et al.* Soluble RAGE and atherosclerosis in youth with type 1 diabetes: a 5-year follow-up study. Cardiovasc Diabetol 2015; 14(1): 126.
[http://dx.doi.org/10.1186/s12933-015-0292-2] [PMID: 26408307]

[64]   Di Pino A, Urbano F, Scicali R, *et al.* 1 h postload glycemia is associated with low endogenous secretory receptor for advanced glycation end product levels and early markers of cardiovascular disease. Cells 2019; 8(8): 910.
[http://dx.doi.org/10.3390/cells8080910] [PMID: 31426413]

[65]   Di Pino A, Urbano F, Zagami RM, *et al.* Low endogenous secretory receptor for advanced glycation end-products levels are associated with inflammation and carotid atherosclerosis in prediabetes. J Clin Endocrinol Metab 2016; 101(4): 1701-9.
[http://dx.doi.org/10.1210/jc.2015-4069] [PMID: 26885882]

[66]   Reichert S, Triebert U, Santos AN, *et al.* Soluble form of receptor for advanced glycation end products and incidence of new cardiovascular events among patients with cardiovascular disease. Atherosclerosis 2017; 266: 234-9.
[http://dx.doi.org/10.1016/j.atherosclerosis.2017.08.015] [PMID: 28864204]

[67]   Park L, Raman KG, Lee KJ, *et al.* Suppression of accelerated diabetic atherosclerosis by the soluble receptor for advanced glycation endproducts. Nat Med 1998; 4(9): 1025-31.
[http://dx.doi.org/10.1038/2012] [PMID: 9734395]

[68]   Sun L, Ishida T, Yasuda T, *et al.* RAGE mediates oxidized LDL-induced pro-inflammatory effects and atherosclerosis in non-diabetic LDL receptor-deficient mice. Cardiovasc Res 2008; 82(2): 371-81.
[http://dx.doi.org/10.1093/cvr/cvp036] [PMID: 19176597]

[69]   Yang S, Zhu L, Han R, Sun L, Li J, Dou J. Pathophysiology of peripheral arterial disease in diabetes mellitus. J Diabetes 2017; 9(2): 133-40.
[http://dx.doi.org/10.1111/1753-0407.12474] [PMID: 27556728]

[70]   Shammas AN, Jeon-Slaughter H, Tsai S, *et al.* Major limb outcomes following lower extremity endovascular revascularization in patients with and without diabetes mellitus. J Endovasc Ther 2017; 24(3): 376-82.
[http://dx.doi.org/10.1177/1526602817705135] [PMID: 28440113]

[71]   Yamagishi S, Matsui T. Role of ligands of receptor for advanced glycation end products (RAGE) in peripheral artery disease. Rejuvenation Res 2018; 21(5): 456-63.
[http://dx.doi.org/10.1089/rej.2017.2025] [PMID: 29644926]

[72]   Falcone C, Bozzini S, Guasti L, *et al.* Soluble RAGE plasma levels in patients with coronary artery disease and peripheral artery disease. ScientificWorldJournal 2013; 2013(1): 584504.
[http://dx.doi.org/10.1155/2013/584504] [PMID: 24228009]

[73]   Malmstedt J, Kärvestedt L, Swedenborg J, Brismar K. The receptor for advanced glycation end products and risk of peripheral arterial disease, amputation or death in type 2 diabetes: a population-based cohort study. Cardiovasc Diabetol 2015; 14(1): 93.
[http://dx.doi.org/10.1186/s12933-015-0257-5] [PMID: 26216409]

[74]   Malmstedt J, Frebelius S, Lengquist M, Jörneskog G, Wang J, Swedenborg J. The receptor for advanced glycation end products (Rage) and its ligands in plasma and infrainguinal bypass vein. Eur J Vasc Endovasc Surg 2016; 51(4): 579-86.
[http://dx.doi.org/10.1016/j.ejvs.2015.12.047] [PMID: 26905625]

[75]   Tsao CF, Huang WT, Liu TT, *et al.* Expression of high-mobility group box protein 1 in diabetic foot atherogenesis. Genet Mol Res 2015; 14(2): 4521-31.
[http://dx.doi.org/10.4238/2015.May.4.10] [PMID: 25966225]

[76]   Oozawa S, Sano S, Nishibori M. Usefulness of high mobility group box 1 protein as a plasma biomarker in patient with peripheral artery disease. Acta Med Okayama 2014; 68(3): 157-62.
[http://dx.doi.org/10.18926/AMO/52656] [PMID: 24942794]

[77]   Konopka CJ, Wozniak M, Hedhli J, *et al.* Multimodal imaging of the receptor for advanced glycation end-products with molecularly targeted nanoparticles. Theranostics 2018; 8(18): 5012-24.
[http://dx.doi.org/10.7150/thno.24791] [PMID: 30429883]

[78]   Tekabe Y, Kollaros M, Li C, Zhang G, Schmidt AM, Johnson L. Imaging receptor for advanced glycation end product expression in mouse model of hind limb ischemia. EJNMMI Res 2013; 3(1): 37.
[http://dx.doi.org/10.1186/2191-219X-3-37] [PMID: 23663412]

[79]   López-Díez R, Shen X, Daffu G, *et al.* Ager deletion enhances ischemic muscle inflammation, angiogenesis, and blood flow recovery in diabetic mice. Arterioscler Thromb Vasc Biol 2017; 37(8): 1536-47.
[http://dx.doi.org/10.1161/ATVBAHA.117.309714] [PMID: 28642238]

[80]   Camm AJ, Lip GYH, De Caterina R, *et al.* 2012 focused update of the ESC Guidelines for the management of atrial fibrillation. Eur Heart J 2012; 33(21): 2719-47.
[http://dx.doi.org/10.1093/eurheartj/ehs253] [PMID: 22922413]

[81]   Yamagishi S, Sotokawauchi A, Matsui T. Pathological role of advanced glycation end products (AGEs) and their receptor axis in atrial fibrillation. Mini Rev Med Chem 2019; 19(13): 1040-8.
[http://dx.doi.org/10.2174/1389557519666190311140737] [PMID: 30854960]

[82]   Raposeiras-Roubín S, Rodiño-Janeiro BK, Grigorian-Shamagian L, *et al.* Evidence for a role of advanced glycation end products in atrial fibrillation. Int J Cardiol 2012; 157(3): 397-402.
[http://dx.doi.org/10.1016/j.ijcard.2011.05.072] [PMID: 21652096]

[83]   Lancefield TF, Patel SK, Freeman M, *et al.* The receptor for advanced glycation end products (RAGE) is associated with persistent atrial fibrillation. PLoS One 2016; 11(9): e0161715.
[http://dx.doi.org/10.1371/journal.pone.0161715] [PMID: 27627677]

[84]   Yang PS, Kim T, Uhm JS, *et al.* High plasma level of soluble RAGE is independently associated with a low recurrence of atrial fibrillation after catheter ablation in diabetic patient. Europace 2016; 18(11): 1711-8.
[http://dx.doi.org/10.1093/europace/euv449] [PMID: 26838688]

[85]   Al Rifai M, Schneider ALC, Alonso A, *et al.* sRAGE, inflammation, and risk of atrial fibrillation: results from the Atherosclerosis Risk in Communities (ARIC) Study. J Diabetes Complications 2015; 29(2): 180-5.
[http://dx.doi.org/10.1016/j.jdiacomp.2014.11.008] [PMID: 25499973]

[86]   Manganelli V, Truglia S, Capozzi A, *et al.* Alarmin HMGB1 and soluble RAGE as new tools to evaluate the risk stratification in patients with the antiphospholipid syndrome. Front Immunol 2019; 10: 460.
[http://dx.doi.org/10.3389/fimmu.2019.00460] [PMID: 30923525]

[87]   Bode M, Haenel D, Hagemeyer CE, *et al.* HMGB1 binds to activated platelets *via* the receptor for advanced glycation end products and is present in platelet rich human coronary artery thrombi. Thromb Haemost 2015; 114(11): 994-1003.
[http://dx.doi.org/10.1160/TH14-12-1073] [PMID: 26202300]

[88]   Spadaccio C, De Marco F, Di Domenico F, *et al.* Simvastatin attenuates the endothelial pro-thrombotic shift in saphenous vein grafts induced by Advanced glycation endproducts. Thromb Res 2014; 133(3): 418-25.
[http://dx.doi.org/10.1016/j.thromres.2013.12.023] [PMID: 24388572]

[89]   Rath D, Geisler T, Gawaz M, Vogel S. HMGB1 expression level in circulating platelets is not significantly associated with outcomes in symptomatic coronary artery disease. Cell Physiol Biochem 2017; 43(4): 1627-33.

[http://dx.doi.org/10.1159/000482026] [PMID: 29041001]

[90] Vogel S, Rath D, Borst O, *et al*. Platelet-derived high-mobility group box 1 promotes recruitment and suppresses apoptosis of monocytes. Biochem Biophys Res Commun 2016; 478(1): 143-8.
[http://dx.doi.org/10.1016/j.bbrc.2016.07.078] [PMID: 27449608]

[91] Kraakman MJ, Lee MKS, Al-Sharea A, *et al*. Neutrophil-derived S100 calcium-binding proteins A8/A9 promote reticulated thrombocytosis and atherogenesis in diabetes. J Clin Invest 2017; 127(6): 2133-47.
[http://dx.doi.org/10.1172/JCI92450] [PMID: 28504650]

[92] Stark K, Philippi V, Stockhausen S, *et al*. Disulfide HMGB1 derived from platelets coordinates venous thrombosis in mice. Blood 2016; 128(20): 2435-49.
[http://dx.doi.org/10.1182/blood-2016-04-710632] [PMID: 27574188]

[93] Fujisawa K, Katakami N, Kaneto H, *et al*. Circulating soluble RAGE as a predictive biomarker of cardiovascular event risk in patients with type 2 diabetes. Atherosclerosis 2013; 227(2): 425-8.
[http://dx.doi.org/10.1016/j.atherosclerosis.2013.01.016] [PMID: 23384720]

[94] Grauen Larsen H, Yndigegn T, Marinkovic G, *et al*. The soluble receptor for advanced glycation end-products (sRAGE) has a dual phase-dependent association with residual cardiovascular risk after an acute coronary event. Atherosclerosis 2019; 287: 16-23.
[http://dx.doi.org/10.1016/j.atherosclerosis.2019.05.020] [PMID: 31181415]

[95] Paradela-Dobarro B, Agra RM, Álvarez L, *et al*. The different roles for the advanced glycation end products axis in heart failure and acute coronary syndrome settings. Nutr Metab Cardiovasc Dis 2019; 29(10): 1050-60.
[http://dx.doi.org/10.1016/j.numecd.2019.06.014] [PMID: 31371263]

[96] Selejan SR, Hewera L, Hohl M, *et al*. Suppressed MMP-9 activity in myocardial infarction-related cardiogenic shock implies diminished rage degradation. Shock 2017; 48(1): 18-28.
[http://dx.doi.org/10.1097/SHK.0000000000000829] [PMID: 28608784]

[97] Bucciarelli LG, Ananthakrishnan R, Hwang YC, *et al*. RAGE and modulation of ischemic injury in the diabetic myocardium. Diabetes 2008; 57(7): 1941-51.
[http://dx.doi.org/10.2337/db07-0326] [PMID: 18420491]

[198] Aleshin A, Ananthakrishnan R, Li Q, *et al*. RAGE modulates myocardial injury consequent to LAD infarction *via* impact on JNK and STAT signaling in a murine model. Am J Physiol Heart Circ Physiol 2008; 294(4): H1823-32.
[http://dx.doi.org/10.1152/ajpheart.01210.2007] [PMID: 18245563]

[99] Ramasamy R, Schmidt AM. Receptor for advanced glycation end products (RAGE) and implications for the pathophysiology of heart failure. Curr Heart Fail Rep 2012; 9(2): 107-16.
[http://dx.doi.org/10.1007/s11897-012-0089-5] [PMID: 22457230]

[100] Shang L, Ananthakrishnan R, Li Q, *et al*. RAGE modulates hypoxia/reoxygenation injury in adult murine cardiomyocytes *via* JNK and GSK-3beta signaling pathways. PLoS One 2010; 5(4): e10092.
[http://dx.doi.org/10.1371/journal.pone.0010092] [PMID: 20404919]

[101] Tsoporis JN, Izhar S, Leong-Poi H, Desjardins JF, Huttunen HJ, Parker TG. S100B interaction with the receptor for advanced glycation end products (RAGE): a novel receptor-mediated mechanism for myocyte apoptosis postinfarction. Circ Res 2010; 106(1): 93-101.
[http://dx.doi.org/10.1161/CIRCRESAHA.109.195834] [PMID: 19910580]

[102] Mohammadzadeh F, Desjardins JF, Tsoporis JN, Proteau G, Leong-Poi H, Parker TG. S100B: Role in cardiac remodeling and function following myocardial infarction in diabetes. Life Sci 2013; 92(11): 639-47.
[http://dx.doi.org/10.1016/j.lfs.2012.09.011] [PMID: 23000886]

[103] Volz H, Kaya Z, Katus H, Andrassy M. The role of HMGB1/RAGE in inflammatory cardiomyopathy. Semin Thromb Hemost 2010; 36(2): 185-94.

[http://dx.doi.org/10.1055/s-0030-1251503] [PMID: 20414834]

[104] Kay AM, Simpson CL, Stewart JA Jr. The role of AGE/RAGE signaling in diabetes-mediated vascular calcification. J Diabetes Res 2016; 2016: 1-8.
[http://dx.doi.org/10.1155/2016/6809703] [PMID: 27547766]

[105] Kim HS, Chung W, Kim AJ, *et al.* Circulating levels of soluble receptor for advanced glycation end product are inversely associated with vascular calcification in patients on haemodialysis independent of S 100 A 12 (EN-RAGE) levels. Nephrology (Carlton) 2013; 18(12): 777-82.
[http://dx.doi.org/10.1111/nep.12166] [PMID: 24124651]

[106] NasrAllah MM, El-Shehaby AR, Osman NA, Salem MM, Nassef A, El Din UAAS. Endogenous soluble receptor of advanced glycation end-products (esRAGE) is negatively associated with vascular calcification in non-diabetic hemodialysis patients. Int Urol Nephrol 2012; 44(4): 1193-9.
[http://dx.doi.org/10.1007/s11255-011-0007-x] [PMID: 21643645]

[107] Wang Y, Shan J, Yang W, Zheng H, Xue S. High mobility group box 1 (HMGB1) mediates high-glucose-induced calcification in vascular smooth muscle cells of saphenous veins. Inflammation 2013; 36(6): 1592-604.
[http://dx.doi.org/10.1007/s10753-013-9704-1] [PMID: 23928875]

[108] Ren X, Shao H, Wei Q, Sun Z, Liu N. Advanced glycation end-products enhance calcification in vascular smooth muscle cells. J Int Med Res 2009; 37(3): 847-54.
[http://dx.doi.org/10.1177/147323000903700329] [PMID: 19589269]

[109] Tanikawa T, Okada Y, Tanikawa R, Tanaka Y. Advanced glycation end products induce calcification of vascular smooth muscle cells through RAGE/p38 MAPK. J Vasc Res 2009; 46(6): 572-80.
[http://dx.doi.org/10.1159/000226225] [PMID: 19571577]

[110] Belmokhtar K, Ortillon J, Jaisson S, *et al.* Receptor for advanced glycation end products: a key molecule in the genesis of chronic kidney disease vascular calcification and a potential modulator of sodium phosphate co-transporter PIT-1 expression. Nephrol Dial Transplant 2019; 34(12): 2018-30.
[http://dx.doi.org/10.1093/ndt/gfz012] [PMID: 30778553]

[111] Gawdzik J, Mathew L, Kim G, Puri TS, Hofmann Bowman MA. Vascular remodeling and arterial calcification are directly mediated by S100A12 (EN-RAGE) in chronic kidney disease. Am J Nephrol 2011; 33(3): 250-9.
[http://dx.doi.org/10.1159/000324693] [PMID: 21372560]

[112] Yan L, Bowman MAH. Research Highlight: Chronic sustained inflammation links to left ventricular hypertrophy and aortic valve sclerosis: a new link between S100/RAGE and FGF23. Inflamm Cell Signal 2014; 1(5): e279.
[http://dx.doi.org/10.14800/ics.279] [PMID: 26082935]

[113] Potenza MA, Nacci C, Sgarra L, Leo V, De Salvia MA, Montagnani M. Endothelial Dysfunction in Diabetes: An Update on Mechanisms and Therapeutic Targets. Pharmacological research Journal 2016; 8: 136-70.
[http://dx.doi.org/10.2174/9781681081755116080006]

[114] Yang DR, Wang MY, Zhang CL, Wang Y. Endothelial dysfunction in vascular complications of diabetes: a comprehensive review of mechanisms and implications. Front Endocrinol (Lausanne) 2024; 15: 1359255.
[http://dx.doi.org/10.3389/fendo.2024.1359255] [PMID: 38645427]

[115] Martín-Carro B, Martín-Vírgala J, Fernández-Villabrille S, *et al.* Role of Klotho and AGE/RAGE-Wnt/β-catenin signalling pathway on the development of cardiac and renal fibrosis in diabetes. Int J Mol Sci 2023; 24(6): 5241.
[http://dx.doi.org/10.3390/ijms24065241] [PMID: 36982322]

[116] Liang B, Zhou Z, Yang Z, *et al.* AGEs–RAGE axis mediates myocardial fibrosis *via* activation of cardiac fibroblasts induced by autophagy in heart failure. Exp Physiol 2022; 107(8): 879-91.
[http://dx.doi.org/10.1113/EP090042] [PMID: 35598104]

[117] Monden M, Koyama H, Otsuka Y, *et al.* Receptor for advanced glycation end products regulates adipocyte hypertrophy and insulin sensitivity in mice: involvement of Toll-like receptor 2. Diabetes 2013; 62(2): 478-89.
[http://dx.doi.org/10.2337/db11-1116] [PMID: 23011593]

[118] Song F, Hurtado del Pozo C, Rosario R, *et al.* RAGE regulates the metabolic and inflammatory response to high-fat feeding in mice. Diabetes 2014; 63(6): 1948-65.
[http://dx.doi.org/10.2337/db13-1636] [PMID: 24520121]

[119] Asadipooya K, Lankarani KB, Raj R, Kalantarhormozi M. RAGE is a potential cause of onset and progression of nonalcoholic fatty liver disease. Int J Endocrinol 2019; 2019: 1-11.
[http://dx.doi.org/10.1155/2019/2151302] [PMID: 31641351]

[120] Schmidt AM. RAGE and implications for the pathogenesis and treatment of cardiometabolic disorders – spotlight on the macrophage. Arterioscler Thromb Vasc Biol 2017; 37(4): 613-21.
[http://dx.doi.org/10.1161/ATVBAHA.117.307263] [PMID: 28183700]

[121] Shimizu T, Yamakuchi M, Biswas KK, *et al.* HMGB1 is secreted by 3T3-L1 adipocytes through JNK signaling and the secretion is partially inhibited by adiponectin. Obesity (Silver Spring) 2016; 24(9): 1913-21.
[http://dx.doi.org/10.1002/oby.21549] [PMID: 27430164]

[122] Gunasekaran MK, Viranaicken W, Girard AC, *et al.* Inflammation triggers high mobility group box 1 (HMGB1) secretion in adipose tissue, a potential link to obesity. Cytokine 2013; 64(1): 103-11.
[http://dx.doi.org/10.1016/j.cyto.2013.07.017] [PMID: 23938155]

[123] Ahima RS. Adipose tissue as an endocrine organ. Obesity (Silver Spring) 2006; 14(S8) (Suppl. 5): 242S-9S.
[http://dx.doi.org/10.1038/oby.2006.317] [PMID: 17021375]

[124] Catrysse L, van Loo G. Adipose tissue macrophages and their polarization in health and obesity. Cell Immunol 2018; 330: 114-9.
[http://dx.doi.org/10.1016/j.cellimm.2018.03.001] [PMID: 29526353]

[125] Feng Z, Zhu L, Wu J. RAGE signalling in obesity and diabetes: focus on the adipose tissue macrophage. Adipocyte 2020; 9(1): 563-6.
[http://dx.doi.org/10.1080/21623945.2020.1817278] [PMID: 32892690]

[126] Feng Z, Du Z, Shu X, *et al.* Role of RAGE in obesity-induced adipose tissue inflammation and insulin resistance. Cell Death Discov 2021; 7(1): 305.
[http://dx.doi.org/10.1038/s41420-021-00711-w] [PMID: 34686659]

[127] Hidmark A, Fleming T, Vittas S, *et al.* A new paradigm to understand and treat diabetic neuropathy. Exp Clin Endocrinol Diabetes 2014; 122(4): 201-7.
[http://dx.doi.org/10.1055/s-0034-1367023] [PMID: 24623503]

[128] Callaghan BC, Little AA, Feldman EL, Hughes RAC. Enhanced glucose control for preventing and treating diabetic neuropathy. Cochrane Libr 2012; 2012(6): CD007543.
[http://dx.doi.org/10.1002/14651858.CD007543.pub2] [PMID: 22696371]

[129] Ismail-Beigi F, Craven T, Banerji MA, *et al.* Effect of intensive treatment of hyperglycaemia on microvascular outcomes in type 2 diabetes: an analysis of the ACCORD randomised trial. Lancet 2010; 376(9739): 419-30.
[http://dx.doi.org/10.1016/S0140-6736(10)60576-4] [PMID: 20594588]

[130] Bongaerts BWC, Rathmann W, Kowall B, *et al.* Postchallenge hyperglycemia is positively associated with diabetic polyneuropathy: the KORA F4 study. Diabetes Care 2012; 35(9): 1891-3.
[http://dx.doi.org/10.2337/dc11-2028] [PMID: 22751964]

[131] Lu B, Hu J, Wen J, *et al.* Determination of peripheral neuropathy prevalence and associated factors in Chinese subjects with diabetes and pre-diabetes - ShangHai Diabetic neuRopathy Epidemiology and Molecular Genetics Study (SH-DREAMS). PLoS One 2013; 8(4): e61053.

[http://dx.doi.org/10.1371/journal.pone.0061053] [PMID: 23613782]

[132]   Beisswenger PJ. Methylglyoxal in diabetes: link to treatment, glycaemic control and biomarkers of complications. Biochem Soc Trans 2014; 42(2): 450-6.
[http://dx.doi.org/10.1042/BST20130275] [PMID: 24646259]

[133]   Pham M, Oikonomou D, Bäumer P, *et al.* Proximal neuropathic lesions in distal symmetric diabetic polyneuropathy: findings of high-resolution magnetic resonance neurography. Diabetes Care 2011; 34(3): 721-3.
[http://dx.doi.org/10.2337/dc10-1491] [PMID: 21266652]

[134]   Juranek JK, Geddis MS, Song F, *et al.* RAGE deficiency improves postinjury sciatic nerve regeneration in type 1 diabetic mice. Diabetes 2013; 62(3): 931-43.
[http://dx.doi.org/10.2337/db12-0632] [PMID: 23172920]

[135]   Thakur V, Sadanandan J, Chattopadhyay M. High-mobility group box 1 protein signaling in painful diabetic neuropathy. Int J Mol Sci 2020; 21(3): 881.
[http://dx.doi.org/10.3390/ijms21030881] [PMID: 32019145]

[136]   Osonoi S, Mizukami H, Takeuchi Y, *et al.* RAGE activation in macrophages and development of experimental diabetic polyneuropathy. JCI Insight 2022; 7(23): e160555.
[http://dx.doi.org/10.1172/jci.insight.160555] [PMID: 36477360]

[137]   Wong TY, Cheung CMG, Larsen M, Sharma S, Simó R. Diabetic retinopathy. Nat Rev Dis Primers 2016; 2(1): 16012.
[http://dx.doi.org/10.1038/nrdp.2016.12] [PMID: 27159554]

[138]   Barile G, Schmidt A. RAGE and its ligands in retinal disease. Curr Mol Med 2007; 7(8): 758-65.
[http://dx.doi.org/10.2174/156652407783220778] [PMID: 18331234]

[139]   Dvoriantchikova G, Hernandez E, Grant J, Santos ARC, Yang H, Ivanov D. The high-mobility group box-1 nuclear factor mediates retinal injury after ischemia reperfusion. Invest Ophthalmol Vis Sci 2011; 52(10): 7187-94.
[http://dx.doi.org/10.1167/iovs.11-7793] [PMID: 21828158]

[140]   Sanajou D, Ghorbani Haghjo A, Argani H, Aslani S. AGE-RAGE axis blockade in diabetic nephropathy: Current status and future directions. Eur J Pharmacol 2018; 833: 158-64.
[http://dx.doi.org/10.1016/j.ejphar.2018.06.001] [PMID: 29883668]

[141]   Gong QY, Yu S-Q, Qian T-W, Xu X, Xu X. Comprehensive assessment of growth factors, inflammatory mediators, and cytokines in vitreous from patients with proliferative diabetic retinopathy. Int J Ophthalmol 2022; 15(11): 1736-42.
[http://dx.doi.org/10.18240/ijo.2022.11.02] [PMID: 36404978]

[142]   Kim HJ, Jeong MS, Jang SB. Molecular characteristics of RAGE and advances in small-molecule inhibitors. Int J Mol Sci 2021; 22(13): 6904.
[http://dx.doi.org/10.3390/ijms22136904] [PMID: 34199060]

[143]   Bianchi R, Giambanco I, Donato R. S100B/RAGE-dependent activation of microglia *via* NF-kappaB and AP-1 Co-regulation of COX-2 expression by S100B, IL-1beta and TNF-alpha. Neurobiol Aging 2010; 31(4): 665-77.
[http://dx.doi.org/10.1016/j.neurobiolaging.2008.05.017] [PMID: 18599158]

[144]   Chen M, Glenn JV, Dasari S, *et al.* RAGE regulates immune cell infiltration and angiogenesis in choroidal neovascularization. PLoS One 2014; 9(2): e89548.
[http://dx.doi.org/10.1371/journal.pone.0089548] [PMID: 24586862]

[145]   Yang X, Wu S, Feng Z, Yi G, Zheng Y, Xia Z. Combination therapy with semaglutide and rosiglitazone as a synergistic treatment for diabetic retinopathy in rodent animals. Life Sci 2021; 269: 119013.
[http://dx.doi.org/10.1016/j.lfs.2020.119013] [PMID: 33417950]

[146]   Zhang Y, Wang W, Yang A. The involvement of ACO3 protein in diabetic retinopathy through the

PI3k/Akt signaling pathway. Adv Clin Exp Med 2022; 31(4): 407-16.
[http://dx.doi.org/10.17219/acem/121930] [PMID: 35275447]

[147] Cui J, Gong R, Hu S, Cai L, Chen L. Gambogic acid ameliorates diabetes-induced proliferative retinopathy through inhibition of the HIF-1α/VEGF expression *via* targeting PI3K/AKT pathway. Life Sci 2018; 192: 293-303.
[http://dx.doi.org/10.1016/j.lfs.2017.11.007] [PMID: 29129773]

[148] Ji Z, Luo J, Su T, Chen C, Su Y. miR-7a targets insulin receptor substrate-2 gene and suppresses viability and invasion of cells in diabetic retinopathy mice *via* PI3K-Akt-VEGF pathway. Diabetes Metab Syndr Obes 2021; 14: 719-28.
[http://dx.doi.org/10.2147/DMSO.S288482] [PMID: 33623407]

[149] Li P, Chen D, Cui Y, *et al.* Src plays an important role in AGE-induced endothelial cell proliferation, migration, and tubulogenesis. Front Physiol 2018; 9: 765.
[http://dx.doi.org/10.3389/fphys.2018.00765] [PMID: 29977209]

[150] Yamazaki Y, Wake H, Nishinaka T, *et al.* Involvement of multiple scavenger receptors in advanced glycation end product-induced vessel tube formation in endothelial cells. Exp Cell Res 2021; 408(1): 112857.
[http://dx.doi.org/10.1016/j.yexcr.2021.112857] [PMID: 34600900]

[151] Pachydaki SI, Tari SR, Lee SE, *et al.* Upregulation of RAGE and its ligands in proliferative retinal disease. Exp Eye Res 2006; 82(5): 807-15.
[http://dx.doi.org/10.1016/j.exer.2005.09.022] [PMID: 16364297]

[152] Abu El-Asrar AM, Imtiaz Nawaz M, Kangave D, Siddiquei MM, Geboes K. Osteopontin and other regulators of angiogenesis and fibrogenesis in the vitreous from patients with proliferative vitreoretinal disorders. Mediators Inflamm 2012; 2012: 1-8.
[http://dx.doi.org/10.1155/2012/493043] [PMID: 23055574]

[153] El-Asrar AM, Nawaz MI, Kangave D, *et al.* High-mobility group box-1 and biomarkers of inflammation in the vitreous from patients with proliferative diabetic retinopathy. Mol Vis 2011; 17: 1829-38.
[PMID: 21850157]

[154] Shen Y, Cao H, Chen F, Suo Y, Wang N, Xu X. A cross-sectional study of vitreous and serum high mobility group box-1 levels in proliferative diabetic retinopathy. Acta Ophthalmol 2020; 98(2): e212-6.
[http://dx.doi.org/10.1111/aos.14228] [PMID: 31421026]

[155] Mohammad G, Siddiquei MM, Othman A, Al-Shabrawey M, Abu El-Asrar AM. High-mobility group box-1 protein activates inflammatory signaling pathway components and disrupts retinal vascular-barrier in the diabetic retina. Exp Eye Res 2013; 107: 101-9.
[http://dx.doi.org/10.1016/j.exer.2012.12.009] [PMID: 23261684]

[156] Chang YC, Lin CW, Hsieh MC, *et al.* High mobility group B1 up-regulates angiogenic and fibrogenic factors in human retinal pigment epithelial ARPE-19 cells. Cell Signal 2017; 40: 248-57.
[http://dx.doi.org/10.1016/j.cellsig.2017.09.019] [PMID: 28970183]

[157] Orlova VV, Choi EY, Xie C, *et al.* A novel pathway of HMGB1-mediated inflammatory cell recruitment that requires Mac-1-integrin. EMBO J 2007; 26(4): 1129-39.
[http://dx.doi.org/10.1038/sj.emboj.7601552] [PMID: 17268551]

[158] Lin Q, Yang XP, Fang D, *et al.* High-mobility group box-1 mediates toll-like receptor 4-dependent angiogenesis. Arterioscler Thromb Vasc Biol 2011; 31(5): 1024-32.
[http://dx.doi.org/10.1161/ATVBAHA.111.224048] [PMID: 21372296]

[159] Yang QW, Lu FL, Zhou Y, *et al.* HMBG1 mediates ischemia-reperfusion injury by TRIF-adaptor independent Toll-like receptor 4 signaling. J Cereb Blood Flow Metab 2011; 31(2): 593-605.
[http://dx.doi.org/10.1038/jcbfm.2010.129] [PMID: 20700129]

[160] Yang S, Xu L, Yang T, Wang F. High-mobility group box-1 and its role in angiogenesis. J Leukoc Biol 2014; 95(4): 563-74.
[http://dx.doi.org/10.1189/jlb.0713412] [PMID: 24453275]

[161] Zong H, Ward M, Madden A, *et al.* Hyperglycaemia-induced pro-inflammatory responses by retinal Müller glia are regulated by the receptor for advanced glycation end-products (RAGE). Diabetologia 2010; 53(12): 2656-66.
[http://dx.doi.org/10.1007/s00125-010-1900-z] [PMID: 20835858]

[162] Steinle JJ. Role of HMGB1 signaling in the inflammatory process in diabetic retinopathy. Cell Signal 2020; 73: 109687.
[http://dx.doi.org/10.1016/j.cellsig.2020.109687] [PMID: 32497617]

[163] Li G, Tang J, Du Y, Lee CA, Kern TS. Beneficial effects of a novel RAGE inhibitor on early diabetic retinopathy and tactile allodynia. Mol Vis 2011; 17: 3156-65.
[PMID: 22171162]

[164] Saleh I, Maritska Z, Parisa N, Hidayat R. Receptor for advanced glycation end products: a key molecule in the genesis of chronic kidney disease vascular calcification and a potential modulator of sodium phosphate co-transporter PIT-1 expression. Nephrol Dial Transplant 2019; 34(12): 2018-30.
[http://dx.doi.org/10.3889/oamjms.2019.759]

[165] Natesan V, Kim SJ. Diabetic nephropathy–a review of risk factors, progression, mechanism, and dietary management. Biomol Ther (Seoul) 2021; 29(4): 365-72.
[http://dx.doi.org/10.4062/biomolther.2020.204] [PMID: 33888647]

[166] Elendu C, John Okah M, Fiemotongha KDJ, *et al.* Comprehensive advancements in the prevention and treatment of diabetic nephropathy: A narrative review. Medicine (Baltimore) 2023; 102(40): e35397.
[http://dx.doi.org/10.1097/MD.0000000000035397] [PMID: 37800812]

[167] Wendt TM, Tanji N, Guo J, *et al.* RAGE drives the development of glomerulosclerosis and implicates podocyte activation in the pathogenesis of diabetic nephropathy. Am J Pathol 2003; 162(4): 1123-37.
[http://dx.doi.org/10.1016/S0002-9440(10)63909-0] [PMID: 12651605]

[168] Andersson U, Tracey KJ. HMGB1 is a therapeutic target for sterile inflammation and infection. Annu Rev Immunol 2011; 29(1): 139-62.
[http://dx.doi.org/10.1146/annurev-immunol-030409-101323] [PMID: 21219181]

[169] Yang H, Wang H, Andersson U. Targeting inflammation driven by HMGB1. Front Immunol 2020; 11: 484.
[http://dx.doi.org/10.3389/fimmu.2020.00484] [PMID: 32265930]

[170] Vande Walle L, Kanneganti TD, Lamkanfi M. HMGB1 release by inflammasomes. Virulence 2011; 2(2): 162-5.
[http://dx.doi.org/10.4161/viru.2.2.15480] [PMID: 21422809]

[171] Kelley N, Jeltema D, Duan Y, He Y. The NLRP3 inflammasome: an overview of mechanisms of activation and regulation. Int J Mol Sci 2019; 20(13): 3328.
[http://dx.doi.org/10.3390/ijms20133328] [PMID: 31284572]

[172] Chen Y, Qiao F, Zhao Y, Wang Y, Liu G. HMGB1 is activated in type 2 diabetes mellitus patients and in mesangial cells in response to high glucose. Int J Clin Exp Pathol 2015; 8(6): 6683-91.
[PMID: 26261550]

[173] Kalousová M, Hodková M, Kazderová M, *et al.* Soluble receptor for advanced glycation end products in patients with decreased renal function. Am J Kidney Dis 2006; 47(3): 406-11.
[http://dx.doi.org/10.1053/j.ajkd.2005.12.028] [PMID: 16490618]

[174] Kaňková K, Kalousová M, Hertlová M, Krusová D, Olšovský J, Zima T. Soluble RAGE, diabetic nephropathy and genetic variability in the *AGER* gene. Arch Physiol Biochem 2008; 114(2): 111-9.
[http://dx.doi.org/10.1080/13813450802033818] [PMID: 18615900]

[175] Nin JWM, Jorsal A, Ferreira I, *et al.* Higher plasma soluble Receptor for Advanced Glycation End Products (sRAGE) levels are associated with incident cardiovascular disease and all-cause mortality in type 1 diabetes: a 12-year follow-up study. Diabetes 2010; 59(8): 2027-32.
[http://dx.doi.org/10.2337/db09-1509] [PMID: 20522598]

[176] Tan KCB, Shiu SWM, Chow WS, Leng L, Bucala R, Betteridge DJ. Association between serum levels of soluble receptor for advanced glycation end products and circulating advanced glycation end products in type 2 diabetes. Diabetologia 2006; 49(11): 2756-62.
[http://dx.doi.org/10.1007/s00125-006-0394-1] [PMID: 16969649]

[177] Humpert PM, Djuric Z, Kopf S, *et al.* Soluble RAGE but not endogenous secretory RAGE is associated with albuminuria in patients with type 2 diabetes. Cardiovasc Diabetol 2007; 6(1): 9.
[http://dx.doi.org/10.1186/1475-2840-6-9] [PMID: 17343760]

[178] Nakamura T, Sato E, Fujiwara N, *et al.* Positive association of serum levels of advanced glycation end products and high mobility group box–1 with asymmetric dimethylarginine in nondiabetic chronic kidney disease patients. Metabolism 2009; 58(11): 1624-8.
[http://dx.doi.org/10.1016/j.metabol.2009.05.018] [PMID: 19604520]

[179] Isoyama N, Leurs P, Qureshi AR, *et al.* Plasma S100A12 and soluble receptor of advanced glycation end product levels and mortality in chronic kidney disease Stage 5 patients. Nephrol Dial Transplant 2015; 30(1): 84-91.
[http://dx.doi.org/10.1093/ndt/gfu259] [PMID: 25074436]

[180] Wadén JM, Dahlström EH, Elonen N, *et al.* Soluble receptor for AGE in diabetic nephropathy and its progression in Finnish individuals with type 1 diabetes. Diabetologia 2019; 62(7): 1268-74.
[http://dx.doi.org/10.1007/s00125-019-4883-4] [PMID: 31127314]

[181] Reiniger N, Lau K, McCalla D, *et al.* Deletion of the receptor for advanced glycation end products reduces glomerulosclerosis and preserves renal function in the diabetic OVE26 mouse. Diabetes 2010; 59(8): 2043-54.
[http://dx.doi.org/10.2337/db09-1766] [PMID: 20627935]

[182] Jensen LJN, Denner L, Schrijvers BF, Tilton RG, Rasch R, Flyvbjerg A. Renal effects of a neutralising RAGE-antibody in long-term streptozotocin-diabetic mice. J Endocrinol 2006; 188(3): 493-501.
[http://dx.doi.org/10.1677/joe.1.06524] [PMID: 16522729]

CHAPTER 9

Receptor for Advanced Glycation End Products in Various Types of Cancers

Parth Malik[1,2,†], **Ruma Rani**[3,†] and **Tapan Kumar Mukherjee**[4,*]

¹ School of Chemical Sciences, Central University of Gujarat, Gandhinagar, Gujarat-382030, India

² Swarrnim Startup & Innovation University, Bhoyan-Rathod, Gandhinagar-Gujarat, India

³ ICAR-National Research Centre on Equines, Hisar-125001, Haryana, India

⁴ Amity Institute of Biotechnology, Amity University, New Town, Kolkata, West Bengal 700156, India

Abstract: The receptor for advanced glycation end products (RAGE) was first isolated and characterized in the bovine lungs. Mammalian lungs express a relatively higher level of RAGE than other organs of the mammalian body. Physiologically, RAGE guards from lung cancer development owing to which, a diminished RAGE expression is implicated in the lung cancer complication. Opposed to this, a high-level RAGE expression is associated with the development of various cancers including breast, ovary, prostate, pancreatic, colon and colorectal, hepatocellular, melanoma, and neuronal. Interactions of RAGE and its multiple ligands, namely advanced glycation end products (AGE), high mobility group box protein 1 (HMGB1), S100/calgranulin, Mac 1, amyloid beta (Aβ) peptide, and others are involved in the complications of cancers. Besides their interactions with RAGE, RAGE ligands also independently aggravate the cancer-promoting actions. In cancer cells, the cellular events affected by RAGE include proliferation, survival, angiogenesis, autophagy, invasion, and metastasis. RAGE-ligands interaction aggravates inflammation and oxidative stress, leading to the propagation of various diseases including cancers.

Keywords: Apoptosis, Autophagy, Angiogenesis, Cancer, Invasion, Metastasis, Proliferation, RAGE, Survival.

INTRODUCTION

In 1992, *Anna Marie Schmidt* and *associates* isolated, purified, and biochemically characterized the receptor for advanced glycation end products (RAGE) from the bovine pulmonary tissues [1]. Following this discovery, multiple studies including

---

* **Corresponding author Tapan Kumar Mukherjee:** Amity Institute of Biotechnology, Amity University, New Town, Kolkata, West Bengal 700156, India; E-mail: tapan400@gmail.com
† These authors contributed equally to this work.

Tapan Kumar Mukherjee, Parth Malik & Ruma Rani (Eds.)

those of *Brett and colleagues*, analyzed the RAGE expression in embryonic and adult bovine, rat, and human tissues of various organs. While all major organs of the human body express RAGE, its maximum expression is noticed in the lungs. The study by *Brett and teammates* demonstrated that the bovine tissue's pulmonary vasculature, particularly endothelium and smooth muscle cells (SMCs) express RAGE. This study also detected RAGE expression in cardiomyocytes and mononuclear cells, including monocyte-derived macrophages. Considerable RAGE expression was also scrutinized in neonatal rat cardiomyocytes, neural tissues, and rat PC12 pheochromocytes [2]. The RT-PCR analysis of human cDNA by *Schlueter C and associates* revealed that alternative mRNA splicing is linked to tissue-specific RAGE distribution [3]. Many versions of RAGE mRNA are produced by alternative splicing, one of which is RAGE variant 1 (RAGEv1), which yields the majority of RAGE soluble form (known as sRAGE in the blood) and can bind a variety of RAGE ligands. Consequently, sRAGE functions as a decoy receptor, attenuating RAGE-driven cell signaling in inflammation and possibly resulting in carcinogenesis [4].

RAGE was not only detected for the first time in lung tissues [1] but its highest-level expression was also observed in the lung tissues [2]. Several investigations conducted after *Schmidt and colleagues* work, described the tissue and cell-specific expression of RAGE in the lungs. Although the initial research conducted by *Katsuoka and colleagues* located RAGE in lung epithelial type II cells [5], studies by *Demling, Shirasawa, and Fehrenbach* research groups demonstrated the exclusive RAGE expression in AT-I epithelial cells [6 - 8]. Significantly, neither AT-II epithelial nor capillary endothelial cells showed signs of RAGE expression in the investigations. Using quantitative immunoelectron microscopy, *Fehrenbach and colleagues* [7] for the first time, showed RAGE localization on the basal face of AT-I cell plasma membrane in rat and human lungs. Henceforth, other researchers also reported the RAGE expression in the basolateral membrane of AT-I epithelial cells [6, 8]. It is now widely accepted that the basolateral membrane of lung AT-1 cells exhibits the highest RAGE expression.

Structurally, RAGE is regarded as a cell surface immunoglobulin class of molecule [9] interacting with multiple ligands and is therefore characterized as a multi-ligand receptor. Further investigation herein, demonstrated a significant homology between the human RAGE sequence and the cell surface immunoglobulin class of molecules, including the cytoplasmic domain of B lymphocyte antigen (CD20), cell surface glycoprotein (MUC18, also known as CD146), and neural cell adhesion molecule (NCAM, also known as CD56). RAGE is therefore regarded as a molecule belonging to the cell surface immunoglobulin class [9]. Many physiological and pathological processes are ascribed to the interactions between RAGE and ligand(s). Full-length human

RAGE cDNA-transfected human HEK-293 cells were used by *Demling and colleagues* to examine the physiological role of RAGE in pulmonary tissues. Analysis revealed the transfected cell adhesion and dissemination in the kidney epithelial HEK-293 cell line are controlled by RAGE [8]. Almost 95% of pulmonary epithelial cells are characterized as AT1 type [10]. Studying the RAGE expression in rat lungs, *Lizotte and colleagues* observed progressive fetal growth with enhanced RAGE expression, wherein lung-RAGE expression was inhibited by hyperoxia. Monitoring the stimulation of AT-I epithelial cell adhesion and spreading (for pulmonary homeostasis) [8, 11], the investigators concluded that RAGE played a significant role in pulmonary homeostasis [11].

High-level RAGE expression is implicated in the pathogenesis of manifold pulmonary [12 - 14] and non-pulmonary diseases including various cancers. Altered RAGE expression is implicated in the complication of non-small cell cancers (NSCLCs) [15 - 19]. NSCLCs are considered the major type of lung cancer as almost 85% of lung cancers are characterized as NSCLCs. However, the role of RAGE in less common (~15%), small cell lung cancer (SCLC) complications, is not known. Several studies correlated a suppressed RAGE expression in the pulmonary tissues to aggravated NSCLC conditions. These findings concurred with the hypothesis that RAGE protects lung tissues from tumour susceptibility, strengthening the feasibility that the suppression of RAGE expression encouraged NSCLC development [20, 21]. However, erstwhile reports illustrate the RAGE contribution to pathophysiological events leading to lung cancer complications [22]. Additional investigation is required to delineate the specific role of RAGE in lung cancer development. However, the results of separate, independent investigations ascertained RAGE involvement in the complication of various cancers including those of breast, prostate, melanoma, and pancreatic. These studies examined the role of RAGE in the proliferation, survival, angiogenesis, autophagy, invasion, and metastasis of various cancer cells [23]. Interaction of RAGE with its manifold ligands aggravates inflammation and oxidative stress, resulting in various diseased conditions including cancers [24]. The dual action-mechanism of RAGE (growth-suppressive in the lung to growth-promoting in the breast, prostate, colon, *etc.*) appears to result from variations in the genetic regulation controlling: (a) location-dependent function; (b) spliced variant functional diversity; and (c) tissue-specific abundance of specific ligands.

The dynamic results of ligand-specific RAGE interactions remain a concern for researchers, even though the role of these interactions has been well studied in the control of homeostasis, inflammatory responses, angiogenesis, and carcinogenesis (in various malignancies) [25 - 28]. Augmenting the complexity, recent studies illustrate the diverse functional responses *vis-à-vis* varied ligand extents [29, 30]. The distinct outcomes attributed to genetic polymorphisms have also been demonstrated in RAGE-dependent NSCLC development [31 - 34] and several erstwhile cancers [35 - 38]. Understanding the physiological diversity of RAGE functions *vis-à-vis* ligand interactions in distinct redox environments is essential for efficient NSCLC diagnosis and treatment.

RAGE LIGANDS IN THE DEVELOPMENT AND PROGRESSION OF CANCERS

- RAGE signaling has been the subject of significant interest, owing to the diverse outcomes of its multi-ligand interactions. The major RAGE ligands comprise advanced glycation end products (AGEs), high mobility group box protein 1 (HMGB1), S100/calgranulins, amyloid-beta (Aβ) peptide, macrophage 1 antigen (Mac1), and others [39]. RAGE is classified as a pattern recognition receptor (PRR) molecule based on the chemical diversity of its ligands and its structure-function similarities to toll-like receptors (TLRs) [40]. Enhanced expression of RAGE and its ligands is a frequent observation in many cancers. In this regard, two major studies screened the gene expression of the diseased human population and tumour-affected mice models [41]. Surprisingly, compared to all cancers including breast, ovarian, pancreatic, and others, the NSCLCs exhibit a suppressed RAGE extent [42].
- Despite the RAGE downregulation, RAGE ligands especially S100 family members and HMGB1 are overexpressed in lung cancer cells [43]. Like RAGE, sRAGE level is also diminished in NSCLCs, supporting and consolidating the enhanced reduction in sRAGE extents in lung cancer patients over healthy controls. Analysis revealed a negative correlation with lymph node involvement [19]. For NSCLCs, increased plasma RAGE extents exhibit a protective effect rather than progression, presenting a simple diagnostic tool for specific outcomes post-curative surgery [44].
- Fig. (1) outlines the major RAGE-ligand interactions leading to altered redox environment and aggravated inflammatory conditions whereby, cytoplasmic to nuclear translocation of the transcription factor NF-κB is facilitated. In turn, the activated NF-κB stimulates cell division and progression by energizing the oncogenes and inactivating the tumor suppressor genes. Concomitantly, enhanced pacing of angiogenesis, invasion, and metastasis altogether, aggravates tumorigenesis. This mechanistic correlation is not observed in the lungs, where RAGE actions are essential for adequate developmental and functional sustenance.
- In various cells, the role of ligand-specific RAGE interactions has been rigorously investigated concerning homeostatic regulation of diverse cellular events, as well as inflammatory responses, angiogenesis, and carcinogenesis. Thus, RAGE is involved in physiological and pathophysiological events characterized by aggravated expression of various cell signaling molecules

involved in the propagation of multiple cancers. The inceptive success of the experimental attempts in this direction is credited to *Taguchi and colleagues*. Working on a mouse model, the investigators observed that inhibition of RAGE regulatory and specific HMGB1-RAGE axis suppressed tumor growth. This study provided the first ever, *in vivo* experimental proof for RAGE involvement in the complication of cancers [45]. A recent study concluded that enhanced HMGB1 and M2 macrophage actions, aggravated osteosarcoma migration, invasion, and epithelial-to-mesenchymal transition *via* positive feedback regulation [46]. Yet another recent study demonstrated a substantial increase in serum HMGB1 extent in multiple myeloma patients [47].

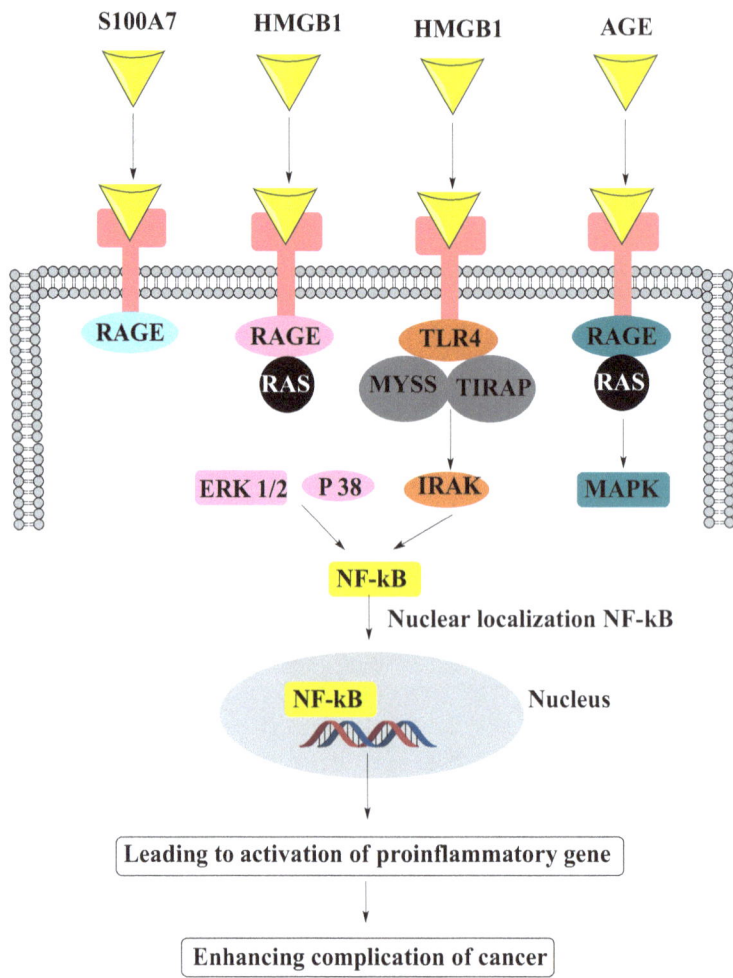

Fig. (1). RAGE-ligand interactions mediated enhanced cytoplasmic to nuclear localization of transcription factor, NF-κB, and consequent activation of pro-inflammatory genes leading to complications of different cancers.

- While it is now well accepted that the interaction of RAGE with its various ligands controls the complications and progression of various cancers, most RAGE ligand activities are exhibited without RAGE involvement. The diversity of ligand-specific RAGE activities continues to excite uncertainty in distinct biochemical environments (Fig. **2**) [25 - 28]. A complex response is expected from the RAGE-ligand interactions with their eventual fate being sensitively influenced by the ligand concentrations [29, 30].

- Further, genetic polymorphisms also affected the role of RAGE in lung and other cancers [31 - 38]. Understanding the dynamics of various RAGE ligand expressions, their polymorphic variants (molecular forms), and tissue-specific functions are highly pertinent for an accurate diagnosis and treatment of different cancers including NSCLCs. Targeting RAGE and its ligands such as S100A6 using various experimental molecules may be an effective therapeutic approach as an anticancer mechanism [48]. Readers are suggested to refer to Chapter 5 of this book which specifically describes the role of various RAGE ligands, *vis-à-vis* diversified biochemical fates.

Fig. (2). Schematic representation of molecular and cellular events, characterized by altered redox and inflammatory environment-driven RAGE-ligand interactions (notably, HMGB1, AGEs, S100), on setting gradually as tumorigenic hallmarks (enhanced angiogenesis, telomere elongation, aggravated metastasis, and invasion, activation of oncogenes and suppression of tumor suppressor genes).

RAGE IN THE PROTECTION AGAINST LUNG CANCER DEVELOPMENT AND PROGRESSION

- In 1992, *Anna Marie Schmidt and colleagues* completed the pioneering works on RAGE, having isolated, purified, and biochemically characterized 35 kDa and 80 kDa AGE binding proteins from bovine pulmonary endothelial cells [1]. The ability of these proteins to bind AGEs enabled their recognition as RAGE [9]. The same group used a human lung library, subsequently isolating a partial clone encoding the human RAGE gene. Its sequence is ~ 90% homologous to bovine RAGE (M_w: 35 kDa).

- In general, RAGE is abundantly expressed in the alveolar type 1 (AT-1) epithelial cells compared to other types of lung cells. However, in various inflammation-driven pathophysiological conditions such as asthma, chronic bronchitis (CB), acute lung injury (ALI), acute respiratory distress syndrome (ARDS), cystic fibrosis (CF), and other pulmonary diseases, RAGE expression continues to rise and worsen the diseased conditions. Yester studies herein discussed the specific role of RAGE in pulmonary physiology and pathophysiology, with an enhanced expression being deciphered as critical in multiple disorders [14, 49].

- The findings published in 1994 and 1997 by *Schraml and associates* demonstrated that viable lung cancer tissues have much lower RAGE expression levels. The researchers showed that the downregulated RAGE gene is a prominent identifying feature of perpetuating lung cancer. Their investigation has shown that RAGE expression was significantly decreased in NSCLC tissues than in normal lungs [15, 16]. Nevertheless, later research showed that RAGE plays a crucial part in the complications associated with lung and other malignancies. The notable study by *Huttunen and colleagues* was the first thorough attempt to address this contradiction when they demonstrated that amphoterin, a common RAGE ligand, harbors a distinctive -COOH-terminal region (amino acids 150-183), normally engaged in RAGE binding. The amphoterin-RAGE interaction dramatically forbids tumor cell migration across the endothelium. This group found that a synthetic peptide corresponding to the amphoterin -COOH terminal motif dramatically inhibited lung cancer cell invasion in further experiments conducted using an *in vivo* model. Human lung cancer tissue specimens subjected to immune-histochemical inspection also showed amphoterin-RAGE binding [50].

- In separate efforts, *Hofmann and colleagues* subsequently used a microarray approach to distinguish the RAGE and cyclin gene expression profile in lung neoplasm. To a surprise, RAGE expression was reduced while that of cyclin B2 (a cell cycle protein) was enhanced in the lung neoplastic tissues. Of note, cyclin B2 regulates the G2/M phase transition of the cell cycle. According to these preliminary investigations, RAGE expression may serve as an implicit diagnostic marker for lung carcinoma [18]. In a 2005 study, *Bartling and colleagues* examined the RAGE expression in normal lung tissues and NSCLC specimens, deciphering a lower RAGE expression in advanced stages. On full-length human RAGE overexpression, the researchers also showed reduced cell proliferation of NSCLC cell line, NCI-H358. It's interesting to see a lesser multiplication of the cells expressing RAGE than those harbouring the cytoplasmic domain alone. In contrast to the cells containing Dcyto RAGE, the deletion mutant (Dcyto) RAGE and mock-transfected NCI-H358 cells similarly developed smaller tumors in spheroid cultures and athymic (nude) animals. Immunofluorescence patterns were used to supplement the distinctions, explaining preferential RAGE localization at the intercellular contact locations irrespective of the cytoplasmic domain prevalence. The results supported reduced RAGE expression for aggravated NSCLC.

- A subsequent study by the same research group supported the negative relationship between RAGE/ligand activity and NSCLC severity, ascertained through resection surgery. The fluorescence associated with plasma AGE was measured. Individuals with AGE-related plasma fluorescence greater than the median extent had a significantly lower chance of developing NSCLC (25% *vs.* 47%, 5-year survival, p = 0.011) than those having a lower than median fluorescence. The analysis also showed that in mice given an AGE-rich diet (to supplement their AGE levels), there was an inverse relationship between the NSCLC growth and the levels of circulating AGEs [21].

- The *Bartling group's* two experiments suggested that inhibited RAGE exacerbated NSCLC growth and metastasis, supporting the idea that RAGE inhibited NSCLC carcinogenesis. These investigations additionally decipher plasma AGE fluorescence as a possible prognostic indicator for screening the NSCLC susceptibility. The findings elucidate an inverse link with expanding tumour spheroids. In a subsequent microarray study, *Stav and colleagues* examined the RAGE, CCNB2 (G2/mitotic-specific cyclin in B2), and CDK5RAP3 (cyclin-dependent kinase 5 regulatory subunit-associated protein 3) expression for a possible screening of NSCLC. On expected lines, the CDK5RAP3 and CCNB2 expressions were enhanced while that of RAGE was significantly decreased, deciphering reduced RAGE expression as an authentic lung adenocarcinoma biomarker [17]. Fig. (3) depicts the multiple sub-links for varied RAGE actions in lung cancer. Exhibiting a prominent role in maintaining normal lung physiology, RAGE expression promotes the spreading of adherent cells on collagen IV in the extracellular matrix (ECM), ensuring an adequate and effective gaseous exchange. While cells harboring RAGE suppressed the proliferation, those with a more than a threshold, adhere rather more aggressively to collagen IV, illustrating the critical RAGE involvement in cell-ECM interactions.

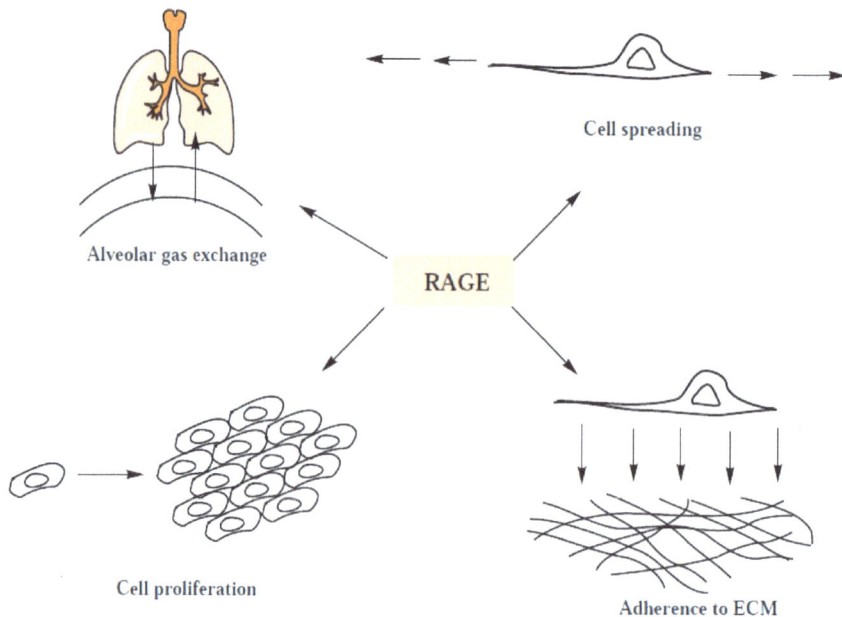

Fig. (3). Multiple interfaces through which RAGE expression affects the functioning of lungs. Contrary to most other organs where RAGE's presence aggravates the inflammatory actions (the reason for anti-RAGE therapies as reliable curative mechanisms), the lungs harbor a threshold RAGE as essential for normal functioning and development.

- Another significant effort by *Kobayashi and associates* (2007) investigated the expression pattern of RAGE splice variants, endogenous secretory RAGE (esRAGE), and its potential significance in the NSCLC prognosis. Notably, esRAGE is a unique splice variant that acts as a decoy receptor to moderate the RAGE-mediated cellular responses. In the study, 182 stage-I NSCLC patients were followed immunohistochemically wherein 137 surgical samples revealed reduced or null cytoplasmic esRAGE expression. The expression level therein was strikingly distinct from the normal lung tissue. Decreased esRAGE cytoplasmic extents were linked with 5-year patient survival on surgical resection. Through laser capture microdissection and quantitative real-time PCR monitored simultaneous esRAGE and full-length RAGE mRNA expression; the researchers discovered all intra-tumoral stromal cells to have generated both mRNAs. On the other hand, only 29% of 36 NSCLC tissues (or 81%) expressed either full-length RAGE or esRAGE. This study extends the distinct full-length or esRAGE expression in NSCLC tissues, as also demonstrated by *Bartling and colleagues* [51].
- A similar study by *Jing and colleagues* reported suppressed serum sRAGE expression during lung cancer progression, supporting the earlier findings. Analysis revealed serum sRAGE significance as an authentic lung cancer

diagnostic biomarker, inferring a negative correlation of its extent with lymph node involvement [19]. The RAGE expression was lower in lung cancer tissues than in adjacent normal lung tissue. Analysis revealed reduced serum sRAGE extents amidst lung cancer progression, relating a decreased tissue and parallel serum sRAGE/RAGE expression as an accurate NSCLC biomarker.

- Significant HMGB1-RAGE interactions in lung tumors were identified in a later investigation by *Chang and colleagues*. Using inverse-variance weighting to aggregate the hazard ratios (HRs) of datasets from the University of Michigan Cancer Centre (n = 178), the Moffit Cancer Centre (n = 79), the Memorial Sloan-Kettering Cancer Centre (MSK, n = 104), and the Dana-Farber Cancer Institute (CAN/DF, n = 82), the investigators examined the biomarkers of 443 lung adenocarcinoma patients. The analysis identified a significant correlation for lung cancer clinical outcomes *vis-à-vis* 21 of 65 HMGB1-RAGE signalling network genes, 22 of 70 genes in the β-adrenergic receptor-regulated extracellular regulated kinase (ERK) pathway, and 31 of 107 in the clathrin-coated vesicle cycle system. Using predictive inspection, the CPBR score of high-risk groups indicated a shorter overall survival and a better lung cancer prognosis. The tumor-susceptibility genes were screened to determine the low *versus* high risk. These findings imply probable reciprocal interactions between differentially expressed genes in every pathway, establishing the basis for the prognosis [52].

- In a follow-up investigation, *Pasquali and colleagues* elaborated that retinol, or vitamin A, modulated RAGE expression in A549-NSCLC cells. The researchers administered retinol (2, 5, 10, and 20 μM) to A549-NSCLC cells in culture, optimizing the dosage through physiological response to 2 μM retinol. Inspection revealed that administering 10 or 20 μM retinol enhanced the production of free radicals, oxidative damage, and anti-oxidative enzyme activity, suggesting a restraint for excessive intake. Moreover, RAGE expression was suppressed on enhanced retinol administration. Co-administration of trolox, a hydrophilic counterpart of α-tocopherol (vitamin E) inhibited the RAGE suppression by retinol. The findings imply that localized redox state influences RAGE expression in NSCLC. Additional research on the underlying mechanisms of the influence of retinol on RAGE expression revealed p38-mediated MAPK activation. The results also showed that NF-κB was a p38 downstream effector in the suppressed RAGE expression by retinol, an observation countered by NF-κB inhibition on SN50 p65 siRNA administration, decreasing the RAGE activity limited by retinol. In conclusion, this work offers a potential strategy for selectively modifying retinol extents to promote NSCLC-specific RAGE expression and thus prevent NSCLC growth [53]. Collectively, the majority of studies established a mechanism for impaired NSCLC growth *via* modulated RAGE expression.

RAGE IN THE PROTECTION VERSUS PROMOTION AGAINST LUNG CANCER

Several research attempts demonstrated RAGE as a protective factor against lung cancer development, suggesting that a reduction in RAGE expression was a contributing factor to the manifested lung malignancies. Nonetheless, a few quite recent investigations asserted that RAGE is accountable for the onset and progression of lung cancer. In one such effort, *Yu and associates* examined the RAGE function in the development and spread of the H1975 NSCLC cell line. As demonstrated by soft-agar colony-forming experiments, RAGE silencing (*via* its particular siRNA) significantly reduced the colony-forming capacity of H1975 cells. RAGE also aided in the proliferation, metastasis, and epithelial to mesenchymal transition (EMT) of H1975 cells *via* controlling intracellular signalling, including the P(I)3K-AKT (pro-proliferative) and RAF-1 (pro-survival) pathways. These findings were complemented by considerably reduced tumour development and expressions of the tumour cell cycle regulator, Ki67 on the injection of RAGE downregulated H1975 cells into the xenograft nude mice. In conclusion, this study showed that RAGE, *via* controlling the P(I)3K-AKT and KRAS-RAF-1 signalling pathways, plays a critical role in NSCLC growth and metastasis. The results contradicted the studies conducted by the *Bartling and Pasquali* groups [20, 21, 53], whereby RAGE expression was deciphered as a potential NSCLC therapeutic target [22]. The precise causes of these differences remain unknown, although one theory is that the used RAGE siRNA concentration was high enough to promote interaction with additional RAGE ligands, such as HMGB1. Consequently, control of the RAGE-siRNA-dependent P(I)3K-AKT and KRAS-RAF-1 signalling pathways infers a simultaneous active state of RAGE and its ligands. Thereby, further investigation is warranted to understand the exact role of RAGE in NSCLC development.

- A significant recent attempt by the *Sanders KA* group screened the RAGE involvement in the induction of oxidative stress, macrophage activation, and lung disease pathogenesis caused *via* exposure to cigarette smoke (CS). Monitoring the corresponding effects on C57BL/6 WT mice (exposed to CS from 1 week to 4 months), the investigators noticed a forbidden development of CS-driven emphysema on pharmacological inactivation of RAGE. Probing the underlying mechanism, the investigators monitored the expression of genes implicated in chorionic obtrusive pulmonary disorder, wherein it was noticed that inactive RAGE altered the expression of antioxidant response genes. Further analysis using immunostaining of lung protein, 4-HNE (4-hydroxynonenal) inferred reduced oxidative stress in RAGE void mice despite a similar duration of CS exposure and lung leukocyte burden as those of normal WT mice. The confirmation of RAGE-mediated aggravation in oxidative stress and pulmonary inflammation was illustrated by the moderated endoplasmic reticulum (ER) stress on monitoring the alveolar macrophages of CS-exposed, RAGE null mice [54]. In this study, the exact reason for NSCLC promotion by RAGE was not elucidated. In such an eventuality, the mouse strain-specific RAGE polymorphic variant may exhibit a key role in NSCLC aggravation. Thus, the confirmation of RAGE involvement in NSCLC mandates a discussion of studies describing RAGE polymorphic variants from NSCLC tissues of various populations.

RAGE GENE POLYMORPHISM IN LUNG CANCER DEVELOPMENT AND PROGRESSION

Polymorphism encompasses two or more forms of a specific gene within a population, the first being single nucleotide polymorphism (SNP) wherein variation of a single nucleotide happens at a specific gene location. Concerning the RAGE gene, multiple polymorphisms with an enhanced frequency have been identified in various cancers. In one important study herein, *Schenk and associates* noted human NSCLC characterized by T–>A in the promoter region, 388 bp upstream of the start codon: T–>A in exon 1 (Ala2Ala), C–>G in exon 3 (Val89Val), C–>T in intron 6, G–>C and C–>G in exon 10 (Arg365Ser and Arg369Gly), a 63 bp deletion in the promoter region (358-421 bp upstream of start codon), besides 429T/C, 374T/A and 82G/S, rs1800625, rs1800624 and rs2070600, 2184A/G, 82G/S, and -374T/A.

- In a notable 2001 effort, RAGE gene expression was monitored in primary NSCLCs and corresponding normal tissues of nine patients. The investigators identified six novel sequence variants, T–>A in the promoter region, 388 bp upstream of the start codon, T–>A in exon 1 (Ala2Ala), C–>G in exon 3 (Val89Val), C–>T in intron 6, G–>C and C–>G in exon 10 (Arg365Ser and Arg369Gly). In addition, one NSCLC patient had a 63 bp deletion in the promoter region (358-421 bp) upstream of the start codon, according to the researchers. The T–>A transversion was found in the promoter region of three patients, and this polymorphic locus was focused on 54 NSCLC patients and 59 unaffected controls. A thorough investigation identified a substantial genotypic distribution setting in NSCLC patients, apart from controls. The AA genotype was found in NSCLC patients (20.8%) more frequently than in controls (3.5%), indicating that an AA variant in NSCLC is a strong risk factor for NSCLC development [34].

- A subsequent investigation by *Wang and colleagues* investigated the association between RAGE genetic polymorphisms and prognosis, chemotherapeutic response, and NSCLC risk. After identifying 429T/C, -374T/A, and 82G/S as RAGE genetic polymorphisms, the researchers treated 432 advanced NSCLC patients with platinum-based chemotherapy. All RAGE polymorphic genotypes were susceptible to NSCLC based on response analysis, with 82G/S being a significantly distinct polymorphism in chemotherapy responders. In addition to raising the NSCLC risk, the 82SS genotype and 82S allele distribution were linked to a poorer prognosis and a lower treatment response rate. According to the findings, the 82G/S RAGE genetic polymorphism may be a genetic marker for predicting the NSCLC patient's prognosis and treatment response [33]. A subsequent effort by *Xia and colleagues* corroborated these findings, wherein 27 RAGE gene polymorphisms were noticed in patients exhibiting lung cancer risk. Their results revealed the 82G/S and 374T/A RAGE gene polymorphisms as correlated with increased and reduced NSCLC risk, respectively [55].

• In a previous study, *Pan and colleagues* assessed five frequent RAGE and apurinic/apyrimidinic endonuclease 1 (APE1) gene polymorphisms in a Han Chinese cohort consisting of 819 lung cancer patients and 803 cancer-free controls. A ligase detection reaction was used to screen the identified polymorphisms in the RAGE (rs1800625, rs1800624, rs2070600) and APE1 (rs1760944, rs1130409) genes. Significant RAGE and APE1 gene connections were discovered as related to the NSCLC incidence [32]. In another effort, *Wang and colleagues* showed that human lung cancer tissue and serum had reduced levels of RAGE and sRAGE, respectively. Particularly, analysis linked the RAGE polymorphisms, 429T/C and 2184A/G to lung cancer incidence and development. These findings supported the credibility to the hypothesis that RAGE polymorphisms could aid in the clinical prognostic evaluation and correct detection of lung cancer, besides serum sRAGE and RAGE tissue levels utility as practical and useful NSCLC biomarkers [31]. Another research by *Wu and colleagues* on RAGE meta-analysis in 2018 revealed that lung cancer tissues had substantially lower RAGE expression than nearby non-tumor tissues. This study also demonstrated that the variant A allele of rs2070600 might decrease RAGE gene expression to aggravate the NSCLC risk [56].

RAGE IN THE DEVELOPMENT AND PROGRESSION OF CANCERS OTHER THAN LUNG CANCERS

The use of RAGE deficient mice (RAGE−/−) as inflammation-mediated carcinogenesis models, including chemically induced carcinogenesis [41] and colitis-associated cancer [57] provides direct evidence for RAGE involvement in various cancers. RAGE expression has been a hallmark of different tumors originating in multiple organs such as breast, ovaries, prostate, colon, colorectal, skin, lymph nodes, and others [58]. Some of the eminent publications in this direction include the following: breast cancer [59], gastric cancer [60], intestinal, colon, and colorectal cancer [61, 62] hepatocellular carcinoma [63], pancreatic and prostate cancers [64], oral squamous cell carcinoma [65] and others. Several

RAGE polymorphisms with increased frequency have been identified in cancers of the breast [35, 36], gastric tissues [37], and pancreas [38].

While there is little molecular evidence available for knowing the RAGE functions during neoplastic transformation and malignant progression, several *in vitro* studies using mouse models have established a direct correlation between RAGE activation and tumor cell proliferation, survival, migration, and invasion [58, 66]. A direct correlation between RAGE activation and tumor cell proliferation, survival, invasion, and migration has been demonstrated by experimental data obtained from *in vivo* examination of mice models [67]. Many investigations on mice models demonstrated a familiarity with RAGE gene expression for a comprehensive understanding of increasing tumor complications. RAGE's role in cell survival and proliferation has been amply illustrated in experimental cell culture settings employing several cancer cell lines [68]. Besides cancer cell proliferation and survival, RAGE is also involved in various cellular events that affect cancer progression, and maintenance. These events include angiogenesis, apoptosis, autophagy, invasion, and metastasis. Fig. (**4**) summarizes possible RAGE-mediated aggravation in these events, *vis-à-vis* cancer progression and complications. The following paragraphs describe these cellular events through which abundantly expressed RAGE modulates proliferation, survival, autophagy, angiogenesis, apoptosis, and metastatic invasion in cancer cells.

CELLULAR PROCESSES INVOLVED IN THE RAGE-DEPENDENT CANCER DEVELOPMENT AND PROGRESSION

Multiple studies have demonstrated the RAGE interaction with its anionic and acidic ligands, prominently AGEs, HMGB1, and S100 regarding their involvement in tumor progression. These interactions encompass tumor progression *via* modulating the activities of anti- and pro-apoptotic proteins amidst concomitant upregulation of phosphatidylinositide 3-kinase (PI3K)/protein kinase B (Akt)/mammalian target of rapamycin (mTOR), mitogen-activated protein kinases (MAPKs), matrix metalloproteinases (MMPs), vascular endothelial growth factor (VEGF), and NF-κB pathways. The specific variations in signaling mediators and enzyme cofactor activities *vis-à-vis* RAGE interactions with AGEs, HMGB1, and S100 ligands are discussed further. The following paragraphs discuss the prominent aspects of studies monitoring the effect of RAGE signaling-mediated tumor cell growth, metastasis, angiogenesis, autophagic activation, and apoptotic inhibition.

RAGE IN CANCER CELL PROLIFERATION

- Cell proliferation is the process through which cells divide to generate further new cells and sustain their propagation. In mitosis, one cell divides to form two daughter cells. Cell proliferation is a highly regulated process, wherein cells divide (on reaching maturity, regulated *via* signaling responses) as and when new cells are required by the body. On the other hand, cancer cells continuously grow and divide, irrespective of the requirement of new cells by the body.

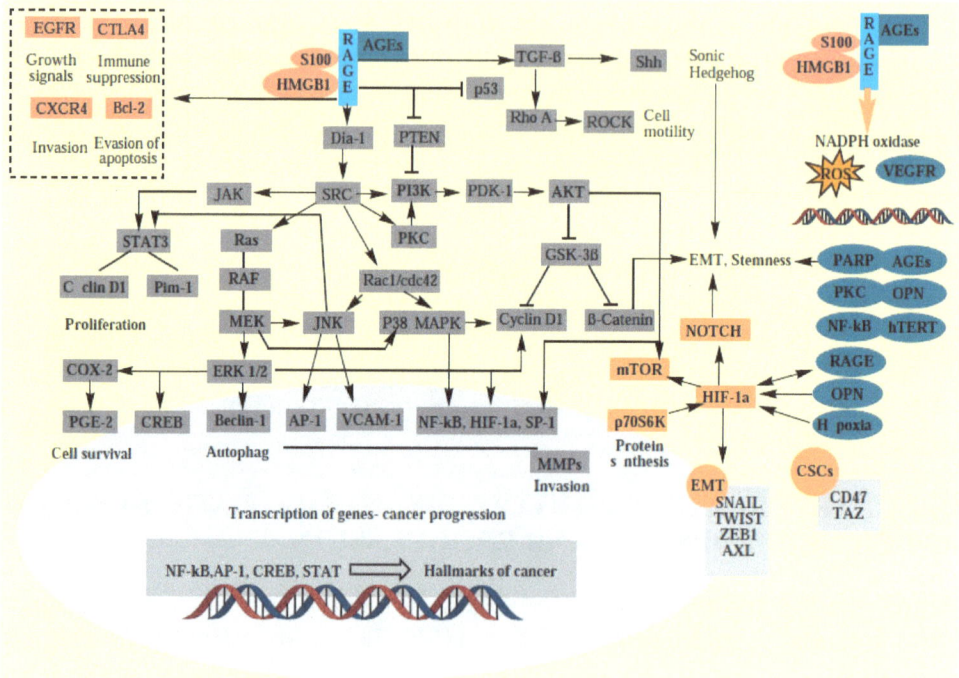

Fig. (4). Schematic representation of the signaling pathways activated by RAGE-ligand (majorly, AGEs, HMGB1, S100) interactions leading to aggravated inflammation and altered α redox conditions triggered angiogenesis, invasion, and metastasis in tumor cells.

- Numerous studies document RAGE's participation in the growth of different cancer cells. For instance, RAGE participation in the complication of pancreatic cancer has been proven through *in vitro* cell culture and *in vivo* animal model experiments. The discovery that the AGE-RAGE interaction induced the proliferation of human pancreatic cancer cells through the autocrine production of platelet-derived growth factor-B was made in two separate investigations [69, 70]. First was the effort of *Kang and colleagues*, who connected the HMGB1-RAGE pathway with bio-energetic dynamics for the first time. Analysis revealed RAGE as predominant in the mitochondria of both primary and cultured tumor cells. RAGE-HMGB1 signaling is critically instrumental in promoting multiple cell growth-sustaining processes, such as the activity of the tumor cell-mitochondrial complex I, the ATP generation, the proliferation and migration of tumor cells, and several others. One important finding was of decreased ATP production driven by HMGB1 inhibition or RAGE deficiency, eventually inhibiting the tumor growth both *in vitro* and *in vivo*. These data reveal a novel mechanism for inflammation and necrosis together to stimulate the growth of tumors [71].

- The RAGE-S100 interaction promotes cell proliferation at lower S100 extents, as per the observations of an *in vitro* investigation conducted on the MCF-7 breast cancer cell line [72 - 74]. A similar effort by *Radia and the group* showed that the treatment of MCF-7, SK-Br-3, and MDA-MB-231 breast cancer cells with RAGE siRNA prevented tumor cells from proliferating past the G1 phase of the cell cycle [75]. In another significant study, *Lata K and colleagues* noticed that 17α-EE treatment of MCF-7 breast cancer cells increased the expression of estrogen receptor-related receptor gamma (ERRγ), a member of the nuclear hormone receptor superfamily. Besides, substantial enhancement in oxidative stress was deciphered, which in turn activated the transcription factor, NF-κB (Fig. **5**). The results showed that RAGE produced *via* 17α-ethiny--estradiol treatment promotes oxidative stress, facilitating the proliferation of MCF-7 breast cells *via* concomitant cyclin D1 activation [59]. Furthermore, RAGE activated mammalian targets of rapamycin (mTOR), a serine/threonine protein kinase. Of note, transcription, cell growth, proliferation, motility, survival, and protein synthesis are all regulated by the mTOR pathway [76]. These studies collectively demonstrate the role of RAGE in the aggravated proliferation of cancer cells.

RAGE IN CANCER CELL SURVIVAL AND AUTOPHAGY

- Several hypothesized mechanisms have been postulated to explain the RAGE-mediated survival of cancer cells. The prominent signaling pathways activated by RAGE are, the (1) activation of pro-survival proteins in cells (like AKT); (2) induction of anti-apoptotic proteins (like Bcl-2); (3) impairing the translocation of pro-apoptotic p53 from the cytoplasm to mitochondria for autophagic induction (programmed cell survival); and (4) angiogenesis (the formation of new blood vessels). As a form of intracellular self-defense, autophagy involves the sequestration of proteins and organelles into autophagic vesicles, which are subsequently eliminated *via* lysosomal-fusion-mediated destruction. The term "programmed" cell survival is often used to describe this process [77]. Autophagy is a process of mediating tumor cell survival *via* degrading the cytoplasmic organelles, proteins, and macromolecules and recycling breakdown products. RAGE, a multi-ligand receptor, suppresses apoptosis *via* oxidative stress, promoting autophagy simultaneously. The AGE-RAGE interaction is a major source of ROS in cancer cells [78].

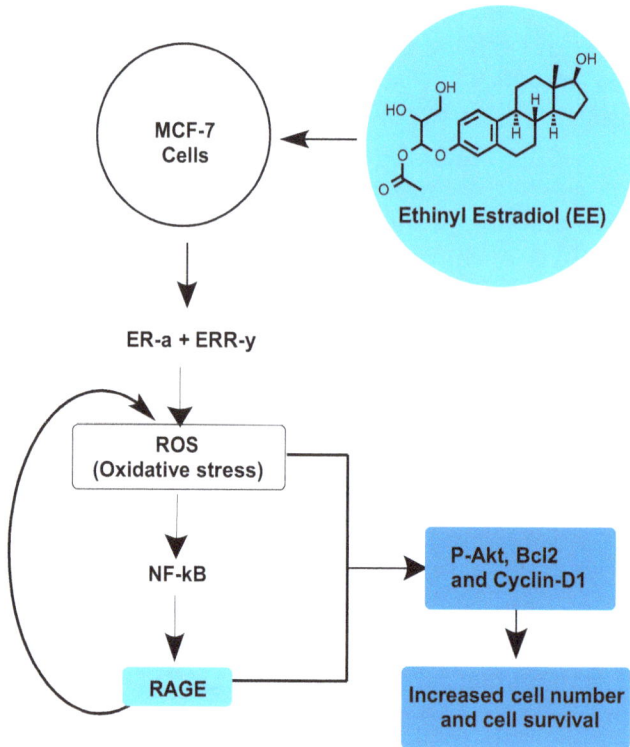

Fig. (5). Mechanistic depiction of the 17α-ethinyl-estradiol generated RAGE end products affected proliferation and survival of MCF-7 breast cancer cells.

- In a similar study, we also demonstrated that treatment of endothelial cells with either pro-inflammatory cytokine-TNFα [79], or steroid hormone, estrogens [80] induces ROS generation. For pancreatic cells, it was found that exposure of tumor cells to H_2O_2 provoked transcription factor NF-κB-aggravated RAGE expression. Further, it is well-known that targeted knockdown of RAGE leads to increased cell death by apoptosis *via* H_2O_2 induced oxidative injury and altered redox balance.
- Erstwhile attempts also decipher RAGE as a positive feedback regulator for transcription factor, NF-κB. This generalization is supported by the RAGE knockdown suppressed H_2O_2-induced NF-κB activity. Collectively, these findings suggest that RAGE is an important regulator of oxidative injury, elucidating the mechanism through which RAGE-generated ROS sustains the apoptosis-autophagy crosstalk [81]. A noteworthy effort by *Kristen M and the group* explained the involvement of the p53-HMGB1 complex in the regulation of autophagy and apoptosis [82]. The HMGB1 released after tumor cell death,

interacts with RAGE to aaggravate autophagy. Hence, aggravated RAGE expression is associated with enhanced autophagy through its interaction with HMGB1 and diminished apoptosis, enabling longer tumor cell survival. This affirmation is well-supported by the results of RAGE-knockdown studies in the tumor cells [83].

RAGE IN CANCER CELL ANGIOGENESIS

- Angiogenesis refers to the formation of new blood vessels that are generated from pre-existing vessels in the earlier phase of *de novo* endothelial cell generation. The process is not restricted to tumor cells although tumor cells do rapidly form new blood vessels due to defunct apoptosis. Tumor angiogenesis, therefore, not only involves enhanced blood vessel formation for nutrients and oxygen provision but also in sustaining the signaling activities triggered by metastasis. Numerous studies demonstrate the involvement of RAGE-ligand(s) axis (HMGB1, AGEs, S100) in aggravated angiogenesis, mediated *via* disruption of ECM constituent's degradation. The modulation of angiogenesis by RAGE is characterized by the activation of VEGF and MMP2, alongside the disruption of the VEGF-cadherin-catenin complex, supporting the capillary formation [18, 84, 85]. Besides this, RAGE activation also increases the macromolecule permeability of endothelial cells, a common aspect in tumor microvasculature. A notable effort by *Tsuji and colleagues* demonstrated the feasibility of angiogenesis *via* lymph node metastasis wherein latent membrane protein 1 trigger enhanced RAGE expression [86]. Similarly, autocrine VEGF has been demonstrated as a major stimulator of AGE-mediated angiogenesis. The AGE-RAGE interaction reduces the pericyte number and subsequently relieves the restriction on endothelial cell replication to facilitate angiogenesis [87]. Hence, collectively, it can be concluded that RAGE-ligand(s) interaction is a considerable factor stimulating tumor angiogenesis.

RAGE IN CANCER CELL APOPTOSIS

- Programmed cell death (apoptosis) is a process through which unwanted or beyond-repair cells are eliminated [88]. Defunct apoptosis implies undisrupted cell division and continual growth potential, making the cells immortal. Studies on RAGE-modulated apoptosis have demonstrated suppressed response leading to enhanced tumor cell survival. The best-known mechanism for RAGE-inhibited apoptosis to date has been *via* p53-dependent mitochondrial pathway [89]. Of note, p53 is a tumor suppressor gene that promotes apoptosis *via* transcription-dependent and independent mechanisms [18, 90]. Interaction of RAGE with its ligands impairs the p53 translocation to the mitochondria, inhibiting apoptosis and aiding the tumor cell survival. Besides p53, RAGE-modulated apoptosis is also triggered *via* RAGE-HMGB1 interaction, aggravating the oxidative and inflammatory stress.

- Experimental evidence supports that RAGE depletion by its specific siRNA aggravates the p53 phosphorylation at Ser392, enhancing the mitochondrial translocation of p53, furthered as mitochondrial cytochrome c release into the cytosol for enhanced apoptosis. A significant 2010 attempt from *Kang and the group* demonstrated RAGE-mediated inflammation *via* NF-κB and MAPK activation that aggravated autophagic response on suppressed apoptosis and enhanced tumor cell viability. The study hereby monitored RAGE expression in pancreatic (Panc1.28, Panc2.03, Panc3.27 and Panc02), breast (4T1), colon (MC38, CT26) and skin (B16V1, B16M05) cancer cells of C57/Bl6 mice and human origin. Analysis revealed the highest RAGE expression in pancreatic cell lines, whereby detailed inspection for mechanism inferred RAGE mediated apoptotic suppression *via* p53 dependent mitochondrial pathway. Apart from this, RAGE-aggravated autophagy is accompanied by suppressed mTOR phosphorylation and enhanced Beclin-1/VPS34 autophagosome formation. The results hereby established RAGE as an inflammation mediator, with a delayed expression in the tumor microenvironment activating the onset of programmed cell death (apoptosis). The observations were corroborated *via* reduced autophagy and suppressed tumor cell survival in RAGE knockdown mice [83].

RAGE IN CANCER CELL INVASION AND METASTASIS

- RAGE exhibits multiple crucial roles in the sustenance of melanoma cell metastasis. A noted study in this regard demonstrated that RAGE aggravation in the dermal cells manifested the formation of a primary melanoma tumor which gradually contributed to metastatic switch [70]. In melanoma cells, the RAGE level was screened to a greater extent in a subset of metastatic tumor strata at the transcription and translation platforms. Experimental data also supported that the metastatic human melanoma cells, G361 harbor higher cellular proliferation and migration extents in the presence of RAGE-activating AGE ligands. Furthermore, a higher RAGE extent implies an altered morphology with an elongated mesenchymal morphology, a typical trait of metastatic cells [90].

MECHANISM OF RAGE-DEPENDENT CANCER CELL DEVELOPMENT AND PROGRESSION

- The RAGE-dependent activation of multiple molecular-inflammatory events further promotes tumor survival, growth and invasion, hypoxia resistance, angiogenesis, tissue remodelling, and inhibition of host antitumor immune response. Based on these observations, a few important therapies are currently underway to destroy tumor cells *via* selective RAGE targeting.

- Myeloid-derived suppressor cells (MDSCs) are crucial in suppressing host immune responses as they produce nitric oxide (NO), arginase, ROS, and immunosuppressive cytokines, together stimulating the immune system and promoting malignant cell proliferation [91]. In the tumor and stromal compartments, overexpressed RAGE ligates to S100A8/A9 generated by MDSCs, assisting the regulation of a tumor gene profile by chemokines and creating a positive feedback loop to further encourage more MDSCs recruitment, stimulating the angiogenesis and tumor invasion [92, 93]. Additionally, it was discovered that RAGE is essential in maintaining tumor growth by releasing interleukin-6 (IL-6) into a pro-inflammatory milieu [70]. Studies on RAGE ligands, specifically S100A8/A9 proteins, have demonstrated their role in the recruitment and retention of MDSCs through a mechanism partially regulated *via* the binding of carboxylated N-glycans produced on the RAGE surface and the consequent activation of the NF-κB signaling cascade [93].
- Apart from the manifold aggravated signaling responses on ligand binding, RAGE could also undergo a ligand-mediated multimodal dimerization or oligomerization, sustained *via* self-association of its specific subunits. Studies provide evidence for RAGE-ligand interactions mediated prolonged cancer cell survival and proliferation [94 - 96]. The binding of ligands on the RAGE surface is known to augment the MAPK and NF-κB mediated cell-signaling activities, resulting in cellular propagation [97, 98]. The AGE-RAGE interaction aggravates cancer cell migration *via* enhanced activities of VEGF and ERK pathways (Table 1). The following paragraphs discuss the major aspects of these targets *vis-à-vis* redox modulation.

Table 1. Signaling targets aggravating tumorigenesis in response to RAGE-AGE interactions.

Name of the Affected Molecule/pathway	Normal Functional Response of the Pathway	Likely Role in Promoting Tumor Cell Survival and Growth
Carbohydrate Response Element Binding Protein, (ChREBP)	Bind to the carbohydrate response element in the pyruvate kinase promoter, the enzyme involved in glycolysis, lipogenesis, and gluconeogenesis	Enhanced anaerobic glucose metabolism, suppressed p53. colorectal and hepatocellular carcinoma being most affected
Janus Kinase/Signal Transducer and Activator of Transcription (JAK/STAT)	Regulates embryonic development, stem cell functioning, hematopoiesis, and inflammatory response. Transduction of cytokines, growth factors, and interleukin signals *via* transmembrane receptors	Helps in the transcription of genes related to the growth of cancer cells. Aggravated STAT3 activities driven by the AGE-RAGE axis have been observed in breast cancer and erythro-leukemia

(Table 1) cont.....

Name of the Affected Molecule/pathway	Normal Functional Response of the Pathway	Likely Role in Promoting Tumor Cell Survival and Growth
Mitogen-activated protein Kinase and Matrix Metalloproteinase 2 (MAPK and MMP2)	Coordinate extracellular signals to intracellular responses, more than a dozen in mammals collectively sustain cell proliferation, differentiation, motility, and survival	Binding of AGEs to RAGE upregulated select MAPK pathways and related family members like ERK, SP1, MMP2, MMP9, and p38. Activated SP1 regulates MMP2 secretion which degrades type IV collagen to boost metastasis of oral, gastric, and colorectal cancers. Also, ERK-aggravated MMP9 induces metastasis in breast and oral cancers
Nuclear factor (erythroid-derived 2)-like 2 (Nrf-2)	Regulates mitochondrial ROS homeostasis *via* (i) detoxifying peroxides, (ii) increasing GSH, and NADPH synthesis, and (iii) inhibiting TXNIP expression	May lead to cancer cell proliferation and metastasis *via* regulating p53 apoptotic pathway. Suppressed Nrf-2 activity inactivated the HO-1, p53 and apoptosis of oral cancer cells
Phosphatidylinositide 3-kinase/protein kinase B (PI3K/Akt)	An intracellular signal transduction pathway that promotes metabolism, proliferation, cell survival, growth, and angiogenesis in response to extracellular signals, phosphorylated Akt (an oncogene) is a hallmark of impaired apoptosis, over-expressed pAkt is a therapeutic target for treating malignant tumors	Activated by multiple factors, wherein Akt plays a key intermediate in survival signaling. Functional Akt phosphorylates and inhibits the pro-apoptotic Bad, Bax, caspase-9, GSK-3 and Foxo1 (pro-apoptotic Bcl-2 family members)

• ChREBP:

A systematic analysis of till date studies reveals colorectal and hepatocellular carcinoma as the major cancers affected by AGE *via* aggravated actions of Carbohydrate-responsive element-binding protein (ChREBP). The ChREBP induces the binding of carbohydrate response elements in the pyruvate kinase promoter, the major enzyme regulating glycolysis, lipogenesis, and gluconeogenesis [99]. Studies link this ChREBP tumor's progressing ability to enhanced anaerobic glucose metabolism and p53 suppression [100].

• JAK/STAT3:

Janus Kinase (JAK)/Signal Transducer and Activator of Transcription (STAT) performs a critical role in the pathologic manifestation of multiple diseases including cancers. Prominently, JAK/STAT signaling activates the transcription of genes responsible for cancer cell proliferation. Studies have correlated the AGE-RAGE axis instigated upregulation of STAT3 in erythroleukemia and breast adenocarcinoma [101, 102].

• MAPKs and MMPs:

AGEs binding to RAGE stimulates certain MAPK pathways and associated family members including ERK, specificity protein 1(Sp1), MMP2, MMP9, and p38 [103 - 108]. The transcription factor Sp1 expression is aggravated on AGE-RAGE binding *via* stimulated ERK pathway. Activated Sp1 thereafter, stimulates multiple tumor regulators such as MMP2 which degrade type IV collagen to assist tumor cell metastasis [109]. Such Sp1 actions *vis-à-vis* the AGE-RAGE axis, have been reported in human oral, gastric, and colorectal cancers [105, 106, 110]. Besides this, activated MMP9 energized by the ERK pathway *via* AGE-RAGE interactions, accelerates the metastatic breast and oral cancer progression [108, 110].

• Nrf-2

Involvement of transcription factor Nrf-2 in tumor proliferation and metastasis is driven *via* regulation of tumor suppressor protein p53 apoptotic pathway. For instance, activation of the AGE-RAGE axis in an oral cancer cell line (SAS) impairs Nrf-2 signaling *via* heme oxygenase (HO-1), eventually resulting in p53 suppression and impaired apoptosis of oral cancer cells [111].

• PI3K/Akt

Based on established PI3K, Akt, NF-κB, VEGF activation, and p53 suppression in multiple cancers, AGE-RAGE interactions aggravate the cancer cells growth by activating PI3K and eventual Akt phosphorylation. Accompanied events with these are activated NF-κB, VEGF, and mouse double minute 2 homolog (MDM2), all together delaying the apoptosis in tumor cells. Aggravated NF-κB triggered tumor cell growth *via* inactivating apoptosis, resulting in defunct caspases-mediated actions [112]. Enhanced VEGF expression stimulates angiogenesis while phosphor-Akt aggravated MDM2 protein impaired p53 (a tumor suppressor gene), terminating both intrinsic and extrinsic apoptotic pathways for proliferating cancer cells.

• The second ligand with which RAGE interacts is HMGB-1 which protects the tumor cells from apoptosis by perturbing telomere stability and stimulating the proteins involved in tumor cell proliferation [113]. Table **2** summarizes the impact of RAGE-HMGB1 interactions on the activation of certain anti-apoptotic molecules and impairment of pro-apoptotic genes/proteins. The major biochemical aspects of these events are as discussed ahead.

• Beclin-1

Known to induce tumor progression *via* regulating growth factor receptor signaling in an autophagy-independent regime, Beclin-1 controlled epidermal growth factor (EGF) and insulin-like growth factor-1 (IGF-1), culminates in Akt and ERK activation driven breast cancer growth [114]. On a similar basis, Beclin-1 activation by AGE-RAGE interaction has been demonstrated to aggravate human pancreatic cancer, Panc2.03 cell lines [115].

Table 2. HMGB1-RAGE interaction mediated tumor growth aggravating controls.

Name of the Affected Molecule/pathway	Regulatory Role in Cancer Cell Development	Effect of HMGB1-RAGE Signaling on the Receptor Action in Tumor Cells
Beclin-1	Associated with cancer cell autophagy, regulates EGF and ILGF-1, activating Akt and ERK in breast cancer cells	Upregulation by AGEs aggravated the tumorigenicity of human pancreatic cancer (Panc2.03) cells
MAPKs	Coordinate external signaling response to intracellular actions, regulate cell division, proliferation, survival differentiation, and motility	HMGB1 aggravated ERK/MAPK pathway aided in survival and proliferation of colon, gastric, hepatocellular, renal cell carcinomas, and breast adenocarcinoma. ERK activation enhanced c-Myc and suppressed p21, assisting cell division *via* enhanced CDK-2 and cyclin D1 expression in hepatocellular carcinoma. HMGB1 upregulated MAPK members (including Cdc42, Rac1, MKK6/p38) aggravated human rhabdomyosarcomas and hepatocellular carcinoma. HMGB1-RAGE activation upregulated SAPK/JNK in glioma, murine Lewis's lung, and hepatocellular carcinoma (HCC).
Micro RNA	Small non-coding RNAs regulate diverse cellular functions, abnormal regulation affects cancer proliferation, invasion, and metastasis	HMGB1-RAGE axis upregulated miR-221/222 in thyroid cancer cells, decreasing p27 protein extents, causing S-phase advance of cancer cells. Similar upregulation of miRNAs *via* HMGB1-RAGE increased cancer progression
Matrix Metallo proteinases (MMPs)	Proteolytic enzymes degrade ECM components to aggravate tumor invasion, angiogenesis, metastasis: deciphering pharmacological utility for cancer therapy	HMGB1 promoted aggravation of MMPs (1,2,3,7,9,10), leading to proliferation of colon, pancreatic, pulmonary, and hepatocellular carcinoma. Upregulated MMPs degraded ECM components to assist tumor progression
Nuclear factor-kappa B (NF-κB)/Snail	Serves as a transcription factor, optimum NF-κB expression kept Snail in control to arrest the N-cadherin and vimentin-mediated EMT	Enhanced EMT *via* HMGB1 aggravated NF-κB expression has been reported in human pancreatic (BxPC-3), hepatocellular (HUH7, HUH22), NSCLC and breast adenocarcinomas (MCF-7) cells

(Table 2) cont.....

Name of the Affected Molecule/pathway	Regulatory Role in Cancer Cell Development	Effect of HMGB1-RAGE Signaling on the Receptor Action in Tumor Cells
Phosphatidylinositide 3-kinase/protein kinase B (PI3K/Akt)	An intracellular signal transduction pathway that supports metabolism, proliferation, cell survival, growth, and angiogenesis in response to extracellular signals	HMGB1-RAGE binding caused VEGF overexpression in human oral squamous cells and bladder carcinomas, aiding angiogenesis. PI3K/Akt stimulation suppressed pro-apoptotic Bax, caspase-3 in human nasopharyngeal carcinoma

Abbreviations: EGF: Epidermal Growth Factor, ILGF-1: Insulin-Like Growth Factor-1, ECM: Extracellular Matrix, NSCLC: Non-small Cell Lung Cancer, VEGF: Vascular Endothelial Growth Factor Receptor; Bax: Bcl-2 associated X protein, Cdc42: Cell Division Control Protein 42 homolog, Rac1: Ras-related C3 botulinum toxin substrate 1 (Rac1), MKK6: Mitogen-activated protein kinase kinase 6.

• MAPK

Multiple studies demonstrate the HMGB1-RAGE interactions aggravated the ERK/MAPK pathway, leading to survival and proliferation of colon, renal, liver, gastric, hepatocellular, and breast cancers [116 - 118]. The ERK activation by HMGB1-RAGE interactions stimulated c-Myc expression, downregulating p21 and activating the cell cycle regulating proteins, cyclin-dependent kinase-2 (CDK-2), and cyclin D1 in hepatocellular carcinoma [119]. Apart from this, ERK activation aggravated MMP2 and MMP9 expressions, leading to cell proliferation and metastasis [120]. Erstwhile MAPK members activated by RAGE-HMGB1 interactions include cell division control protein 42 homolog (Cdc42)/Ras-related C3 botulinum toxin substrate 1 (Rac1)/MAPK kinase6/p38, stimulating the growth of human rhabdomyosarcoma and hepatocellular carcinoma [119, 121]. Stress-activated protein kinases (SAPk)/c-Jun N-terminal kinase (JNK) (an additional member of MAPKs) is upregulated *via* HMGB1-RAGE interaction, in glioma, murine lung, and hepatocellular carcinomas [45, 119].

• MicroRNA

MicroRNAs (miRNAs) are small non-coding RNAs regulating diversified cell functions and are well-versed in their effects on cancer cell proliferation, invasion, and metastasis. HMGB1-RAGE axis aggravates miR-221/222 in thyroid carcinoma, decreasing p27 (Kip1) protein extents, a prominent cell cycle sustaining molecule mediating advancing thyroid carcinoma in S-phase [122]. Another miRNA, miR-155-5p was administered to colorectal cancer cells through macrophage-derived exosomes and impaired BRG1 (one of the two catalytic subunits of the SWI/SNF ATP-dependent chromatin remodeling complex) expression, a decisive factor responsible for regulating metastasis in colorectal cancer [123].

• MMPs

Known to degrade the multiple ECM components *via* proteolytic actions, the tumor invasion, angiogenesis, and metastatic actions of MMPs (MMP1, MMP2, MMP3, MMP7, MMP9, MMP10) make them reliable pharmacological anti-cancer therapeutic targets [68, 124]. HMGB1-aggravated MMP activities have been demonstrated as critical regulators in the proliferation of colon, pancreatic, pulmonary, and hepatocellular carcinoma [125 - 127]. Degradation of ECM components by activated MMPs assists in invading tumor cells.

• NF-κB/Snail

HMGB1-RAGE interactions driven NF-κB expression in human pancreatic (BxPC-3), hepatocellular (HUH7, H22), NSCLC and breast adenocarcinoma (MCF-7) cells resulted in aggravated expression of Snail, thereafter activating N-cadherin and vimentin. Activated N-cadherin and vimentin stimulate EMT and augment the cancer cell migration. Apart from the above, Snail-mediated impaired E-cadherin and phosphatase-tensin homolog (PTEN) fueled cell proliferation and metastasis [128, 129].

• PI3K/Akt

Activation of PI3K/Akt by RAGE-HMGB1 signaling regulated the functioning of multiple proteins associated with aggravated proliferation and suppressed apoptosis in tumor cells, such as Vascular Endothelial Growth Factor (VEGF). The activation of VEGF following HMGB1-RAGE interaction has been demonstrated in human oral squamous cell carcinoma and bladder carcinoma, aiding angiogenesis, and proliferation of tumor cells. On similar lines, PI3K/Akt activation suppressed the activity of pro-apoptotic proteins such as Bcl-2 associated X protein (Bax) and caspase-3 in human nasopharyngeal carcinoma and lung cancer cells [130].

• S100 proteins as RAGE ligands, are involved in multiple regulatory aspects of cancer cells including proliferation, differentiation, apoptosis, and inflammation. Some prominent cancers like human osteosarcoma, melanoma, pancreatic, colorectal carcinoma, and breast adenocarcinoma exhibit a distinct S100 protein activity. The mechanisms and target proteins by which RAGE-S100 signaling aggravates the tumor cell's growth, differentiation, and metastasis are summarized in Table **3**.

Table 3. S100 Family-RAGE axis mediated aggravated tumorigenesis events.

Name of the Affected Molecule/pathway	Regulatory Role in Cancer Cell Development	The Resultant Impact of the S100-RAGE Interaction
Angiogenesis	Formation of new blood vessels from prior existing vessels, an essential process of cancer cell proliferation *via* nutrient and oxygen provision	S100s-RAGE binding in pancreatic, colorectal, and breast cancers resulted in VEGF suppression: retarding vasculogenesis and angiogenesis
MAPKs	Coordination of intracellular responses *via* cell signaling to assist in cell proliferation, differentiation, motility, and survival	S100 members aggravated MAPK family: ERK, Cdc42/p38, SAPK/JNK pathways leading to colorectal, thyroid, nasopharyngeal, breast, and prostate cancers
Matrix Metallo proteinases (MMPs)	Enzymes degrading ECM components for easier access to tumor cells besides aiding angiogenesis, metastasis: deciphering pharmacological utility for cancer therapy	S100A2, S100A4, and S100A7 are involved in breast tumorigenesis and progression in varied extents. S100A7 driven MMP9 upregulation was enhanced in pancreatic carcinoma (BxPC3) and nasopharyngeal carcinoma (C666-1) Alongside S100P triggered MMP2
Nuclear factor-kappa B (NF-κB)	Mediates crosstalk between inflammation and cancer at multiple levels. Aggravated NF-κB activity in tumor cells, fuels the gathering of pro-inflammatory cytokines at the tumor site, establishing the pro-tumorigenic microenvironment	RAGE activation by S100A4, S100A7, S100A8, S100A9, S100A14, S100B, and S100P is accompanied by NF-κB upregulation-induced progression of pancreatic, melanoma, breast, prostate, colon, esophageal cancers. S100A4 stimulated E-cadherin, regulated *via* NF-κB/Snail aggravated cancer metastasis
p53	A tumor suppressor gene whose activity terminates tumor formation. Inheritance of only one functional copy of p53 gene exposes an individual to cancers, with increased vulnerability to independent tumor development in multiple tissues during early age	To date studies have established the effect of S100B on p53, inhibiting the intrinsic and extrinsic apoptotic pathways. Besides recent studies demonstrated that S100A16 reduced p21 and p27 expressions in human prostate cancer, promoting tumor cell propagation.
PI3K/Akt/mTOR	An intracellular pathway regulating tumor cell proliferation. Plays a critical role in the development of endocrine resistance in breast cancer. PI3K is frequently altered in cancers, contributing to tumor growth and survival	Upregulated by S100A4 in colorectal, S100A16 in the prostate, and S100B in neuroblastoma cancers, promoted growth *via* anti-apoptosis and cell-proliferation
Signal Transducer and Activator of Transcription 3 (STAT3)	Transcription factor that regulates the expression of genes related to cell cycle, cell survival, and cancer malignancy	STAT3 activation aided in its nuclear translocation where it promotes translation of target genes. Inhibition of the S100B-RAGE pathway could control certain STAT3 cancer progression

• Angiogenesis

Being a primitive and fundamental process for cancer cell proliferation and viability, angiogenesis is characterized by the formation of new blood vessels from those previously existing (to ensure provision of nutrition and oxygen). RAGE-S100 signaling surprisingly impairs the VEGF mediated vasculogenesis, and angiogenesis, resisting the metastasis of solid tumors [131]. RAGE-S10--mediated impaired VEGF actions have been demonstrated in pancreatic, colorectal, and breast cancers [132, 133].

• MAPKs

S100-RAGE signaling activates the MAPK family members, prominently, ERK, Cdc42/p38 or SAPK/JNK pathways, resulting in progression of colorectal (HCT116, MC38, SW620, DLD-1), thyroid, nasopharyngeal, breast (all in human specimens) along with prostate (LnCaP, PC-3) cancers [92, 134, 135]. The resulting outcomes favored tumor cell progression and proliferation.

• MMPs

The prominent members of the S100 family reported aggravating tumorigenesis *via* RAGE-S100 signaling including S100A2, S100A4, and S100A7, all reported for distinct expressions in breast cancer onset and progression [136]. A noted effort by *Nesser and colleagues* reported S100A7-mediated progression of triple-negative breast cancer (human specimens) *via* upregulated MMP9 activity [73]. Erstwhile studies demonstrated exaggerated MMP9 expression in pancreatic (BxPC3) and nasopharyngeal (C666-1) carcinoma, along with MMP2-activated S100P-RAGE interactions. Thereby, S100A7 and S100P act as authentic therapeutic targets to treat pancreatic and nasopharyngeal carcinomas *via* diverse RAGE proximities [137, 138].

• NF-κB

S100 family members are known to upregulate NF-κB following interactions with RAGE including S100A4, S100A7, S100A8, S100A9, S100A14, S100B, and S100P, resulting in the progression of varied tumors including pancreatic, melanoma, breast, prostate, colon, neuroblastoma and esophageal. The S100A4-activated E-cadherin, regulated by NF-κB/Snail eventually resulted in enhanced cancer metastasis. More conclusive attempts to understand the S100s family members induced cancer cell progression *via* NF-κB signaling are presently underway as concomitant effects on inhibitors of apoptosis members (IAPs), B-cell lymphoma (Bcl-2), and Snail [139].

• p53

RAGE activation by S100B is exhaustively reported for worsened pathology in melanoma, lung, breast, colorectal, and ovarian cancers, leading to p53 impairment suppressed intrinsic and extrinsic apoptosis in cancer cells. Disappointingly, the effect of RAGE-S100B signaling has not yet been validated through actions other than of p53 proteins (caspases 3, 8, and 9). Erstwhile efforts of recent origin do establish substantial decrements in p21 and p27 expressions *via* RAGE-S100A16 interactions in human prostate cancer, causing cell-cycle sustenance [140].

• P13K/Akt/mTOR

RAGE-S100 signaling has been reported to aggravate PI3K/Akt/mTOR pathway in colorectal (*via* S100A4), prostate (S100A16), neuroblastoma (S100B) cancers. The characteristic implications mediated herein include disrupted apoptosis and stimulated cell proliferation.

• STAT3

S100B-RAGE signaling has been reported as instrumental in glioma cells *via* aggravated tumorigenic effect [141, 142]. The transcription factor STAT3 controls the functioning of genes related to the cell cycle, cell survival, and tumor aggressiveness. When STAT3 is activated, it moves from the cytoplasm to the nucleus, where it triggers the translation of target genes linked to angiogenesis, anti-apoptotic processes, and invasion-aggravating actions. Although much remains unknown, the studies to date decipher the S100B-RAGE pathway as a reliable therapeutic target to terminate STAT3-mediated tumor cell progression.

CONCLUSION

In a nutshell, this chapter focused on manifold ligand-diversified tumorigenic actions of RAGE biochemical diversity. The tumor cell aggravating RAGE actions are primarily driven by aggressive inflammatory and pro-oxidative responses, whereby the interactions with AGEs, HMGB1, and the S100A family are critical. Contrary to most cancers where RAGE activities aggravate the inflammatory responses and have been successfully validated as reliable therapeutic targets, the lung cancer actions of RAGE are quite distinct. While a certain minimal extent of RAGE is critical for normal lung functioning, abundant RAGE expression protects against NSCLCs. Specific polymorphic variations of RAGE in the lung tissue may be related to their protection against NSCLCs, which requires further experimental validation.

In most other well-studied cancers (like breast, pancreatic, bladder, hepatocellular), RAGE-ligand interactions aggravate the inflammation and concomitantly alter the redox environment. The resultant dynamic redox conditions activate the oncogenes and inactivate the tumor suppressor genes, to assist in the continued tumor cell survival. Additionally, RAGE-ligand signaling also degrades the ECM proteins (*via* collagen IV disruption), facilitating the tumor cell trafficking of nutrients and blood. Discussions made in this chapter also validate anti-RAGE therapies in synergism with pleiotropic natural polyphenols as a reliable treatment course for tumors, attributed to their multi-targeted action mechanism. Design protocols and studies are however in progress for assured findings wherein computationally assisted protocols may enhance the reliability of targeted therapeutic approaches *via* active drug delivery mechanisms.

REFERENCES

[1]   Schmidt AM, Vianna M, Gerlach M, *et al.* Isolation and characterization of two binding proteins for advanced glycosylation end products from bovine lung which are present on the endothelial cell surface. J Biol Chem 1992; 267(21): 14987-97.
[http://dx.doi.org/10.1016/S0021-9258(18)42137-0] [PMID: 1321822]

[2]   Brett J, Schmidt AM, Yan SD, *et al.* Survey of the distribution of a newly characterized receptor for advanced glycation end products in tissues. Am J Pathol 1993; 143(6): 1699-712.
[PMID: 8256857]

[3]   Schlueter C, Hauke S, Flohr AM, Rogalla P, Bullerdiek J. Tissue-specific expression patterns of the RAGE receptor and its soluble forms—a result of regulated alternative splicing? Biochim Biophys Acta Gene Struct Expr 2003; 1630(1): 1-6.
[http://dx.doi.org/10.1016/j.bbaexp.2003.08.008] [PMID: 14580673]

[4]   Kalea AZ, See F, Harja E, Arriero M, Schmidt AM, Hudson BI. Alternatively spliced RAGEv1 inhibits tumorigenesis through suppression of JNK signaling. Cancer Res 2010; 70(13): 5628-38.
[http://dx.doi.org/10.1158/0008-5472.CAN-10-0595] [PMID: 20570900]

[5]   Katsuoka F, Kawakami Y, Arai T, *et al.* Type II alveolar epithelial cells in lung express receptor for advanced glycation end products (RAGE) gene. Biochem Biophys Res Commun 1997; 238(2): 512-6.
[http://dx.doi.org/10.1006/bbrc.1997.7263] [PMID: 9299542]

[6]   Shirasawa M, Fujiwara N, Hirabayashi S, *et al.* Receptor for advanced glycation end-products is a marker of type I lung alveolar cells. Genes Cells 2004; 9(2): 165-74.
[http://dx.doi.org/10.1111/j.1356-9597.2004.00712.x] [PMID: 15009093]

[7]   Fehrenbach H, Kasper M, Tschernig T, Shearman MS, Schuh D, Müller M. Receptor for advanced glycation endproducts (RAGE) exhibits highly differential cellular and subcellular localisation in rat and human lung. Cell Mol Biol 1998; 44(7): 1147-57.
[PMID: 9846897]

[8]   Demling N, Ehrhardt C, Kasper M, Laue M, Knels L, Rieber EP. Promotion of cell adherence and spreading: a novel function of RAGE, the highly selective differentiation marker of human alveolar epithelial type I cells. Cell Tissue Res 2006; 323(3): 475-88.
[http://dx.doi.org/10.1007/s00441-005-0069-0] [PMID: 16315007]

[9]   Neeper M, Schmidt AM, Brett J, *et al.* Cloning and expression of a cell surface receptor for advanced glycosylation end products of proteins. J Biol Chem 1992; 267(21): 14998-5004.
[http://dx.doi.org/10.1016/S0021-9258(18)42138-2] [PMID: 1378843]

[10] Crapo JD, Barry BE, Gehr P, Bachofen M, Weibel ER. Cell number and cell characteristics of the normal human lung. Am Rev Respir Dis 1982; 126(2): 332-7.
[http://dx.doi.org/10.1164/arrd.1982.126.2.332] [PMID: 7103258]

[11] Lizotte PP, Hanford LE, Enghild JJ, Nozik-Grayck E, Giles BL, Oury TD. Developmental expression of the receptor for advanced glycation end-products (RAGE) and its response to hyperoxia in the neonatal rat lung. BMC Dev Biol 2007; 7(1): 15.
[http://dx.doi.org/10.1186/1471-213X-7-15] [PMID: 17343756]

[12] Hudson BI, Lippman ME. Targeting RAGE signaling in inflammatory diseases. Annu Rev Med 2018; 69(1): 349-64.
[http://dx.doi.org/10.1146/annurev-med-041316-085215] [PMID: 29106804]

[13] Chavakis T, Bierhaus A, Nawroth PP. RAGE (receptor for advanced glycation end products): a central player in the inflammatory response. Microbes Infect 2004; 6(13): 1219-25.
[http://dx.doi.org/10.1016/j.micinf.2004.08.004] [PMID: 15488742]

[14] Mukherjee TK, Mukhopadhyay S, Hoidal JR. Implication of receptor for advanced glycation end product (RAGE) in pulmonary health and pathophysiology. Respir Physiol Neurobiol 2008; 162(3): 210-5.
[http://dx.doi.org/10.1016/j.resp.2008.07.001] [PMID: 18674642]

[15] Schraml P, Bendik I, Ludwig CU. Differential messenger RNA and protein expression of the receptor for advanced glycosylated end products in normal lung and non-small cell lung carcinoma. Cancer Res 1997; 57(17): 3669-71.
[PMID: 9288769]

[16] Schraml P, Shipman R, Colombi M, Ludwig CU. Identification of genes differentially expressed in normal lung and non-small cell lung carcinoma tissue. Cancer Res 1994; 54(19): 5236-40.
[PMID: 7923146]

[17] Stav D, Bar I, Sandbank J. Usefulness of CDK5RAP3, CCNB2, and RAGE genes for the diagnosis of lung adenocarcinoma. Int J Biol Markers 2007; 22(2): 108-13.
[http://dx.doi.org/10.1177/172460080702200204] [PMID: 17549666]

[18] Hofmann HS, Hansen G, Burdach S, Bartling B, Silber RE, Simm A. Discrimination of human lung neoplasm from normal lung by two target genes. Am J Respir Crit Care Med 2004; 170(5): 516-9.
[http://dx.doi.org/10.1164/rccm.200401-127OC] [PMID: 15331390]

[19] Jing R, Cui M, Wang J, Wang H. Receptor for advanced glycation end products (RAGE) soluble form (sRAGE): a new biomarker for lung cance. Neoplasma 2010; 57(1): 55-61.
[http://dx.doi.org/10.4149/neo_2010_01_055] [PMID: 19895173]

[20] Bartling B, Hofmann HS, Weigle B, Silber RE, Simm A. Down-regulation of the receptor for advanced glycation end-products (RAGE) supports non-small cell lung carcinoma. Carcinogenesis 2004; 26(2): 293-301.
[http://dx.doi.org/10.1093/carcin/bgh333] [PMID: 15539404]

[21] Bartling B, Hofmann HS, Sohst A, *et al.* Prognostic potential and tumor growth-inhibiting effect of plasma advanced glycation end products in non-small cell lung carcinoma. Mol Med 2011; 17(9-10): 980-9.
[http://dx.doi.org/10.2119/molmed.2011.00085] [PMID: 21629968]

[22] Yu YX, Pan WC, Cheng YF. Silencing of advanced glycosylation and glycosylation and product-specific receptor (RAGE) inhibits the metastasis and growth of non-small cell lung cancer. Am J Transl Res 2017; 9(6): 2760-74.
[PMID: 28670367]

[23] Malik P, Chaudhry N, Mittal R, Mukherjee TK. Role of receptor for advanced glycation end products in the complication and progression of various types of cancers. Biochim Biophys Acta, Gen Subj 2015; 1850(9): 1898-904.

[http://dx.doi.org/10.1016/j.bbagen.2015.05.020] [PMID: 26028296]

[24]   Palanissami G, Paul SFD. RAGE and its ligands: Molecular interplay between glycation, inflammation, and hallmarks of cancer-a Review. Horm Cancer 2018; 9(5): 295-325.
[http://dx.doi.org/10.1007/s12672-018-0342-9] [PMID: 29987748]

[25]   Koch M, Chitayat S, Dattilo BM, *et al.* Structural basis for ligand recognition and activation of RAGE. Structure 2010; 18(10): 1342-52.
[http://dx.doi.org/10.1016/j.str.2010.05.017] [PMID: 20947022]

[26]   Xiong W-C, Xiong W-C. RAGE and its ligands in bone metabolism. Front Biosci (Schol Ed) 2011; S3(2): 768-76.
[http://dx.doi.org/10.2741/s185] [PMID: 21196410]

[27]   Rai V, Touré F, Chitayat S, *et al.* Lysophosphatidic acid targets vascular and oncogenic pathways *via* RAGE signaling. J Exp Med 2012; 209(13): 2339-50.
[http://dx.doi.org/10.1084/jem.20120873] [PMID: 23209312]

[28]   Kierdorf K, Fritz G. RAGE regulation and signaling in inflammation and beyond. J Leukoc Biol 2013; 94(1): 55-68.
[http://dx.doi.org/10.1189/jlb.1012519] [PMID: 23543766]

[29]   Sims GP, Rowe DC, Rietdijk ST, Herbst R, Coyle AJ. HMGB1 and RAGE in inflammation and cancer. Annu Rev Immunol 2010; 28(1): 367-88.
[http://dx.doi.org/10.1146/annurev.immunol.021908.132603] [PMID: 20192808]

[30]   Armstrong A, Ravichandran KS. Phosphatidylserine receptors: what is the new RAGE? EMBO Rep 2011; 12(4): 287-8.
[http://dx.doi.org/10.1038/embor.2011.41] [PMID: 21399618]

[31]   Wang H, Li Y, Yu W, Ma L, Ji X, Xiao W. Expression of the receptor for advanced glycation end-products and frequency of polymorphism in lung cancer. Oncol Lett 2015; 10(1): 51-60.
[http://dx.doi.org/10.3892/ol.2015.3200] [PMID: 26170976]

[32]   Pan H, Niu W, He L, *et al.* Contributory role of five common polymorphisms of RAGE and APE1 genes in lung cancer among Han Chinese. PLoS One 2013; 8:7: e69018.
[http://dx.doi.org/10.1371/journal.pone.0069018]

[33]   Wang X, Cui E, Zeng H, *et al.* RAGE genetic polymorphisms are associated with risk, chemotherapy response and prognosis in patients with advanced NSCLC. PLoS One 2012; 7(10): e43734.
[http://dx.doi.org/10.1371/journal.pone.0043734] [PMID: 23071492]

[34]   Schenk S, Schraml P, Bendik I, Ludwig CU. A novel polymorphism in the promoter of the RAGE gene is associated with non-small cell lung cancer. Lung Cancer 2001; 32(1): 7-12.
[http://dx.doi.org/10.1016/S0169-5002(00)00209-9] [PMID: 11282423]

[35]   Yue L, Zhang Q, He L, *et al.* Genetic predisposition of six well-defined polymorphisms in HMGB 1/ RAGE pathway to breast cancer in a large Han Chinese population. J Cell Mol Med 2016; 20(10): 1966-73.
[http://dx.doi.org/10.1111/jcmm.12888] [PMID: 27241711]

[36]   Tesařová P, Kalousová M, Jáchymová M, Mestek O, Petruželka L, Zima T. Receptor for advanced glycation end products (RAGE)--soluble form (sRAGE) and gene polymorphisms in patients with breast cancer. Cancer Invest 2007; 25(8): 720-5.
[http://dx.doi.org/10.1080/07357900701560521] [PMID: 18058469]

[37]   Gu H, Yang L, Sun Q, *et al.* Gly82Ser polymorphism of the receptor for advanced glycation end products is associated with an increased risk of gastric cancer in a Chinese population. Clin Cancer Res 2008; 14(11): 3627-32.
[http://dx.doi.org/10.1158/1078-0432.CCR-07-4808] [PMID: 18519797]

[38]   Krechler T, Jáchymová M, Mestek O, Žák A, Zima T, Kalousová M. Soluble receptor for advanced glycation end-products (sRAGE) and polymorphisms of RAGE and glyoxalase I genes in patients with

pancreas cancer. Clin Biochem 2010; 43(10-11): 882-6.
[http://dx.doi.org/10.1016/j.clinbiochem.2010.04.004] [PMID: 20398646]

[39]   Fritz G. RAGE: a single receptor fits multiple ligands. Trends Biochem Sci 2011; 36(12): 625-32.
[http://dx.doi.org/10.1016/j.tibs.2011.08.008] [PMID: 22019011]

[40]   Lin L. RAGE on the toll road? Cell Mol Immunol 2006; 3(5): 351-8.
[PMID: 17092432]

[41]   Gebhardt C, Riehl A, Durchdewald M, *et al.* RAGE signaling sustains inflammation and promotes
tumor development. J Exp Med 2008; 205(2): 275-85.
[http://dx.doi.org/10.1084/jem.20070679] [PMID: 18208974]

[42]   Mukherjee TK, Malik P, Hoidal JR. Receptor for Advanced Glycation End (RAGE) products and its
polymorphic variants as predictive diagnostic and prognostic markers of NSCLCs: a perspective. Curr
Oncol Rep 2021; 23(1): 12.
[http://dx.doi.org/10.1007/s11912-020-00992-x] [PMID: 33399986]

[43]   Miyazaki N, Abe Y, Oida Y, *et al.* Poor outcome of patients with pulmonary adenocarcinoma showing
decreased E-cadherin combined with increased S100A4 expression. Int J Oncol 2006; 28(6): 1369-74.
[http://dx.doi.org/10.3892/ijo.28.6.1369] [PMID: 16685438]

[44]   Zhang YB, He FL, Fang M, *et al.* Increased expression of Toll-like receptors 4 and 9 in human lung
cancer. Mol Biol Rep 2009; 36(6): 1475-81.
[http://dx.doi.org/10.1007/s11033-008-9338-9] [PMID: 18763053]

[45]   Taguchi A, Blood DC, del Toro G, *et al.* Blockade of RAGE–amphoterin signalling suppresses tumour
growth and metastases. Nature 2000; 405(6784): 354-60.
[http://dx.doi.org/10.1038/35012626] [PMID: 10830965]

[46]   Hou C, Lu M, Lei Z, *et al.* HMGB1 positive feedback loop between cancer cells and tumor-associated
macrophages promotes osteosarcoma migration and invasion. Lab Invest 2023; 103(5): 100054.
[http://dx.doi.org/10.1016/j.labinv.2022.100054] [PMID: 36801636]

[47]   Yuan S, Liu Z, Xu Z, Liu J, Zhang J. High mobility group box 1 (HMGB1): a pivotal regulator of
hematopoietic malignancies. J Hematol Oncol 2020; 13(1): 91.
[http://dx.doi.org/10.1186/s13045-020-00920-3] [PMID: 32660524]

[48]   Faruqui T, Singh G, Khan S, Khan MS, Akhter Y. Differential gene expression analysis of RAGE-
S100A6 complex for target selection and the design of novel inhibitors for anticancer drug discovery. J
Cell Biochem 2023; 124(2): 205-20.
[http://dx.doi.org/10.1002/jcb.30356] [PMID: 36502516]

[49]   Marinakis E, Bagkos G, Piperi C, Roussou P, Diamanti-Kandarakis E. Critical role of RAGE in lung
physiology and tumorigenesis: a potential target of therapeutic intervention? Clinical Chemistry and
Laboratory Medicine (CCLM) 2014; 52(2): 189-200.
[http://dx.doi.org/10.1515/cclm-2013-0578] [PMID: 24108211]

[50]   Huttunen HJ, Fages C, Kuja-Panula J, Ridley AJ, Rauvala H. Receptor for advanced glycation end
products-binding COOH-terminal motif of amphoterin inhibits invasive migration and metastasis.
Cancer Res 2002; 62(16): 4805-11.
[PMID: 12183440]

[51]   Kobayashi S, Kubo H, Suzuki T, *et al.* Endogenous secretory receptor for advanced glycation end
products in non-small cell lung carcinoma. Am J Respir Crit Care Med 2007; 175(2): 184-9.
[http://dx.doi.org/10.1164/rccm.200602-212OC] [PMID: 17023736]

[52]   Chang YH, Chen CM, Chen HY, Yang PC. Pathway-based gene signatures predicting clinical
outcome of lung adenocarcinoma. Sci Rep 2015; 5(1): 10979.
[http://dx.doi.org/10.1038/srep10979] [PMID: 26042604]

[53]   de Bittencourt Pasquali MA, Gelain DP, Zeidán-Chuliá F, *et al.* Vitamin A (retinol) downregulates the
receptor for advanced glycation endproducts (RAGE) by oxidant-dependent activation of p38 MAPK

and NF-kB in human lung cancer A549 cells. Cell Signal 2013; 25(4): 939-54.
[http://dx.doi.org/10.1016/j.cellsig.2013.01.013] [PMID: 23333461]

[54]　Sanders KA, Delker DA, Huecksteadt T, *et al.* RAGE is a critical mediator of pulmonary oxidative stress, alveolar macrophage activation and emphysema in response to cigarette smoke. Sci Rep 2019; 9(1): 231.
[http://dx.doi.org/10.1038/s41598-018-36163-z] [PMID: 30659203]

[55]　Xia W, Xu Y, Mao Q, *et al.* Association of RAGE polymorphisms and cancer risk: a meta-analysis of 27 studies. Med Oncol 2015; 32(2): 33.
[http://dx.doi.org/10.1007/s12032-014-0442-5] [PMID: 25603950]

[56]　Wu S, Mao L, Li Y, *et al.* RAGE may act as a tumour suppressor to regulate lung cancer development. Gene 2018; 651: 86-93.
[http://dx.doi.org/10.1016/j.gene.2018.02.009] [PMID: 29421442]

[57]　Turovskaya O, Foell D, Sinha P, *et al.* RAGE, carboxylated glycans and S100A8/A9 play essential roles in colitis-associated carcinogenesis. Carcinogenesis 2008; 29(10): 2035-43.
[http://dx.doi.org/10.1093/carcin/bgn188] [PMID: 18689872]

[58]　Onyeagucha BC, Mercado-Pimentel ME, Hutchison J, Flemington EK, Nelson MA. S100P/RAGE signaling regulates microRNA-155 expression *via* AP-1 activation in colon cancer. Exp Cell Res 2013; 319(13): 2081-90.
[http://dx.doi.org/10.1016/j.yexcr.2013.05.009] [PMID: 23693020]

[59]　Lata K, Mukherjee TK. Knockdown of receptor for advanced glycation end products attenuate 17α-ethinyl-estradiol dependent proliferation and survival of MCF-7 breast cancer cells. Biochim Biophys Acta, Gen Subj 2014; 1840(3): 1083-91.
[http://dx.doi.org/10.1016/j.bbagen.2013.11.014] [PMID: 24252278]

[60]　Kuniyasu H, Oue N, Wakikawa A, *et al.* Expression of receptors for advanced glycation end-products (RAGE) is closely associated with the invasive and metastatic activity of gastric cancer. J Pathol 2002; 196(2): 163-70.
[http://dx.doi.org/10.1002/path.1031] [PMID: 11793367]

[61]　Heijmans J, Büller NVJA, Hoff E, *et al.* Rage signalling promotes intestinal tumourigenesis. Oncogene 2013; 32(9): 1202-6.
[http://dx.doi.org/10.1038/onc.2012.119] [PMID: 22469986]

[62]　Liang H, Zhong Y, Zhou S, Peng L. Knockdown of RAGE expression inhibits colorectal cancer cell invasion and suppresses angiogenesis *in vitro* and *in vivo*. Cancer Lett 2011; 313(1): 91-8.
[http://dx.doi.org/10.1016/j.canlet.2011.08.028] [PMID: 21945853]

[63]　Hiwatashi K, Ueno S, Abeyama K, *et al.* A novel function of the receptor for advanced glycation end-products (RAGE) in association with tumorigenesis and tumor differentiation of HCC. Ann Surg Oncol 2008; 15(3): 923-33.
[http://dx.doi.org/10.1245/s10434-007-9698-8] [PMID: 18080716]

[64]　Ishiguro H, Nakaigawa N, Miyoshi Y, Fujinami K, Kubota Y, Uemura H. Receptor for advanced glycation end products (RAGE) and its ligand, amphoterin are overexpressed and associated with prostate cancer development. Prostate 2005; 64(1): 92-100.
[http://dx.doi.org/10.1002/pros.20219] [PMID: 15666359]

[65]　Bhawal UK, Ozaki Y, Nishimura M, *et al.* Association of expression of receptor for advanced glycation end products and invasive activity of oral squamous cell carcinoma. Oncology 2005; 69(3): 246-55.
[http://dx.doi.org/10.1159/000087910] [PMID: 16127291]

[66]　Shimomoto T, Luo Y, Ohmori H, *et al.* Advanced glycation end products (AGE) induce the receptor for AGE in the colonic mucosa of azoxymethane-injected Fischer 344 rats fed with a high-linoleic acid and high-glucose diet. J Gastroenterol 2012; 47(10): 1073-83.
[http://dx.doi.org/10.1007/s00535-012-0572-5] [PMID: 22467055]

[67] Logsdon C, Fuentes M, Huang E, Arumugam T. RAGE and RAGE ligands in cancer. Curr Mol Med 2007; 7(8): 777-89.
[http://dx.doi.org/10.2174/156652407783220697] [PMID: 18331236]

[68] Kang R, Tang D, Lotze MT, Zeh HJ III. RAGE regulates autophagy and apoptosis following oxidative injury. Autophagy 2011; 7(4): 442-4.
[http://dx.doi.org/10.4161/auto.7.4.14681] [PMID: 21317562]

[69] Yamagishi S, Nakamura K, Inoue H, Kikuchi S, Takeuchi M. Possible participation of advanced glycation end products in the pathogenesis of colorectal cancer in diabetic patients. Med Hypotheses 2005; 64(6): 1208-10.
[http://dx.doi.org/10.1016/j.mehy.2005.01.015] [PMID: 15823719]

[70] Kang R, Loux T, Tang D, *et al.* The expression of the receptor for advanced glycation endproducts (RAGE) is permissive for early pancreatic neoplasia. Proc Natl Acad Sci USA 2012; 109(18): 7031-6.
[http://dx.doi.org/10.1073/pnas.1113865109] [PMID: 22509024]

[71] Kang R, Tang D, Schapiro NE, *et al.* The HMGB1/RAGE inflammatory pathway promotes pancreatic tumor growth by regulating mitochondrial bioenergetics. Oncogene 2014; 33(5): 567-77.
[http://dx.doi.org/10.1038/onc.2012.631] [PMID: 23318458]

[72] Ghavami S, Rashedi I, Dattilo BM, *et al.* S100A8/A9 at low concentration promotes tumor cell growth *via* RAGE ligation and MAP kinase-dependent pathway. J Leukoc Biol 2008; 83(6): 1484-92.
[http://dx.doi.org/10.1189/jlb.0607397] [PMID: 18339893]

[73] Nasser MW, Wani NA, Ahirwar DK, *et al.* RAGE mediates S100A7-induced breast cancer growth and metastasis by modulating the tumor microenvironment. Cancer Res 2015; 75(6): 974-85.
[http://dx.doi.org/10.1158/0008-5472.CAN-14-2161] [PMID: 25572331]

[74] Nasser MW, Qamri Z, Deol YS, *et al.* S100A7 enhances mammary tumorigenesis through upregulation of inflammatory pathways. Cancer Res 2012; 72(3): 604-15.
[http://dx.doi.org/10.1158/0008-5472.CAN-11-0669] [PMID: 22158945]

[75] Radia ALM, Yaser ALM, Ma X, *et al.* Specific siRNA targeting receptor for advanced glycation end products (RAGE) decreases proliferation in human breast cancer cell lines. Int J Mol Sci 2013; 14(4): 7959-78.
[http://dx.doi.org/10.3390/ijms14047959] [PMID: 23579957]

[76] Hou X, Hu Z, Xu H, *et al.* Advanced glycation endproducts trigger autophagy in cadiomyocyte Via RAGE/PI3K/AKT/mTOR pathway. Cardiovasc Diabetol 2014; 13(1): 78.
[http://dx.doi.org/10.1186/1475-2840-13-78] [PMID: 24725502]

[77] Amaravadi RK, Lippincott-Schwartz J, Yin XM, *et al.* Principles and current strategies for targeting autophagy for cancer treatment. Clin Cancer Res 2011; 17(4): 654-66.
[http://dx.doi.org/10.1158/1078-0432.CCR-10-2634] [PMID: 21325294]

[78] Wautier MP, Chappey O, Corda S, Stern DM, Schmidt AM, Wautier JL. Activation of NADPH oxidase by AGE links oxidant stress to altered gene expression *via* RAGE. Am J Physiol Endocrinol Metab 2001; 280(5): E685-94.
[http://dx.doi.org/10.1152/ajpendo.2001.280.5.E685] [PMID: 11287350]

[79] Mukherjee TK, Mukhopadhyay S, Hoidal JR. The role of reactive oxygen species in TNFα-dependent expression of the receptor for advanced glycation end products in human umbilical vein endothelial cells. Biochim Biophys Acta Mol Cell Res 2005; 1744(2): 213-23.
[http://dx.doi.org/10.1016/j.bbamcr.2005.03.007] [PMID: 15893388]

[80] Mukherjee TK, Reynolds PR, Hoidal JR. Differential effect of estrogen receptor alpha and beta agonists on the receptor for advanced glycation end product expression in human microvascular endothelial cells. Biochim Biophys Acta Mol Cell Res 2005; 1745(3): 300-9.
[http://dx.doi.org/10.1016/j.bbamcr.2005.03.012] [PMID: 15878629]

[81] Kang R, Tang D, Livesey KM, Schapiro NE, Lotze MT, Zeh HJ III. The Receptor for Advanced

Glycation End-products (RAGE) protects pancreatic tumor cells against oxidative injury. Antioxid Redox Signal 2011; 15(8): 2175-84.
[http://dx.doi.org/10.1089/ars.2010.3378] [PMID: 21126167]

[82]　Livesey KM, Kang R, Vernon P, *et al.* p53/HMGB1 complexes regulate autophagy and apoptosis. Cancer Res 2012; 72(8): 1996-2005.
[http://dx.doi.org/10.1158/0008-5472.CAN-11-2291] [PMID: 22345153]

[83]　Kang R, Tang D, Schapiro NE, *et al.* The receptor for advanced glycation end products (RAGE) sustains autophagy and limits apoptosis, promoting pancreatic tumor cell survival. Cell Death Differ 2010; 17(4): 666-76.
[http://dx.doi.org/10.1038/cdd.2009.149] [PMID: 19834494]

[84]　Otero K, Martínez F, Beltrán A, *et al.* Albumin-derived advanced glycation end-products trigger the disruption of the vascular endothelial cadherin complex in cultured human and murine endothelial cells. Biochem J 2001; 359(3): 567-74.
[http://dx.doi.org/10.1042/bj3590567] [PMID: 11672430]

[85]　Yamagishi S, Yonekura H, Yamamoto Y, *et al.* Advanced glycation end products-driven angiogenesis *in vitro*. Induction of the growth and tube formation of human microvascular endothelial cells through autocrine vascular endothelial growth factor. J Biol Chem 1997; 272(13): 8723-30.
[http://dx.doi.org/10.1074/jbc.272.13.8723] [PMID: 9079706]

[86]　Tsuji A, Wakisaka N, Kondo S, Murono S, Furukawa M, Yoshizaki T. Induction of receptor for advanced glycation end products by EBV latent membrane protein 1 and its correlation with angiogenesis and cervical lymph node metastasis in nasopharyngeal carcinoma. Clin Cancer Res 2008; 14(17): 5368-75.
[http://dx.doi.org/10.1158/1078-0432.CCR-08-0198] [PMID: 18765528]

[87]　Rojas A, Morales MA. Advanced glycation and endothelial functions: A link towards vascular complications in diabetes. Life Sci 2004; 76(7): 715-30.
[http://dx.doi.org/10.1016/j.lfs.2004.09.011] [PMID: 15581904]

[88]　Chipuk JE, Green DR. Do inducers of apoptosis trigger caspase-independent cell death? Nat Rev Mol Cell Biol 2005; 6(3): 268-75.
[http://dx.doi.org/10.1038/nrm1573] [PMID: 15714200]

[89]　Chipuk JE, Kuwana T, Bouchier-Hayes L, *et al.* Direct activation of Bax by p53 mediates mitochondrial membrane permeabilization and apoptosis. Science 2004; 303(5660): 1010-4.
[http://dx.doi.org/10.1126/science.1092734] [PMID: 14963330]

[90]　Meghnani V, Vetter SW, Leclerc E. RAGE overexpression confers a metastatic phenotype to the WM115 human primary melanoma cell line. Biochim Biophys Acta Mol Basis Dis 2014; 1842(7): 1017-27.
[http://dx.doi.org/10.1016/j.bbadis.2014.02.013] [PMID: 24613454]

[91]　Clark CE, Hingorani SR, Mick R, Combs C, Tuveson DA, Vonderheide RH. Dynamics of the immune reaction to pancreatic cancer from inception to invasion. Cancer Res 2007; 67(19): 9518-27.
[http://dx.doi.org/10.1158/0008-5472.CAN-07-0175] [PMID: 17909062]

[92]　Ichikawa M, Williams R, Wang L, Vogl T, Srikrishna G. S100A8/A9 activate key genes and pathways in colon tumor progression. Mol Cancer Res 2011; 9(2): 133-48.
[http://dx.doi.org/10.1158/1541-7786.MCR-10-0394] [PMID: 21228116]

[93]　Sinha P, Okoro C, Foell D, Freeze HH, Ostrand-Rosenberg S, Srikrishna G. Proinflammatory S100 proteins regulate the accumulation of myeloid-derived suppressor cells. J Immunol 2008; 181(7): 4666-75.
[http://dx.doi.org/10.4049/jimmunol.181.7.4666] [PMID: 18802069]

[94]　He M, Kubo H, Morimoto K, *et al.* Receptor for advanced glycation end products binds to phosphatidylserine and assists in the clearance of apoptotic cells. EMBO Rep 2011; 12(4): 358-64.
[http://dx.doi.org/10.1038/embor.2011.28] [PMID: 21399623]

[95] Win MTT, Yamamoto Y, Munesue S, *et al.* Regulation of RAGE for attenuating progression of diabetic vascular complications. Exp Diabetes Res 2012; 2012: 1-8.
[http://dx.doi.org/10.1155/2012/894605] [PMID: 22110482]

[96] Yan SD, Chen X, Fu J, *et al.* RAGE and amyloid-β peptide neurotoxicity in Alzheimer's disease. Nature 1996; 382(6593): 685-91.
[http://dx.doi.org/10.1038/382685a0] [PMID: 8751438]

[97] Huttunen HJ, Fages C, Rauvala H. Receptor for advanced glycation end products (RAGE)-mediated neurite outgrowth and activation of NF-kappaB require the cytoplasmic domain of the receptor but different downstream signaling pathways. J Biol Chem 1999; 274(28): 19919-24.
[http://dx.doi.org/10.1074/jbc.274.28.19919] [PMID: 10391939]

[98] Lander HM, Tauras JM, Ogiste JS, Hori O, Moss RA, Schmidt AM. Activation of the receptor for advanced glycation end products triggers a p21(ras)-dependent mitogen-activated protein kinase pathway regulated by oxidant stress. J Biol Chem 1997; 272(28): 17810-4.
[http://dx.doi.org/10.1074/jbc.272.28.17810] [PMID: 9211935]

[99] Iizuka K, Horikawa Y. ChREBP: A Glucose-activated Transcription Factor Involved in the Development of Metabolic Syndrome Endocr J 2008; 55(4): 617-24.
[http://dx.doi.org/10.1507/endocrj.K07E-110]

[100] Tong X, Zhao F, Mancuso A, Gruber JJ, Thompson CB. The glucose-responsive transcription factor ChREBP contributes to glucose-dependent anabolic synthesis and cell proliferation. Proc Natl Acad Sci USA 2009; 106(51): 21660-5.
[http://dx.doi.org/10.1073/pnas.0911316106] [PMID: 19995986]

[101] Sharaf H, Matou-Nasri S, Wang Q, *et al.* Advanced glycation endproducts increase proliferation, migration and invasion of the breast cancer cell line MDA-MB-231. Biochim Biophys Acta Mol Basis Dis 2015; 1852(3): 429-41.
[http://dx.doi.org/10.1016/j.bbadis.2014.12.009] [PMID: 25514746]

[102] Kim JY, Park HK, Yoon JS, *et al.* Advanced glycation end product (AGE)-induced proliferation of HEL cells *via* receptor for AGE-related signal pathways. Int J Oncol 2008; 33(3): 493-501.
[PMID: 18695878]

[103] Yaser AM, Huang Y, Zhou RR, *et al.* The Role of receptor for Advanced Glycation End Products (RAGE) in the proliferation of hepatocellular carcinoma. Int J Mol Sci 2012; 13(5): 5982-97.
[http://dx.doi.org/10.3390/ijms13055982] [PMID: 22754344]

[104] Zill H, Günther R, Erbersdobler HF, Fölsch UR, Faist V. RAGE expression and AGE-induced MAP kinase activation in Caco-2 cells. Biochem Biophys Res Commun 2001; 288(5): 1108-11.
[http://dx.doi.org/10.1006/bbrc.2001.5901] [PMID: 11700025]

[105] Deng R, Mo F, Chang B, *et al.* Glucose-derived AGEs enhance human gastric cancer metastasis through RAGE/ERK/Sp1/MMP2 cascade. Oncotarget 2017; 8(61): 104216-26.
[http://dx.doi.org/10.18632/oncotarget.22185] [PMID: 29262634]

[106] Deng R, Wu H, Ran H, *et al.* Glucose-derived AGEs promote migration and invasion of colorectal cancer by up-regulating Sp1 expression. Biochim Biophys Acta Gen Subj 1861; 1861(5PtA): 1065-74.
[http://dx.doi.org/10.1016/j.bbagen.2017.02.024]

[107] Hudson BI, Kalea AZ, del Mar Arriero M, *et al.* Interaction of the RAGE cytoplasmic domain with diaphanous-1 is required for ligand-stimulated cellular migration through activation of Rac1 and Cdc42. J Biol Chem 2008; 283(49): 34457-68.
[http://dx.doi.org/10.1074/jbc.M801465200] [PMID: 18922799]

[108] Matou-Nasri S, Sharaf H, Wang Q, *et al.* Biological impact of advanced glycation endproducts on estrogen receptor-positive MCF-7 breast cancer cells. Biochim Biophys Acta Mol Basis Dis 2017; 1863(11): 2808-20.
[http://dx.doi.org/10.1016/j.bbadis.2017.07.011] [PMID: 28712835]

[109] Mook ORF, Frederiks WM, Van Noorden CJF. The role of gelatinases in colorectal cancer progression and metastasis. Biochim Biophys Acta Rev Cancer 2004; 1705(2): 69-89.
[http://dx.doi.org/10.1016/j.bbcan.2004.09.006] [PMID: 15588763]

[110] Ko SY, Ko HA, Shieh TM, *et al.* Cell migration is regulated by AGE-RAGE interaction in human oral cancer cells *in vitro*. PLoS One 2014; 9(10): e110542.
[http://dx.doi.org/10.1371/journal.pone.0110542] [PMID: 25330185]

[111] Lee YM, Auh QS, Lee DW, *et al.* Involvement of Nrf2-mediated upregulation of heme oxygenase-1 in mollugin-induced growth inhibition and apoptosis in human oral cancer cells. BioMed Res Int 2013; 2013: 1-14.
[http://dx.doi.org/10.1155/2013/210604] [PMID: 23738323]

[112] Sheikh MS, Huang Y. Death receptor activation complexes: it takes two to activate TNF receptor 1. Cell Cycle 2003; 2(6): 549-51.
[http://dx.doi.org/10.4161/cc.2.6.566] [PMID: 14504472]

[113] Smolarczyk R, Cichoń T, Jarosz M, Szala S. HMGB1 – its role in tumor progression and anticancer therapy. Postepy Hig Med Dosw 2012; 66: 913-20.
[http://dx.doi.org/10.5604/17322693.1021108] [PMID: 23175347]

[114] Rohatgi RA, Janusis J, Leonard D, *et al.* Beclin 1 regulates growth factor receptor signaling in breast cancer. Oncogene 2015; 34(42): 5352-62.
[http://dx.doi.org/10.1038/onc.2014.454] [PMID: 25639875]

[115] Tang D, Kang R, Cheh C-W, *et al.* HMGB1 release and redox regulates autophagy and apoptosis in cancer cells. Oncogene 2010; 29(38): 5299-310.
[http://dx.doi.org/10.1038/onc.2010.261] [PMID: 20622903]

[116] Sharma S, Evans A, Hemers E. Mesenchymal-epithelial signalling in tumour microenvironment: role of high-mobility group Box 1. Cell Tissue Res 2016; 365(2): 357-66.
[http://dx.doi.org/10.1007/s00441-016-2389-7] [PMID: 26979829]

[117] Lin L, Zhong K, Sun Z, Wu G, Ding G. Receptor for advanced glycation end products (RAGE) partially mediates HMGB1-ERKs activation in clear cell renal cell carcinoma. J Cancer Res Clin Oncol 2012; 138(1): 11-22.
[http://dx.doi.org/10.1007/s00432-011-1067-0] [PMID: 21947243]

[118] Pusterla T, Nèmeth J, Stein I, *et al.* Receptor for advanced glycation endproducts (RAGE) is a key regulator of oval cell activation and inflammation-associated liver carcinogenesis in mice. Hepatology 2013; 58(1): 363-73.
[http://dx.doi.org/10.1002/hep.26395] [PMID: 23504974]

[119] Chen Y, Lin C, Liu Y, Jiang Y. HMGB1 promotes HCC progression partly by downregulating p21 *via* ERK/c-Myc pathway and upregulating MMP-2. Tumour Biol 2016; 37(4): 4399-408.
[http://dx.doi.org/10.1007/s13277-015-4049-z] [PMID: 26499944]

[120] Wang C, Fei G, Liu Z, Li Q, Xu Z, Ren T. HMGB1 was a pivotal synergistic effecor for CpG oligonucleotide to enhance the progression of human lung cancer cells. Cancer Biol Ther 2012; 13(9): 727-36.
[http://dx.doi.org/10.4161/cbt.20555] [PMID: 22617774]

[121] Riuzzi F, Sorci G, Donato R. RAGE expression in rhabdomyosarcoma cells results in myogenic differentiation and reduced proliferation, migration, invasiveness, and tumor growth. Am J Pathol 2007; 171(3): 947-61.
[http://dx.doi.org/10.2353/ajpath.2007.070049] [PMID: 17640970]

[122] Jikuzono T, Kawamoto M, Yoshitake H, *et al.* The miR-221/222 cluster, miR-10b and miR-92a are highly upregulated in metastatic minimally invasive follicular thyroid carcinoma. Int J Oncol 2013; 42(6): 1858-68.
[http://dx.doi.org/10.3892/ijo.2013.1879] [PMID: 23563786]

[123] Lan J, Sun L, Xu F, *et al.* M2 macrophage-derived exosomes promote cell migration and invasion in colon cancer. Cancer Res 2019; 79(1): 146-58.
[http://dx.doi.org/10.1158/0008-5472.CAN-18-0014] [PMID: 30401711]

[124] Winer A, Adams S, Mignatti P. Matrix metalloproteinase inhibitors in cancer therapy: Turning past failures into future successes. Mol Cancer Ther 2018; 17(6): 1147-55.
[http://dx.doi.org/10.1158/1535-7163.MCT-17-0646] [PMID: 29735645]

[125] Zhu L, Li X, Chen Y, Fang J, Ge Z. High-mobility group Box 1: A novel inducer of the epithelial–mesenchymal transition in colorectal carcinoma. Cancer Lett 2015; 357(2): 527-34.
[http://dx.doi.org/10.1016/j.canlet.2014.12.012] [PMID: 25511739]

[126] Takada M, Hirata K, Ajiki T, Suzuki Y, Kuroda Y. Expression of receptor for advanced glycation end products (RAGE) and MMP-9 in human pancreatic cancer cells. Hepatogastroenterology 2004; 51(58): 928-30.
[PMID: 15239215]

[127] Gong W, Wang ZY, Chen GX, Liu YQ, Gu XY, Liu WW. Invasion potential of H22 hepatocarcinoma cells is increased by HMGB1-induced tumor NF-κB signaling *via* initiation of HSP70. Oncol Rep 2013; 30(3): 1249-56.
[http://dx.doi.org/10.3892/or.2013.2595] [PMID: 23836405]

[128] Mardente S, Mari E, Massimi I, *et al.* HMGB1-Induced cross talk between PTEN and miRs 221/222 in thyroid cancer. BioMed Res Int 2015; 2015: 1-7.
[http://dx.doi.org/10.1155/2015/512027] [PMID: 26106610]

[129] Mardente S, Mari E, Consorti F, *et al.* HMGB1 induces the overexpression of miR-222 and miR-221 and increases growth and motility in papillary thyroid cancer cells. Oncol Rep 2012; 28(6): 2285-9.
[http://dx.doi.org/10.3892/or.2012.2058] [PMID: 23023232]

[130] Xu X, Zhu H, Wang T, *et al.* Exogenous high-mobility group box 1 inhibits apoptosis and promotes the proliferation of lewis cells *via* RAGE/TLR4-dependent signal pathways. Scand J Immunol 2014; 79(6): 386-94.
[http://dx.doi.org/10.1111/sji.12174] [PMID: 24673192]

[131] Ferrara N. Vascular endothelial growth factor as a target for anticancer therapy. Oncologist 2004; 9(S1) (Suppl. 1): 2-10.
[http://dx.doi.org/10.1634/theoncologist.9-suppl_1-2] [PMID: 15178810]

[132] Shubbar E, Vegfors J, Carlström M, Petersson S, Enerbäck C. Psoriasin (S100A7) increases the expression of ROS and VEGF and acts through RAGE to promote endothelial cell proliferation. Breast Cancer Res Treat 2012; 134(1): 71-80.
[http://dx.doi.org/10.1007/s10549-011-1920-5] [PMID: 22189627]

[133] Wang H, Duan L, Zou Z, *et al.* Activation of the PI3K/Akt/mTOR/p70S6K pathway is involved in S100A4-induced viability and migration in colorectal cancer cells. Int J Med Sci 2014; 11(8): 841-9.
[http://dx.doi.org/10.7150/ijms.8128] [PMID: 24936148]

[134] Dahlmann M, Okhrimenko A, Marcinkowski P, *et al.* RAGE mediates S100A4-induced cell motility *via* MAPK/ERK and hypoxia signaling and is a prognostic biomarker for human colorectal cancer metastasis. Oncotarget 2014; 5(10): 3220-33.
[http://dx.doi.org/10.18632/oncotarget.1908] [PMID: 24952599]

[135] Hermani A, Deservi B, Medunjanin S, Tessier P, Mayer D. S100A8 and S100A9 activate MAP kinase and NF-κB signaling pathways and trigger translocation of RAGE in human prostate cancer cells. Exp Cell Res 2006; 312(2): 184-97.
[http://dx.doi.org/10.1016/j.yexcr.2005.10.013] [PMID: 16297907]

[136] Emberley ED, Murphy LC, Watson PH. S100A7 and the progression of breast cancer. Breast Cancer Res 2004; 6(4): 153-9.
[http://dx.doi.org/10.1186/bcr816] [PMID: 15217486]

[137] Dakhel S, Padilla L, Adan J, *et al.* S100P antibody-mediated therapy as a new promising strategy for the treatment of pancreatic cancer. Oncogenesis 2014; 3(3): e92.
[http://dx.doi.org/10.1038/oncsis.2014.7] [PMID: 24637492]

[138] Liu Y, Wang C, Shan X, *et al.* S100P is associated with proliferation and migration in nasopharyngeal carcinoma. Oncol Lett 2017; 14(1): 525-32.
[http://dx.doi.org/10.3892/ol.2017.6198] [PMID: 28693201]

[139] El-Far AH, Sroga G, Al Jaouni SK, Mousa SA. Role and mechanisms of RAGE-ligand complexes and RAGE-inhibitors in cancer progression. Int J Mol Sci 2020; 21(10): 3613.
[http://dx.doi.org/10.3390/ijms21103613] [PMID: 32443845]

[140] Takino J, Yamagishi S, Takeuchi M. Glycer-AGEs-RAGE signaling enhances the angiogenic potential of hepatocellular carcinoma by upregulating VEGF expression. World J Gastroenterol 2012; 18(15): 1781-8.
[http://dx.doi.org/10.3748/wjg.v18.i15.1781] [PMID: 22553402]

[141] Tan B, Shen L, Yang K, *et al.* C6 glioma-conditioned medium induces malignant transformation of mesenchymal stem cells: Possible role of S100B/RAGE pathway. Biochem Biophys Res Commun 2018; 495(1): 78-85.
[http://dx.doi.org/10.1016/j.bbrc.2017.10.071] [PMID: 29050939]

[142] Zhang F, Banker G, Liu X, *et al.* The novel function of advanced glycation end products in regulation of MMP-9 production. J Surg Res 2011; 171(2): 871-6.
[http://dx.doi.org/10.1016/j.jss.2010.04.027] [PMID: 20638679]

Receptor for Advanced Glycation End Products in Neuronal Pathophysiology

Parth Malik[1,2,†], **Ruma Rani**[3,†] and **Tapan Kumar Mukherjee**[4,*]

[1] *School of Chemical Sciences, Central University of Gujarat, Gandhinagar, Gujarat-382030, India*

[2] *Swarrnim Startup & Innovation University, Bhoyan-Rathod, Gandhinagar-Gujarat, India*

[3] *ICAR-National Research Centre on Equines, Hisar-125001, Haryana, India*

[4] *Amity Institute of Biotechnology, Amity University, New Town, Kolkata, West Bengal 700156, India*

Abstract: The receptor for advanced glycation end products (RAGE) is a multi-ligand receptor molecule expressed in the cells of the nervous system (neurons and glial cells). Compared to embryonic cells, RAGE expression is significantly decreased within the adult tissues, including the nervous system. Various RAGE ligands such as amyloid-beta peptide (Aβ-peptide), high mobility group box protein 1 (HMGB1), S100/calgranulin, and advanced glycation end products (AGEs) are expressed by the cells of the nervous system. Several studies have predicted the role of RAGE in neurogenesis. Interaction of RAGE with its various ligands has been demonstrated as the responsible factor for complicating multiple diseased conditions such as Neuronal Differentiation and Outgrowth, Alzheimer's Disease (AD), Parkinson's Disease (PD), Huntington's Disease (HD), Amyotrophic Lateral Sclerosis (ALS), Creutzfeldt-Jakob's Disease (CJD), Peripheral Neuropathies, Familial Amyloid Polyneuropathy (FAP), Spinal Cord Injury (SCI), and epilepsy. The interactions of RAGE with its ligands are critically dependent on the relative extents of inflammation and oxidative stress, controlling the various neurological disease manifestations. Redox sensitivity of such interactions is inferred by their treatment using targeted and sustainable antioxidant delivery at the affected regions. Besides targeting RAGE-ligand interactions *via* blocking RAGE expression may be useful against various neurological diseases.

Keywords: AGEs, Aβ-peptide, HMGB1, Inflammatory and Oxidative stress, Multi-ligand receptor molecule, Neurogenesis, RAGE-ligand interactions.

---

* **Corresponding author Tapan Kumar Mukherjee:** Amity Institute of Biotechnology, Amity University, New Town, Kolkata, West Bengal 700156, India; E-mail: tapan400@gmail.com
† These authors contributed equally to this work.

Tapan Kumar Mukherjee, Parth Malik & Ruma Rani (Eds.)

INTRODUCTION

The nervous system is the network of nerve cells (neurons) and fibers that transmit nerve impulses between various parts of the body surrounded by many glial cells (the supporting cells of neurons). While neurons or nerve cells with their fibers are associated with the transmission of nerve signals, glial cells are implicated in supporting and nourishing neuronal cells. The receptor for advanced glycation end products (RAGE) is expressed in neuronal and glial cells both in the embryonic and adult stages, albeit to a lower extent in the latter. The multi-ligand receptor molecule RAGE binds with several ligands in neuronal cells such as amyloid-beta (Aβ)-peptide, high mobility group box protein 1 (HMGB1), S100/calgranulin, and advanced glycation end products (AGEs). The prevalence of RAGE and its ligands in the cells of the nervous system has been studied quite keenly from physiological and pathophysiological considerations. Several studies have predicted the involvement of RAGE in neurogenesis. However, a high-level expression of RAGE or its various ligands and subsequent enhanced extent of RAGE-ligand interactions, have been screened as the causative factors of various neurological disorders (NDs). For instance, accumulation and aggregation of Aβ peptide, one of the major RAGE ligands in the neuronal tissues induces the hyperphosphorylation of tau (τ) proteins, promoting the pro-inflammatory activation of microglia and supplementary neurological cells leading to synaptic loss and the consequent manifestation of Alzheimer's disease (AD). Similarly, investigations at multiple platforms have demonstrated aggravated AGE and RAGE extents in the frontal cortex and other brain regions, exhibiting enhanced vulnerability in the sufferers of Parkinson's disease (PD). The RAGE-ligand interactions are predicted to complicate PD. In Huntington's disease (HD), RAGE is upregulated in many affected brain regions, confining with the mutant Huntingtin (Htt), a prominent protein involved in HD pathogenesis. The other major neurological disorders (NDs) where RAGE and its various ligands are claimed to be involved in the propagation and complication, are Amyotrophic Lateral Sclerosis (ALS), Creutzfeldt-Jakob's Disease (CJD), Peripheral Neuropathies, Familial Amyloid polyneuropathy (FAP), Spinal Cord Injury (SCI), and epilepsy. High-level inflammation and oxidative stress generated by RAGE-ligand interactions are claimed as one of the major reasons for the enhanced severity of various NDs. Thus, several studies examined the efficacy of anti-RAGE molecules, including inhibitors against various RAGE ligands, besides suppressing the RAGE-ligand interactions as the treatment strategies for various NDs. This chapter discusses RAGE expression in neuronal cells, various RAGE ligands active in the neurological cells, and the NDs complicated by RAGE-ligand interactions. The experimental observations probing the efficacy of anti-RAGE/anti-RAGE ligand molecules affecting the RAGE-ligand interactions as neuronal disease treatment are also recalled. However, these results are not conclusive since no clinical trials have been conductedbased on them.

RAGE EXPRESSION IN NEURONAL AND SUPPORTING CELLS

The mammalian nervous system consists of the brain, spinal cord, a complex network of cells, and associated fibers (dendrites, dendrons, axons, *etc.*). Neurons (also called nerve cells) and glial cells (providing physical and chemical support to neurons) constitute the nervous system. Neurons are associated with the transmission of nerve signals, through which these cells command a mammalian body, wherein abundant glial cells support and protect neuronal cells. In estimation, the human nervous system consists of nearly 360 billion non-neural glial cells and 90 billion nerve cells. There is a substantial level of interaction between neurons and glial cells, maintaining all biological functions of the nervous system [1 - 3].

RAGE expression varies with the specified cell nature and the corresponding embryogenic-developmental stage. In convention, RAGE is abundantly expressed in the developing embryo, where it serves as a receptor for amphoterin. The RAGE expression is downregulated in adult life. In a typical adult, while lungs and skin are the major locations of RAGE expression, a low-level RAGE expression is also detected in the brain [4, 5]. Several other studies subsequently confirmed a low-level RAGE expression in the neuronal tissues, particularly in those associated with various adult neuronal tissues. Certainly, excluding the skin and lungs where RAGE is exceedingly expressed throughout life, in the physiological environment, RAGE is expressed at meager extents in the adult central nervous system (CNS), glia, and neurons [6 - 9]. Based on the results of several investigations, it can now be confirmed that certain regions of CNS and PNS express RAGE. In the CNS, RAGE prevalence has been demonstrated in neurons, microglia, astrocytes, and pyramidal cells, with numerous studies establishing a decisive role of RAGE across the CNS [10, 11]. In peripheral nerves, RAGE is present in nerve bundles associated with blood vessels [12] and in axons [13]. In a notable effort, *Qin and colleagues* detected RAGE expression in the oligodendrocytes in response to oxidative stress. In this study, the investigators used primers and antibodies specific for rat RAGE, detected mRNA, and a 55-kDa RAGE protein using PCR and Western blotting respectively. The analysis further revealed stronger staining for membrane-localized RAGE oligodendrocytes in neonatal rats wherein 100 μM hydrogen peroxide (H_2O_2) treatment caused 55-kDa RAGE loss from the cell membrane, with the appearance of "soluble" 45-kDa RAGE in the culture medium, followed by restoration of RAGE expression to normal extents [9].

Through yet another elegant effort, *Kamide and associates* noticed a likely enhanced RAGE expression in the cerebrovascular endothelium and hippocampal neurons on transient brain ischemia during bilateral common carotid artery occlusion (BCCAO) [14]. Another significant attempt by *Camila and associates* examined the content and specific prevalence of RAGE in Wistar rat's brains, manifested with complete inflammation on treatment with a single lipopolysaccharide (LPS, 5 mg•kg^{-1}, administered intraperitoneally) dose. In their study, investigators ascertained the RAGE content post 15 days of LPS administration, in the prefrontal cortex (PFC), hippocampus (HIPP), cerebellum (CB), and substantia nigra (SN). Inspection revealed enhanced RAGE extents in all locations, except HIPP. Nevertheless, immunohistochemistry screening revealed altered cellular location of RAGE expression from blood vessel-like structures to neuronal cells in all brain regions.

Apart from this, the maximum extent of RAGE expression was noticed in SN, immunofluorescence screening of which assured exclusive RAGE co-localization within the endothelial cells (RAGE/PECAM-1 co-staining) in untreated animals. Contrary to this, the LPS-administered animals mainly exhibited RAGE prevalence in dopaminergic neurons (RAGE/TH co-staining). Impaired TH extents, besides enhanced pro-inflammatory markers (TNF-α, IL-1β, GFAP, and phosphorylated ERK1/2) in SN, were observed concomitant to RAGE enhancement in identical locations. Such observations suggested a role of RAGE in establishing a neuroinflammatory-degeneration axis that manifests as a chronic retort to systemic LPS inflammation [15]. Fig. (**1**) briefly summarizes the above observations to establish the LPS-mediated pro-inflammatory role of RAGE in rat models, prevailing substantially within dopaminergic neurons.

Numerous erstwhile attempts also demonstrated the role of RAGE in neurogenesis, amongst which 2010 and 2013 studies by *Meneghini and accomplices* noticed an exclusive RAGE expression in the neural stem/progenitor cells (NS/PCs) along the neurogenic sub-ventricular region of adult mouse brain, being co-expressed with Sox2 [16]. The 2013 attempt by the same research group noticed that interactions of RAGE (and concomitant NF-κB axis) with ligands, HMGB1 and Aβ fueled the neuronal differentiation of adult hippocampal neural progenitors, as the presumptive cause of neurodegenerative AD [17]. However, RAGE expression in mature Tbr1 positive neurons is not reported, hinting at the likely possibility of RAGE involvement in the early proceedings of adult neurogenesis. Additionally, it has been established that HMGB1 released from reactive astrocytes fuels the NS/PC propagation *via* RAGE stimulation and subsequent JNK phosphorylation [18]. Erstwhile studies confirmed the significance of RAGE in sustaining brain repair and nerve renewal, aiding crosstalk with inflammatory pathways *via* transgenic mice [19] or S100 B-induced neuronal regeneration [20].

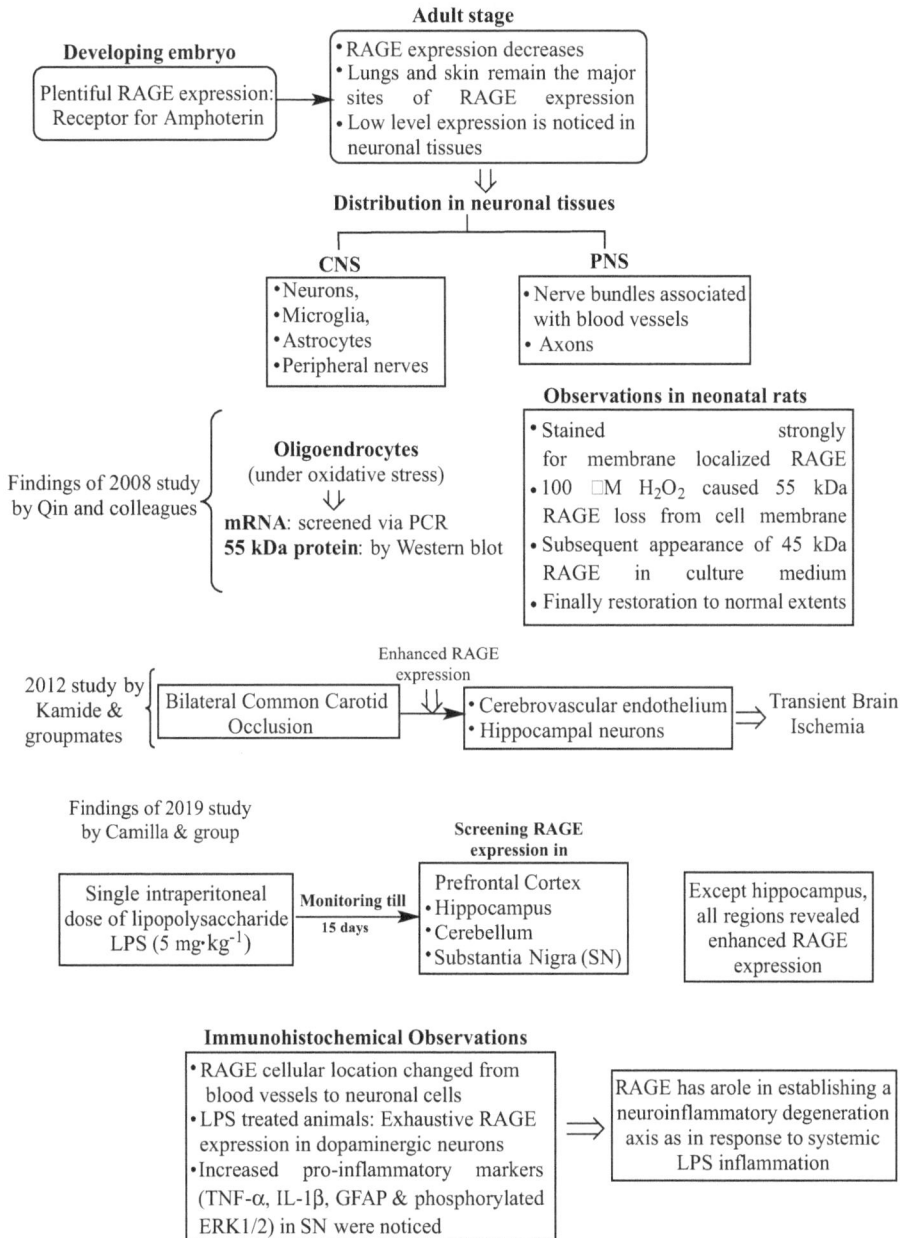

Fig. (1). Summarized findings of 2008, 2012, and 2019 studies by Qin, Kamide, and Camilla groups, illustrating varied RAGE expression in neuronal tissues under inflammatory stress conditions.

RAGE LIGANDS IN NEURONAL AND SUPPORTING CELLS

The interaction of RAGE with multiple ligands leads to its recognition as a multiligand receptor molecule. Several RAGE isoforms have been identified in neuronal cells [21]. The aggravated activities of RAGE have been reported in the presence of its ligands [10]. The major RAGE ligands known to date are advanced glycation end products (AGEs), high mobility group box protein 1 (HMGB1)/amphoterin, S100/calgranulin, and Aβ-peptide. A noteworthy effort by *Hori and colleagues* discussed RAGE expression in the developing nervous system where it serves as an amphoterin receptor [4].

Of note, amphoterin is abundantly expressed in embryonic and transformed cells and stimulates neurite outgrowth [22 - 24]. Amphoterin and RAGE co-localize at the leading edge of advancing neurites in the developing nervous system wherein anti-RAGE antibodies suppress the neurite outgrowth of cortical neurons on amphoterin-coated substrates [4]. Henceforth, in 1999, *Hofmann and group* identified the S100 protein family in the nervous system [25]. It is pertinent to mention that S100 proteins form a multigenic family of Ca^{2+} binding, EF-hand type proteins exhibiting intracellular and extracellular functions [24]. Another major ligand of RAGE in the nervous system with significant clinical importance is Aβ-peptide, the accumulation of which has been demonstrated to play a decisive role in Alzheimer's disease (AD) pathogenesis. The β-amyloid clumps into plaques within the neurons, with a swift spread of *tau* throughout the brain on its attainment of a threshold extent. Multiple studies have elucidated the Aβ role in several physiological functions, such as modulated synaptic functioning, facilitating neuronal growth and survival, and guarding against oxidative stress, neuroactive toxins, and pathogens [26 - 28].

Ahead is a brief discussion of some major RAGE ligands expressed in the cells of the nervous system:

AMYLOID BETA PEPTIDE IN NEURONAL CELLS

The build-up and aggregation of Aβ-peptide onsets hyper-phosphorylation of tau protein, causing subsequent activation of (pro-inflammatory) microglia and other neurological cells and culminates in synaptic loss and AD onset. Of note, AD is the most evident form of dementia in the elderly, characterized by accretion of senile plaques, tau tangles, neurodegeneration, and neuroinflammation in the brain tissues. Of note, AD progression is a pathological cascade commencing with the amyloid hypothesis. During the early stages of AD, soluble Aβ-peptide probably exhibits a prominent role by perturbing synaptic functioning and cognitive processes.

The involvement of RAGE in neurogenesis has been identified in the Aβ-peptide induced neuronal impairment [29]. In 1997, further study by the same group showed that binding of Aβ-peptide to neuronal RAGE induces macrophage-colony stimulating factor (M-CSF) *via* pro-inflammatory, pro-oxidative transcription factor, nuclear factor kappa B (NF-κB)-dependent pathway [30]. A nearly similar study by *Lue and colleagues* demonstrated the expression of RAGE by microglial cells amidst AD complications. The results of this study revealed that Aβ-peptide-RAGE-mediated microglial activation stimulates the M-CSF and RAGE expression [31]. Experimental evidence from several groups elucidated the RAGE-mediated effects of Aβ on target cells (neurons, glia, endothelial cells) [25, 31 - 34]. Furthermore, the introduction of wild-type (WT) RAGE gene in neurons of the AD-transgenic (T_g) mice expressing mutant human amyloid precursor protein (mAPP), exhibited a quick Aβ-mediated neuronal alteration [35]. Another significant effort by *Origlia and the team* noticed that Aβ impaired long-ter--potentiation (LTP) in the entorhinal cortex through neuronal RAGE-mediated p38 MAPK activation [36]. Apical-to-basolateral transport of Aβ-peptides *via the* blood-brain barrier (BBB) is regulated *via* RAGE and is restricted by p-glycoprotein [37]. A further noteworthy attempt from *Chaney and colleagues* explored the molecular level of RAGE-Aβ-peptide interaction in the AD-affected brain. Analysis revealed elevated quantities of soluble and insoluble Aβ-peptides to aggravate the expression of membrane-bound RAGE (mRAGE) and soluble (sRAGE), respectively. The binding of soluble Aβ to sRAGE inhibits Aβ-peptide aggregation, while mRAGE-Aβ-peptide interactions activate the transcription factor, NF-κB, sustaining chronic inflammation.

Analysis using atomic force microscopy (AFM) revealed the N-terminal domain of RAGE in the interacting site with Aβ-peptide. This study further explored RAGE-Aβ peptide structural interactions *via* chronological computational chemistry algorithms. The study model suggested that a soluble dimeric RAGE assembly generates a cationic well for docking the dimeric, N-terminal domain negative charges [38]. Subsequent studies by *Kim and accomplices* demonstrated the detailed molecular aspects of RAGE interaction with Aβ peptide, wherein observations in a mouse model revealed two β-strands of RAGE interacting with Aβ peptide. Serial deletion analysis of the RAGE V domain deciphered a need for third and eighth β-strands for interacting with Aβ-peptide. Site-directed mutagenesis of amino acids residing in these strands ruled out the feasibility of RAGE-Aβ peptide interactions. It was noticed that wild-type RAGE stimulated the NF-κB signaling as an aftermath of Aβ peptide treatment, whereas a mutant RAGE having impaired Aβ-peptide proximity failed to do so. Furthermore, investigation in mice models using a third β-strand peptide or a RAGE monoclonal antibody targeting RAGE-Aβ interaction weakened the Aβ-peptide trafficking across the BBB. These findings together established RAGE-Aβ interactions in AD pathology, validating RAGE inhibition as a promising therapeutic strategy for AD treatment. Readers are suggested to follow (Fig. **2a**) for a prompt follow-through of decisive findings [39].

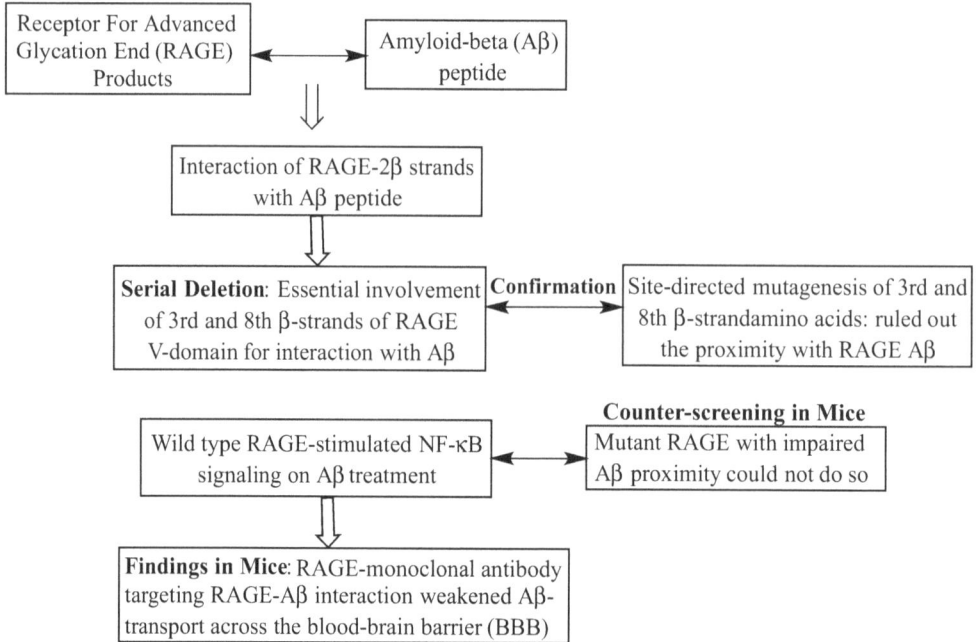

Fig. (2a). Summary of the 2013 study by Kim and colleagues, demonstrating RAGE-Aβ interaction as critical in Aβ transport across the blood-brain barrier (BBB).

Yet another recent attempt by *Tolstova and colleagues* probed the prominent interacting interfaces between RAGE and Aβ-phosphorylated (Aβ$_{40}$, Aβ$_{42}$) as well as isomeric (pS8-Aβ$_{42}$, isoD7-Aβ$_{42}$) isoforms. Analysis revealed Aβ transcytosis *via* Aβ-RAGE complex formation amidst blood-to-brain trafficking across the BBB. The noteworthy 2022 attempt took note of no attention being given to the Aβ isoform's interactions with RAGE and screened two potential interfaces *via* docking [40]. The interfaces are distinguished by an elongated region at the junction of the V and C1 RAGE domains and another C1-C2 linking confined region. Simulation studies using molecular dynamics (MD) revealed all Aβ isoforms as stable and equipped to form strongly bound complexes. Besides, it was also observed that all Aβ isoforms could be trafficked across the cells, *via* RAGE-complexation. Further scrutiny of possible RAGE-Aβ interactions assisted identifying the potential chemical compounds that can suppress this interaction and obstruct the related pathogenic cycle of events.

Hereby, RAP, FPS-ZM1, and RP-1 were identified as potent RAGE inhibitors, *via* Aβ-RAGE interactions probed through docking and concurrently using MD. Owing to a coincidence of RP-1 and Aβ interacting domains, the RP-1's ability to suppress RAGE-Aβ interaction is quite reliable. For a robust summary, readers are suggested to refer to (Fig. **2b**), summarizing the sequential findings [41].

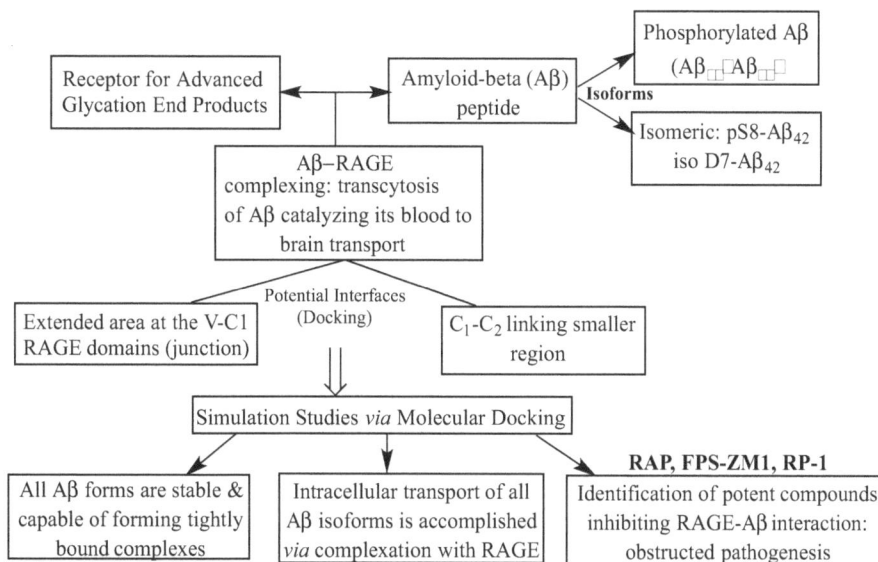

Fig. (2b). Summary of 2022 study by *Tolstova and groupmates*, noting the significance of RAGE-Aβ interaction mediated Aβ blood to brain transport. The findings herein provided insights for screening novel RAGE-Aβ interaction inhibitors for possible impaired Aβ actions within the cell.

HMGB1 IN NEURONAL CELLS

HMGB1 is perhaps the most copious nuclear non-histone protein engaged in the regulation of transcription. This protein is richly involved in developing CNS across varied species [42 - 44]. Quite intriguingly, HMGB1 exhibits a dynamic expression regime in the embryonic mouse brain, profusely expressed in the dorsal telencephalon until E16 (Embryonic day 16) but decreasing radically post E18 [45]. These distinctions indicated a likely HMGB1 involvement in neurogenesis during early brain development. Under certain conditions, RAGE is secreted from neuronal and non-neuronal cells [46] and mainly *via* signals generated from the RAGE-ligand interactions [47]. The extracellular HMGB1 was initially incepted as a heparin-binding protein, expressed with ease in the embryonic rat brain. The study by *Merenmies and teammates* comprised the generation of a cDNA library from rat brain mRNA, encoding a 30 kDa heparin-binding RAGE ligand (amphoterin). Analysis revealed encoded amphoterin as regulated in the rat brain, adjacently augmenting the neurite outgrowth in cerebral neurons. The mRNA instructing amphoterin release is substantially reduced in quantitative expression on the rapid perinatal growth phase of a rat brain. Parallel studies revealed anti-synthetic peptide antibodies raised in line with the amphoterin sequence as capable of binding specifically to the protein derived from rat brain alongside perceiving sub-cellular fractions and cellular immunostaining. Multiple events were thereafter elucidated as regulated *via* amphoterin expression including the homeostatic compositional balance of cell soma-cytoplasm. Besides this, amphoterin expression also correlated well with the plasma membrane filopadia expression in the cells at the vibrant stage of expansion and cytoplasmic growth. Distinguished localization of amphoterin to the filopodia of advancing plasma membrane inferred the involvement of endogenous amphoterin in the extension of neurite resembling cytoplasmic events in growing cells. Further support of such realizations was provided by the impaired outgrowth of cytoplasmic events on supplementation of culture media with anti-amphoterin and anti-synthetic peptide antibodies against amphoterin [23].

RAGE is essential for neurite outgrowth and neuronal migration in a budding nervous system [48]. Over the past few years, HMGB1 extracellular functions have been studied with significant interest. HMGB1 is released by the immune cells and platelets under specific conditions, *viz.* infection or injury, and is highly critical for cell migration, differentiation, and activation [47, 48]. The protein is widely regarded as a damage-associated molecular pattern (DAMP) molecule due to its involvement in inflammation and regulation of immune responses against tumor formation [49]. A recent study by *Zhao and groupmates* confirmed the HMGB1 requirement for the proliferation and differentiation of neuronal stem cells/progenitor cells. Besides, HMGB1-devoid neuronal cells exhibited enhanced apoptosis. Analysis revealed that a weakened HMGB1 activity disrupts Wnt/β-catenin signaling wide transcription factor expression in the developing cortex, including Foxg1, Tbr2, Emx2, and Lhx6. Finally, HMGB1 null mice revealed aberrant CXCL12/CXCR4 expression on impaired RAGE signaling [50].

A significant 2022 study elucidated the critical role of HMGB1 in Glioma stem cell (GSCs) self-renewal potential and tumorigenicity. In this investigation, the researchers observed over-expression of HMGB1 in human GBM specimens, wherein corresponding expression and secretion in GSCs are mediated *via* hypoxia induction. The HMGB1 involvement in the development of tumor cells was complemented by its silencing-driven loss of stem cell markers, besides the reduced self-renewal ability of GSCs. Besides, HMGB1 knockdown impaired the activation of the RAGE-mediated ERK1/2 signaling pathway, leading to cell cycle arrest in GSCs [51]. The above findings were further validated in studies on xenograft mice models, wherein treatment with FPS-ZM1 (a potent RAGE inhibitor) reduced the HMGB1 expression (and concurrent ERK1/2 phosphorylation), resulting in marred tumor growth and enhanced mice survival. Altogether the results elucidated a decisive HMGB1 involvement in sustaining GSC self-renewal and tumorigenicity. Fig. (**3**) summarizes regulatory events coordinated by HMGB1, characterized by their progressive entry from ECM into Glioma cells *via* phosphorylated ERK expression. Subsequent intracellular modulation *via* hypoxia and concomitant HIF-1α expression paved the way for self-renewal-driven increased tumorigenic behavior.

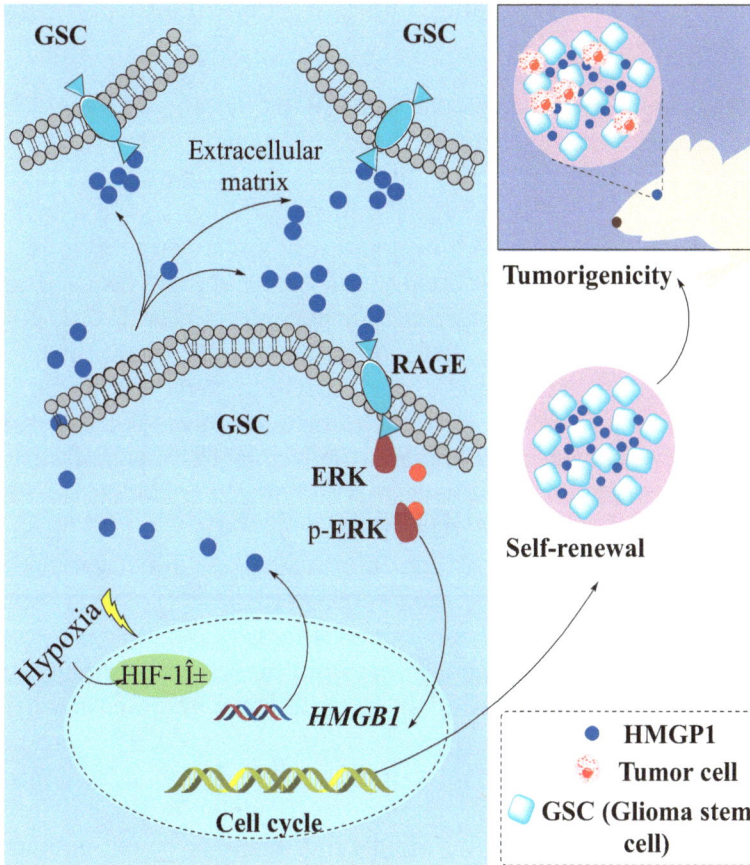

Fig. (3). Hypoxia-mediated enhanced High Mobility Group Box-1 (HMGB1: a potent RAGE ligand) expression in Glioma stem cells (GSCs), paving the way for their prolonged self-renewal. The cycle of events elucidates arrested HMGB1 expression as a mechanism to limit the GSC's plasticity.

S100. IN NEURONAL CELLS

The S100B protein prevails as one of a 25-member mutagenic family (like calmodulin/parvalbumin/troponin C) and got its name due to its solubility in a 100% saturated solution of ammonium sulfate at pH = 7. Perhaps, the initially screened member of this family was a non-fractionated S100A1 and S100B proteins mixture [52]. In the nervous system, S100B majorly prevails in astrocytes and other glial cells, such as oligodendrocytes, Schwann cells, ependymal cells, retinal Muller cells, and others. Investigations also reported S100B prevailing within the discrete neuron sub-populations [53 - 57].

The S100B protein was first detected in the late 1970s, in the extracellular compartment through its enhanced extents in the cerebrospinal fluid (CSF) of acute phase multiple sclerosis (MS) sufferers, contrary to the lower extents in the stationary phase of the disease [58]. As a cellular injury hallmark in the nervous system, measurements of S100B extents in biological fluids were suggested [59]. Thereafter, research on S100B as an indicator of brain injury focused on biological fluids other than that of CSF, prominently peripheral blood [60], cord blood [61], amniotic fluid [62], urine [63], and saliva [64]. These locations were analyzed to express the S100B detectable extents, which were observed to enhance in several nervous system pathologies. The prominent diseases therein include acute brain injury (cardiovascular disorders and traumatic injury), neurodegenerative diseases (Alzheimer's disease-AD, Parkinson's disease-PD, amyotrophic lateral sclerosis-ALS, MS), congenital/perinatal disorders (Trisomy 21, pre- and full-term asphyxiated newborns, intrauterine growth underdeveloped fetuses), and psychiatric challenges (schizophrenia, mood disorders, *etc.*). Studies on S100B prevalence in biological fluids consolidate this protein as a primitive biomarker of active neural distress, even though manifold diversified pathological conditions involving it significantly impaired its specificity. The assessment of S100B protein extents in biological fluids concurrently emerged as a crucial remedy for clinical diagnosis with concurrent symptotic prevalence, in screening the trend of disorders and eventual clinical outcomes. Extendable utilities include assistance in therapeutic decision-making by identifying responsive patients and monitoring therapeutic interventions in response.

STRUCTURAL DISTINCTIONS OF S100B-RAGE LIGANDS

S100B is an acidic protein with Zn^{+2} and Ca^{+2}, being restricted to the nucleus and cytoplasm of multiple cells. The S100 protein genes consist of 13 members, in a clustered state on chromosome 1q21 [65]. Structurally, this protein consists of twin EF-hand type binding regions, with a helix-loop-helix motif conjoined with a central hinge region [66]. Of note, EF-hand is an assorted motif class comprising

30 aa, which folds into a helix-loop-helix structure (resembling a right-hand fist, with an extended thumb and index finger). Biochemically, the protein consists of 91-amino acid (aa) polypeptide, 2 identical chains, further containing 2 EF-hands, the typical helix-loop-helix Ca^{+2} binding domains. These regions are exclusively implicated in cytoskeleton organization and cellular proliferation [67]. Each S100B subunit harbors 4 helices (helix 1: E2-R20; helix 2: K29-N38, helix 3: Q50-D61, helix 4: F70-A83) besides an anti-parallel β-sheet (strand 1: K26-K28, strand 2: E67-D69). The helices and sheets generate normal and pseudo-E-hands, wherein the C-terminal contains 12, canonical aa binding loops, the classical FF-hand while the *N*-terminal harbors 14 S100B explicit aa-binding loops.

The AA sequence is known to have areas of rigorously distinguishing lipophilic, basic, and acidic aa besides having a Ca^{+2} binding domain within the acidic domain. With a moderate affinity, S100B (2-20 M) binds 2 Ca^{+2} per subunit (Fig. 4). Target protein interactions are mediated by Ca^{+2} binding to EF-hand-driven structural changes. The distinct Ca^{+2} binding ability of S100B proteins in their $-NH_2$ terminal binding sites renders them different from erstwhile helix-loo-helix EF-hand proteins. Furthermore, the potential of a transition metal (such as Cu, Zn, Mn) binding at the dimer interface is higher than in the histidine-rich domains. The prevalence of Mg and K marks a protein's affinity towards Ca^{+2}, occurring normally at α, and β sites and is potently antagonized by potassium, K. A striking aspect of such a protein is its interaction with target proteins, bringing the cysteine residues (one in S100A1 and two in S100B) in proximity. For the specific studies of this description, readers are suggested to refer to the 2021 Current Neuropharmacology review article by *Langeh and colleagues.*

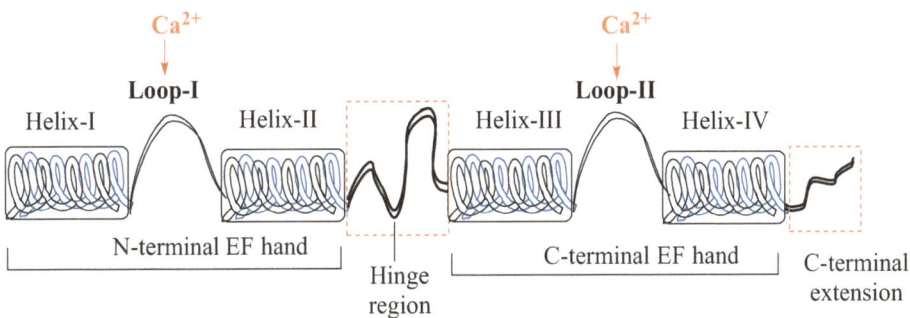

Fig. (4). Functional structure of S100B protein, characterized by EF-Hand 1 and EF-Hand 2, Ca^{+2} binding regions in pseudo and canonical configurations. Overactivation of this protein aggravates the release of pro-inflammatory markers, causing neurodegeneration.

Functionally, the S100 proteins are divided into those, (i) exerting merely intracellular regulatory effects, (ii) exhibiting intracellular and extracellular actions, and (iii) having exclusive extracellular regulatory effects. The S100 proteins are expressed sheerly in vertebrates and illustrate cell-specific expression dynamics. In some pathological events, a specific S100 protein expression could be triggered in the cells albeit no such expression is noticed in normal conditions. At the intracellular platform, S100 proteins regulate proliferation, differentiation, apoptosis, Ca^{2+} homeostasis, energy metabolism, inflammation, and migration/invasion by interacting with the various target proteins including enzymes, cytoskeletal subunits, receptors, transcription factors, and nucleic acids. Certain S100 proteins are secreted for regulating cellular functions in an autocrine and paracrine regime *via* surface receptor(s) activation (*e.g.* RAGE and TL4), G-protein-coupled receptors, scavenger receptors, or heparan sulfate proteoglycans and N-glycans. Extracellular S100A4 and S100B too, interact with epidermal growth factor (EGF) and basic fibroblast growth factor (BFGF), respectively, for sustaining the corresponding cell-to-cell intercommunication. Thus, extracellular S100 proteins perform multiple regulatory activities on various cells, participating in innate and adaptive immune responses, cell migration and chemotaxis, tissue development and repair, and leukocyte and tumor cell invasion [68]. In a comprehensive review article, *Langeh and colleagues* summarized the role of S100B in various NDs, discussing the therapeutic measures to reduce their existence. The 2021 contribution from India describes NDs as the characteristic CNS complications resulting from oxidative stress, mitochondrial impairment, and overexpressed state of proteins, such as S100B. The authors have taken note of the S100B protein's Ca^{+2} binding domain as a decisive mediator of NDs *via* activation of the MAPK pathway and NF-κB expression, contributing to cell survival, proliferation, and upregulated gene expression. Typical implication of S100B protein in AD, Parkinson's disease, multiple sclerosis, schizophrenia, and epilepsy (prominent NDs) is mediated *via* its aggressive expression besides astrocytes targeting aggravated neuroinflammation. Interactions with RAGE are a potent attribute of S100B's pathophysiological outcomes, exhibited under stressful conditions. Erstwhile notable traits include its neuroprotective ability (*via* arresting microgliosis and TNF-α expression) with a concentration-dependent frequency. The authors concluded that S100B enhanced expression is a positive factor for predicting the release of inflammatory hallmarks, nitric oxide, and excitotoxicity manifested neuronal loss [69].

The most notable intracellular activity regulated by S100B is its ability to function as a Ca^{+2} sensing protein, adjacently regulating diversified tasks besides transferring signals from secondary messengers and ensuring cell-to-cell communication *via* communicating with surface receptors. Most notably, S100B intercedes in cell proliferation, survival, and differentiation, regulating cellular

Ca^{+2} homeostasis and enzymatic activities besides interacting with the cytoskeleton proteins. These S100B functions are well-studied and experimentally established but an eluding factor despite this, leaves considerable ambiguity regarding a clearly defined and established S100B functioning. Readers are suggested to refer to the 2019 review article by *Michetti and groupmates*, for the relevant studies, based on which the authors have comprehensively discussed the role of S100B as a neural biomarker and its related aspects [70]. Quite unexpectedly, evidence for extracellular functioning of S100B is significantly higher than for intracellular tasks. About the biological fluid's prevalence of S100B, its formidable role as a neural injury hallmark is deciphered. However, the involvement of this protein in diversified pathological backgrounds diminishes its specificity. Screening the S100B prevalence in extracellular fluids is considered a reliable diagnostic positivity, assisting in the pathological manifestation of the disorder, the efficacy of the treatment administered, and guiding subsequent therapeutic decisions, *vis-à-vis* monitoring the patient's response to therapeutic interventions. The reliability and pathological validation of S100B as a diagnostic marker for neural complications has been the subject of numerous post-2010 comprehensive literature sources (could be tracked in the Journal of Neurochemistry 2019 Michetti's review article).

The centered theme of most studies probing S100B's role as a neural injury biomarker in the physiological environment has construed the prevalence of this protein owing to its leakage from damaged cells. The release of S100B by various cells under stressful conditions has been a well-versed fact since 1984. The involvement of RAGE gradually emerged as a central discourse of S100B (in extracellular fluids) interaction with different cells in the vicinity [7, 25]. In this reference, a 2016 study elucidated a capture of extracellular S100B by the vesicles and subsequent uptake by astrocytes in a manner depending on RAGE activity [71]. Many of the S100B *in vivo* functions are the results of its polymeric existence, wherein the tetrameric form has a higher RAGE binding proximity over the dimeric texture [72]. On a similar note, several S100B responses are mediated *via* its interaction with receptors other than RAGE. Prominent among these actions include S100B actions in non-neural cultured myoblasts, wherein S100B interacts with the basic fibroblast growth factor (bFGF)/FGF receptor 1 (FGFR1) system. For example, at higher extents, S100B is known to stimulate the mitogenic bFGF/FGFR1 signaling in low-density myoblasts, subject to the EGF presence. Herein, a new function of S100B is deciphered, wherein it functions as a signal generated from injured muscle cells to facilitate the regeneration of skeletal muscles *via* activation of promyogenic RAGE (or the mitogenic bFGF/FGFR1), as per the native concentration, bFGF presence and the respective myoblast intensity [73, 74]. The findings elucidate the interactions of S100B receptor system with RAGE as a reliable therapeutic target. More studied and well-versed extracellular S100B actions have been elucidated as modulated, concerning varying S100B concentrations. For instance, nanomolar extents were reported and subsequently confirmed as physiologic wherein outcomes of neurite extension, long-term potentiation, guarded neuron survival, counteractive response to neurotoxicant insults, and enhanced scavenging of reactive oxygen species (ROS) are reported in several studies from the 1980s to 2010 (details can be tracked in 2019 Michetti's review article). The nanomolar amounts of S100B are known to generate small quantities of signaling oxygen radicals that in turn activate the anti-apoptotic molecule Bcl-2 [20]. Contrary to this, the micromolar ranges are known to exhibit toxic and pro-inflammatory outcomes [75 - 77].

Prolonged RAGE activation by micromolar S100B extents generates an enhanced quantity of oxygen radicals, likely to result in mitochondrial impairment and apoptotic induction. Furthermore, the related signaling pathways converge to the transcription of the pro-apoptotic gene, often referred to as the Hyde side of the S100B protein (source of recent unceremonious attention). The micromolar amounts of S100B have been of specific interest, aggravating the inducible nitric oxide synthase for NO release, succeeded by impaired actions of neurons and glial cells. This cycle of events encourages multiple detrimental roles, like glutamate-triggered neuronal death, aggressive activation in astrocytes, and upregulated cyclooxygenase expression in microglia together fueling ROS generation in neurons, interfering with the lipid homeostasis and the cell cycle [78, 79].

Because astrocytes are the prominent sites of extracellular S100B functions, S100B is known to activate a RAGE-mediated autocrine loop in them, conferring a pro-inflammatory and neurodegenerative phenotype, assisting in the moderation of "Hyde side" S100B activity [80]. Thereby, like other S100 family proteins, S100B could be integrated with damage-associated molecular patterns (DAMP) molecules or alarmins. DAMPs are generated in the endogenous microenvironment to induce damage-responsive tissue reaction [81]. Quite intriguingly, certain S100B specialties such as its non-canonical secretion modality that provides an alternative to the conventional Golgi processing, its interaction with RAGE, and stimulated microglial migration, are shared with DAMPs [82]. Knowing that DAMPs are released in the extracellular region in the early stage of disease with the extents that often correlate with their intensity, they can be used as reliable biomarkers for accurate diagnosis and prognosis of varied pathologies. This generalization is best established for HMGB1 for its pathogenic and biomarker significance of the related diseases [83]. Similar generalization almost fits well for S100B although concentration threshold is indeed a concern. This has led to the consensus that a high S100B concentration could function squarely in the pathological cycle of neural injury besides aiding in accurate diagnosis and planning future treatments [84]. In a nutshell, plentiful experimental evidence established S100B's role in the pathogenesis of several nervous system disorders. The climax is that irrespective of the initiating sites herein, inflammatory activities have unanimously exhibited a key role in neural disorder manifestation, making anti-inflammatory drugs a fitting recourse (Fig. **5**).

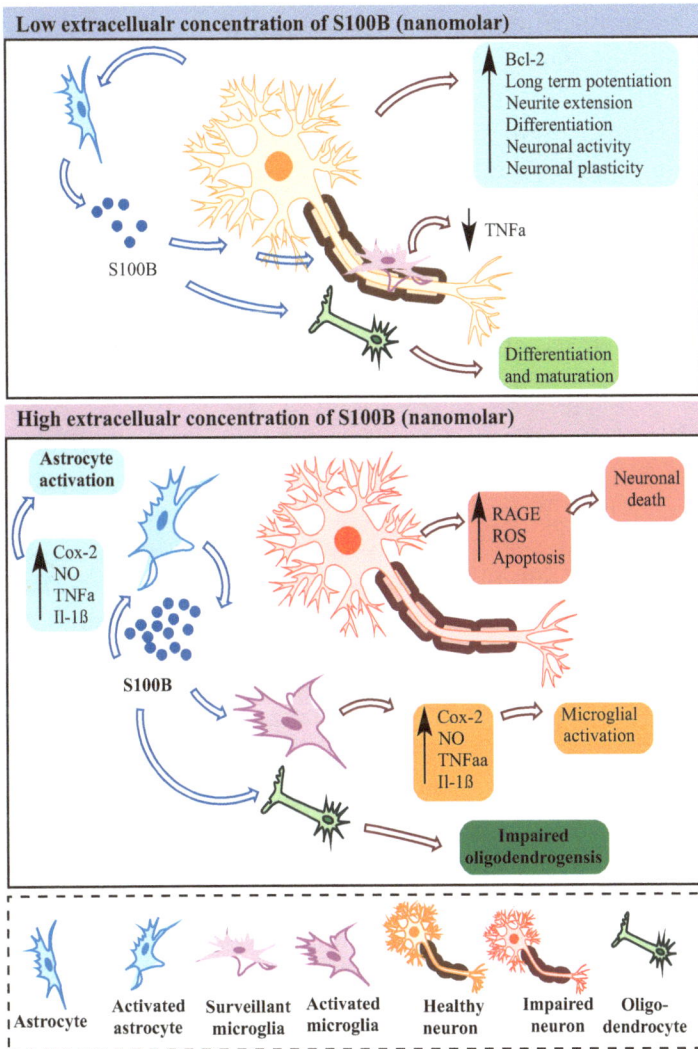

Fig. (5). Concentration-dependent biological actions of extracellular S100B in the cells of the central nervous system. Aggravated inflammation and oxidative stress emerge as major culprits for a diseased state.

ADVANCED GLYCATION END PRODUCTS IN NEURONAL CELLS

AGE is the most prominent RAGE ligand in the nervous system (neurons and glial cells). A noted study by *Luth HJ and groupmates* demonstrated that the percentage of AGE-positive neurons (and astroglia), both increase with age and, in AD patients, with gradual disease progression. Interestingly, nearly all neurons undergoing diffused cytosolic AGE immune reactivity contain hyper-phosphorylated tau protein. This suggests a possible correlation between AGE accumulation and the formation of early neurofibrillary tangles (NFTs). Many, but not all, neurons exhibit a co-localization of AGEs with other neurodegenerative markers, such as neuronal nitric oxide synthetase (nNOS) and caspase-3 [85]. In another study by *Takeuchi and colleagues*, cortical neurons were incubated with 5 immunochemically distinct AGEs, designated as 1 to 5. Analysis of treated neuron cells revealed a dose-dependent increase in neuronal cell death, as assessed by MTT assay, Trypan blue, and Hoechst 33258 staining. The structural epitope, AGE-2 exhibited the highest cytopathic effect with the corresponding neurotoxicity getting nullified on treatment with an anti-AGE-2-specific antibody and not by others. Based on these observations, the investigators proposed structural epitope, AGE-2 as a toxic moiety for neuronal cells [86].

Several lines of evidence demonstrate the involvement of RAGE-AGE interaction in AD pathogenesis. In one such effort, *Sasaki and colleagues* examined the RAGE, AGE, and Aβ-peptide distributions in neurons, astrocytes of AD, Diabetes Mellitus (DM) patients, and healthy subjects. The anti-RAGE molecules generated by investigators therein recognized full-length (50 kDa) and N-terminal RAGE (35 kDa) in human brain tissues. The Aβ-, AGE-, and RAGE-positive granules were screened in the perikaryon of hippocampal neurons (especially from CA3 and CA4) in all subjects. In AD, most astrocytes contained both AGE and RAGE-positive granules with a nearly unchanged distribution. The Aβ-positive granules were less common, while those for Aβ-, AGE-, and RAGE- co-localized in one region of a single astrocyte. Besides, in DM patients and control models, AGE- and RAGE-positive astrocytes rarely featured. These findings support the theory that in astrocytes, glycated Aβ is taken up *via* RAGE and is degraded *via the* lysosomal pathway. The receptor-mediated reactions could also contribute to neuronal impairment, aggravating AD progression [87].

The mechanism through which AGE-RAGE interaction complicates manifold diseased conditions has been rigorously explored. The inevitable outcome of AGE-RAGE interaction generates ROS that damages the brain. Enhanced AGE and ROS formation aggravate the synthesis of amyloid β (Aβ) leading to the Aβ deposition and phosphorylation of tau, culminating in the formation of plaques and NFTs. Accumulating ROS aggravates the formation of Aβ, HMGB1, and S100 RAGE-ligands, generating additional ROS on inducing RAGE expression through a vicious self-propagation cyclic mechanism. The enhanced ROS complicates AD pathology by elevating the Aβ plaque formation. With a progressive clarity on AGE-RAGE involvement in AD pathology, the treatment strategies are judiciously programmed to moderate the AGE build-up *via* reduced AGE consumption and formation, decreasing the RAGE expression to hinder the ligand binding, increasing the sRAGE extent and use of antioxidants [88]. Therefore, the aggressive and prompt interaction of RAGE with its various ligands aggravates the vulnerability to various pathophysiological conditions. The following paragraphs briefly describe the role of RAGE in NDs, signifying their impairment as a likely therapeutic strategy.

ROLE OF RAGE IN NEURONAL GROWTH AND DIFFERENTIATION

RAGE is widely expressed in neurons and astrocytes, exhibiting an important role in axon regeneration and neurite outgrowth interacting with HMGB1 [7, 89, 90]. HMGB1-RAGE interaction is well-demonstrated in promoting neurite outgrowth and is highly co-expressed in developing CNS. Several studies have demonstrated that in the neuronal cells, HMGB1 and S100B RAGE ligands are prominently involved in neuronal differentiation besides RAGE participation in embryonic and adult neuronal cells differentiation, peripheral nerve regeneration, and neurite elongation [4, 7, 91]. One of the first major efforts demonstrating RAGE involvement in neuro-differentiation was made during the early 2000s, wherein RAGE signaling induced embryonic stem cell demarcation for neuronal phenotype, impairing proliferation arrest and neurodifferentiation stages in N18 neuroblastoma cells [92]. Subsequent attempts focused further on the in-depth mechanisms of RAGE-mediated neuronal involvement, validating RAGE signaling contribution to the neuronal segregation in adult sensory neurons *via* partially interrelated pathways [93]. A noted study by *Wang and colleagues* revealed that in retinoic acid-induced P19 embryonic carcinoma stem cells, impaired RAGE expression using RNA interference (RNAi) suppressed the delineation of P19 cells into neuronal cells, enhancing the generation of vimentin-positive fibroblast-like cells. The investigators observed that RAGE knockdown inhibited retinoic acid-induced activation, hindering the NF-κB nuclear translocation, and deciphering the RAGE-regulated NF-κB activation. Analysis revealed RAGE involvement in the prevalence of P19 cells amidst retinoic acid-induced differentiation. Apart from this, RAGE knockdown considerably reduced the neurite outgrowth in retinoic acid-differentiated P19 cells, suggesting a RAGE requirement for neurite outgrowth of delineated P19 cells. The RAGE-devoid, retinoic acid-treated P19 cells activated GTPases, Rac1, and Cdc42. A key distinction in RAGE-impaired P19 cells was the GTPases activation. Additionally, RAGE inactivation in primary cerebellar granule neurons inhibited the neurite outgrowth. In these cells, Rac1 and Cdc42 (dominant-negative forms) over-expression impaired the neurite outgrowth, whereas excessive expression of constitutively active forms in RAGE-deficient neurons restored neurite outgrowth.

These distinctions inferred a role of the Rac1/Cdc42 pathway in RAGE-mediated neurite outgrowth. The study was the first major attempt to describe the role of RAGE in the cell lines and primary neurons, ascertained through RNAi knockdown [94]. Another study screening Stem-Cell Injury (SCI)-experiments established that HMGB1-RAGE interaction modulates endogenous neural stem cell differentiation model. Analysis revealed no direct pro-inflammatory induction by HMGB1 although it did promote the growth of MAP-2-positive mature neurons. Investigators concluded that HMGB1-RAGE signaling aided in neurogenesis and functional recovery in endogenous stem cells post-SCI [95].

THE PATHOPHYSIOLOGICAL ROLE OF RAGE IN THE NERVOUS SYSTEM

Despite a long history of understanding RAGE signaling-driven diversified responses and physiological outcomes, a clear-cut understanding of ligand-specific RAGE interactions has eluded biochemists and molecular biologists. Focusing on some regulatory actions *vis-à-vis* RAGE-ligand interactions in the neuronal tissues, the RAGE involvement in neuronal differentiation has been demonstrated in mouse-derived, adult neuronal progenitor cells. The investigations documented a possible RAGE involvement in neurite outgrowth, and activated neuronal differentiation *via* proximity with S100B, HMGB1, glycated bovine serum albumin, and subsequent participation of NF-κB signaling [13]. Another set of studies established the coupling of RAGE-ligand interaction with the upregulated secretion of chromogranin, a structural constituent of secretory vesicles engaged in neurosecretion at chemical synapses. Investigations have established the consequence of RAGE-signaling in the nervous system mediated through interactions with S1000B, HMGB1, and Aβ1-42. In neuroblastoma cells, RAGE-Aβ1-42 interactions have been screened to trigger an enhancement of AMIGO (amphoterin-induced gene and ORF) proteins in the neurite elongation and fasciculation, resulting in neuronal differentiation. The regulatory significance of RAGE in the nervous system is well-known in the non-physiological, post-injury conditions, as elucidated in studies deciphering the role of RAGE in spinal cord and peripheral nerve injury. For the spinal cord, the pro-regenerative essence of RAGE signaling is inferred through its HMGB1-signaling, superseding its pro-inflammatory functioning to stimulate neuronal stem cell delineation and a swift spinal cord normalcy. In the peripheral nerve insults, RAGE-S100B interaction activates the pathways promoting Schwann cell migration, enabling the injured peripheral nerve recovery [13, 96-99].

Opposed to the less-reported positive effects of RAGE-signaling-driven molecular activities, more studies focus on RAGE-signaling's detrimental impact on the nervous system. Fig. (6) summarizes the double-edged sword-like behavior of

RAGE signaling events having positive and negative impacts on neurological development. It is established hereby that the unfavorable impact of RAGE is sustained *via* its transduction receptor functioning on proximity with manifold, pro-inflammatory, and pro-oxidative stress-inducing ligands, culminating in the activation of inflammatory and oxidative stress pathways. Such activities result in cellular impairment and degeneration, leading to cell death. In general, for a healthy cellular environment, RAGE expression in the adult nervous system is quite abysmal and suddenly shoots up in proportion with inflammatory and oxidative stress-inducing ligands such as AGEs, HMGB1, S100B or Aβ-1-42. It is pertinent to mention here that the positive effect of RAGE signaling in the nervous system is due to its coordinated activities with threshold HMGB1, S100B, and Aβ1-42 expression. Herein, a higher than minimal extent (of these interactions) is directly correlated with detrimental outcomes. An element of interest herein relates to the uncommon interacting pattern of AGEs, RAGE binding of which results in aggressive oxidative and inflammatory stress. Negative impacts of RAGE-mediated signaling have also been reported on interactions with erstwhile ligands, such as AOPP, complement factors, heat shock proteins, and HSP-70, all being the critical regulators of oxidative and inflammatory stress-causing signaling events. The outcome of RAGE signaling *vis-à-vis* diverse ligands, also depends on specific nervous system cells, such as microglia, resident macrophages, and astrocytes, fueling the pathological events in the nerve cells and tissues [100 - 103].

Despite many studies demonstrating the detrimental role of RAGE in multiple neurodegenerative disorders, no clarity prevails for its involvement in diseased conditions or sustaining normal physiology. The consensus aspects as per the best understanding establish that RAGE itself does not instigate neurodegeneration although it does exacerbate it *via* oxidative and inflammatory stress. Aggravated oxidative stress could vary from one cell to another, while the aggressive inflammatory response gets aid partly from the hyper-immune activities. Herein, immunological molecules like cytokines and other protecting proteins, and growth factors synergistically contribute to their guarding activities. For the specific actions of RAGE in the nervous system, microglia of CNS and resident macrophages of the PNS are the initial responders to the excessive gathering of pro-oxidative and pro-inflammatory vulnerable RAGE ligands [104].

The specific functioning of RAGE on the surface of immune cells is aided by the activation of ligand binding besides manifesting downstream pathological pathways. This corresponds to the activated expression of neuronal and astrocyte RAGE activity, comprising an integrated loop with the progression of pathological alterations culminating in neuronal impairment and degeneration [105]. The major reason for the dual sensitivity of RAGE cellular responses correlates to its wider tissue distribution. Based on the ligand-specific interactions, the prominent outcomes of their explicit communications in different biochemical environments accounting for inflammatory stress-worsened pathology of NDs are discussed ahead.

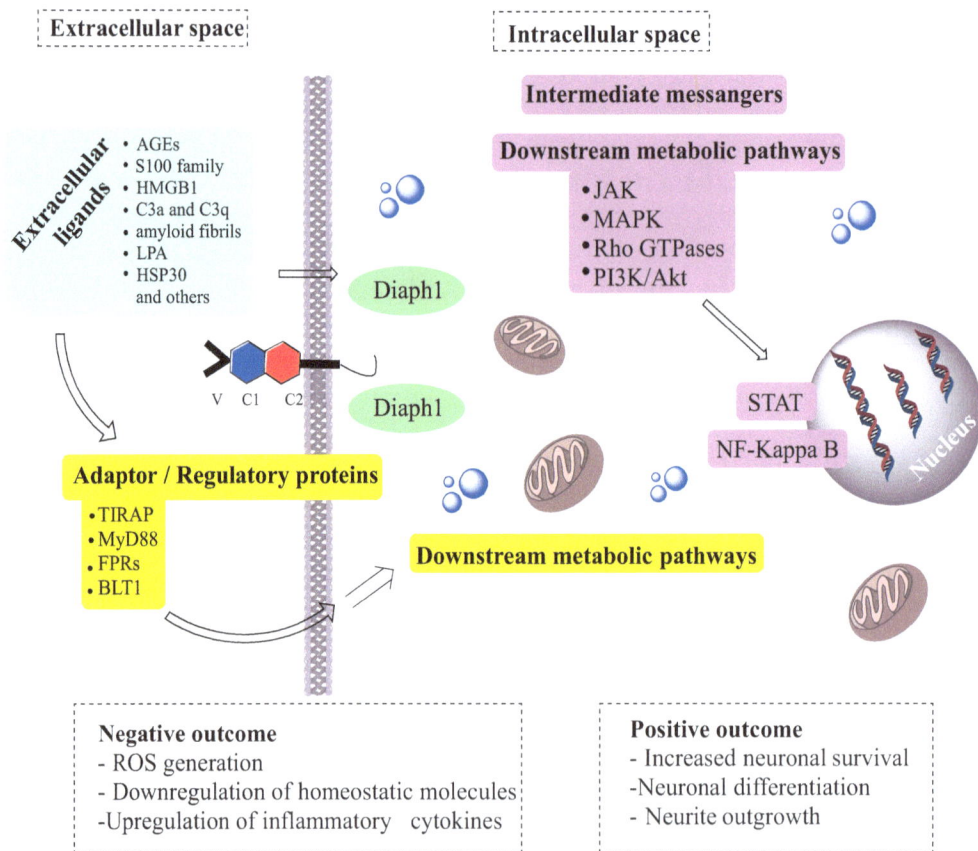

Fig. (6). Positive and negative outcomes of the receptor for advanced glycation end products, RAGE signaling pathways, ligands such as AGEs, S100, HMGB1, and others along with regulatory proteins (such as TIRAM, MyD88, *etc.*). The interactions with RAGE stimulate manifold transcription factors and signaling pathways, which destabilize neuronal cells *via* enhanced oxidative and inflammatory stress. Positive outcomes support neuronal survival, differentiation, and neurite outgrowth.

ROLE OF RAGE IN THE COMPLICATION OF ALZHEIMER'S DISEASE

Alzheimer's disease (AD) is recognized as a progressive ND that causes atrophy (shrinking) followed by possible death of the brain cells. On a histopathological platform, abnormal deposition of proteins/peptides such as Aβ and hyperphosphorylated, τ proteins, leads to the formation of amyloid plaques and NFTs in the vicinity of neuronal cells. Thereafter, shrinking of neuronal cells, loss of memory (dementia), and cognitive power are the eventual outcomes. Besides other factors, increasing age seems an additive prospect in the AD manifestation [106]. Experimental observations infer a decisive role of glycation in the creation of amyloid proteins. For example, a noted effort reported glycation-mediated albumin (a globular protein) conversion (initially having a large α-helical morphology) into a β-pleated sheet and quaternary structural configuration known as cross-conformation [107]. Of note, AGEs are known to prevail in β-amyloid plaques and NFTs [108, 109] with higher extents in the former than samples from age-matched controls [110]. Furthermore, immunohistochemical approaches have convincingly demonstrated AGE prevalence in NFTs and senile plaques [111]. Although some investigations have opined that AGEs are very late markers of AD [114], it is now a consensus that these are active participants in AD progression [112, 113]. Hereby, a noted effort by *Li and colleagues* speculated that Aβ-AGE aggregation may aggravate neurotoxicity *vis-à-vis* upregulated RAGE and stimulated glycogen synthase kinase-3 (GSK-3) [113, 114]. The variations thereof culminated in tau protein hyperphosphorylation, impaired synapse signaling, and memory, in the rats.

The potential involvement of RAGE in AD molecular pathogenesis was first highlighted in 1996 by *Yan and associates* when they noticed that RAGE binding to Aβ mediates the Aβ-induced cellular cascade of pathological happenings, involving oxidative stress and cytotoxicity triggered neuronal degeneration and apoptosis. The investigators observed that RAGE not merely binds to Aβ but also to its soluble monomeric form, Aβ (1-40) peptide, (sAβ1-40) that can invade the BBB and bind to brain monolayer endothelium, resulting in toxicity therein. The study noticed that blocking RAGE using anti-RAGE antibodies in human brain endothelial cell cultures, reduced sAβ1-40 transcytosis, improving the overall conditions of cells therein [30, 115]. Knowing its involvement in early AD, RAGE may be considered a more potent biomarker than Aβ. Screening using Positron Emission Tomography (PET) provides valuable information regarding AD pathology, several years before the clinical symptoms. Thus, to further elucidate the role of RAGE in AD pathology, investigators opined the need for an exclusively RAGE-targeting tracer [116].

Yet another effort explaining RAGE involvement in AD noticed its enhanced expression, being upregulated in microglia of AD patients, stimulating RAGE-Aβ-induced liberation of MCSF *via* spirally coordinated pathological events, culminating in neurodegeneration with an exacerbated disease progress [31]. The 2012 attempt by *Liu and associates* involved mice with intra-cerebroventricular Aβ fusion. Administration of pinocembrin (a neuroprotective agent) provided further proof for RAGE involvement in AD pathogenesis, after which Aβ formation is effectively inhibited leading to decreased RAGE expression and subsequent moderation of mitochondrial stress, relieving the diseased state [117]. Another study reported RAGE-signaling in AD as the source of neurodegeneration, deciphering RAGE as an interface mediating the effects of AD and instigating risk factors [118]. Fig. (7) depicts the chemical structure of pinocembrin (a dihydroxyflavanone), having an abysmal water solubility but well-demonstrated immunoprotective functions (including as an antioxidant, antineoplastic, and neuroprotective agent) having vasodilator traits. The terminal –OH groups indeed make it interactive in an aqueous environment, but the integrated, bulk hydrophobicity persistently impairs its adequate aqueous dissolution. With a functional essence similar to edible and readily used secondary polyphenols (curcumin, genistein, and others), there is a substantial possibility of a synergistic impact on the co-delivery mode of pinocembrin with stoichiometry varied proportion of erstwhile phytochemicals. Apart from co-delivery, the administration *via* different nanocarriers (such as nanoliposomes, nanoemulsions, functionalized nanoparticles, and integrated nano assemblies) could be an even better option for maximizing its functional impact.

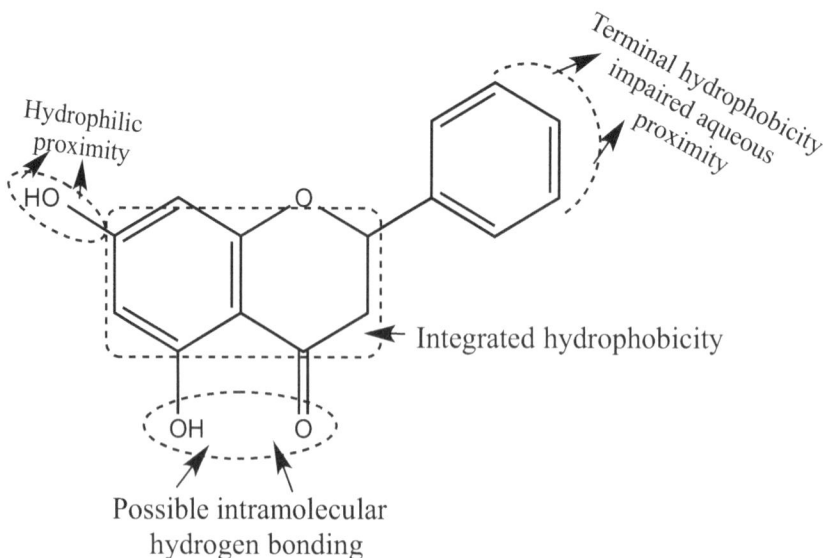

Fig. (7). Chemical structure of pinocembrin, demarcating the segregated terminal hydrophilicity from the integrated bulk hydrophobic domain for its low aqueous solubility.

Several studies have probed the mechanisms by which RAGE aggravates AD manifestation, the prominent of which involves recognition and binding with Aβ-peptide. A decisive role of RAGE in AD is the characteristic generation and consequent build-up of Aβ plaques, NFTs, adjacent to the failure of synaptic transmission and neuronal degeneration. The steady-state extent of Aβ depends on the balance between its formation and clearance. RAGE plays an important role in Aβ clearance, acting as a prominent transporter by regulating the brain influx of circulating Aβ. The efflux of accumulated Aβ into the circulation *via* BBB is facilitated by low-density lipoprotein receptor-related protein 1 (LRP1). RAGE could be a significant contributor to Aβ generation *via* aggravated functions of β- and/or γ-secretases, promoting the inflammatory response and oxidative stress. However, sRAGE-Aβ interactions could inhibit Aβ neurotoxicity, paving way for its clearance from the brain. Meanwhile, RAGE could promote synaptic and neuronal circuit dysfunction, the major aspects of cognition alongside physiological and pathological regulation. Besides, RAGE could trigger the pathogenesis of Aβ and tau hyper-phosphorylation, both of which indulge in cognitive impairment. Preclinical and clinical studies have demonstrated the efficacy of RAGE impairment agents in AD treatment, establishing RAGE expression as a novel therapeutic target [119 - 122].

Compelling evidence infers RAGE functioning as an inflammation intermediator as well as a critical oxidative stress inducer, through which it manifests the pathophysiological changes leading to the onset of AD and its further aggravation. For example, a study conducted on retinal ganglion cell line RCG-5 by *Lee and colleagues* reported that Aβ-RAGE interaction mediated aggressive pro-inflammatory response *via* Toll-like receptor (TLR)-4 signaling pathway in the RGC-5 cell line. Subsequent analysis revealed that treatment of these cells with AGEs stimulated the expression of amyloid precursor protein (APP) alongwith enhanced Aβ formation [120]. This study also noticed the inhibition of p65 (subunit of NF-κB) nuclear translocation and tumor necrosis factor-α (TNF-α, a pro-inflammatory cytokine) transcription *via* small interfering RNA-mediated impaired actions of RAGE or TLR-4. RAGE-dependent p38 MAPK activation contributes to Aβ peptide-mediated synaptic plasticity [36]. Of note, MAPKs are involved in neurodegenerative processes, characterized *via* Aβ modulated phosphorylated states besides synaptic dysfunction, cognitive decline, and the onset of inflammatory responses in AD [121].

ROLE OF RAGE IN THE COMPLICATION OF PARKINSON'S DISEASE

Parkinson's disease (PD) is another familiar form of neurodegenerative disease, distinguished *via* inactive tremors, rigidity, sluggish movements, and postural and autonomic vulnerability. The disease involves a degeneration of dopaminergic neurons in the SN of mid-brain and erstwhile, monoaminergic neurons in the brain stem [122]. The build-up of intracellular proteinaceous deposits such as Lewy bodies and neuromelanin is recognized as a PD pathological hallmark [123, 124]. Lewy bodies majorly comprise neurofilament proteins including α-synuclein (α-syn). In a matching context to the other neurodegenerative diseases including AD, the PD etiology is complex, involving genetic, and environmental factors and their mutual crosstalk. Shockingly, though, very few cases of entirely genetic or environmental PD have been reported [125]. Genes such as α-syn, Parkin, DJ-1, and Pink1 are recognized as key players in PD [126]. The prominent environmental factors contributing to PD onset include brief exposure to pesticides, herbicides, heavy metals, increased stress besides brain injuries. Selected, autosomal dominant PD genes are SNCA (α-synuclein) and LRRK2 (leucine-rich repeat kinase 2). Aggravated inflammation and RAGE expression in PD infer α-syn and glycation as hallmarks of diseased conditions, leading to neuronal cell death [127 - 129].

Several *in vitro* studies have demonstrated α-syn aggregation induced by AGEs. RAGE is considered one of the potential PD contributors, being noticed in neuronal Lewy bodies, the propagation of which signifies oxidative stress and neuro-inflammation [130 - 134]. Histochemical staining conducted using *in vivo* studies demonstrated enhanced AGE and RAGE levels in the PD patient's frontal cortex as compared to control subjects [131, 132]. Another experiment on MPTP mice (a laboratory model of PD), revealed that the RAGE deletion moderates the MPTP-triggered cell toxicity, reducing the pro-inflammatory impact on the affected cells to impair the disease progress [133]. Similarly, one study screening the role of S100B (a pro-inflammatory RAGE ligand), observed its deficiency as directly correlated to lowered concentration and activation of RAGE and TNF-α respectively, in MPTP mice. These hallmarks were subsequently screened as critical interrupting agents in the RAGE signaling-driven PD aggravation [130]. Yet another research on PD patients from the Han population, revealed a certain RAGE gene polymorphism(s) as being related to the increased probability of PD development, deciphering RAGE involvement in PD pathogenesis at the proteinaceous and genetic platform [131].

ROLE OF RAGE IN THE COMPLICATION OF HUNTINGTON'S DISEASE

Huntington's disease (HD), also known as Huntington's chorea is a rare, inherited disorder caused by genetic mutations that result in progressive degeneration of nerve cells in the brain. This hereditary, neurodegenerative disorder affects muscle coordination, often indicated *via* cognitive decline and psychiatric problems [135, 136]. Although the genetic background of this disorder is well-versed, currently there is no cure for it, and the available options merely moderate the symptoms without targeting the underlying pathologies. The mean age of onset for HD is approximately 45 years. RAGE over-expression was established in the striatal neuron of an HD-positive animal model [136].

Immunohistochemistry and double labeling by *Ma and accomplices* revealed localized RAGE expression in HD caudate nucleus (CN). Analysis deciphered RAGE expression in at least two cell types, medium spiny projection neurons, and astrocytes, with stronger staining in astrocytes [11]. Studies on an R6/2 mouse HD model revealed enhanced RAGE expression in multiple affected brain regions, localizing with mutant Huntingtin (Htt), a key protein involved in HD pathogenesis [137]. The data support the hypothesis that RAGE is upregulated in the neurodegenerative process of HD, with its activation being related to the sufferer's vulnerability *vis-à-vis* the striatal neuronal subtype. Likewise, both RAGE and its ligands were found to be richly prevalent in the SN and frontal cortex of PD patients and MPTP-induced PD models. Of note, it has been discussed earlier that increased binding of AGEs to RAGE instigates oxidative stress and inflammation, leading to mitochondrial dysfunction and neuronal cell death. The analysis by *Shi and accomplices* showed that the RAGE expression and its co-localization with ligands were aggravated in the striatum of HD patients, exhibiting further aggravation with grade severity alongside a coincidence of mediolateral neurodegeneration [138]. Detailed studies establishing the role of RAGE in HD pathogenesis are currently in progress.

ROLE OF RAGE IN AMYOTROPHIC LATERAL SCLEROSIS

Amyotrophic Lateral Sclerosis (ALS), also known as Lou Gehrig's disease, is a universal motor neuron disorder. It is the third most frequent neurological issue of human casualties [139]. ALS is polygenic, highly diverse, *vis-à-vis* its origin. The prominent identifying features of ALS include discerning failure of upper and lower motor neurons in the brain and spinal cord, progressive muscle weakness, atrophy, and spasticity [140 - 144]. ALS can be sporadic or inherited and its etiology remains largely unknown, although it is established that 5-10% of cases are familial, and 15% of these exhibit mutations in the gene encoding Cu/Zn superoxide dismutase (SOD-1) [145].

The mutation in the SOD-1 gene not merely results in the impairment of its biological function but also interrupts the homeostasis besides enhanced toxicity caused by accumulating and aggregating, mutated SOD-1 and other RAGE ligands, such as AGEs. It is conjectured that AGE accumulation further aggravates cellular functioning, triggering RAGE activation besides inducing RAGE-mediated neuronal stress and microglial stimulation. A 2011 study aimed at screening TLR/RAGE signaling involvement in ALS inflammatory pathways deciphered that RAGE was likewise over-expressed in ALS brains to TTL (Tubulin-Tyrosine Ligase, needed to repair injured axons) [143], establishing a link for an unswerving RAGE/TTL pathway involvement in ALS mediated inflammation. Reduced extents of circulating soluble RAGE (sRAGE), natural decoy RAGE, in the sufferer's blood, has been correlated with the disease severity in ALS patients [144]. Furthermore, the correlation between glycation end-product amount and Cu/Zn-SOD-1 mutation has also been observed [145], wherein, Lys122 and Lys128 were identified as protein-specific glycation loci [146]. Apart from this, ALS patients exhibited glycation in the neurons and spinal cord, likely to affect the Lys-Ser-Pro sequences of neurofilaments' subunit. The process suppresses the generation of neurofilament protein and promotes intramolecular cross-linking, manifesting finally as ALS [147, 148].

Recent studies pinpoint oxidative stress and neuro-inflammation as major contributors to ALS progression and aggravation. Mounting evidence implicates RAGE as a significant pathogenic mediator of multiple neurodegenerative diseases and chronic conditions. It is hypothesized that detrimental actions of RAGE are triggered by ligand binding, such as AGEs, S100/calgranulin family members, and HMGB1 proteins. To validate the success of RAGE-ligand signaling inhibition in the human ALS, tissue samples from age-matched human controls and ALS spinal cords were examined for ligand(s) expression (AGE, S100B, and HMGB1), through varied immunofluorescent and immunoblotting signals intensity. Analysis revealed vigorously enhanced RAGE and its ligands expression in the ALS sufferer's spinal cords rather than the age-matched control subjects. This observation corresponded to the first breakthrough ever about the RAGE and its ligands co-expression in human ALS spinal cords, although detailed and reproducible attempts are needed to probe the rationality of RAGE involvement in human ALS neurodegeneration [146, 147].

ROLE OF RAGE IN CREUTZFELDT-JAKOB'S DISEASE

Creutzfeldt-Jakob's disease (CJD) is a prion-mediated disastrous, swiftly progressive, and invariably fatal neurodegenerative brain disorder. The histopathological signatures of CJD share similarities with those of AD, such as the formation of prion protein (PrP)-amyloid plaques. The observed CJD and AD histopathological resemblance creates interest in screening the role of RAGE in CJD pathology. A study conducted by *Makita and associates*, on PrP plaques from CJD postmortem brain samples revealed enhanced RAGE and AGE actions in CJD brain samples, suggesting a role of RAGE in CJD pathogenesis. It was anticipated that AGEs fuel plaque generation, and in this case, RAGE could indeed act as a scavenger protein to reduce the circulating AGEs pool, accounting for PrP degradation and improved disease outcome [91]. An identical study assessed the oxidative stress in affected tissues of CJD patients, and supported the proven findings, with enhanced RAGE concentration aggravating the expression of multiple oxidative stress hallmarks in CJD brain tissues. The findings indicated RAGE-targeting as a therapeutic resource for CJD [148].

ROLE OF RAGE IN PERIPHERAL NEUROPATHIES

Peripheral neuropathies are progressive, tangential nerve-related disorders of usually unknown etiology. The most common form of peripheral neuropathy is "Diabetic Sensorimotor Neuropathy (DPN)" in the lower limbs, affecting almost 60% of global diabetic sufferers. DPN is succeeded by idiopathic neuropathy, several toxic/drug, infectious disease-related, and inheritable neuropathies such as Charcot-Marie-Tooth disease, vasculitis polyneuropathy, and several others [149, 150]. A study by *Sbai and groupmates* elucidated that RAGE and its ligand S100B, secreted by Schwann cells (SC), are needed for the *in vivo* revamp of peripheral nerve. The study demonstrated that RAGE stimulates the expression of thioredoxin interacting protein (TXNIP) in SC-injured sciatic nerves. Consequent RAGE-TXNIP assists in the refurbishing of damaged peripheral nerves by stimulating the p38 MAPK, CREB, and NF-κB pathways in SCs [99]. Separate investigations herein observed that RAGE and erstwhile pro-inflammatory markers (CML, AGE representative, S100B, HMGB1), over-expressed in human, porcine, and mouse diabetic nerves, during diabetic neuropathy (DN) [151, 152]. Detailed analysis aimed towards screening the molecular mechanisms by which RAGE induces DPN succession, in RAGE-deficient mice after acute peripheral nerve injury. Inspection revealed that besides the onset of multiple metabolic injuries, RAGE contributes to nerve pathology by activating a switch in macrophage phenotype from an anti-to pro-inflammatory regime, at the injury site [153]. Two erstwhile attempts focused on prognostic screening, coordinated AGE-RAGE expression in skin biopsies and DN neurological scores in diabetic patients revealed a considerable rise in dermal biopsy. The AGE and RAGE extents significantly correlated with neuropathy severity, deciphering RAGE's role in DPN pathogenesis [154, 155]. Furthermore, research efforts screening the pathogenesis of several neuropathies (except diabetes), such as inflammatory demyelinating polyneuropathy, Charcot-Marie-Tooth neuropathy, alcohol-related neuropathy, B12 deficiency neuropathy, or vasculitis neuropathy, inferred RAGE-triggered NF-κB activation and RAGE-AGE aggravated inflammation, as major factors involved the pathogenesis. The investigators noticed increased RAGE-AGE interaction and NF-κB activation, elucidating a prominent involvement of RAGE-ligand interactions and RAGE-triggered NF-κB activation in neuronal dysfunctions and nerve degeneration [150].

ROLE OF RAGE IN FAMILIAL AMYLOID NEUROPATHY

Familial Amyloid Polyneuropathy (FAP) is an autosomal dominant, inherited disorder. Pathological traits of FAP typically comprise the amyloid fibril gatherings, majorly constituted of mutant transthyretin. Of note, transthyretin is a protein usually involved in the transport of thyroxin hormone and retinol (vitamin

A) [156, 157]. The defective genotype is characterized by nearly 80-point mutations, but still, there is an acute paucity of their phenotypic association. Typical onset age varies by decades in patients harboring similar mutations contrary to which, some sufferers may remain asymptomatic for their entire life [158]. A noted effort herein by *Matsunaga and colleagues* reported the prevalence of methylglyoxal-derived AGE in FAP patients. The glycated transthyretin could be vulnerable *via* oxidative stress and may result in cytotoxicity. Another likelihood is the transthyretin-RAGE interaction, resulting in NF-κB nuclear translocation and stimulating TNF-α and interleukin-1β [156]. Another worthy effort by *Sousa and groupmates* revealed enhanced RAGE expression in FAP tissues, aiding in fibrillar transthyretin binding besides an increased NF-κB expression [159]. Yet another effort deciphered the RAGE involvement in amyloidosis biology, wherein RAGE was deciphered as bound with Aβ. RAGE inactivation suppressed the splenic prevalence of pro-inflammatory cytokines, NF-κB stimulation, and amyloid build-up in a systemic amyloidosis model [160].

ROLE OF RAGE IN SPINAL CORD INJURY

Spinal cord injury (SCI) is exhaustively related to impaired motor neuron function, triggering the activation of diverse CNS mechanisms in a bid to correct the injured spinal cord tissue. Recently, the potential remedies for SCI together with stem cell therapy, Schwann cell transplantation, restoration of neurotrophin growth factors, inflammatory response regulation have been studied. SCI instigates an inflammatory outcome, causing subsequent tissue damage and neurodegeneration. The inflammatory conditions play a key role in aggravating secondary insult to the spinal cord neuronal tissues amidst traumatic injury, regulating the pathological improvement amidst SCI. Later-stage outcomes of SCI include apoptosis of neurons and oligodendrocytes besides the Walleriande generation (WD) of white matter. Of note, WD sums up the morphological traits in axons, Schwann cells, and macrophages, in a location other than of nerve injury, establishing a microenvironment-driven axonal regeneration [161, 162].

Multiple mechanisms out of RAGE-ligand interactions are known to affect the SCI in a distinctive manner, wherein prominent aspects modulating the role of RAGE in SCI, are (1) binding of RAGE and its ligands seems crucial to the SCI inflammatory response caused *via* cytokine, chemokine release, and modulated apoptotic signaling; (2) RAGE and its ligands may aggravate functional impairment post SCI manifestation by augmenting the neurite outgrowth, critical for regeneration post-CNS injury; (3) RAGE and its ligands could stimulate the axons myelination and regeneration in SCI by activating Schwann cells [162, 163]. RAGE, thereby, emerges as a therapeutic target to promote recovery from SCI. Genetic ablation of RAGE is reported to assist functional recovery in a mouse model of SCI. Additionally, RAGE suppression attenuated neuronal survival at the ventral horn post-SCI. Analysis revealed the role of RAGE in maintaining oligodendrocyte autophagy to promote neuronal regeneration.

ROLE OF RAGE IN EPILEPSY

Epilepsy is a chronic and devastating ND characterized by recurrent, unprovoked seizures. A substantial proportion (~30%) of epilepsy patients are rendered refractory to carefully optimized pharmacological treatment. The core physiological feature of epileptic seizures is the hyperexcitability of CNS neurons. When enough neurons synchronously depolarize and generate action potentials, a seizure initiates. Focal epilepsy is a condition in which a localized brain region induces spontaneous and recurrent seizures. Underlying this condition is a process called epileptogenesis, inducing epilepsy in a rather normal brain. Microglia and astrocytes are the major nervous system cells affected by epilepsy. Glia-mediated inflammation induced by various brain insults can promote seizures and epileptogenesis, particularly for excessive inflammation. Mutations in SCN2A and SCN1B are known to cause febrile seizures while those in SCN9A, GPA6, and GPR98 result in familial febrile seizures. The mutations in GABRG2 are rather generalized outcomes with febrile and familial febrile seizures [163, 164].

Studies have demonstrated the AGE involvement in neuropsychiatric and inflammatory diseases *via* activation of AGE receptors. Epilepsy is characterized by spontaneous synchronized discharge as well as hyper-perfusion and metabolic abnormalities of epileptogenic foci amidst epileptic seizures. Moreover, over the past decade, brain inflammation has been considered a familiar hallmark. Based on this evidence, the AGE-RAGE interaction is deciphered as an adverse signal transduction pathway for epileptic seizures. The potential contribution of the HMGB1-TLR-RAGE axis to seizures and epileptogenesis has been extensively investigated in two models of acute seizures involving unilateral intra-hippocampal administration of kainite and bicuculline (a light-sensitive competitive antagonist of GABAA receptors), respectively, and a model of chronic epilepsy wherein spontaneous seizures arise 1-week post-kainite-induced epilepsy [165, 166]. In a significant attempt, *Lori V and colleagues* examined wild-type RAGE in an experimental epilepsy model, observing RAGE-TLR4 involvement in the manifestation of seizures and HMGB1 progestogenic effects. The analysis also revealed increased RAGE expression in experimental mesial temporal lobe epilepsy (mTLE) hippocampi of mice and humans. These observations made the investigators conclude that activation of the HMGB1-RAGE axis is a new molecular target for seizure inhibition [164]. Functional polymorphisms in the regulatory elements and ligand-binding RAGE regions may alter RAGE expression and function, affecting susceptibility towards epilepsy.

The G82S polymorphism in *RAGE* is associated with increased DRE risk in the Chinese population [165]. Another study observed a critical role of RAGE-TLR4 signaling in post-traumatic epileptogenesis, wherein prompt impairment of post-traumatic brain injury competently prevented post-traumatic epilepsy [166].

THE ANTI-RAGE MOLECULES AGAINST NEUROLOGICAL DISEASES

The onset of NDs *vis-à-vis* RAGE signaling actions relates to ligand-specific RAGE biochemical actions, wherein AGEs, HMGB1 and S100 proteins are the most reported signaling intermediates. The crux factor of cumulative interactions is the sum total of biochemical events leading to enhanced inflammatory responses accompanied by varied redox conditions (although to a minor extent). The primary sources of AGEs in the physiological environment are the reducing sugars' non-enzymatic reactions with primary aa $-NH_2$ groups (proteins and nucleic acids), succeeded by oxidative degradation. Manifested NDs resulting from AGEs-RAGE signaling are typically mediated *via* their diversely modulated cell-damaging effects. The prominent factor amongst these actions is the activated intracellular signaling and the expression of pro-inflammatory transcription factors alongwith inflammatory cytokines. This inflammatory signaling cascade is implicated in multiple NDs, such as AD, secondary effects of traumatic brain injury, ALS, and DN. Besides aggravated inflammation in the brain vicinity, AGEs-RAGE interaction also exhibits a critical role in changing the gut microbiota composition, concomitantly enhancing the permeability and cytokine-modulating actions. The inequity of gut microbiota along with intestinal irritation is squarely implicated with endothelial impairment, suppressed BBB communication, and increasing vulnerability to progressive AD and similar NDs. The observations from several matching studies have elucidated inhibited AGE-RAGE signaling using small molecule-based therapeutics to arrest inflammation-mediated cellular events. The RAGE antagonist, Apeirogon, is presently in the clinical trial of development [167-169].

Investigations have also elucidated RAGE involvement in the onset of DN, as revealed by enhanced localized inflammation and oxidative stress in hyperglycemia-sensitive tissues in the murine models. A further aspect of interest herein relates to the equivalent RAGE expression and concurrent activation squarely in the diabetic and non-diabetic neuropathies (in animal models and human sufferers) [154, 168]. Another concern and limiting aspect is the action of trusted chemotherapeutic drugs (CDs, long in use) such as paclitaxel enhanced RAGE expression tendency in dorsal root ganglia, leading to diabetic neuropathies [169]. A major source of paclitaxel toxicity *vis-à-vis* RAGE signaling actions is the drug's tendency to release the ligand, HMGB1 from macrophage proximity to the peripheral nerves [170]. A similar increment in capsaicin-triggered HMGB1 liberation from macrophages has been testified of late, in response to proteasome inhibiting CD, bortzomib administration in a mouse model. Observations for paclitaxel and bortzomib for aggravated RAGE actions, inferred altered RAGE signaling activities in most chemotherapy-triggered peripheral neuropathies [171]. A coincident observation in oxaliplatin-administered rodents herein is the involvement of non-macrophage cells released HMGB1 in chemotherapy-induced peripheral neuropathy [171]. The pathological sensitivity of peripheral neuropathies distinguishes them as painful and painless, the former being vulnerable to generations of allodynia (aberrant pain). Explicit RAGE involvement in such a troubling pathophysiological state meets justification from the moderated allodynia on the administration of RAGE antagonists (studies in animal models) [170]. Likewise, thrombomodulin-mediated HMGB1 degradation moderates allodynia in oxaliplatin-instigated peripheral neuropathy.

Several studies have explored the anti-HMGB1 molecules inhibiting RAGE-HMGB1 interactions as possible therapeutic targets to treat NDs. For instance, Glycyrrhizin (GLCN) is a biologically vibrant triterpenoid (saponin, ingredient of *Glycyrrhiza glabra*), a conventional medication harboring significantly potent antioxidant, anti-aging, and anti-inflammatory attributes. In a comprehensive review article, *Paudel and colleagues* have discussed GLCN ability to inhibit HMGB1, as a possible cure for multiple NDs [172]. The investigators have raised potential concerns for the extracellular release-mediated intracellular signaling on binding with RAGE and TLR-4. The worthy contribution herein has highlighted manifold emerging aspects of GLCN neuroprotective effects against multiple HMGB1-triggered NDs such as traumatic brain injury, neuro-inflammation, and associated conditions comprising epileptic seizures, Parkinson's disease, and multiple sclerosis. Scientists have elucidated the therapeutic activities of GLCN as mediated *via* attenuated HMGB1 expression, translocation, and suppressed inflammatory cytokine expression. Multiple pre-clinical studies support the GLCN actions against the mainstream therapeutic strategies for NDs, prominently *via* arresting the disease progress. Fig. (**8a**) depicts the chemical structure of GLCN, whose HMGB1 binding mechanism has been studied *via* nuclear magnetic resonance (NMR) and fluorescence analysis. Multiple –OH groups at terminals make this molecule hydrophilically sensitive with closely spaced in and out of the plane orientations inducing a residual stress. The screening revealed a direct HMGB1 binding with GLCN *via* proximity with two shallow concave surfaces, constituted of the two HMG box arms, with an unaltered secondary structure. The GLCN binding with HMGB1 interferes with the latter's chemo-attractant and mitogenic activity. The native affinity of GlCN towards HMGB1 is quite modest with the dissociation constant, K_d being ~150 μM (a concentration-dependent regime), inferring a need for further improved therapeutics. Of further note is the potent ability of glycyrrhetinic acid (formed *via* GLCN hydrolysis) to inhibit HMGB1 *via* interfering with Lys90, Arg91, Ser1011, Tyr149, C230 and C231 residues in the HMGB1-DNA complex. Apart from this, GLCN also impairs the HMGB1-TLR4 and RAGE-NF-κB signaling axis, impairing the cytokine-like HMGB1 actions (prominently displayed by the sulfide form therein). Thus, GLCN inhibition of HMGB1 is mediated by its direct suppressed discharge from injured cells prominently *via* impaired HMGB1 phosphorylation and in a small proportion by reduced expression. A further distinction underlined by authors is the stronger HMGB1 inhibition of GLCN than the cytokine release inhibiting drugs (CRIDs, such as ethyl pyruvate, nicotine, acetylcholine, stearoyl lysophosphatidylcholine, and tanshinone IIA). Of note, CRIDs operate *via* modulating HMGB1 nuclear release from activated inflammatory cells. Thereby, these drugs are rendered unable to obstruct HMGB1 extracellular functions. Readers are suggested to track the cross-cited studies in the 2020 review article by *Paudel and colleagues*, featured in ACS Chemical Neuroscience.

Numerous other natural compounds (of polyphenolic background) have been demonstrated as useful in controlling the RAGE actions by suppressing the inflammatory and oxidative aggravations. Such options provide a sustainable alternative to CDs and synthetic compounds and are encouraged by the pleiotropic essence of most secondary plant metabolites. For instance, epigallocatechin--gallate (EGCG), quercetin, and lycopene represent some alternatives that have been validated for moderating the lethal systemic inflammation resulting from endotoxemia. Prevailing as a major ingredient of tea and brewed from the leaves of the plant *Camellia sinensis*, EGCG has been reported to protect mice from lethal endotoxemia, rescuing them from sepsis *via* moderating HMGB1 systemic accumulation in macrophage/monocyte cultures and forbidding the clustering of exogenous HMGB1 on the macrophages surface, which otherwise contribute to aggressive inflammatory conditions [173]. Structurally, EGCG is distinguished *via* its manifold, closely positioned –OH groups generating a likelihood for intramolecular hydrogen bonding (HB) and a higher sensitivity towards hydrophilic solvents. However, in and out of the plane asymmetry in the molecule generates residual stress to hinder its aqueous solubilization (Fig. **8b**). Similarly, quercetin is another plant polyphenol obtained from fruits, vegetables, leaves, and grains, widely reported for its antioxidant and anticancer efficacy. Studies on quercetin-modulated HMGB1 activity reported an impaired release and cytokine actions of HMGB1, along with the suppressed activation of mitogen-activated protein kinase (MAPKs) [174]. Fig. (**8c**) represents the chemical structure of quercetin, exhibiting multiple –OH groups in close vicinity. These closely positioned –OH groups confer a high possibility of HB-controlled interactions, deciphering the suitability of robust H^+ generation (due to the higher stability of conjugate phenoxide ion) and the consequent antioxidant effect.

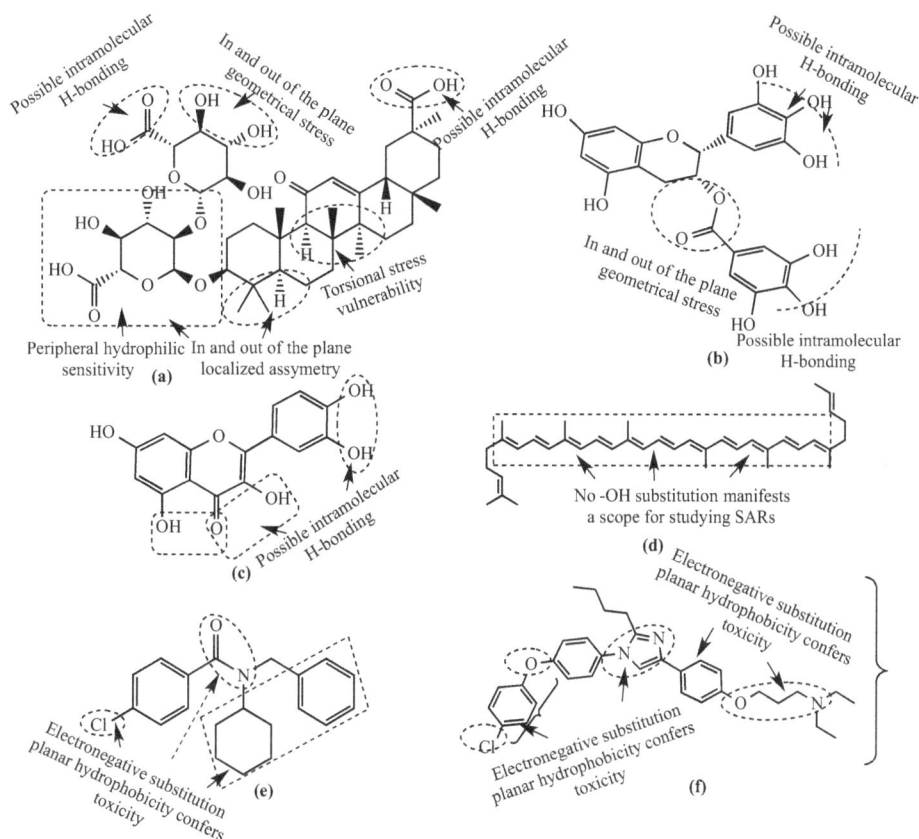

Fig. (8). Structural distinctions of natural polyphenols (a-d) and synthetically made potent HMGB1 inhibitors (e-f). Missing –OH functionality and electronegative N, O, Cl in synthetic compounds make them strongly reactive and toxic, cautioning for concentration optimization.

Likewise, lycopene (a carotene present in tomatoes and red fruits) is well-versed for its antioxidant and anti-inflammatory activities, together inhibiting the lipopolysaccharide triggered HMGB1 release besides the HMGB1 expression driven TNF secretory phospholipase A2 and pro-inflammatory signaling in endothelial cells. Known to operate *via* suppressing the cell surface expression of cell adhesion molecules (CAMs) and HMGB1 receptors (TLR-2,4 and RAGE), this polyphenolic compound is devoid of any –OH group (Fig. **8d**) [175]. This structural distinction opens a new gate to probe the mechanism for the anti-inflammatory essence as of other phytochemicals described in Fig. (**6**). The understanding of structure-activity-relationships (SAR) could open new avenues for a possible synergism with CDs or compatible hydrophobicity moderators.

Similarly, a significant study from 2021 elucidated curcumin's ability to suppress the RAGE-aggravated inflammatory response in AD-positive mice models. Analysis revealed curcumin's relieving ability from memory shortfalls *via* inhibiting HMGB1-RAGE/TLR4-NF-κB signaling [176]. Accelerated cognitive impairment has been notified in several studies with a weakened activity of gut microbiota resulting in enhanced oxidative stress and aggravated actions of inflammatory cytokines. Edible polyphenols and certain other nutraceutical grade materials (oils containing PUFAs) exhibit potent anti-oxidative sensitivity to relieve the consequent oxidative stress, reported as a causative factor of AD in mice models [177]. A constant incentive behind the use of edible polyphenols is their biocompatible essence owing to which, the intake of their major extent with small extent of CDs moderates the CDs intake and toxicity by their specific actions within the diseased cells only (*via* pleiotropic essence of polyphenols, enhanced permeation and retention hypothesis). Our earlier contributions have reported such possibilities as accomplishable for combinatorial delivery of CDs with flavonoids having polyphenolic sensitivity and curcumin resembling pleiotropic functional mechanisms, like hesperetin and genistein [178 - 180].

Studies on the development of RAGE antagonists have distinguished them as endogenous or exogenous (typical small molecule synthetic compounds), which on binding to RAGE suppress the ligand-RAGE interactions and pro-inflammatory signaling to arrest the inflammatory and oxidative stress complications. Recent studies on two such small molecule RAGE inhibitors, FPS-ZM1 (4-chloro-N-cyclohexyl-N-(phenylmethyl) benzamide) and Azeliragon (TTP488)   (3-[4-[2-butyl-1,   [4-(4-chlorophenoxy)phenyl]-1H-imida-ol-4-yl]phenoxy]-N,N-diethyl-1-proponamine), have progressed to clinical trials.

Fig. (**8**) depicts the chemical structures of FPS-ZM1 and TTP488, wherein substitutions of N and Cl are deciphered as common aspects, attributing to toxicity induction. Both N and Cl are electronegative and could attract a shared electron towards themselves in a chemical bond, thereby serving as the basis of antioxidant functioning. Earlier studies on FPS-ZM1 have revealed its selective binding of the RAGE V-domain (having cationic residues) and subsequent inhibition of concurrent Aβ40 and Aβ2 formations, besides inhibiting β-secretase activity (enzyme implicated in Aβ formation). Of note, α-synuclein is the interlink facilitating the interaction of its acidic, C-terminal, and cationic residues at its V-domain to provoke the signaling activities implicated in neuroinflammation. Studies on FPS-ZM1's working have revealed its selective binding to suppress the α-synuclein fibril's RAGE proximity, moderating the microglial inflammatory actions. Studies on animal models have revealed FPS-ZM1 efficacy in treating ischemia-reperfusion-triggered hippocampal damage only at higher doses. The risk of higher toxicity could be reduced *via* its combinatorial delivery with compatible polyphenols or nanocarrier-assisted delivery [181]. Similarly, Azeliragon is currently undergoing phase 3 trials for AD treatment, with phase2b studies having successfully revealed suppressed Aβ plaque formation, accompanying the increase in plasma Aβ extents besides reduced extents of inflammatory cytokines and a moderated cognitive suppression in mild AD sufferers [182]. Readers are suggested to refer to the 2016 research effort of *Schmidt and teammates*, wherein 58,000 small molecules competitive, RAGE cytoplasmic terminal inhibitors as RAGE-DIAPH1 antagonists have been screened and 13 of the tested compounds were eventually identified as potent for impairing the AGE-RAGE signaling outcomes. Subsequently, the investigators, through fluorescence titration, demonstrated the nanomolar affinity of these compounds to the RAGE cytoplasmic terminal (carboxy terminus), offering hope for specificity with low toxicity risk. 12 of the screened 13 lead molecules significantly impaired CML (an AGE) triggered IL-6 upregulation while all the compounds restrained the ML-mediated TNF-α enhanced activity. Thereby, the anti-inflammatory essence of optimized AGE-RAGE signaling inhibitors is inferred, analogous to the extracellular RAGE antagonists (having benzamide moiety like FPS-ZM1) as possible therapeutic agents to treat AGE-mediated NDs. The findings of some other close efforts are discussed ahead, providing further anticipation for a better analysis *via* accurate *in silico* simulation models and combinatorial formulation.

A noteworthy 2013 effort from *Li and colleagues* reported impaired Aβ glycation (in mice) on aminoguanidine administration for up to 3 months, an analysis of which deciphered improved cognitive function and a higher toxicity for glycated Aβ [113]. More recently, *in vitro* studies deciphered a regulation of amyloid precursor protein (APP) expression by AGEs, leading to increased Aβ levels. This aggravated Aβ-APP proximity could be impaired using ROS inhibitor, N-acety--L-cysteine (signifying it as a result of oxidative stress) [183]. For instance, the use of soluble RAGE substantially decreased the infarct size, exhibiting neuroprotective essence on cortical neurons in mouse-MCAO models [184, 185]. Besides, the expression of inflammatory molecules, *viz.* TNF-α or inducible nitric oxide synthase was attenuated at the mRNA level in the hippocampal CA1 region of esRAGE transgenic (T_g) and RAGE knockout (KO) mice, than the wild-type (WT) [14]. Analysis revealed esRAGE conferred neuroprotection wherein examined mice were engineered to generate human esRAGE in the liver using albumin promoter [186]. Disappointingly, human esRAGE was also present in the brain of this mouse model. Therefore, to ascertain the outcome of bloodstream esRAGE in solace, a parabiotic way out was employed, demonstrating a stable blood esRAGE extent. However, a pilot study of esRAGE intravenous injection is needed to ascertain the underlying mechanism for neuroprotection.

Abnormal Aβ influx across the BBB by RAGE inhibitor (FPS-ZM1) at a higher dosage, contributes to reduced neuronal apoptosis, improved hippocampal plasticity, and cognitive impairment in db/db mice [187]. A recent study exploring the therapeutic targets in AD demonstrated RAGE binding to monomeric and non-cellular toxic Aβ revealed aggressive neurite formation in neuroblastoma cells on all-trans-retinoic acid neurogenesis-induced treatments [96]. Nevertheless, it must be noted that, despite promising results of *in vitro* and *in vivo* mice models, the findings of 10-week and 1-year clinical trials of PF-04494700, an oral RAGE inhibitor, in mild to moderate AD sufferers remained inconclusive, exhibiting no significant improvements for the overall well-being. The follow-up inspection after the trials indicated a likelihood of belated positive cognitive effects due to the relatively short duration of both trials [188]. Currently, the preparation strategies for revoking the trial have been proclaimed by Transtech Pharma (www.alzforum.org/therapeutics/ttp488).

CONCLUSION

To sum up, this chapter discussed the role of RAGE in various neuronal cells and related diseases; wherein distinctive ligand-mediated interactions assume prime significance. The abnormal expression and trafficking of proteins arising from altered redox balance is the prominent aspect of deleterious RAGE-ligand interactions. Prominent RAGE ligands attributing to NDs include Aβ peptide, HMGB1, S100, and AGEs. Understanding the structural modulation of various RAGE ligands thereby assumes paramount significance in their role for NDs. While RAGE ligands are natively the proteins with functions and activities important for normal functioning, conditions resulting in the manifold neuronal disorders are ascribed by the defects in post-translational modifications or maybe high-level expression under various pathological conditions. Many of the actions of RAGE ligands are without any involvement of the RAGE molecule. Various RAGE ligands with or without interaction with RAGE aggravate the altered redox balance, manifested *via* enhanced oxidative and inflammatory stress. The gradually impaired coordination in neurodegenerative diseases, prominently in Alzheimer's and Parkinson's disorders, is characterized by the manifested aging of local tissues due to their continuous use and improper nourishment, along with enhanced levels of oxidative and inflammatory stress. Therefore, dietary sources of RAGE ligands such as AGE must be moderated along with processed carbohydrate intake which could minimize the *in vivo* RAGE expression and its pro-inflammatory and pro-oxidative activities. An urgent caution and safeguard measure to moderate the vulnerability of RAGE-ligand interactions is the proper maintenance of physiological antioxidants, such as glutathione, thioredoxin (non-enzymatic antioxidants), and superoxide dismutase (SOD, an enzymatic antioxidant). Thus, RAGE-ligand interactions are detrimental to normal physiology with oxidative and inflammatory stress. The reliability of such observations has paved the way for RAGE-targeted therapeutic strategies as treatment options. In this scenario, a thorough understanding of ligand biochemistry with the diversity of structure-function aspects is highly urgent. Ongoing research and trials in this direction have faced challenges *vis-à-vis* cumbersome access to the intricate blood-brain barrier (BBB) regions. Understanding the comparative significance of contributory factors in the RAGE-ligand interactions triggered neuronal diseases is an aspect where computational simulations and support from case-history stereotypes could be used for vitalizing the remedies. The pegging challenge in all RAGE-ligand interactions is to discover the specific inhibitors for RAGE-ligand interactions. Since RAGE ligands fulfill functions independent of binding with RAGE, understanding the inhibitor that specifically inhibits ligands is also a necessity of the hour. Ingestion of plentiful natural products in the food and minimizing the intake of free radical-generating sources alongside a healthy working routine with adequate physical activities are in themselves, major remedies, since this will limit the production of AGE and expression of other RAGE ligands. Genetic vulnerability and polymorphisms of RAGE and its ligands form an area mandating pattern identification with a low success rate, although it may be improved *via* rigorous genetic identification of personalized polymorphic genes.

REFERENCES

[1]   Sousa AMM, Meyer KA, Santpere G, Gulden FO, Sestan N. Evolution of the human nervous system function, structure and development. Cell 2017; 170(2): 226-47.
        [http://dx.doi.org/10.1016/j.cell.2017.06.036] [PMID: 28708995]

[2]   Laming PR, Kimelberg H, Robinson S, *et al.* Neuronal–glial interactions and behaviour. Neurosci Biobehav Rev 2000; 24(3): 295-340.
        [http://dx.doi.org/10.1016/S0149-7634(99)00080-9] [PMID: 10781693]

[3]   Kettenmann H, Faissner A, Trotter J. Neuron-glia interactions in homeostasis and degeneration.Comprehensive human physiology. Berlin, Heidelberg: Springer 1996; pp. 533-43.
        [http://dx.doi.org/10.1007/978-3-642-60946-6_27]

[4]   Hori O, Brett J, Slattery T, *et al.* The receptor for advanced glycation end products (RAGE) is a cellular binding site for amphoterin. Mediation of neurite outgrowth and co-expression of rage and amphoterin in the developing nervous system. J Biol Chem 1995; 270(43): 25752-61.
        [http://dx.doi.org/10.1074/jbc.270.43.25752] [PMID: 7592757]

[5]   Brett J, Schmidt AM, Yan SD, *et al.* Survey of the distribution of a newly characterized receptor for advanced glycation end products in tissues. Am J Pathol 1993; 143(6): 1699-712.
        [PMID: 8256857]

[6]   Ott C, Jacobs K, Haucke E, Navarrete Santos A, Grune T, Simm A. Role of advanced glycation end products in cellular signaling. Redox Biol 2014; 2(1): 411-29.
        [http://dx.doi.org/10.1016/j.redox.2013.12.016] [PMID: 24624331]

[7] Huttunen HJ, Kuja-Panula J, Sorci G, Agneletti AL, Donato R, Rauvala H. Coregulation of neurite outgrowth and cell survival by amphoterin and S100 proteins through receptor for advanced glycation end products (RAGE) activation. J Biol Chem 2000; 275(51): 40096-105.
[http://dx.doi.org/10.1074/jbc.M006993200] [PMID: 11007787]

[8] Schmidt A, Kuhla B, Bigl K, Münch G, Arendt T. Cell cycle related signaling in neuro2a cells proceeds *via* the receptor for advanced glycation end products. J Neural Transm (Vienna) 2007; 114(11): 1413-24.
[http://dx.doi.org/10.1007/s00702-007-0770-0] [PMID: 17564756]

[9] Qin J, Goswami R, Dawson S, Dawson G. Expression of the receptor for advanced glycation end products in oligodendrocytes in response to oxidative stress. J Neurosci Res 2008; 86(11): 2414-22.
[http://dx.doi.org/10.1002/jnr.21692] [PMID: 18438937]

[10] Ding Q, Keller JN. Evaluation of rage isoforms, ligands, and signaling in the brain. Biochim Biophys Acta Mol Cell Res 2005; 1746(1): 18-27.
[http://dx.doi.org/10.1016/j.bbamcr.2005.08.006] [PMID: 16214242]

[11] Ma L, Carter RJ, Morton AJ, Nicholson LFB. RAGE is expressed in pyramidal cells of the hippocampus following moderate hypoxic–ischemic brain injury in rats. Brain Res 2003; 966(2): 167-74.
[http://dx.doi.org/10.1016/S0006-8993(02)04149-5] [PMID: 12618340]

[12] Haslbeck KM, Bierhaus A, Erwin S, *et al.* Receptor for advanced glycation endproduct (RAGE)–mediated nuclear factor-κB activation in vasculitic neuropathy. Muscle Nerve 2004; 29(6): 853-60.
[http://dx.doi.org/10.1002/mus.20039] [PMID: 15170618]

[13] Rong LL, Yan SF, Wendt T, *et al.* RAGE modulates peripheral nerve regeneration *via* recruitment of both inflammatory and axonal outgrowth pathways. FASEB J 2004; 18(15): 1818-25.
[http://dx.doi.org/10.1096/fj.04-1900com] [PMID: 15576485]

[14] Kamide T, Kitao Y, Takeichi T, *et al.* RAGE mediates vascular injury and inflammation after global cerebral ischemia. Neurochem Int 2012; 60(3): 220-8.
[http://dx.doi.org/10.1016/j.neuint.2011.12.008] [PMID: 22202666]

[15] Gasparotto J, Ribeiro CT, da Rosa-Silva HT, *et al.* Systemic inflammation changes the site of rage expression from endothelial cells to neurons in different brain areas. Mol Neurobiol 2019; 56(5): 3079-89.
[http://dx.doi.org/10.1007/s12035-018-1291-6] [PMID: 30094805]

[16] Meneghini V, Francese MT, Carraro L, Grilli M. A novel role for the Receptor for Advanced Glycation End-products in neural progenitor cells derived from adult SubVentricular Zone. Mol Cell Neurosci 2010; 45(2): 139-50.
[http://dx.doi.org/10.1016/j.mcn.2010.06.005] [PMID: 20600932]

[17] Meneghini V, Bortolotto V, Francese MT, *et al.* High-mobility group box-1 protein and β-amyloid oligomers promote neuronal differentiation of adult hippocampal neural progenitors *via* receptor for advanced glycation end products/nuclear factor-κB axis: relevance for Alzheimer's disease. J Neurosci 2013; 33(14): 6047-59.
[http://dx.doi.org/10.1523/JNEUROSCI.2052-12.2013] [PMID: 23554486]

[18] Li M, Sun L, Luo Y, Xie C, Pang Y, Li Y. High-mobility group box 1 released from astrocytes promotes the proliferation of cultured neural stem/progenitor cells. Int J Mol Med 2014; 34(3): 705-14.
[http://dx.doi.org/10.3892/ijmm.2014.1820] [PMID: 24970310]

[19] Dong H, Zhang Y, Huang Y, Deng H. Pathophysiology of RAGE in inflammatory diseases. Front Immunol 2022; 13: 931473.
[http://dx.doi.org/10.3389/fimmu.2022.931473] [PMID: 35967420]

[20] Donato R, Sorci G, Riuzzi F, *et al.* S100B's double life: Intracellular regulator and extracellular signal.

Biochim Biophys Acta Mol Cell Res 2009; 1793(6): 1008-22.
[http://dx.doi.org/10.1016/j.bbamcr.2008.11.009] [PMID: 19110011]

[21]   C RC, Lukose B, Rani P. G82S RAGE polymorphism influences amyloid-RAGE interactions relevant in Alzheimer's disease pathology. PLoS One 2020; 15(10): e0225487.
[http://dx.doi.org/10.1371/journal.pone.0225487] [PMID: 33119615]

[22]   Rauvala H, Merenmies J, Pihlaskari R, Korkolainen M, Huhtala ML, Panula P. The adhesive and neurite-promoting molecule p30: analysis of the amino-terminal sequence and production of antipeptide antibodies that detect p30 at the surface of neuroblastoma cells and of brain neurons. J Cell Biol 1988; 107(6): 2293-305.
[http://dx.doi.org/10.1083/jcb.107.6.2293] [PMID: 2461949]

[23]   Merenmies J, Pihlaskari R, Laitinen J, Wartiovaara J, Rauvala H. 30-kDa heparin-binding protein of brain (amphoterin) involved in neurite outgrowth. Amino acid sequence and localization in the filopodia of the advancing plasma membrane. J Biol Chem 1991; 266(25): 16722-9.
[http://dx.doi.org/10.1016/S0021-9258(18)55361-8] [PMID: 1885601]

[24]   Donato R. Functional roles of S100 proteins, calcium-binding proteins of the EF-hand type. Biochim Biophys Acta Mol Cell Res 1999; 1450(3): 191-231.
[http://dx.doi.org/10.1016/S0167-4889(99)00058-0] [PMID: 10395934]

[25]   Hofmann MA, Drury S, Fu C, et al. RAGE mediates a novel proinflammatory axis: a central cell surface receptor for S100/calgranulin polypeptides. Cell 1999; 97(7): 889-901.
[http://dx.doi.org/10.1016/S0092-8674(00)80801-6] [PMID: 10399917]

[26]   Bishop GM, Robinson SR. Physiological roles of amyloid-beta and implications for its removal in Alzheimer's disease. Drugs Aging 2004; 21(10): 621-30.
[http://dx.doi.org/10.2165/00002512-200421100-00001] [PMID: 15287821]

[27]   Whitson JS, Selkoe DJ, Cotman CW. Amyloid beta protein enhances the survival of hippocampal neurons *in vitro*. Science 1989; 243(4897): 1488-90.
[http://dx.doi.org/10.1126/science.2928783] [PMID: 2928783]

[28]   Cai W, Li L, Sang S, Pan X, Zhong C. Physiological roles of β-amyloid in regulating synaptic function: Implications for AD pathophysiology. Neurosci Bull 2023; 39(8): 1289-308.
[http://dx.doi.org/10.1007/s12264-022-00985-9] [PMID: 36443453]

[29]   Yan SD, Chen X, Fu J, et al. RAGE and amyloid-β peptide neurotoxicity in Alzheimer's disease. Nature 1996; 382(6593): 685-91.
[http://dx.doi.org/10.1038/382685a0] [PMID: 8751438]

[30]   Du Yan S, Zhu H, Fu J, et al. Amyloid-β peptide–Receptor for Advanced Glycation Endproduct interaction elicits neuronal expression of macrophage-colony stimulating factor: A proinflammatory pathway in Alzheimer disease. Proc Natl Acad Sci USA 1997; 94(10): 5296-301.
[http://dx.doi.org/10.1073/pnas.94.10.5296] [PMID: 9144231]

[31]   Lue LF, Walker DG, Brachova L, et al. Involvement of microglial receptor for advanced glycation endproducts (RAGE) in Alzheimer's disease: identification of a cellular activation mechanism. Exp Neurol 2001; 171(1): 29-45.
[http://dx.doi.org/10.1006/exnr.2001.7732] [PMID: 11520119]

[32]   Yan SD, Yan SF, Chen X, et al. Non-enzymatically glycated tau in Alzheimer's disease induces neuronal oxidant stress resulting in cytokine gene expression and release of amyloid β-peptide. Nat Med 1995; 1(7): 693-9.
[http://dx.doi.org/10.1038/nm0795-693] [PMID: 7585153]

[33]   Deane R, Du Yan S, Submamaryan RK, et al. RAGE mediates amyloid-β peptide transport across the blood-brain barrier and accumulation in brain. Nat Med 2003; 9(7): 907-13.
[http://dx.doi.org/10.1038/nm890] [PMID: 12808450]

[34]   Lue L, Yan S, Stern D, Walker D. Preventing activation of receptor for advanced glycation

endproducts in Alzheimer's disease. Curr Drug Targets CNS Neurol Disord 2005; 4(3): 249-66.
[http://dx.doi.org/10.2174/1568007054038210] [PMID: 15975028]

[35]   Arancio O, Zhang HP, Chen X, *et al.* RAGE potentiates Aβ-induced perturbation of neuronal function in transgenic mice. EMBO J 2004; 23(20): 4096-105.
[http://dx.doi.org/10.1038/sj.emboj.7600415] [PMID: 15457210]

[36]   Origlia N, Righi M, Capsoni S, *et al.* Receptor for advanced glycation end product-dependent activation of p38 mitogen-activated protein kinase contributes to amyloid-beta-mediated cortical synaptic dysfunction. J Neurosci 2008; 28(13): 3521-30.
[http://dx.doi.org/10.1523/JNEUROSCI.0204-08.2008] [PMID: 18367618]

[37]   Candela P, Gosselet F, Saint-Pol J, *et al.* Apical-to-basolateral transport of amyloid-β peptides through blood-brain barrier cells is mediated by the receptor for advanced glycation end-products and is restricted by P-glycoprotein. J Alzheimers Dis 2010; 22(3): 849-59.
[http://dx.doi.org/10.3233/JAD-2010-100462] [PMID: 20858979]

[38]   Chaney MO, Stine WB, Kokjohn TA, *et al.* RAGE and amyloid beta interactions: Atomic force microscopy and molecular modeling. Biochim Biophys Acta Mol Basis Dis 2005; 1741(1-2): 199-205.
[http://dx.doi.org/10.1016/j.bbadis.2005.03.014] [PMID: 15882940]

[39]   Kim SJ, Ahn JW, Kim H, *et al.* Two β-strands of RAGE participate in the recognition and transport of amyloid-β peptide across the blood brain barrier. Biochem Biophys Res Commun 2013; 439(2): 252-7.
[http://dx.doi.org/10.1016/j.bbrc.2013.08.047] [PMID: 23973487]

[40]   Tolstova AP, Adzhubei AA, Mitkevich VA, Petrushanko IY, Makarov AA. Docking and molecular dynamics-based identification of interaction between various beta-amyloid isoforms and RAGE receptor. Int J Mol Sci 2022; 23(19): 11816.
[http://dx.doi.org/10.3390/ijms231911816] [PMID: 36233130]

[41]   Huang X, Wang L, Zhang H. Developmental expression of the High Mobility Group B gene in the amphioxus, *Branchiostoma belcheri* tsingtauense. Int J Dev Biol 2005; 49(1): 49-6.
[http://dx.doi.org/10.1387/ijdb.041915xh] [PMID: 15744666]

[42]   Kinoshita M, Hatada S, Asashima M, Noda M. *HMG-X*, a *Xenopus* gene encoding an HMG1 homolog, is abundantly expressed in the developing nervous system. FEBS Lett 1994; 352(2): 191-6.
[http://dx.doi.org/10.1016/0014-5793(94)00909-0] [PMID: 7925972]

[43]   Zhao X, Kuja-Panula J, Rouhiainen A, Chen Y, Panula P, Rauvala H. High mobility group box-1 (HMGB1; amphoterin) is required for zebrafish brain development. J Biol Chem 2011; 286(26): 23200-13.
[http://dx.doi.org/10.1074/jbc.M111.223834] [PMID: 21527633]

[44]   Guazzi S, Strangio A, Franzi AT, Bianchi ME. HMGB1, an architectural chromatin protein and extracellular signalling factor, has a spatially and temporally restricted expression pattern in mouse brain. Gene Expr Patterns 2003; 3(1): 29-33.
[http://dx.doi.org/10.1016/S1567-133X(02)00093-5] [PMID: 12609598]

[45]   Fages C, Nolo R, Huttunen HJ, Eskelinen EL, Rauvala H. Regulation of cell migration by amphoterin. J Cell Sci 2000; 113(4): 611-20.
[http://dx.doi.org/10.1242/jcs.113.4.611] [PMID: 10652254]

[46]   Rouhiainen A, Kuja-Panula J, Tumova S, Rauvala H. RAGE-mediated cell signaling. Methods Mol Biol 2013; 963: 239-63.
[http://dx.doi.org/10.1007/978-1-62703-230-8_15] [PMID: 23296615]

[47]   Rouhiainen A, Imai S, Rauvala H, Parkkinen J. Occurrence of amphoterin (HMG1) as an endogenous protein of human platelets that is exported to the cell surface upon platelet activation. Thromb Haemost 2000; 84(6): 1087-94.
[PMID: 11154118]

[48]   Rouhiainen A, Kuja-Panula J, Wilkman E, *et al.* Regulation of monocyte migration by amphoterin

(HMGB1). Blood 2004; 104(4): 1174-82.
[http://dx.doi.org/10.1182/blood-2003-10-3536] [PMID: 15130941]

[49] Campana L, Bosurgi L, Bianchi ME, Manfredi AA, Rovere-Querini P. Requirement of HMGB1 for stromal cell–derived factor–1/CXCL12–dependent migration of macrophages and dendritic cells. J Leukoc Biol 2009; 86(3): 609-15.
[http://dx.doi.org/10.1189/jlb.0908576] [PMID: 19414537]

[50] Zhao X, Rouhiainen A, Li Z, Guo S, Rauvala H. Regulation of neurogenesis in mouse brain by HMGB1. Cells 2020; 9(7): 1714.
[http://dx.doi.org/10.3390/cells9071714] [PMID: 32708917]

[51] Ye C, Li H, Li Y, *et al.* Hypoxia-induced HMGB1 promotes glioma stem cells self-renewal and tumorigenicity *via* RAGE. iScience 2022; 25(9): 104872.
[http://dx.doi.org/10.1016/j.isci.2022.104872] [PMID: 36034219]

[52] Yardan T, Erenler AK, Baydin A, Aydin K, Cokluk C. Usefulness of S100B protein in neurological disorders. J Pak Med Assoc 2011; 61(3): 276-81.
[PMID: 21465945]

[53] Ludwin SK, Kosek JC, Eng LF. The topographical distribution of S-100 and GFA proteins in the adult rat brain: An immunohistochemical study using horseradish peroxidase-labelled antibodies. J Comp Neurol 1976; 165(2): 197-207.
[http://dx.doi.org/10.1002/cne.901650206] [PMID: 1107363]

[54] Brockes JP, Fields KL, Raff MC. Studies on cultured rat Schwann cells. I. Establishment of purified populations from cultures of peripheral nerve. Brain Res 1979; 165(1): 105-18.
[http://dx.doi.org/10.1016/0006-8993(79)90048-9] [PMID: 371755]

[55] Ferri GL, Probert L, Cocchia D, Michetti F, Marangos PJ, Polak JM. Evidence for the presence of S-100 protein in the glial component of the human enteric nervous system. Nature 1982; 297(5865): 409-10.
[http://dx.doi.org/10.1038/297409a0] [PMID: 7043279]

[56] Didier M, Harandi M, Aguera M, *et al.* Differential immunocytochemical staining for glial fibrillary acidic (GFA) protein, S-100 protein and glutamine synthetase in the rat subcommissural organ, nonspecialized ventricular ependyma and adjacent neuropil. Cell Tissue Res 1986; 245(2): 343-51.
[http://dx.doi.org/10.1007/BF00213941] [PMID: 2874885]

[57] Rickmann M, Wolff JR. S100 protein expression in subpopulations of neurons of rat brain. Neuroscience 1995; 67(4): 977-91.
[http://dx.doi.org/10.1016/0306-4522(94)00615-C] [PMID: 7675218]

[58] Michetti F, Massaro A, Murazio M. The nervous system-specific S-100 antigen in cerebrospinal fluid of multiple sclerosis patients. Neurosci Lett 1979; 11(2): 171-5.
[http://dx.doi.org/10.1016/0304-3940(79)90122-8] [PMID: 460686]

[59] Michetti F, Massaro A, Russo G, Rigon G. The S-100 antigen in cerebrospinal fluid as a possible index of cell injury in the nervous system. J Neurol Sci 1980; 44(2-3): 259-63.
[http://dx.doi.org/10.1016/0022-510X(80)90133-1] [PMID: 7354371]

[60] Kato K, Kimura S, Semba R, Suzuki F, Nakajima T. Increase in S-100 protein levels in blood plasma by epinephrine. J Biochem 1983; 94(3): 1009-11.
[http://dx.doi.org/10.1093/oxfordjournals.jbchem.a134397] [PMID: 6358201]

[61] Gazzolo D, Vinesi P, Marinoni E, *et al.* S100B protein concentrations in cord blood: correlations with gestational age in term and preterm deliveries. Clin Chem 2000; 46(7): 998-1000.
[http://dx.doi.org/10.1093/clinchem/46.7.998] [PMID: 10894846]

[62] Gazzolo D, Bruschettini M, Corvino V, *et al.* S100b protein concentrations in amniotic fluid correlate with gestational age and with cerebral ultrasound scanning results in healthy fetuses. Clin Chem 2001; 47(5): 954-6.

[http://dx.doi.org/10.1093/clinchem/47.5.954] [PMID: 11325908]

[63] Gazzolo D, Florio P, Ciotti S, *et al.* S100B protein in urine of preterm newborns with ominous outcome. Pediatr Res 2005; 58(6): 1170-4. a
[http://dx.doi.org/10.1203/01.pdr.0000185131.22985.30] [PMID: 16306188]

[64] Gazzolo D, Lituania M, Bruschettini M, *et al.* S100B protein levels in saliva: correlation with gestational age in normal term and preterm newborns. Clin Biochem 2005; 38(3): 229-33. b
[http://dx.doi.org/10.1016/j.clinbiochem.2004.12.006] [PMID: 15708543]

[65] Abboud T, Mende KC, Jung R, *et al.* Prognostic value of early S100 calcium binding protein B and neuron-specific enolase in patients with poor-grade aneurysmal subarachnoid hemorrhage: a pilot study. World Neurosurg 2017; 108: 669-75.
[http://dx.doi.org/10.1016/j.wneu.2017.09.074] [PMID: 28943424]

[66] Leclerc E, Fritz G, Vetter SW, Heizmann CW. Binding of S100 proteins to RAGE: An update. Biochim Biophys Acta Mol Cell Res 2009; 1793(6): 993-1007.
[http://dx.doi.org/10.1016/j.bbamcr.2008.11.016]

[67] Grzybowska EA. Calcium-binding proteins with disordered structure and their role in secretion, storage, and cellular signaling. Biomolecules 2018; 8(2): 42.
[http://dx.doi.org/10.3390/biom8020042] [PMID: 29921816]

[68] Donato R, Cannon BR, Sorci G, *et al.* Functions of S100 proteins. Curr Mol Med 2013; 13(1): 24-57.
[http://dx.doi.org/10.2174/156652413804486214] [PMID: 22834835]

[69] Langeh U, Singh S. Targeting S100B protein as a surrogate biomarker and its role in various neurological disorders. Curr Neuropharmacol 2021; 19(2): 265-77.
[http://dx.doi.org/10.2174/18756190MTA44NjEs3] [PMID: 32727332]

[70] Michetti F, D'Ambrosi N, Toesca A, *et al.* The S100B story: from biomarker to active factor in neural injury. J Neurochem 2019; 148(2): 168-87.
[http://dx.doi.org/10.1111/jnc.14574] [PMID: 30144068]

[71] Lasič E, Galland F, Vardjan N, *et al.* Time-dependent uptake and trafficking of vesicles capturing extracellular S100B in cultured rat astrocytes. J Neurochem 2016; 139(2): 309-23.
[http://dx.doi.org/10.1111/jnc.13754] [PMID: 27488079]

[72] Ostendorp T, Heizmann CW, Kroneck PMH, Fritz G. Purification, crystallization and preliminary X-ray diffraction studies on human Ca $^{2+}$ -binding protein S100B. Acta Crystallogr Sect F Struct Biol Cryst Commun 2005; 61(7): 673-5.
[http://dx.doi.org/10.1107/S1744309105018014] [PMID: 16511125]

[73] Riuzzi F, Sorci G, Donato R. S100B protein regulates myoblast proliferation and differentiation by activating FGFR1 in a bFGF-dependent manner. J Cell Sci 2011; 124(14): 2389-400.
[http://dx.doi.org/10.1242/jcs.084491] [PMID: 21693575]

[74] Riuzzi F, Sorci G, Beccafico S, Donato R. S100B engages RAGE or bFGF/FGFR1 in myoblasts depending on its own concentration and myoblast density. Implications for muscle regeneration. PLoS One 2012; 7(1): e28700.
[http://dx.doi.org/10.1371/journal.pone.0028700] [PMID: 22276098]

[75] Koppal T, Lam AGM, Guo L, Van Eldik LJ. S100B proteins that lack one or both cysteine residues can induce inflammatory responses in astrocytes and microglia. Neurochem Int 2001; 39(5-6): 401-7.
[http://dx.doi.org/10.1016/S0197-0186(01)00047-X] [PMID: 11578775]

[76] Valencia JV, Mone M, Zhang J, Weetall M, Buxton FP, Hughes TE. Divergent pathways of gene expression are activated by the RAGE ligands S100b and AGE-BSA. Diabetes 2004; 53(3): 743-51.
[http://dx.doi.org/10.2337/diabetes.53.3.743] [PMID: 14988260]

[77] Schmitt KRL, Kern C, Lange PE, Berger F, Abdul-Khaliq H, Hendrix S. S100B modulates IL-6 release and cytotoxicity from hypothermic brain cells and inhibits hypothermia-induced axonal outgrowth. Neurosci Res 2007; 59(1): 68-73.

[http://dx.doi.org/10.1016/j.neures.2007.05.011] [PMID: 17604861]

[78]   Hu J, Castets F, Guevara JL, Van Eldik LJ. S100 beta stimulates inducible nitric oxide synthase activity and mRNA levels in rat cortical astrocytes. J Biol Chem 1996; 271(5): 2543-7.
[http://dx.doi.org/10.1074/jbc.271.5.2543] [PMID: 8576219]

[79]   Hu J, Ferreira A, Van Eldik LJ. S100beta induces neuronal cell death through nitric oxide release from astrocytes. J Neurochem 1997; 69(6): 2294-301.
[http://dx.doi.org/10.1046/j.1471-4159.1997.69062294.x] [PMID: 9375660]

[80]   Villarreal A, Seoane R, González Torres A, *et al.* S100B protein activates a RAGE-dependent autocrine loop in astrocytes: implications for its role in the propagation of reactive gliosis. J Neurochem 2014; 131(2): 190-205.
[http://dx.doi.org/10.1111/jnc.12790] [PMID: 24923428]

[81]   Braun M, Vaibhav K, Saad NM, *et al.* White matter damage after traumatic brain injury: A role for damage associated molecular patterns. Biochim Biophys Acta Mol Basis Dis 2017; 1863(10): 2614-26.
[http://dx.doi.org/10.1016/j.bbadis.2017.05.020] [PMID: 28533056]

[82]   Sorci G, Bianchi R, Riuzzi F, *et al.* S100B protein, a damage-associated molecular pattern protein in the brain and heart, and beyond. Cardiovasc Psychiatry Neurol 2010; 2010: 1-13.
[http://dx.doi.org/10.1155/2010/656481] [PMID: 20827421]

[83]   Walker LE, Frigerio F, Ravizza T, *et al.* Molecular isoforms of high-mobility group box 1 are mechanistic biomarkers for epilepsy. J Clin Invest 2017; 127(6): 2118-32.
[http://dx.doi.org/10.1172/JCI92001] [PMID: 28504645]

[84]   Michetti F, Corvino V, Geloso MC, *et al.* The S100B protein in biological fluids: more than a lifelong biomarker of brain distress. J Neurochem 2012; 120(5): 644-59.
[http://dx.doi.org/10.1111/j.1471-4159.2011.07612.x] [PMID: 22145907]

[85]   Lüth HJ, Ogunlade V, Kuhla B, *et al.* Age- and stage-dependent accumulation of advanced glycation end products in intracellular deposits in normal and Alzheimer's disease brains. Cereb Cortex 2004; 15(2): 211-20.
[http://dx.doi.org/10.1093/cercor/bhh123] [PMID: 15238435]

[86]   Takeuchi M, Bucala R, Suzuki T, *et al.* Neurotoxicity of advanced glycation end-products for cultured cortical neurons. J Neuropathol Exp Neurol 2000; 59(12): 1094-105.
[http://dx.doi.org/10.1093/jnen/59.12.1094] [PMID: 11138929]

[87]   Sasaki N, Toki S, Chowei H, *et al.* Immunohistochemical distribution of the receptor for advanced glycation end products in neurons and astrocytes in Alzheimer's disease. Brain Res 2001; 888(2): 256-62.
[http://dx.doi.org/10.1016/S0006-8993(00)03075-4] [PMID: 11150482]

[88]   Prasad K. AGE–RAGE stress: a changing landscape in pathology and treatment of Alzheimer's disease. Mol Cell Biochem 2019; 459(1-2): 95-112.
[http://dx.doi.org/10.1007/s11010-019-03553-4] [PMID: 31079281]

[89]   Huttunen HJ, Fages C, Rauvala H. Receptor for advanced glycation end products (RAGE)-mediated neurite outgrowth and activation of NF-kappaB require the cytoplasmic domain of the receptor but different downstream signaling pathways. J Biol Chem 1999; 274(28): 19919-24.
[http://dx.doi.org/10.1074/jbc.274.28.19919] [PMID: 10391939]

[90]   Huttunen HJ, Rauvala H. Amphoterin as an extracellular regulator of cell motility: from discovery to disease. J Intern Med 2004; 255(3): 351-66.
[http://dx.doi.org/10.1111/j.1365-2796.2003.01301.x] [PMID: 14871459]

[91]   Chou DKH, Zhang J, Smith FI, McCaffery P, Jungalwala FB. Developmental expression of receptor for advanced glycation end products (RAGE), amphoterin and sulfoglucuronyl (HNK-1) carbohydrate in mouse cerebellum and their role in neurite outgrowth and cell migration. J Neurochem 2004; 90(6):

1389-401.
[http://dx.doi.org/10.1111/j.1471-4159.2004.02609.x] [PMID: 15341523]

[92]　Huttunen HJ, Kuja-Panula J, Rauvala H. Receptor for advanced glycation end products (RAGE) signaling induces CREB-dependent chromogranin expression during neuronal differentiation. J Biol Chem 2002; 277(41): 38635-46.
[http://dx.doi.org/10.1074/jbc.M202515200] [PMID: 12167613]

[93]　Saleh A, Smith DR, Tessler L, *et al.* Receptor for advanced glycation end-products (RAGE) activates divergent signaling pathways to augment neurite outgrowth of adult sensory neurons. Exp Neurol 2013; 249: 149-59.
[http://dx.doi.org/10.1016/j.expneurol.2013.08.018] [PMID: 24029001]

[94]　Wang L, Li S, Jungalwala FB. Receptor for advanced glycation end products (RAGE) mediates neuronal differentiation and neurite outgrowth. J Neurosci Res 2008; 86(6): 1254-66.
[http://dx.doi.org/10.1002/jnr.21578] [PMID: 18058943]

[95]　Wang H, Mei X, Cao Y, *et al.* HMGB1/Advanced Glycation End Products (RAGE) does not aggravate inflammation but promote endogenous neural stem cells differentiation in spinal cord injury. Sci Rep 2017; 7(1): 10332.
[http://dx.doi.org/10.1038/s41598-017-10611-8] [PMID: 28871209]

[96]　Kim J, Wan CK, J O'Carroll S, Shaikh SB, Nicholson LFB. The role of receptor for advanced glycation end products (RAGE) in neuronal differentiation. J Neurosci Res 2012; 90(6): 1136-47.
[http://dx.doi.org/10.1002/jnr.23014] [PMID: 22344976]

[97]　Piras S, Furfaro AL, Piccini A, *et al.* Monomeric Aβ1–42 and RAGE: key players in neuronal differentiation. Neurobiol Aging 2014; 35(6): 1301-8.
[http://dx.doi.org/10.1016/j.neurobiolaging.2014.01.002] [PMID: 24484607]

[98]　Song J, Lee W, Park K, Lee J. Receptor for advanced glycation end products (RAGE) and its ligands: focus on spinal cord injury. Int J Mol Sci 2014; 15(8): 13172-91.
[http://dx.doi.org/10.3390/ijms150813172] [PMID: 25068700]

[99]　Sbai O, Devi TS, Melone MAB, *et al.* RAGE–TXNIP axis is required for S100B-promoted Schwann cell migration, fibronectin expression and cytokine secretion. J Cell Sci 2010; 123(24): 4332-9.
[http://dx.doi.org/10.1242/jcs.074674] [PMID: 21098642]

[100]　Tóbon-Velasco J, Cuevas E, Torres-Ramos M. Receptor for AGEs (RAGE) as mediator of NF-kB pathway activation in neuroinflammation and oxidative stress. CNS Neurol Disord Drug Targets 2014; 13(9): 1615-26.
[http://dx.doi.org/10.2174/1871527313666140806144831] [PMID: 25106630]

[101]　Daffu G, Del Pozo C, O'Shea K, Ananthakrishnan R, Ramasamy R, Schmidt A. Radical roles for RAGE in the pathogenesis of oxidative stress in cardiovascular diseases and beyond. Int J Mol Sci 2013; 14(10): 19891-910.
[http://dx.doi.org/10.3390/ijms141019891] [PMID: 24084731]

[102]　Kalea AZ, Reiniger N, Yang H, Arriero M, Schmidt AM, Hudson BI. Alternative splicing of the murine receptor for advanced glycation end-products (RAGE) gene. FASEB J 2009; 23(6): 1766-74.
[http://dx.doi.org/10.1096/fj.08-117739] [PMID: 19164451]

[103]　Hudson BI, Lippman ME. Targeting RAGE signaling in inflammatory disease. Annu Rev Med 2018; 69(1): 349-64.
[http://dx.doi.org/10.1146/annurev-med-041316-085215] [PMID: 29106804]

[104]　Corica D, Aversa T, Ruggeri RM, *et al.* Could AGE/RAGE-related oxidative homeostasis dysregulation enhance susceptibility to pathogenesis of cardiometabolic complications in childhood obesity? Front Endocrinol (Lausanne) 2019; 10: 426.
[http://dx.doi.org/10.3389/fendo.2019.00426] [PMID: 31316471]

[105]　Derk J, MacLean M, Juranek J, Schmidt AM. The receptor for advanced glycation end products

(RAGE) and mediation of inflammatory neurodegeneration. J Alzheimers Dis Parkinsonism 2018; 8(1): 421.
[http://dx.doi.org/10.4172/2161-0460.1000421] [PMID: 30560011]

[106] Knopman DS, Amieva H, Petersen RC, *et al.* Alzheimer disease. Nat Rev Dis Primers 2021; 7(1): 33.
[http://dx.doi.org/10.1038/s41572-021-00269-y] [PMID: 33986301]

[107] Obrenovich ME, Monnier VM. Glycation stimulates amyloid formation. Sci SAGE KE 2004; 2004(2): pe3.
[http://dx.doi.org/10.1126/sageke.2004.2.pe3] [PMID: 14724325]

[108] Wong A, Lüth HJ, Deuther-Conrad W, *et al.* Advanced glycation endproducts co-localize with inducible nitric oxide synthase in Alzheimer's disease. Brain Res 2001; 920(1-2): 32-40.
[http://dx.doi.org/10.1016/S0006-8993(01)02872-4] [PMID: 11716809]

[109] Vitek MP, Bhattacharya K, Glendening JM, *et al.* Advanced glycation end products contribute to amyloidosis in Alzheimer disease. Proc Natl Acad Sci USA 1994; 91(11): 4766-70.
[http://dx.doi.org/10.1073/pnas.91.11.4766] [PMID: 8197133]

[110] Smith MA, Taneda S, Richey PL, *et al.* Advanced Maillard reaction end products are associated with Alzheimer disease pathology. Proc Natl Acad Sci USA 1994; 91(12): 5710-4.
[http://dx.doi.org/10.1073/pnas.91.12.5710] [PMID: 8202552]

[111] Mattson MP, Carney JW, Butterfield DA. A tombstone in Alzheimer's? Nature 1995; 373(6514): 481.
[http://dx.doi.org/10.1038/373481a0] [PMID: 7845457]

[112] Smith MA, Say re LM, Vitek MP, Monnier VM, Perry G. Early AGEing and Alzheimer's. Nature 1995; 374(6520): 316.
[http://dx.doi.org/10.1038/374316b0] [PMID: 7885469]

[113] Li X-H, Du L-L, Cheng X-S, *et al.* Glycation exacerbates the neuronal toxicity of β-amyloid. Cell Death Dis 2013; 4(6): e673.
[http://dx.doi.org/10.1038/cddis.2013.180] [PMID: 23764854]

[114] Li XH, Lv BL, Xie JZ, Liu J, Zhou XW, Wang JZ. AGEs induce Alzheimer-like tau pathology and memory deficit *via* RAGE-mediated GSK-3 activation. Neurobiol Aging 2012; 33(7): 1400-10.
[http://dx.doi.org/10.1016/j.neurobiolaging.2011.02.003] [PMID: 21450369]

[115] Mackic JB, Stins M, McComb JG, *et al.* Human blood-brain barrier receptors for Alzheimer's amyloid-β 1- 40. Asymmetrical binding, endocytosis, and transcytosis at the apical side of brain microvascular endothelial cell monolayer. J Clin Invest 1998; 102(4): 734-43.
[http://dx.doi.org/10.1172/JCI2029] [PMID: 9710442]

[116] Kong Y, Liu C, Zhou Y, *et al.* Progress of RAGE molecular imaging in Alzheimer's Disease. Front Aging Neurosci 2020; 12: 227.
[http://dx.doi.org/10.3389/fnagi.2020.00227] [PMID: 32848706]

[117] Liu R, Wu C, Zhou D, *et al.* Pinocembrin protects against β-amyloid-induced toxicity in neurons through inhibiting receptor for advanced glycation end products (RAGE)-independent signaling pathways and regulating mitochondrion-mediated apoptosis. BMC Med 2012; 10(1): 105.
[http://dx.doi.org/10.1186/1741-7015-10-105] [PMID: 22989295]

[118] Matrone C, Djelloul M, Taglialatela G, Perrone L. Inflammatory risk factors and pathologies promoting Alzheimer's disease progression: is RAGE the key? Histol Histopathol 2015; 30(2): 125-39.
[http://dx.doi.org/10.14670/HH-30.125] [PMID: 25014735]

[119] Cai Z, Liu N, Wang C, *et al.* Role of RAGE in Alzheimer's Disease. Cell Mol Neurobiol 2016; 36(4): 483-95.
[http://dx.doi.org/10.1007/s10571-015-0233-3] [PMID: 26175217]

[120] Lee JJ, Wang PW, Yang IH, Wu CL, Chuang JH. Amyloid-β mediates the receptor of advanced glycation end product-induced pro-inflammatory response *via* toll-like receptor 4 signaling pathway in

retinal ganglion cell line RGC-5. Int J Biochem Cell Biol 2015; 64: 1-10.
[http://dx.doi.org/10.1016/j.biocel.2015.03.002] [PMID: 25783987]

[121] Origlia N, Arancio O, Domenici L, Yan SS. MAPK, β-amyloid and synaptic dysfunction: the role of RAGE. Expert Rev Neurother 2009; 9(11): 1635-45.
[http://dx.doi.org/10.1586/ern.09.107] [PMID: 19903023]

[122] Guerrero E, Vasudevaraju P, Hegde ML, Britton GB, Rao KS. Recent advances in α-synuclein functions, advanced glycation, and toxicity: implications for Parkinson's disease. Mol Neurobiol 2013; 47(2): 525-36.
[http://dx.doi.org/10.1007/s12035-012-8328-z] [PMID: 22923367]

[123] Braak H, Tredici KD, Rüb U, de Vos RAI, Jansen Steur ENH, Braak E. Staging of brain pathology related to sporadic Parkinson's disease. Neurobiol Aging 2003; 24(2): 197-211.
[http://dx.doi.org/10.1016/S0197-4580(02)00065-9] [PMID: 12498954]

[124] Defebvre L. [Parkinson's disease: Role of genetic and environment factors. Involvement in everyday clinical practice]. Rev Neurol (Paris) 2010; 166(10): 764-9.
[http://dx.doi.org/10.1016/j.neurol.2010.07.014] [PMID: 20817232]

[125] Jankovic J. Parkinson's disease: clinical features and diagnosis. J Neurol Neurosurg Psychiatry 2008; 79(4): 368-76.
[http://dx.doi.org/10.1136/jnnp.2007.131045] [PMID: 18344392]

[126] Nuytemans K, Theuns J, Cruts M, Van Broeckhoven C. Genetic etiology of Parkinson disease associated with mutations in the SNCA, PARK2, PINK1, PARK7, and LRRK2 genes: a mutation update. Hum Mutat 2010; 31(7): 763-80.
[http://dx.doi.org/10.1002/humu.21277] [PMID: 20506312]

[127] Hegde ML, Vasudevaraju P, Rao KJ. DNA induced folding/fibrillation of alpha-synuclein: new insights in Parkinson's disease. Front Biosci 2010; 15(1): 418-36.
[http://dx.doi.org/10.2741/3628] [PMID: 20036828]

[128] Lee D, Park CW, Paik SR, Choi KY. The modification of α-synuclein by dicarbonyl compounds inhibits its fibril-forming process. Biochim Biophys Acta Proteins Proteomics 2009; 1794(3): 421-30.
[http://dx.doi.org/10.1016/j.bbapap.2008.11.016] [PMID: 19103312]

[129] Padmaraju V, Bhaskar JJ, Prasada Rao UJS, Salimath PV, Rao KS. Role of advanced glycation on aggregation and DNA binding properties of α-synuclein. J Alzheimers Dis 2011; 24(s2) (Suppl. 2): 211-21.
[http://dx.doi.org/10.3233/JAD-2011-101965] [PMID: 21441659]

[130] Teismann P, Sathe K, Bierhaus A, *et al.* Receptor for advanced glycation endproducts (RAGE) deficiency protects against MPTP toxicity. Neurobiol Aging 2012; 33(10): 2478-90.
[http://dx.doi.org/10.1016/j.neurobiolaging.2011.12.006] [PMID: 22227007]

[131] Gao J, Teng J, Liu H, Han X, Chen B, Xie A. Association of RAGE gene polymorphisms with sporadic Parkinson's disease in Chinese Han population. Neurosci Lett 2014; 559: 158-62.
[http://dx.doi.org/10.1016/j.neulet.2013.11.038] [PMID: 24304868]

[132] Dalfó E, Portero-Otín M, Ayala V, Martínez A, Pamplona R, Ferrer I. Evidence of oxidative stress in the neocortex in incidental Lewy body disease. J Neuropathol Exp Neurol 2005; 64(9): 816-30.
[http://dx.doi.org/10.1097/01.jnen.0000179050.54522.5a] [PMID: 16141792]

[133] Sathe K, Maetzler W, Lang JD, *et al.* S100B is increased in Parkinson's disease and ablation protects against MPTP-induced toxicity through the RAGE and TNF-α pathway. Brain 2012; 135(11): 3336-47.
[http://dx.doi.org/10.1093/brain/aws250] [PMID: 23169921]

[134] Caron NS, Wright GEB, Hayden MR. Huntington Disease. In: Adam MP, Feldman J, Mirzaa GM, Pagon RA, Wallace SE, Bean LJH, Gripp KW, Amemiya A, editors. Gene Reviews Seattle (WA): University of Washington, Seattle; 1998; pp. 1993-2024.

[PMID: 202301482]

[135]   Franciosi S, Shim Y, Lau M, Hayden MR, Leavitt BR. A systematic review and meta-analysis of clinical variables used in Huntington disease research. Mov Disord 2013; 28(14): 1987-94.
[http://dx.doi.org/10.1002/mds.25663] [PMID: 24142393]

[136]   Bates GP, Dorsey R, Gusella JF, *et al.* Huntington disease. Nat Rev Dis Primers 2015; 1(1): 15005.
[http://dx.doi.org/10.1038/nrdp.2015.5] [PMID: 27188817]

[137]   Anzilotti S, Giampà C, Laurenti D, *et al.* Immunohistochemical localization of receptor for advanced glycation end (RAGE) products in the R6/2 mouse model of Huntington's disease. Brain Res Bull 2012; 87(2-3): 350-8.
[http://dx.doi.org/10.1016/j.brainresbull.2011.01.009] [PMID: 21272617]

[138]   Shi D, Chang JW, Choi J, *et al.* Receptor for Advanced Glycation End Products (RAGE) is expressed predominantly in medium spiny neurons of tgHD rat Striatum. Neuroscience 2018; 380: 146-51.
[http://dx.doi.org/10.1016/j.neuroscience.2018.03.042] [PMID: 29625216]

[139]   Redler RL, Dokholyan NV. The complex molecular biology of amyotrophic lateral sclerosis (ALS). Prog Mol Biol Transl Sci 2012; 107: 215-62.
[http://dx.doi.org/10.1016/B978-0-12-385883-2.00002-3] [PMID: 22482452]

[140]   Saez-Atienzar S, Bandres-Ciga S, Langston RG, *et al.* Genetic analysis of amyotrophic lateral sclerosis identifies contributing pathways and cell types. Sci Adv 2021; 7(3): eabd9036.
[http://dx.doi.org/10.1126/sciadv.abd9036] [PMID: 33523907]

[141]   Antoniadi AM, Galvin M, Heverin M, *et al.* Identifying features that are predictive of quality of life in people with amyotrophic lateral sclerosis. IEEE International Conference on Healthcare Informatics (ICHI).
[http://dx.doi.org/10.1109/ICHI48887.2020.9374298]

[142]   Casula M, Iyer AM, Spliet WGM, *et al.* Toll-like receptor signaling in amyotrophic lateral sclerosis spinal cord tissue. Neuroscience 2011; 179: 233-43.
[http://dx.doi.org/10.1016/j.neuroscience.2011.02.001] [PMID: 21303685]

[143]   Iłżecka J. Serum-soluble receptor for advanced glycation end product levels in patients with amyotrophic lateral sclerosis. Acta Neurol Scand 2009; 120(2): 119-22.
[http://dx.doi.org/10.1111/j.1600-0404.2008.01133.x] [PMID: 19053950]

[144]   Shibata N, Hirano A, Hedley-Whyte TE, *et al.* Selective formation of certain advanced glycation end products in spinal cord astrocytes of humans and mice with superoxide dismutase-1 mutation. Acta Neuropathol 2002; 104(2): 171-8.
[http://dx.doi.org/10.1007/s00401-002-0537-5] [PMID: 12111360]

[145]   Takamiya R, Takahashi M, Myint T, *et al.* Glycation proceeds faster in mutated Cu, Zn-superoxide dismutases related to familial amyotrophic lateral sclerosis. FASEB J 2003; 17(8): 1-18.
[http://dx.doi.org/10.1096/fj.02-0768fje] [PMID: 12626432]

[146]   Chou SM, Wang HS, Taniguchi A, Bucala R. Advanced glycation endproducts in neurofilament conglomeration of motoneurons in familial and sporadic amyotrophic lateral sclerosis. Mol Med 1998; 4(5): 324-32.
[http://dx.doi.org/10.1007/BF03401739] [PMID: 9642682]

[147]   Juranek JK, Daffu GK, Wojtkiewicz J, Lacomis D, Kofler J, Schmidt AM. Receptor for Advanced Glycation End Products and its inflammatory ligands are upregulated in amyotrophic lateral sclerosis. Front Cell Neurosci 2015; 9: 485.
[http://dx.doi.org/10.3389/fncel.2015.00485] [PMID: 26733811]

[148]   Freixes M, Rodríguez A, Dalfó E, Ferrer I. Oxidation, glycoxidation, lipoxidation, nitration, and responses to oxidative stress in the cerebral cortex in Creutzfeldt–Jakob disease. Neurobiol Aging 2006; 27(12): 1807-15.
[http://dx.doi.org/10.1016/j.neurobiolaging.2005.10.006] [PMID: 16310893]

[149]   Hughes RAC. Regular review: Peripheral neuropathy. BMJ 2002; 324(7335): 466-9.
[http://dx.doi.org/10.1136/bmj.324.7335.466] [PMID: 11859051]

[150]   Haslbeck KM, Neundörfer B, Schlötzer-Schrehardt U, *et al.* Activation of the RAGE pathway: a general mechanism in the pathogenesis of polyneuropathies? Neurol Res 2007; 29(1): 103-10.
[http://dx.doi.org/10.1179/174313206X152564] [PMID: 17427284]

[151]   Toth C, Rong LL, Yang C, *et al.* Receptor for advanced glycation end products (RAGEs) and experimental diabetic neuropathy. Diabetes 2008; 57(4): 1002-17.
[http://dx.doi.org/10.2337/db07-0339] [PMID: 18039814]

[152]   Juranek JK, Geddis MS, Song F, *et al.* RAGE deficiency improves postinjury sciatic nerve regeneration in type 1 diabetic mice. Diabetes 2013; 62(3): 931-43. a
[http://dx.doi.org/10.2337/db12-0632] [PMID: 23172920]

[153]   Juranek JK, Kothary P, Mehra A, Hays A, Brannagan TH III, Schmidt AM. Increased expression of the receptor for advanced glycation end-products in human peripheral neuropathies. Brain Behav 2013; 3(6): 701-9. b
[http://dx.doi.org/10.1002/brb3.176] [PMID: 24363972]

[154]   Bekircan-Kurt CE, Üçeyler N, Sommer C. Cutaneous activation of rage in nonsystemic vasculitic and diabetic neuropathy. Muscle Nerve 2014; 50(3): 377-83.
[http://dx.doi.org/10.1002/mus.24164] [PMID: 24395344]

[155]   Park SY, Kim YA, Hong YH, Moon MK, Koo BK, Kim TW. Up-regulation of the receptor for advanced glycation end products in the skin biopsy specimens of patients with severe diabetic neuropathy. J Clin Neurol 2014; 10(4): 334-41.
[http://dx.doi.org/10.3988/jcn.2014.10.4.334] [PMID: 25324883]

[156]   Saraiva MJM, Birken S, Costa PP, Goodman DS. Family studies of the genetic abnormality in transthyretin (prealbumin) in Portuguese patients with familial amyloidotic polyneuropathy. Ann N Y Acad Sci 1984; 435(1): 86-100.
[http://dx.doi.org/10.1111/j.1749-6632.1984.tb13742.x] [PMID: 6099706]

[157]   da Costa G, Gomes RA, Guerreiro A, *et al.* Beyond genetic factors in familial amyloidotic polyneuropathy: protein glycation and the loss of fibrinogen's chaperone activity. PLoS One 2011; 6(10): e24850.
[http://dx.doi.org/10.1371/journal.pone.0024850] [PMID: 22053176]

[158]   Shorter J, Lindquist S. Prions as adaptive conduits of memory and inheritance. Nat Rev Genet 2005; 6(6): 435-50.
[http://dx.doi.org/10.1038/nrg1616] [PMID: 15931169]

[159]   Sousa MM, Yan SD, Stern D, Saraiva MJ. Interaction of the receptor for advanced glycation end products (RAGE) with transthyretin triggers nuclear transcription factor kB (NF-kB) activation. Lab Invest 2000; 80(7): 1101-10.
[http://dx.doi.org/10.1038/labinvest.3780116] [PMID: 10908156]

[160]   Yan SD, Zhu H, Zhu A, *et al.* Receptor-dependent cell stress and amyloid accumulation in systemic amyloidosis. Nat Med 2000; 6(6): 643-51.
[http://dx.doi.org/10.1038/76216] [PMID: 10835680]

[161]   Guo JD, Li L, Shi Y, *et al.* Genetic ablation of receptor for advanced glycation end products promotes functional recovery in mouse model of spinal cord injury. Mol Cell Biochem 2014; 390(1-2): 215-23.
[http://dx.doi.org/10.1007/s11010-014-1972-z] [PMID: 24526523]

[162]   Mei X, Wang H, Zhang H, *et al.* Blockade of receptor for advanced glycation end products promotes oligodendrocyte autophagy in spinal cord injury. Neurosci Lett 2019; 698: 198-203.
[http://dx.doi.org/10.1016/j.neulet.2019.01.030] [PMID: 30660637]

[163]   Maroso M, Balosso S, Ravizza T, *et al.* Toll-like receptor 4 and high-mobility group box-1 are involved in ictogenesis and can be targeted to reduce seizures. Nat Med 2010; 16(4): 413-9.

[http://dx.doi.org/10.1038/nm.2127] [PMID: 20348922]

[164] Iori V, Maroso M, Rizzi M, *et al.* Receptor for Advanced Glycation Endproducts is upregulated in temporal lobe epilepsy and contributes to experimental seizures. Neurobiol Dis 2013; 58: 102-14.
[http://dx.doi.org/10.1016/j.nbd.2013.03.006] [PMID: 23523633]

[165] Guo M, Wang J, Qi H, *et al.* Polymorphisms in the receptor for advanced glycation end products gene are associated with susceptibility to drug-resistant epilepsy. Neurosci Lett 2016; 619: 137-41.
[http://dx.doi.org/10.1016/j.neulet.2016.01.043] [PMID: 26828298]

[166] Ping X, Chai Z, Wang W, Ma C, White FA, Jin X. Blocking receptor for advanced glycation end products (RAGE) or toll-like receptor 4 (TLR4) prevents posttraumatic epileptogenesis in mice. Epilepsia 2021; 62(12): 3105-16.
[http://dx.doi.org/10.1111/epi.17069] [PMID: 34535891]

[167] Reddy VP, Aryal P, Soni P. RAGE inhibitors in neurodegenerative diseases. Biomedicines 2023; 11(4): 1131.
[http://dx.doi.org/10.3390/biomedicines11041131] [PMID: 37189749]

[168] Bekircan Kurt CE, Tan E, Erdem Özdamar S. The activation of RAGE and NF-κB in nerve biopsies of patients with axonal and vasculitic neuropathy. Noro Psikiyatri Arsivi 2015; 52(3): 279-82.
[http://dx.doi.org/10.5152/npa.2015.8801] [PMID: 28360724]

[169] Klein I, Lehmann H. Pathomechanisms of paclitaxel-induced peripheral neuropathy. Toxics 2021; 9(10): 229.
[http://dx.doi.org/10.3390/toxics9100229] [PMID: 34678925]

[170] Sekiguchi F, Domoto R, Nakashima K, *et al.* Paclitaxel-induced HMGB1 release from macrophages and its implication for peripheral neuropathy in mice: Evidence for a neuroimmune crosstalk. Neuropharmacology 2018; 141: 201-13.
[http://dx.doi.org/10.1016/j.neuropharm.2018.08.040] [PMID: 30179591]

[171] Tsubota M, Miyazaki T, Ikeda Y, *et al.* Caspase-dependent HMGB1 release from macrophages participates in peripheral neuropathy caused by bortezomib, a proteasome inhibiting chemotherapeutic agent, in mice. Cells 2021; 10(10): 2550.
[http://dx.doi.org/10.3390/cells10102550] [PMID: 34685531]

[172] Paudel YN, Angelopoulou E, Semple B, Piperi C, Othman I, Shaikh MF. Potential Neuroprotective Effect of the HMGB1 inhibitor Glycyrrhizin in neurological disorders. ACS Chem Neurosci 2020; 11(4): 485-500.
[http://dx.doi.org/10.1021/acschemneuro.9b00640] [PMID: 31972087]

[173] Li Y, Peng Y, Shen Y, Zhang Y, Liu L, Yang X. Dietary polyphenols: regulate the advanced glycation end products-RAGE axis and the microbiota-gut-brain axis to prevent neurodegenerative diseases. Crit Rev Food Sci Nutr 2023; 63(29): 9816-42.
[http://dx.doi.org/10.1080/10408398.2022.2076064] [PMID: 35587161]

[174] Tang D, Kang R, Xiao W, *et al.* Quercetin prevents LPS-induced high-mobility group box 1 release and proinflammatory function. Am J Respir Cell Mol Biol 2009; 41(6): 651-60.
[http://dx.doi.org/10.1165/rcmb.2008-0119OC] [PMID: 19265175]

[175] Lee W, Ku SK, Bae JW, Bae JS. Inhibitory effects of lycopene on HMGB1-mediated pro-inflammatory responses in both cellular and animal models. Food Chem Toxicol 2012; 50(6): 1826-33.
[http://dx.doi.org/10.1016/j.fct.2012.03.003] [PMID: 22429818]

[176] Han Y. Chen, Lin Q, Liu Y, Ge W, Cao H, Li J. Curcumin improves memory deficits by inhibiting HMGB1-RAGE/TLR4-NF-B signaling pathway in APPswe/PS1dE9 transgenic mice hippocampus. J Cell Mol Med 2021; 25: 8947-56.
[http://dx.doi.org/10.1111/jcmm.16855] [PMID: 34405526]

[177] Weng MH, Chen SY, Li ZY, Yen GC. Camellia oil alleviates the progression of Alzheimer's disease

in aluminum chloride-treated rats. Free Radic Biol Med 2020; 152: 411-21.
[http://dx.doi.org/10.1016/j.freeradbiomed.2020.04.004] [PMID: 32294510]

[178] Malik P, Malik P, Hoidal JR, Hoidal JR, Mukherjee TK, Mukherjee TK. Recent advances in curcumin treated non-small cell lung cancers: an impetus of pleiotropic traits and nanocarrier aided delivery. Curr Med Chem 2021; 28(16): 3061-106.
[http://dx.doi.org/10.2174/0929867327666200824110332] [PMID: 32838707]

[179] Malik P, Singh R, Kumar M, Malik A, Mukherjee TK. Understanding the phytoestrogen genistein actions on breast cancer: Insights on estrogen receptor equivalence, pleiotropic essence and emerging paradigms in bioavailability modulation. Curr Top Med Chem 2023; 23(15): 1395-413.
[http://dx.doi.org/10.2174/1568026623666230103163023] [PMID: 36597609]

[180] Malik P, Bernela M, Seth M, Kaushal P, Mukherjee TK. Recent progress in the Hesperetin delivery regimes: Significance of pleiotropic actions and synergistic anticancer efficacy. Curr Pharm Des 2023; 29(37): 2954-76.
[http://dx.doi.org/10.2174/0113816128253609231030070414] [PMID: 38173051]

[181] Ren L, Yan H. Targeting AGEs-RAGE pathway inhibits inflammation and presents neuroprotective effect against hepatic ischemia-reperfusion induced hippocampus damage. Clin Res Hepatol Gastroenterol 2022; 46(2): 101792.
[http://dx.doi.org/10.1016/j.clinre.2021.101792] [PMID: 34400367]

[182] Burstein AH, Sabbagh M, Andrews R, Valcarce C, Dunn I, Altstiel L. Development of Azeliragon, an oral small molecule antagonist of the receptor for advanced glycation end products, for the potential slowing of loss of cognition in mild Alzheimer's Disease. J Prev Alzheimers Dis 2018; 5(2): 1-6.
[http://dx.doi.org/10.14283/jpad.2018.18] [PMID: 29616709]

[183] Ko SY, Lin YP, Lin YS, Chang SS. Advanced glycation end products enhance amyloid precursor protein expression by inducing reactive oxygen species. Free Radic Biol Med 2010; 49(3): 474-80.
[http://dx.doi.org/10.1016/j.freeradbiomed.2010.05.005] [PMID: 20471471]

[184] Shimizu Y, Harashima A, Munesue S, *et al.* Neuroprotective effects of endogenous secretory receptor for advanced glycation end-products in brain ischemia. Aging Dis 2020; 11(3): 547-58.
[http://dx.doi.org/10.14336/AD.2019.0715] [PMID: 32489701]

[185] Rao NL, Kotian GB, Shetty JK, *et al.* Receptor for Advanced Glycation End Product, organ crosstalk, and pathomechanism targets for comprehensive molecular therapeutics in diabetic ischemic stroke. Biomolecules 2022; 12(11): 1712.
[http://dx.doi.org/10.3390/biom12111712] [PMID: 36421725]

[186] Sugihara T, Munesue S, Yamamoto Y, *et al.* Endogenous secretory receptor for advanced glycation end-products inhibits amyloid-β1-42 uptake into mouse brain. J Alzheimers Dis 2012; 28(3): 709-20.
[http://dx.doi.org/10.3233/JAD-2011-110776] [PMID: 22064071]

[187] Wang H, Chen F, Du YF, *et al.* Targeted inhibition of RAGE reduces amyloid-β influx across the blood-brain barrier and improves cognitive deficits in db/db mice. Neuropharmacology 2018; 131: 143-53.
[http://dx.doi.org/10.1016/j.neuropharm.2017.12.026] [PMID: 29248482]

[188] Sabbagh MN, Agro A, Bell J, Aisen PS, Schweizer E, Galasko D. PF-04494700, an oral inhibitor of receptor for advanced glycation end products (RAGE), in Alzheimer disease. Alzheimer Dis Assoc Disord 2011; 25(3): 206-12.
[http://dx.doi.org/10.1097/WAD.0b013e318204b550] [PMID: 21192237]

SUBJECT INDEX

A

Acids 49, 106, 170, 296, 310, 311, 313, 441
  glycyrrhetinic 441
  hyaluronic 49, 106
Actions 270, 369, 393, 396
  cancer-promoting 369
  ligand-diversified tumorigenic 396
  maleylacetoacetate isomerize 270
  metastatic 393
  proteolytic 393
Activating 211, 343
    immune cells 211
    signaling pathways 343
Activation, pro-inflammatory 409
Activator protein 160
Activities 97, 161, 201, 214, 215, 220, 222,
      258, 264, 268, 270, 277, 329, 382, 388,
      422, 424, 442
  anti-inflammatory 442
  citrate synthase 258
  cytokine 201
  enzymatic 97, 161, 220, 264, 329, 422
  enzyme cofactor 382
  glutathione reductase 268
  growth factor 277
  inflammatory 424
  mediated cell-signaling 388
  monomethyl arsenate reductase 270
  neuronal 222
  pro-inflammatory 215
  telomerase 214
  transcriptional 214
Acute 192, 229, 241, 244, 245, 286, 287, 294,
      295, 297, 300, 310, 313, 315, 317, 341,
      375
  coronary syndrome (ACS) 341
  inflammation 244, 300
  lung injury 192, 229, 241, 245, 286, 287,
      294, 295, 375

respiratory distress syndrome (ARDS) 192,
      229, 241, 245, 286, 294, 295, 297, 310,
      313, 315, 317, 375
AD-associated neuropathology 168
Adeno-associated virus (AAV) 315
Adhesion 42, 48, 67, 154, 230, 248, 273, 317
  microbial 67
  monocyte 273
  pathogenic microbial 42
Advanced 57, 275, 276, 277, 287, 428
    oxidation protein products (AOPPs) 275,
      276, 277, 287, 428
    therapeutic medicinal products (ATMPs)
      57
Age 131, 150, 153, 154, 155, 156, 158, 163,
      164, 168, 169
  -altered protein damage 155
  -altered proteins 150, 154, 168
  -based chronic complications 164
  -binding protein 153
  -induced damage 168
  -lipoprotein 131
  -mediated autophagy 158
  -mediated disorders 169
  -modulated proteins 156, 163
  proteins 169
Aggressive frictional forces 277
Aging-mediated complications 302
Airway inflammation 230, 292, 293, 304
  neutrophilic 230
  smoke-induced 293
Airway neutrophilia 294
Albumin, serum protein 290
Alzheimer's disease 164, 168, 174, 192, 198,
      245, 254, 408, 409, 413, 429
Amino acid 4, 121
  nucleophilic 121
  tryptophan 4
Amyloid 198, 222, 223, 414, 432, 436, 444
  fibril gatherings 436
  precursor protein (APP) 198, 222, 223,
      414, 432, 444

www.ingramcontent.com/pod-product-compliance
Lightning Source LLC
Chambersburg PA
CBHW050758220326
41598CB00006B/51